Modern Technical Management Techniques

for Engineers in Management, and for Those Who Want to Get There

Edited by

Herbert Popper

and the Staff of *Chemical Engineering*
and other McGraw-Hill Publications

McGraw-Hill Book Company

New York St. Louis San Francisco Düsseldorf Johannesburg
Kuala Lumpur London Mexico Montreal New Delhi Panama
Rio de Janeiro Singapore Sydney Toronto

Modern Technical Management Techniques

Library of Congress Catalog Card Number 70–133809

ISBN 07-050529-2

234567890 HDBP 7654321

Preface

It is debatable whether technical management is necessarily a nobler career pursuit than straight engineering. But one cannot really argue with facts such as these:

- On the average, technical managers receive more pay than engineers—they tend to escape the flattening-out of salary curves common among nonsupervisory professionals.
- Technical managers have a wider range of opportunities because managerial skills are more readily transferable than specialized technical skills and may be less prone to obsolescence.
- In general, technical managers have a greater sense of participation in their company's decisions and enjoy higher status (notwithstanding the success some firms have had in giving comparable status to outstanding engineers or scientists via the "dual ladder" approach).

Despite such inducements, it really makes little sense for a good engineer to become a technical manager unless he develops into a really good one—i.e., one who is in command of his job and can advance beyond the first rung of the management ladder.

This is where this book should prove helpful. It is aimed not only at engineers who are interested in management and who want to get in, but also at technical managers who want to improve their skills and get ideas or reference information for new ways of handling certain aspects of their job.

The six main sections are devoted to:

1. Careers and career-building in technical management, from group leader to vice-president.
2. Comprehensive advice on effective communications: writing, public speaking, gathering information.
3. A hard look at the appraisal, development, and rewarding of professional employees.
4. An examination of business law as related to technical management, with emphasis on areas such as job mobility, equipment performance, product liability, patent requirements, and contract enforcement.
5. Guidelines on exploiting operations-research and economic-evaluation techniques—e.g., critical-path methods, linear programming, and cost and profitability studies (particularly with regard to R&D projects).
6. Techniques for successful plant and project management.

Incidentally, I believe that much of this material should interest any engineer or scientist who wants to do a better job of managing his own activities, regardless of whether his sights are set on the managerial or technical/scientific career ladder. For instance, one certainly does not have to become a manager to profit from skills in the areas of communications, human relations and economics.

Most of the material comes from *Chemical Engineering,* but some has been drawn from other McGraw-Hill publications. Acknowledgment is gratefully made to *Product Engineering* for the article on p. 225, and to *Modern Manufacturing* (formerly *Factory*) for articles on pp. 337 and 339.

Herbert Popper

Contents

1

Careers and Career-Building in Technical Management

Here is a methodical approach to building a career as a technical executive, and advice on coping with problems and responsibilities at various managerial levels. Included are self-examination sections to help you decide where you want to go and how (and how fast) you want to get there.

Custom-Building Your Management Career

Part I: Establishing Your Performance Record

This important, six-part series represents a methodical approach to a successful career in technical management. Part I provides a general perspective, and examines how the nonsupervisory engineer can build up his managerial qualifications.

E. M. KIPP, Sun Oil Co.

If you are looking for magic shortcuts to "success," or for startling new revelations on how to succeed in business without hardly trying, you are not likely to find them in this series.

Rather, what you will find is a career planning and implementation program aimed at helping you become a technical manager of superior skill—one who can advance to higher and higher levels of responsibility. If that happens to be your career goal, and if you are willing to put in some extra effort in order to reach it, then please read on.

In constructing this career program, I will begin with the young engineer who has recently accepted his first job in industry, and who has to establish his credentials for advancement. Later parts of this series will carry him through progressively higher phases of his management development.

Hopefully, this approach will also be useful to the engineer who already has managerial responsibilities and is looking to take on bigger ones; he can "jump in" where he sees fit as the series progresses, skipping such sections as he considers self-evident, and focusing on the ones that help him take stock of his present position and of those that lie ahead.

Representative Responsibilities and Salaries

	Novice Manager	Apprentice Manager
Typical position	Group leader	Section chief
Years of experience	2 to 10	5 to 20
Annual salary range (1969)	$14,000 to $23,000 (depends on degree)	$18,000 to $31,000*
Time devoted to administration	10 to 20%	40 to 50%
Time devoted to technology	80 to 90%	50 to 60%
Number of subordinates	Technical: 2 to 10; Nontechnical: 2 to 10	Technical: 5 to 25; Nontechnical: 5 to 25
Typical responsibilities	Match work assignments with human resources.	Plan and direct limited programs, and organize the workforce to meet objectives.
	Measure progress and quality of work.	Evaluate performance of subordinates; administer salaries.
	Distribute budgetary resources.	Sell ideas and programs to supervisors, colleagues and subordinates.
	Develop effective communications with his group and his supervisor.	Evaluate and control the projects under jurisdiction.
	Recommend salary changes, and further the growth of members of his group.	Broaden his section's assignments and responsibilities in support of other departments.
	Coordinate his group's assignments with other activities.	Determine his subordinates' training and personal-development needs.

*Exclusive of additional compensation in the form of bonuses, profit-sharing plans, or stock options.

For the sake of simplicity, I have selected an organizational model that consists of only five levels, i.e.:

1. Nonsupervisory engineer.
2. Novice manager. This phase gives the engineer his introductory exposure to management, via assignments as a group leader (or equivalent).
3. Apprentice manager. The engineer is now at the section-chief level, and is preparing himself for a role in middle management and possibly above.
4. Middle manager. This category includes senior division managers, directors of R&D, managers of substantial plants, etc.
5. Top level manager—e.g., a vice president in a large company, or the top executive in a small one.

In considering each of these levels, we will examine their pivotal success factors from the viewpoint of the engineer, as well as from that of his supervisors.

The table on this page summarizes some of the pivotal factors, shows typical salary progressions, and provides other data. At this point, the reader is urged not to jump to the conclusion that he must necessarily set his goal on a top-level managerial position, no matter how alluring he finds the salary level and responsibilities shown in the table. Later in the series, after we examine the apprentice manager category, we will take a closer look at the factors that should be evaluated in deciding where to set one's goals; this section will be entitled "At the Crossroads."

The Managerial Challenge

The industrial establishment is undergoing a fundamental reassessment of the complex environmental trends with which it must learn to cope successfully. The superior technical manager must develop an impressive and almost frightening array of skills and resources, not only in the technical areas but, even more significantly, in the social and human behavioral sciences. Many widely and firmly held premises must be modified or discarded where no longer suitable, and many new viewpoints and sets of values must be developed.

Outstanding performance in any endeavor requires a high degree of dedication and self-discipline. Chances for success will be greatly enhanced if the journey is guided by a carefully developed and implemented program—modified and improved as time goes by. The original goals may or may not be reached; nevertheless, more realistic but equally challenging objectives will have been secured. With this in mind, let us proceed to the first step in custom-building your management career.

Establishment of Performance Record

Our lives can be more effective and purposeful if we continually evaluate these factors:

for Five Levels of Managerial Jobs

Middle Manager Director of R&D (small to medium-size firm); Technical Director or Division Manager in larger firm	Top-Level Manager Vice President (large organization) President or top executive (small company)
15 to 25	20 to 30
$28,000 to $42,000*	$40,000 to $60,000*
50 to 80%	80 to 90%
20 to 50%	10 to 20%
50 to several hundred	Several hundred to several thousand
Structure comprehensive progress and budgetary proposals. Structure and administer line-management positions. Define and monitor staff- and service-functions needed. Further develop the technical resources and capabilities of his people. Identify and train managerial candidates. Establish effective relationships with heads of other divisions. Maintain progress-balance within division. Effectively delegate authority and responsibility; apply participative management principles. Develop profitable applications of technology and engineering output.	Develop his departments for optimum profits, keeping costs in line. Interpret his firm's needs and opportunities to his subordinates; develop mutually acceptable work objectives. Serve on major intra- and intercorporate committees. Maintain sound corporate relationships with official agencies, etc. Make final recommendations on major proposed courses of action (e.g., product realignments, internal diversification, mergers, acquisitions, joint ventures, major licensing agreements) to president or board of directors. Maintain optimum progress-balance among departments; match the resources of his departments to the needs and opportunities of the company as a whole. Seek to improve his company's contributions to national and world progress.

- Where do I want to go?
- When do I want to arrive?
- How do I wish to travel?
- What will it cost me?

The quality of these four primary evaluations should improve with time, as we benefit from the knowledge and wisdom that develop with living. For example, a newly graduated chemical engineer may not, and most probably does not, have adequate basis for judging whether he wishes to become a manager. Nor does he yet know whether a career as a highly skilled technical specialist, for instance, might be more appropriate and possibly even more lucrative.

At this beginning phase, his concept of the four factors above can thus be extremely tentative. However, there is one objective that is primary: that of building a performance record of substance and of quality. This is the best possible foundation for subsequent phases in his career, whether they turn out to be in the direction of management or in some other direction.

In seeking to establish and develop his competency, and his ability to apply it usefully, the young engineer will recognize through first-hand experience those traits that will promote and those that will impede his ability to get results (see box on next page).

The most common and tragic example of the importance of these impeding and promoting factors can be seen frequently in those engineers who though technically competent may irritate and antagonize their associates or supervisors to such an extent that emotional considerations completely obscure the actual merits of their work.

In this regard, the old saying that a good leader must first learn to be a good follower is still true if we make it clear that "good follower" refers to the man who has learned how to make his leader successful. This demands the skill to recognize when constructive reasoning with his leader is timely, and when the leader must be alerted to things of which he may not be aware, but which are important to his success. In the "follower," this will often involve the skill of handling the situation so that the leader feels that if he did not originate the idea, at least he improved it considerably.

Perhaps above all else, a really effective follower must have the courage, skills and judgment to confront his boss with distasteful information when the need arises.

These are the things the "follower" will also require of his subordinates when he becomes a leader.

Technical Competency

Although technical proficiency does not in itself guarantee success as a technical manager, there is

Factors That Promote or Impede the Engineer's Ability to Get Results

Promotors	Impeders
Technical proficiency	Technical weak spots
Enthusiasm	Self-centeredness; prima donna-ism
Ability to clearly recognize and define the problem	Negative or isolationist attitudes about associates
Objective approach to solutions to the problem	Tendency to fight the problem rather than solve it
Communications skills	Undue conformity (e.g., to the way it was taught in school)
Intellectual and moral integrity	Cynicism
Willingness to seek and to use help	Tendency toward "politicking"
Graceful acknowledgement of contributions of others	Inability to be a good listener
Calm approach to opposing views and criticisms	Tendency to pass a problem on, rather than to think things through
Ability to sell one's work in terms that will relate to the needs of others	

no question that such proficiency is essential to the construction of the performance record. By and large, this is as it should be, because even the man who supervises 100 engineers does not spend all his time on administrative matters (see the Job Table). Unless he himself is technically competent, it is hard for him to tell how technically competent his subordinates are, and to guide them if some sort of impasse is reached: He may turn down a poorly presented but technically brilliant proposal in favor of one that is brilliantly presented but technically unsound.

Most companies therefore insist that no matter how much managerial promise a young engineer displays, he must also display technical proficiency in order to earn a promotion into technical management. And this means that the young engineer must assign a high initial priority to rounding out his technical skills —on the job, and off the job via carefully selected reading and continuing education.

Specifically, even M.S.-level chemical engineering graduates may find that their college education was relatively weak in certain fields such as:

- Selection, specification and troubleshooting of equipment.
- Some aspects of practical design work.
- Materials sciences; safety engineering.
- Modern, instrumented, analytical chemistry.

If you have gaps such as these, and if some of them are likely to impede your performance on your present assignment, your work is cut out for you.

Responsibility

Building of the performance record requires development and demonstration of responsibility. I use responsibility in the broad sense of an awareness and dedication to accepted commitments. A strong sense of responsibility is an indication not only of personal self-respect, but also of concern for the responsibilities and needs of associates and of superiors.

A man who cannot be relied upon to carry out his own assignments and commitments cannot aspire to positions of trust and importance.

Another key attribute—one that overlaps many of the others—is the use of good judgment. We will merely mention it now, and provide examples later.

Innovation

An outstanding work-performance record will reflect latent or partially developed skills in the areas of technological creativity and innovation. These skills are important to the future high-quality technical manager, because the ability to effectively innovate in managerial functions is becoming increasingly prized.

Even though technological creativity differs from managerial creativity, they share many common skills and drives. I recall a man with whom I was associated for a number of years who was truly an exceptionally creative inventor. He was not inhibited by available knowledge; he was not afraid of failure; he delighted in the skillful construction of a variety of broad and novel associations of ideas and knowledge; he was comfortable in ambiguous technical situations; and above all, he was adept at devising and carrying out ingenious experimentation to confirm or further develop his creative concepts. His perceptive boss recognized these rare assets. He further appreciated that these assets are also those needed for effective managerial creativity. Through training in "people engineering," as well as appropriate managerial assignments, this engineer made a very natural transition whereby the varied skills that defined his technical creativity were applied to the management area, with outstanding success.

The skills and characteristics that are demonstrated in creative innovation during the early years will also provide a valuable awareness of basic personality characteristics, such as risk-taking attitudes, independence, promotional skills (ethical propaganda and salesmanship), etc.

Implementation

The manner in which the young engineer discharges his working assignments develops important self-understanding. It is also evidence upon which others base their impressions of his general capabilities and promise.

For example, some engineers will:

- Attempt to execute or implement assignments in a very literal fashion.
- Operate under limitations of real or imagined restrictions.
- Attempt to do precisely what is asked and not one iota more.
- Stop for new instructions every time problems arise that seem to preclude completion of the assignment.

However, other engineers will:

- Tend to deviate from original work assignments.
- Seek own opportunities for broadening work assignments.
- Attempt to resolve unexpected problems to the limit of their resources before asking for new instructions.
- Reflect greater confidence, drive, and willingness to accept the risks of failure.

Thus, the construction of the performance record is a most valuable time to make progress in the important requirement of "knowing thyself" as well as in benefiting from the counsel, advice, and other inputs from supervisors and associates. An outstanding technical manager must enjoy the challenge of risk-taking; of making his own job, rather than the reverse; and of regarding as the only true security that which is based on internally generated strengths and drives.

Group and Individual Loyalties

A person with strong tendencies to self-centeredness —one who is so obsessed with personal goals that he cannot appreciate group or task-related goals—will have great difficulty in becoming a superior manager. On the other hand, certain types of loyalties, such as strong personal friendships, are not necessarily desirable assets in managerial capacities if such loyalties make it difficult to be impartial and objective.

The early working experiences provide a basis for self-understanding and development in these important areas by virtue of the manner in which we interact with other persons. Above all, let not loyalty to the company and to one's associates be equated to conformity and yesmanship.

Work Productivity: The Role of Counseling

In the final analysis, most managements tend to judge the individual engineer by what is sometimes referred to as total or over-all productivity—a label

How to Get Off to a Fast Start in Establishing Your Performance Record

Get as much information as possible on the background and reasons for the work assignment. Seek to understand how your work results can be put to use if successful.

Seek every opportunity for constructive initiative.

Be considerate of and dedicated to the boss's needs.

Seek timely advice and help in the proper places.

Learn all you can about the company and its key people.

Clarify with your supervisor what is expected of you and how your performance will be measured.

Spend as much time as possible on improving and sharpening your communications skills.

Maintain and further develop your technical skills.

Plan and implement a balanced program of reading in the nontechnical area; specifically, in the human behavioral sciences, business administration as applied in the chemical process industries, industrial relations, sociology, and government. A minimum of four hours a week should be budgeted for this reading (but not at the expense of remedying technical weak-spots).

Be an active member of your technical society and seek committee assignments.

Join and actively support at least one civic or community organization.

Do your best at all times to meet your commitments.

that can mean different things to different people.

Success in the following objectives is related to quality work-productivity:

- Establishing and meeting significant objectives.
- Efficiently using and coordinating resources.
- Developing appropriate interpretations of the work results, and profitable implementation.

In any event, productivity defined as "useful or quality results" is the definition I have in mind.

Although quality of work is the dominant consideration, quantity is also a factor. Obviously, two men judged to be doing work at the same quality level would be rated by the amount of output.

The way the young engineer performs his work can guide the planning and implementation of suitable training programs. Here is an example:

As a supervisor to a technically competent and bright young man in his work (which was to optimize the economics of an entirely new process concept—working as a professional assistant to a project leader), I noted that he was well liked by his coworkers. He was cooperative and flexible without being merely agreeable. He enjoyed hard work and put in long hours; he had a quick and a good mind, as was clearly shown by his innovative suggestions, his ability to discern pivotal problems and devise effective experimental approaches to their solution. However, even more significantly, he was people-oriented with far greater than usual perceptivity of the complexities of people interactions. He had already established a managerial career as a personal goal. It soon developed, however, that his over-all effectiveness was seriously compromised by an unwarranted inferiority complex, in that he:

- Would not make a decision on his own.
- Was very much afraid of making a mistake.
- Was reluctant to present controversial suggestions for fear of giving offense.
- Had great difficulty in presenting results of his work and his ideas to those in authority.

Identification of these weaknesses provided the basis upon which a series of training and counseling activities were programmed and carried out. These were done with the full involvement of the young man. In less than two years, his self-confidence was restored and his communications skills greatly improved. He now knew that his company would make all reasonable efforts to continue his growth and development if he would do his part. He is today, and has been for some time, a highly successful vice president in the technical management area.

The above example shows how the engineer's work record during the initial years can be used as a basis for skilled performance evaluation and career counseling; I would strongly urge every young engineer to give this particular aspect great emphasis. He should make sure he knows how his performance is evaluated in terms of his resources (actual and potential), including strengths and weaknesses. He should make every attempt to relate his job performance to his career planning in terms of professional growth and development. He should do so in the closest possible collaboration with his boss. If this evaluation is properly handled, it will become apparent to both the supervisor and the young engineer himself, whether or not he should now be seriously considered as a candidate for a technical management career.

Applying Economic and Mathematical Concepts

In establishing his initial professional reputation through an outstanding work record, the chemical engineer will usually have the opportunity to acquire management-related skills—i.e., in applying economic measures and controls to his work; in making reliable cost estimates for processes and plants; and in adopting new, mathematical, computer-based, techniques.

He must take the initiative in making the most of the foregoing opportunities. He can do so by actively seeking to view his work assignment in economic terms at this early stage in his career—for instance, by identifying and measuring the cost elements associated with the sub-units or component steps of his process. After ascertaining which steps are the most dominant economically, the engineer can allocate his own time and resources so that he works on the most important problems first, rather than attempting to work on all of the component parts. At the same time, he should periodically test the development of his process by trying to work up simple mathematical models as a basis for more-effective guidance of his experimental work. He should continue to look into the feasibility of developing simple computer programs that could greatly increase his productivity.

In doing these things, he will seek out the appropriate persons and other resources he can call on for assistance and education. In this way, he will at the very least have learned a great deal about the kinds of input data that managers must deal with, and the problems involved in obtaining and using such information. He will develop a fuller understanding of the short-range and long-range economic factors that govern the commitment of company resources; and this sort of understanding can help him establish a reputation for using good judgment in decision-making, particularly in areas where benefits must be weighed against costs.

Resiliency and Salesmanship

An outstanding leader of men, particularly in the technical arena, must develop a high degree of resiliency if he is to deal successfully with the frustrations and setbacks that will characterize the endeavors of any motivated and ambitious person. He must learn to recognize when the objective may still be sound but requires alternative approaches or temporary compromises. In particular, the quality of resiliency is essential if he is to successfully maintain a spirit of enthusiasm in all that he does. The quality of enthusiasm—one that embodies discerning conviction—is one of the most essential resources of an effective person.

Making the Most of Your Time

Building an outstanding performance record need not mean neglecting your family if you use your working, reading and educational time to best advantage. Here are a few hints:

1. Organize your working hours in line with logical priorities. Take a few minutes at the start of each day to set some tentative goals and plans of action, and then start with a full head of steam. Think along "critical path" lines; this means that when you are given a deadline, you set a timetable for key activities that must take place by a certain time in order for you to meet your deadline. If you fall into the trap of favoring easy-to-handle trivia over more-important matters that involve tougher decisions, you are setting up a "stay late because of crisis" work pattern.

2. Try to combine relaxation with intellectual stimulation during lunch hours and other breaks. It may not be relaxing to talk about very specific aspects of your job over lunch, but there are broader aspects that may represent enough of a break, and discussing these with stimulating colleagues can be more fruitful in the long run than spending the lunch hour trying to get a $3 radio tube for $2.49.

3. When reading something for information purposes (rather than for literary enjoyment), read it quickly and keep the goal of your reading in mind—e.g., jot down salient points, decide whether this or that page is going to be relevant before you read every line, read the results or conclusions of a research study before you become bogged down in detail, and so on. A good book on efficient reading is "How to Manage Your Reading," by Thomas E. Anastasi, Executive Development Center, Massachusetts (1965).

4. When documenting your work, plan ahead. For example, try writing parts of your final report before a project is completed. Certainly, such detailed information as tables of data, description of procedures, etc., can be prepared well in advance. With a little courage, you can even try writing the summary and tentative conclusions before all the results are in. These may have to be changed, but probably with less effort than starting from scratch. You will also be making some typist or secretary much happier.

5. Defeat that tired feeling. Don't stop working at the first feeling of being tired. Push yourself out of the fatigue stage by eliminating distractions, sitting in an alert posture, varying your activities, and plunging forward. Soon you will acquire the habit of not tiring easily.

6. Eliminate nonconstructive mind wandering. Consciously jerk yourself back to the business at hand whenever your attention wanes. We are particularly prone to this problem when concentrating on an unpleasant task or listening to a boring speaker. However, unless you prevent mind wandering, you have wasted your time. In the case of a speaker, even a rambling, undynamic presentation may contain little nuggets of information. You won't know unless you have been an attentive listener.

7. Develop an excellent memory; in the long run, this can be a great time-saver. Associate new facts or ideas with your previous experience to make them meaningful. Review often and soon after hearing new information. Preferably, tell someone verbally. Overlearn material for permanent recollection. Continue to review material even when it seems to be thoroughly learned.

8. Make sure your "communicating" time is spent effectively. For instance, if you are a very slow writer, don't evade the issue by shirking writing chores, but take every opportunity to write letters and reports where you can experiment with ways of speeding up your pace without sacrificing quality. Be natural, and try to write as you would speak. Always remember who your readers are, and tailor communications to meet their specific needs: How much do they need to know? Can a phone call fill their need? Would they prefer a short, informal memo today to a long, formal report a week from now? Or is the matter one that does require a lot of detail and documentation to show that all practical alternatives have been considered? Another example: If it appears likely that the results of your work may not be what your boss is expecting, give him some inkling of this in advance; if you spring a big surprise on him at the last minute, you and he may both end up like the man in the drawing above, reevaluating the techniques or deciding how to change course quickly.

I recollect quite vividly a past associate who had one serious fault—he never was able to deal effectively with frustrations and setbacks. He was unable to understand human nature sufficiently to appreciate that the great majority of persons are most interested in their own problems and in their own ideas, and don't always respond to other people's ideas at first hearing. Instead of using salesmanship in formulating and reformulating his ideas, he would send up trial balloons and, if they were shot down, he would become dejected and abandon further effort on their behalf. Since many of his proposals were of high quality, the penalties paid by him and by his employers because of his easily induced pessimism were substantial.

Resiliency, therefore, is the ability to maintain confidence and conviction under adverse circumstances, recognizing that there are many ways in which problems can be approached. Thus, so long as he maintains his convictions, the engineer should not abandon his goals.

As for salesmanship, the engineer should waste no time learning effective techniques for selling his ideas: His work output has no value until someone believes in it. Salesmanship requires skills in oral and written communications. It also requires the ability to recognize appropriate timing. An effective salesman of technology learns how to present his work output in terms that appeal to the interests of his organization and/or of his superiors or colleagues. His salesmanship will be most effective when it includes well-thought-out suggestions as to implementation.

Individual vs. Team Performance

The establishment of a performance record provides the engineer his best opportunity to develop further understanding of his career objectives in terms of ability to perform effectively on an individual effort basis, as well as on a collaborative group- or team-effort basis. Many engineers discover that they are most effective when working on independent assignments for which they are primarily accountable. Others find this to be the case only when involved as one of a group. A smaller number find equal satisfaction and skills in working on either basis. These are the ones who tend to be most effective as technical managers.

Emotional Maturity

No person should develop such rigid self-control that he loses his warm human qualities; in fact, there are times when emotional intensity is constructive and useful. However, we are often harmed by things said that better had been left unsaid. It is extremely important in our emotional development to learn how to filter out overtones of annoyance, patronization, gloating or chips-on-the-shoulder by the way we modulate and use our speaking voices. It is important to learn how to consider and to deal with conflict and opposition without losing the battle in debilitating emotional skirmishes.

What to Do, and How to Do It

The "fast start" table on p. 7 summarizes some of the what-to-do suggestions we have made, while the table on p. 9 indicates how to follow these suggestions without becoming a slave to your work. If you analyze your own particular situation with a critical eye on how you are spending your time, you can probably add many efficiency pointers to this second table.

By keeping such pointers in mind in your day-to-day work (and assuming your work load is not completely unreasonable), you should have the time to do full justice to your immediate assignments while broadening your interests and skills. Thus you can refine your viewpoint on the four factors (where do you want to go, etc.) with which we started this discussion, and prepare yourself for your first managerial post. ■

Meet the Author and Contributors

E. M. Kipp, author of this series, is shown at the left. As Associate Director of Research and Development for Sun Oil Co. (P.O. Box 426, Marcus Hook, Pa.), he has direct supervision of the firm's corporate basic research, process development and research service divisions. He joined Sun Oil in 1959, was appointed Manager of Basic Research in 1960, and assumed his present position in 1962. His earlier experience was obtained with Aluminum Co. of America and with Foote Mineral Co., where he was Director of Research and Development during 1957-1959. Dr. Kipp has 13 patents to his credit. He has a Ph.D. degree in physical chemistry from Pennsylvania State University, an A.M. from Boston University, and a B.S. from Iowa Wesleyan College. The latter awarded him an honorary D.Sc. degree in 1961.

S. L. Thompson (center) and **A. W. Pollock** (right) made several contributions to this series. Mr. Thompson is a Research Chemical Engineer at Sun Oil, having joined the firm in 1962. He has B.S. and M.S. degrees in chemical engineering from the University of Minnesota. Mr. Pollock is concerned with process development work (process models, mathematical applications, optimization, etc.) He joined Sun Oil in 1952, after obtaining a B.S. in chemical engineering at Columbia University.

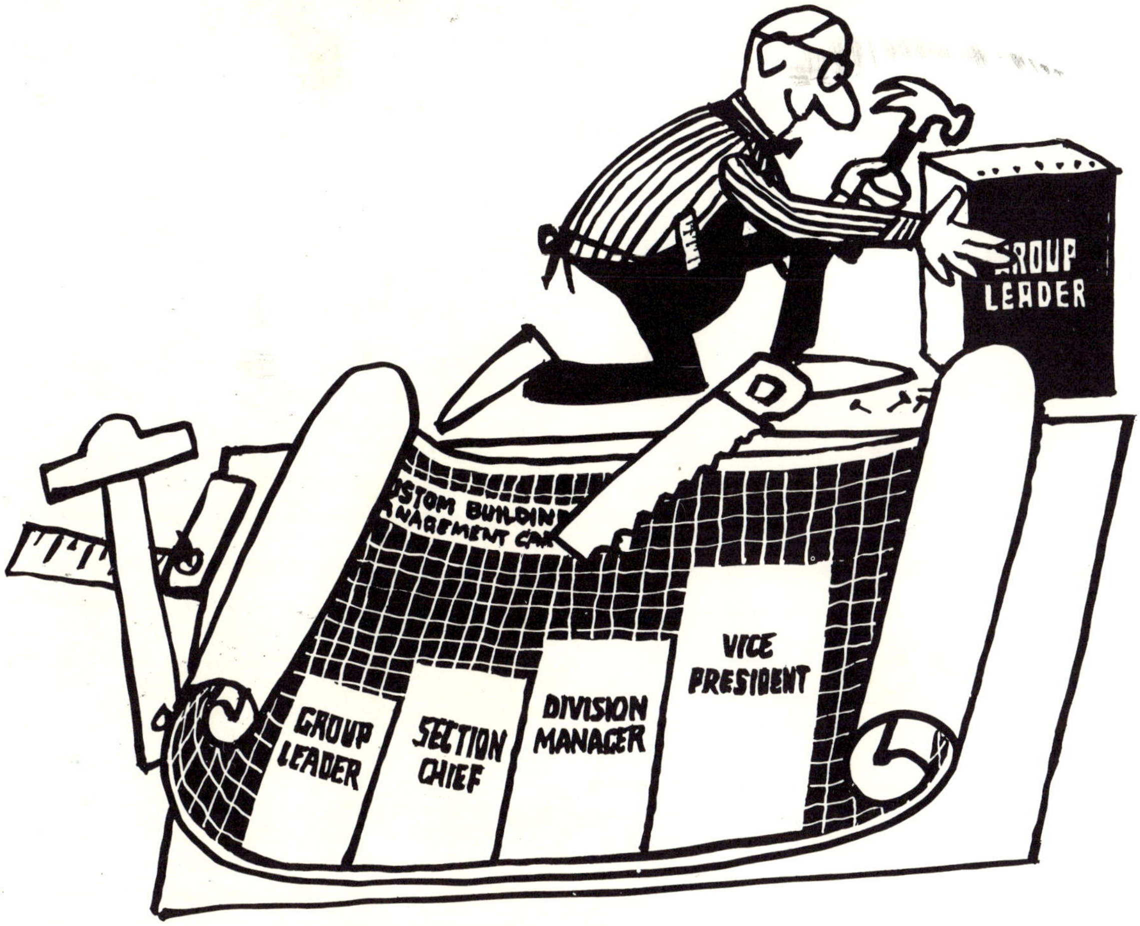

Custom-Building Your Management Career

Part II: The Novice Manager

Here is practical advice on how the engineer can evaluate his own performance as group leader or as some other type of "novice manager"—how he can stay clear of some occupational hazards, and make the most of his opportunities.

E. M. KIPP, Sun Oil Co.

In the previous article, we examined how the nonsupervisory chemical engineer can establish an outstanding performance record. We are now ready to consider the engineer who has already done so.

Let us say that in addition to technical competency, he has displayed promise of having at least some of the following "management characteristics":

- Commitment
- Creativeness
- Decisiveness
- Diplomacy
- Drive
- Honesty
- Intellectual ability
- Judgment
- Motivation
- Objectivity
- Sincerity
- Sociability
- Stability
- Versatility

The next move is to provide the engineer with an assignment where he can put these characteristics to the test, and develop them further. This usually involves a position in the group-leader category, where about 80% of the man's time will be spent on do-it-yourself technical matters, and 20% on managerial matters.

The assignment should not be made lightly, but should be based on careful performance evaluation. Clearly, it is not advantageous to either the engineer or the company to have him embark upon a course of events that might simply end in the destruction of an outstanding engineer and the production of an inadequate manager. (One must also remember that many of the desirable management characteristics listed earlier are also highly desirable for, say, a nonmanagerial senior technologist.)

Some Pointers on the Parceling-Up and Delegation of Work

Crucial to the novice manager's success is how well he delegates work, and how well he integrates his own activities with those of his group. Here are a few suggestions:

1. If you don't know very much about people in your group, familiarize yourself with their backgrounds and interests as soon as possible. What problems has each man handled recently, and how well has he handled them? Does he appear ready to take on bigger problems? What are his goals? Does he appear to be a self-starter? Is he a "loner," or does he prefer to work with a partner? Some of this information can be gleaned from the man's employment record, his past supervisor, and from the man himself. Remember, of course, that this information is just a starting point, and that you will want to refine or revise the initial impressions, based on how well the man works for you.
2. Next, analyze the project you are to manage. Break it into discrete component objectives.
3. Make a preliminary estimate of the needed resources in terms of:
   a. Individual skills and experience.
   b. Equipment and materials.
   c. Space.
   d. Budget.
4. Relate your inventory of individual skills and experience to information you have developed about the people in your group, then:
5. Decide tentatively whether you will make the necessary individual work assignments or whether you will do so by a delegation-of-work meeting.
6. Call your group together for a detailed briefing on the project, being sure to include all background information, justification for the project, probable implementation of favorable results, etc. Encourage questions, then:
7. Present your proposed detailed work-program in terms of objectives, relating this to specific workers, or:
8. Present your detailed work programs, but indicate that final assignment of personnel is open and will be determined at this time through joint discussion. Indicate that preferences will be acted upon to the extent that you can do so without jeopardizing the productivity of the group.
9. Discuss and identify your communications needs thoroughly. In particular, emphasize the individual responsibilities for the various communications segments, specified deadlines, frequencies of reports, circulation listings, and the relative part to be played by written vs. oral communications. Clarify those communications activities reserved for yourself.
10. Agree upon guides for responding to the line-management hierarchy. What should your people do when they are approached by members of higher line management, and what are their responsibilities to you in such cases?
11. Be sure to establish effective communications with your boss(es). Identify their needs; determine the limit of your authority and responsibility, finding out the things your boss does and does not want to be involved in.
12. Translate the output of your group into specific recommendations, based upon your interpretation of the work results. Wherever possible, make suggestions as to the implementation of your recommendations.

Fortunately, getting an assignment as group leader need not mean that the engineer is putting all his career eggs in the line-management basket. If it turns out that he prefers the technical side of the job to the managerial, the wise company makes it easy for him to rejoin the purely technical mainstream without loss of prestige. Chances are that such an engineer will have profited from his experience as group leader; he will have picked up skills and perspectives that he can use in managing and optimizing his own activities.

Getting the New Job Done

Generally, the new job will involve responsibility for the management of a relatively small group of technical people—perhaps two to five, plus a few nonprofessional assistants. The novice manager will be entrusted with conducting a program of work whose objectives are clearly stated and thought to be in keeping with the size and nature of the group. A budget will be assigned, and tentative dates given for completion of the work objectives.

If the company philosophy stresses "participative management" on all levels, our novice manager will have considerable leeway as to how he reaches the work objectives. He must:

1. Determine how to fit the work assignments to the human resources committed to his care: How shall the work objectives be divided, and how explicitly shall they be stated? To what extent will each individual have freedom of decision in planning and carrying out his individual part of the total work? In effect, how will the young manager and the individuals on his staff agree upon their individual work plans?
2. Decide what agreements or arrangements are to be developed for measuring work progress and quality of the work.
3. Begin to view his budget appropriations with a personal sense of responsibility. Therefore, he must learn how to distribute the budgetary resources —including capital equipment, supplemental hardware, etc.—in a realistic and equitable manner among his group, and he must maintain control.
4. Develop mutual understanding and agreement of the degree of initiative and independent decisions allocated to each of his group.
5. Make proper arrangements regarding the always present and critical problem of communications. His main concern will be to develop effective communications within his group and with his own supervisor.
6. Prepare to become involved in salary administration recommendations for his subordinates, and thus begin to contribute to their further professional and personal growth.
7. If he is really an outstanding managerial candidate, he must begin to recognize and accept responsibility for examining the activities of his small group in order to determine whether work assignments can be improved or enlarged. He will make appropriate recommendations to his section chief.
8. If the novice manager is in fact a managerial candidate of outstanding potential, he will not be content to limit himself to his specific assignment. He will, on his own initiative if necessary, seek to view his assignment within the context of the activities of the larger organization of which his group is but a part. This may lead to suggestions, and he will discuss these with his immediate supervisor to see if they call for action—e.g., for lateral communications, or changes in the work assignment.

The novice manager's progress in dealing with responsibilities such as the foregoing will depend almost equally upon his immediate supervisor and himself.

Case Study of a Group Leader

Let us consider the specific case of one Stanley Westbrook. (The name is fictitious, but the facts are as stated.) Stanley was advised at the end of his annual performance counselling and appraisal that management had its eye on him as a potential manager. Accordingly, it had been decided that he was to be promoted to the position of group leader. This is the first recognized line-management position in his company. In this position, he would have two Ph.D.'s, a B.S. degree man, and three technicians reporting to him. The group was assigned to develop the process technology for a new chemical, upon which process economics could be determined and the engineering design for the plant could be based.

The input data to the group were in the form of a detailed technical report from the basic-research section. The work of this section was bench-scale in nature and had been limited to establishing technical feasibility of the underlying process chemistry and identifying the primary reaction parameters in order of technical but not economic importance. The process chemistry had not been optimized in terms of factors such as feedstocks, recycle, byproducts, etc. Stanley was advised that the assignment was expected to be completed within 10 or 12 months and that the human resources assigned to him were considered adequate for the task.

Stanley, being somewhat overwhelmed, excused himself and left the discussion with a mixture of elation and trepidation. The following Monday, the first day in his new position, he called his small group together and told them that they were to work on a special project assignment. He outlined this assignment to them, indicating that their real task was to translate the basic-research technology into a commercially economic process and distributed copies of the basic-research report to them. It was tentatively agreed that on Friday morning the group would meet again, to receive work assignments. No other information was given to, or volunteered by, any of the group.

Stanley spent the rest of the week studying the

report and its function as a base line from which to launch his new project. In fact, it took considerable homework before he understood the information in the report. He did not initiate informal discussions with the basic-research personnel involved, or with anyone else.

On Friday, when the group reassembled, Stanley's first question was, "Do any of you have suggestions as to how we should approach our assignment?"

After a few moments of embarrassed silence, one of the men hesitatingly suggested that he would personally like to work on the problem of feedstock preparation and purity, to determine the optimum specifications for this part of the process. A second man then suggested that he was rather interested in optimizing the catalyst utilization part of the process, as he had considerable experience in catalyst preparation and regeneration. The rest of the group remained silent.

Stanley then said, "Well, O.K." and arbitrarily assigned one of the three technicians to the person who volunteered to work on feedstocks, and the remainder of the group to the person who was to carry out the catalyst utilization study.

For the time being, Stanley assumed the responsibility of considering the other component technologies of the process, reserving the right to select what work he would actually carry out personally, as well as the right to then reassign one or more of the group to work directly with him. He also advised the entire group that if they needed any particular equipment or supplies, he would attempt to provide them. The meeting was then adjourned.

During the next several days, Stanley made it a point to discuss the work of his two key men with them on an individual basis every day. These discussions were characterized by frequent disagreements on matters of technology and design of the proposed experiments. More often than not, the discussions would terminate with a request by Stanley that the work be carried out his way.

In the meantime, Stanley decided that he wished to personally tackle that part of the process concerned with working up of the crude reaction product to recover the major primary product in high purity, and to develop the best use of the other by-products whose primary value would be as recycle feedstock. He determined that he would like to have two technicians to assist him. However, if he did so, one of his key people would have no technician at all. He therefore decided to make do with one technician, and advised the key man involved that a specific one of the two technicians was henceforth to work directly with him (Stanley). This decision was not discussed in advance with the key man involved, but made known to him as a request.

After about three months, evidence of conflict began to be increasingly apparent. The professional man working on the feedback portion of the problem began to complain that he was not being kept properly advised by the man in charge of catalyst utilization as to the progress of his work. Because the feedstock optimization had to be closely tied in with the catalyst work, he could not be certain that his own work would be useful or consistent with the other. Before long, the man in charge of catalyst utilization had a similar complaint.

Finally, both men began to grumble to Stanley, telling him that he was not keeping them advised of his work and progress on product "workup," and that they felt he should do a better job of communicating with them in this regard.

In the meantime, Stanley had made it a practice to report on a biweekly basis to his immediate supervisor on the progress of the various segments of the work—primarily in terms of new information, but without relating this to time and budget schedules of the original assignment. Stanley's boss was overworked and had elected to rely upon Stanley to indicate whether his assistance was required.

At the end of the first three months, Stanley's boss advised him that he would like to have a formal review of the work, and a date was set for this purpose. In preparing for the review, Stanley began to confirm his growing feeling that all was not well. For the first time, he reviewed the status of the budget and whether work progress was on schedule. He recognized that he was in trouble and in need of help.

Growing in the Basic Areas

In moving to the novice manager stage, the engineer needs to expand his experience and skills in these seven basic areas previously discussed in Part I of this series:

Responsibility	There is now more temptation to postpone or avoid unpleasant decisions that have to be made.
Judgment	Human factors as well as technical and economic ones must now be weighed.
Implementation	The novice manager has more latitude in setting objectives.
Group vs. Individual Loyalty	Loyalty to subordinates may sometimes clash with loyalty to others.
Resiliency	Not only the novice manager's own morale but the morale of his subordinates is now a factor.
Salesmanship	He must be able to sell the work of his subordinates as well as his own.
Emotional Maturity	Exerting authority poses new challenges.

So, for the first time, he called his group together, put the cards on the table, and asked for suggestions as to how the review would be handled the types of problems that required identification and elaboration, and the nature of the recommendations that would be in order. It was agreed that the professional group would have to set all else aside to work on this matter. In so doing, a number of things became apparent—the most important of which was that the main problems had not been appropriately identified, orders of priority had not been established, and intercommunications within the group were virtually nonexistent.

Accordingly, the group addressed itself to these needs, and developed recommendations to minimize the adverse effects of this prior neglect. Included was a plan to reorganize the work in a more appropriate fashion to meet the scheduled completion dates within the budget allotments.

Stanley immediately communicated these matters to his supervisor so that the progress to be reported at the scheduled review would not come as a complete surprise. Also, Stanley hoped that his supervisor would be able to give him counseling and advice that was obviously needed.

Stanley's Kind of Group Therapy

As a result of the review presentation, the deficiencies of the approach applied to date were recognized and Stanley agreed to implement corrective actions. His first step was to call another meeting of his small group, stating that the members were to jointly determine the significant component elements of the total assignment (problem); attempt to set some order of importance or sequence to the problems; and develop an over-all PERT-type diagram for coordination of individual efforts within the group.

It was also decided that the group would meet on a weekly basis for a detailed discussion, to assure that there was effective intercommunication. Each man could make suggestions on any aspect of the work.

In addition, Stanley continued daily discussions with each of his group, these discussions being characterized by frequent disagreements on technical matters. Although Stanley defended his own opinions strongly in some of these discussions, he no longer insisted that his opinions prevail in all instances.

Within about four weeks' time, it became apparent that the project was on the upswing; morale was improving and the almost daily gripe sessions with Stanley, wherein each man complained about the deficiencies of the rest of the group, had virtually ended.

Stanley also discussed his problems much more fully with his supervisor, seeking his counsel and advice, but making his own decisions in the final analysis.

At the six-month formal review period of the project, progress was found satisfactory. Barring some

"How Am I Doing?" Self-Evaluation Sheet

The ability to critically evaluate one's own parformance is a valuable skill. As a first step in sharpening up and applying this skill, I suggest that at fairly frequent intervals during your "novice manager" tenure, you retire to some quiet, private corner and work through this self-evaluation list as conscientiously and searchingly as possible.

1. **Work Objectives:**
   Are activities on schedule? Within budget?
   If not, could this have been avoided or at least foreseen?
2. **Quality of Work:**
   Putting yourself in your boss' place, would you say the quality of your group's work is outstanding, good, marginal or poor?
   If less than outstanding, do you have some viable plan to upgrade it?
3. **Interpretation of Work Results:**
   Do they confirm original work objectives?
   If not, what recommendations have you made for appropriate adjustments in your assignments, including budget, target dates, objectives, etc.?
4. **Communications:**
   Does each subordinate feel well informed about the group project and any related matters? Does he freely propose and defend his ideas?
   Does your supervisor feel you are keeping him adequately informed and involved?
   Have people in the organization (other than your subordinates) who are qualified to help been given the opportunity to do so?
5. **Morale and Growth:**
   Would you rate your group's morale as good, marginal or poor?
   In the latter two cases, do you have a program to improve it? For instance, are you helping your people to develop their full potential by encouraging their educational and keep-informed activities? If you have been in charge of the group for some time, are you giving your people a chance for increasingly challenging or at least varied assignments, rather than have them do the same sort of thing over and over?
   How is your own morale? If it is low, are you sure the reasons are valid and beyond your control? For instance, if you think other group leaders have more-promising assignments than yours, are you sure you are not just comparing the good points of their job with the bad points in yours?
6. **Challenges and Achievements:**
   In considering your stewardship of the group, of how many achievements are you really proud? Do you feel you have invested your time to best advantage, or have you dissipated a lot of your time on relatively unimportant aspects of your job? Have you developed any other bad habits—e.g., falling prey to unconstructive worry instead of thinking through a plan for contingencies?
   When you examine your image, are you sure that you appear to enjoy the challenges of your work and are eager to take on bigger ones, or do you project either a "barely holding on," "martyr" or "in a rut" image?

unforeseen incident, the technical targets would be met on schedule. However, it was necessary to request an incremental budget allocation of 30% of the amount originally approved for Stanley's group. Otherwise, the project was successfully completed.

However, there was one important sour note associated with the conclusion of the project. At the final report, a number of other supervisors and executives from the R&D department had been invited to be present. The ensuing discussion clearly indicated that important additional work could have been done that would have made the process considerably more useful in terms of related opportunities than was the case. It also became evident that had Stanley or his supervisor recognized the importance of lateral discussions with others throughout the R&D group, these additional opportunities would have been more likely recognized.

After submitting his final report, Stanley was asked by his supervisor for salary recommendations for his small group. He was also asked to submit an analysis of their performance in terms of respective strengths and weaknesses, with suggestions as to further training or other assignments that might be appropriate to their actual and potential skills. Stanley was quite unprepared for this request, and was able to handle it only on a minimal basis.

Stanley's immediate supervisor did not initiate a review of Stanley's performance in his first line-management assignment, nor did Stanley request such a review. Fortunately, the division manager was alert to this omission, and was able to develop a meaningful and fair appraisal of Stanley's performance, evaluating it in terms of potential for higher management assignments.

Perhaps the reader would wish to analyze Stanley's performance in the light of the guidelines given previously, and develop his own assessment not only of Stanley himself, but also of his supervisor.

Opportunities of the Novice Manager

Contrary to what many young men seem to believe, the individual subordinate can and often does have as much, and sometimes greater, control of his professional destiny as does his supervisor. Therefore, if the young engineer is suited to a managerial career, he must begin to demonstrate at this initial phase the qualities of his motivation and drive that will be so essential in his further development.

If we assume that the young engineer's supervisor is competent, the two men will be successful in working out the guidance, training, or counseling appropriate to helping the engineer meet the new challenges of his first step up the managerial ladder.

Although not every engineer goes through the group-leader phase as he moves up this ladder, such a phase can be very fruitful for him and his company because:

1. He has the opportunity to learn without undue risks to the company or himself. The limited size of the group he is supervising provides sensible restrictions in this content. Experience based upon the commitment of honest or intelligent mistakes can be a very effective teacher. In my opinion, there continues to be too much managerial restraint upon innovativeness and leadership (both technical and administrative) because of an attitude toward mistakes that seems to be more punitive in nature than constructive.

2. He can further develop his technological competence. Generally, the managerial and administrative aspects of the job as described should not require more than from 10-20% of the novice manager's time, as mentioned earlier.

3. He is challenged to develop his leadership potential early in his career. Many younger managers of great potential are often discouraged by what appears to be many interminable years ahead of them before present incumbents of higher managerial positions are ready for retirement or are themselves promoted to larger opportunities. If advance in management depends on these kinds of logistics, both the motivated young manager and the company suffer. The group-leader position provides him with an early outlet for his managerial drive, because he can make the work of his group so useful that it will expand in both scope and function. In this way, he can literally create his own promotional opportunities.

For example, consider an innovative group leader who is ambitious and highly motivated. If he can broaden the scope of the work assigned to him by virtue of outstanding performance, this will justify his being given additional resources. If all goes well, the group will become a section, and the young manager will have achieved his promotion without the necessity of a vacancy in a section chief's job. (By the same token, an aggressive and productive section can grow into the scope of a division.)

Input for Career Decisions

The final virtue of structuring a group-leader rung into the managerial-career ladder is that this rung provides information to enable the "climber"— and his superiors—to determine whether this type of ladder is the one he is best fitted for.

In Stanley's case, his performance was certainly imperfect. However, in view of his rather inadequate supervisor and the marginal nature of on-the-job training that was provided, I would say that he had passed his initial "first rung" test quite adequately. The way he turned the project around gave evidence that he could profit from rocky experiences and apply his new managerial insights in a purposeful manner. If he continues along this path in his next projects, and if his desire to cope with the challenges of management remain intact, he will soon be ready to tackle the next managerial rung—i.e., the Section Chief type of position, which we will examine in the next part of this series. ■

Custom-Building Your Management Career

Part III: Section Chiefs and Other Apprentice Managers

The "section chief" level can be a dignified plateau for anyone who wants to divide his time evenly between management and technology—or it can be a stepping-stone to higher administrative levels. For either case, here are stimulating guidelines.

E. M. KIPP, Sun Oil Co.

The positions we will be discussing in this section generally encompass a salary range of \$18,000 to \$30,000. Although at least ten years of prior experience may usually be required, some engineers have made it with less. The work is 40 to 50% administrative, and 50 to 60% technical.

In the R&D field, the most common title would be section chief, and the position is one or possibly two rungs below that of director of R&D (depending on the size of the research organization.)

In the engineering section of an operating company, the most common title would be assistant chief (i.e., of process, design or construction engineering), and the position may be one or two rungs below that of chief engineer or manager of engineering.

In an engineering-construction company, the comparable job is generally one level below that of project manager, and can have a number of titles involving project engineering, field engineering, or some other function.

Typical titles in a manufacturing division would include assistant superintendent, assistant plant manager, and production supervisor; the position is either one rung (small plant) or two rungs (large plant) below the plant manager.

The above positions are the first ones on the ladder that cover most of the basic managerial functions in a primary way (rather than in the adjunct way that is more characteristic of the group leader). I will therefore refer to this level as that of an apprentice manager. The apprenticeship is completed when the job is mastered and when a certain degree of self-discovery has taken place; the engineer can then decide whether to strive for higher managerial levels that involve less technology and more administration, whether to try to grow within the section-chief level because it strikes the optimum balance for him, or whether to pursue some other career path. (Part IV will take up this decision in detail.)

Let us now assume you have just been promoted from group leader to section chief. Although our discussion will take place in an R&D setting, most of it will apply to the parallel positions cited earlier.

As a section chief, your managerial and administrative functions and responsibilities now escalate by a significant amount. In the preceding "novice" phase, you were responsible for the work of a group consisting typically of from two to five persons; now you will be entrusted with the direct line supervision and leadership of several such groups. These may most often comprise from two to five groups, representing a total staff of from 10 to 25 professionals and perhaps a similar number of technicians.

What additional skills and resources must you now acquire? I believe the factors below are the more critical ones, arranged in an approximate order of significance to outstanding performance.

Establishing and Meeting Work Objectives

It is not enough merely to identify the work objectives. To be meaningful, they must be subjected to certain constraints and associated requirements; the most important being the total resources to be committed to the programs, and an appropriate schedule of target dates for completion of the various components of the work assignments.

In effect, then, we are talking about a program and budget. These will be most meaningful when worked out jointly with your supervisor in a manner that ensures that the final workplan is fully understood and supported by both of you.

The workplan, as reflected in the program and budget, will be seriously compromised if human resources are not commensurate in quality and quantity, or if these resources—engineers, technicians, chemists, etc.—are not skillfully allocated to the various components of the program.

The human-resource-allocation guidelines that were offered to the group leader in Part II (particularly the "Pointers on the Parceling-Up and Delegation of Work" on p. 108 of the Feb. 10 issue) also apply to the section chief. Some of these guidelines take on added significance: Whereas the group leader who is promoted from within the group is likely to be familiar with everyone's capabilities and idiosyncrasies, the newly promoted section chief may encoun-

Developing Your Natural Managerial Style

As a section chief, engineering supervisor, or similar type of technical manager, you have the choice of basing your managerial style on what has been called* "Theory X" or "Theory Y," or somewhere in between.

In essence, the assumptions of Theory X are that most people inherently dislike work, and will try to avoid as much of it as possible. They must therefore be directed, controlled—and threatened with punishment for failure—in order to put forth adequate efforts toward meeting company objectives. Because most people wish to avoid responsibility, and yet want security, they prefer strong external direction to individual decision-making.

Theory Y, on the other hand, basically assumes that:

- Most people do not inherently dislike work; it is as natural as play or rest.
- People will exercise self-direction and self-control to meet objectives to which they are committed.
- Commitment to objectives is a function of the rewards associated with their achievement, and is materially affected by management policies and practices.
- Most people can learn not only to accept but also to seek responsibility.
- Imagination and ingenuity in the solution of organizational problems are widely, not narrowly, distributed traits in the population.
- In modern industrial life, the intellectual potential of most people is only partially used.

For most technical managers, Theory Y has much more to offer than Theory X, particularly in managing fellow professionals. Within Theory Y, there is considerable leeway for individuality in choosing the response to a given situation. Even the occasional blunt use of authority is not necessarily inconsistent with this theory, where commitment to organizational objectives has proved unobtainable by other means. But once this commitment has been obtained and obstacles resulting from complacency or internal dissension have been removed, reliance on authority should give way to participative management—e.g., in getting your men directly involved in the process of defining their responsibilities, identifying the meaningful indicators of their performance, establishing their performance goals and formulating their own self-improvement targets.

* McGregor, D., "The Human Side of Enterprise," McGraw-Hill, New York, 1960.

ter many a surprise from his larger and more-varied staff, and can take much less for granted.

He can no longer merely philosophize in a detached manner about the great complexities and enigmas of the human organism; he now has to organize his men into an effective team. Once the workplan is developed, this involves the following three tasks:

Definition—Here, you must classify the work to be done in such a manner that each person can see how he fits into the total picture. Jobs should be so defined that there is little overlap of individual functions.

Delegation—Once a job is delegated, you must be careful to avoid trespassing upon the delegated responsibilities. Individuals should be encouraged to make decisions as much as possible on their own, leaving a minimum of detail for you.

Direction—This important function involves recognizing situations that need improvement and correcting them in an effective manner.

In carrying out these team-building functions, you will encounter the question of managerial style; this is amplified in the box on the opposite page. Try to develop a sensitivity for how people react to your managerial style. This does not mean that you necessarily have to change your style every time you sense an objection, but you should be aware of the extent of the objection, and of its impact on productivity.

As a new manager you must realize that you have yet to earn the respect and confidence of your subordinates. You are starting out with a clean slate, because even the people who knew you as an effective group leader will usually have an open mind as to what sort of section chief you will make. Earning their loyalty represents a challenge and an opportunity—one that can be met via high ethical standards, fair play, consistency (i.e., practicing what you preach), effective communication, and personal productivity.

Performance Evaluation and Counseling

On accepting your managerial assignment, take the initiative and request counseling and performance evaluation from your superior on a more frequent basis than the usual annual review.

Such an evaluation can be very helpful if related to mutually agreed upon performance targets. These must be suitably qualified in terms of limitations beyond your control, unforeseen events, etc.:

- Underlying premises and qualifications must be spelled out and understood.
- Control and review procedures must be established.

The performance evaluation then has specific meaning in terms of whether the agreed upon objectives have been attained. In examining this aspect, it will also be useful for both you and your supervisor to analyze whether the actual performance might have been even more successful than it was. Constructive understanding can then be developed as to the reasons for less than complete success.

The analysis should be carried out on a positive note, with the major intent being to learn from experience. The discussion can lead in a very natural way to an evaluation of your performance in terms of your personal skills, resources and weaknesses.

Based on such inputs, one of your major objectives should be the selection of personal training programs that will help you grow in your job. For the most part, suitable courses would be quite similar to those offered by organizations such as the American Management Assn., and would more often than not be introductory rather than advanced. For an R&D section chief, such courses should emphasize project selection and evaluation, product and process development, and salary administration.

With this counseling and training approach, supplemented by outside reading, you can develop a sensitivity for new developments or trends, both within and outside the company, that ought to have an immediate bearing upon the workplan agreed to by you and your supervisor. Remember that *any* plan, *any* program, *any* budget, should be a living agreement flexible enough to adjust to new development: Steering a straight course does not mean locking the rudder.

Having thus committed yourself to continued personal growth and development, you must now try to get the same sort of commitment from your subordinates, via performance evaluation and counseling that you carry out yourself. (More on this later.)

Project Evaluation

In accepting the assignments given to your section, you must develop the skills required to properly determine when a project should be terminated, changed in orientation, escalated or de-escalated.

A wide variety of economic tools are available for this purpose; and in larger organizations, these resources will be available as a staff service or can be hired as a consultant service. A typical grouping of economic yardsticks for R&D projects will include:

- Interest Rate of Return (IRR). Otherwise known as the Profitability Index, Discounted Cash Flow or True Rate of Return, this yardstick has been well described in the literature.
- Present Worth (PW). This is an estimate of the cash flow, discounted back to the present time, over and above that needed to earn 15% and to recover the projected investment in a potential venture. The PW can be used in conjunction with a factor that reflects the estimated probability of accomplishing the technical and marketing goals set for the project, and any other pertinent contingencies.
- Project Incentive Ratio (PIR). This ratio represents the PW divided by anticipated research expenditures that would still be necessary for commercialization (put on an aftertax basis and discounted at 15%). For instance, a PIR of 1.0 would mean that project is estimated to yield an aftertax return of 15% over and above the recovery of R&D costs.

The First Hundred Days on the Job

Getting off on the right foot is a trite yet appropriate admonition. What you do or do not do during the first three to six months in your new managerial post will shape the course of subsequent events. Here are some thoughts that may be helpful:

1. Do not play the role of a "big wheel." A reputation for arrogance and vanity is easy to acquire, but hard to shake off.

2. A slow start and a fast finish is better than a fast start and a disastrous finish. Take the time for an effective countdown to be sure that all systems are "go." This will let you check and adjust vital matters such as:

- Do you properly understand your assignments and responsibilities and their implications? This is usually best done by constructing your personal workplan for review and evaluation by your boss.
- Do your subordinates clearly understand the role that they are to play and your expectations as to what will constitute satisfactory or unsatisfactory performance? Again, this is often effectively done by seeing that they submit their individual workplans to you for review and discussion.
- Are the resources, human and otherwise, entrusted to you commensurate with your assignments? A good approach to this question is the preparation of a PERT-type arrow diagram to check out target dates, flow of work results, specific work assignments, etc.
- Have you mapped out your communications strategy? Poor communications are too often an excuse for poor performance; therefore, decide in quite specific terms how you want to handle this. A good way to construct your communications chart is to do so in terms of the key people involved. Each of these people will have their rather specific needs and interests. Thus, what will your superior need; what will others in the organization who have need to know require; what will each of your subordinate groups need for most effective performance; and, finally, what are the miscellaneous communication needs? Having prepared this assessment, you should find it relatively easy to establish an appropriate charting of your communications needs and the manner in which you propose to respond to them. This chart should be specific as to dates of deadlines for specific communications events—which ones are to be written, oral, or both. In particular, the individuals whom you will ask to be responsible for particular communications events should be identified and you should discuss with each of these how they are to handle this responsibility.

3. Establish your personal priorities in budgeting your time for the different phases of your job. This will help you to identify those parts of your job that can be delegated to subordinates. Do not permit yourself to become a bottleneck in decision-making, flow of paperwork, or in the solving of important personnel or related matters. Above all, tackle the tough jobs first and get these out of the way.

4. Be attentive to the attitudes, needs and problems of your people. Listen to them. A good start here is important.

5. Be sure that your actions are consistent with your expressed philosophy. Try to see yourself as others see you during the sizing-up period. For instance, let us say that one of your initial statements was that you welcome suggestions from your group leaders on any phase of the section's operations. You then let a group leader's suggestion memo languish on your desk for a month because things have been too busy to discuss the memo with him even on an interim basis. While you may just see yourself as a man with a lot of demands on his time, he may see you as insincere or autocratic.

The minimum acceptable numerical value for an economic yardstick such as IRR or PW that a research project must meet if it is to be continued, varies from company to company. One firm, for instance, may require that the IRR be at least 15%, on the theory that unforeseen problems will inevitably develop that will tend to erode the project's profitability by the time it is ready to be fully commercialized. On the other hand, some companies will accept a lower IRR value, say 5 to 10%, on the theory that once a commitment has been made and the project commercialized, opportunities will present themselves for further process and product improvements, and cost cutting.

Good managers learn rather quickly that the value of elementary economic yardsticks such as IRR and PW is their usefulness in pinpointing factors that will have the most significant economic leverage upon the attractiveness of the project. (For example, a drop in raw-material cost has more impact on the IRR of ventures with low investment/volume ratios than on those with high ones.) In this way, the main problems can be more readily recognized, and available resources applied to their solution—rather than in assigning manpower to problems that, even if solved, would not be controlling.

Beyond the Economic Yardsticks

Of course, there is a rather large variety of other managerial decision-making ideas and techniques. You normally need not become skilled in applying all of them, but you should be aware of their existence and, in a general way, of their chief limitations and strengths. Among these are managerial tools such as decision trees; various varieties of PERT diagrams; planning by objectives in the larger corporate sense, etc.

Most of these tools—as well as large-scale optimization methods such as linear programming, stochastic or competitive models to simulate random or competitive influences on company operation, and comprehensive management information systems—are computer-implemented.

Advances in computer hardware and related skills are among the most significant developments of our time. Although computer resources are by no means panaceas, they are potentially one of the most powerful tools available to researchers, designers, and managers alike. The apprentice manager phase is probably the most appropriate time to become personally involved in the managerial applications of the computer.

If you have already developed familiarity with computer resources in programming and solving your engineering problems, you should have little difficulty in understanding and assessing these new techniques for solving management problems—particularly if you avail yourself of one or several of the following opportunities:

- Renew acquaintance with the computer, possibly through a time-sharing remote terminal where you can converse with the computer in writing, debugging and executing programs.
- Read some of the many articles on advanced computer-based management techniques frequently appearing in *The Harvard Business Review, Datamation, Management Science*, etc. (Some process-industry-oriented articles of this kind have also appeared in CHEMICAL ENGINEERING, particularly in the investment-appraisal area.)
- Consider taking some of the short courses (one week or less) on these techniques offered by the American Management Assn., various computer manufacturers and software houses, or consultants, etc.
- Acquaint yourself with how your own company is applying or planning to apply computer-based problem solving—perhaps by having informal lunch-time chats with appropriate staff specialists.

One thing should be remembered in using these managerial approaches: the computer solution obtained should be subjected to the same scrutiny as the solution to any engineering problem, namely:

1. The models used by the computer should be tested, where possible, against their real-life corporate counterparts.
2. Parameter values and input data fed to the computer should be clearly specified and understood.
3. The sensitivity of the solution to the more uncertain parameters and inputs should be known.
4. The reason *why* a given solution was obtained should make sense, based on judgment and experience.

Salary Administration

The apprentice manager must be fully conversant with the salary administration philosophies and

Coping With Crises

Adversity comes to all of us in many packages. Times of crisis are inevitable—in fact, they often constitute the real challenges of a job and, when properly surmounted, can provide considerable satisfaction. Crises can involve technical, administrative or personality aspects of your job. Your handling of these inevitable moments of tension should be effective if they include most of the following guides:

- Don't lose your cool. Most crises in retrospect are relatively inconsequential. A touch of humor at the right time can sometimes work wonders.
- Define the crisis in terms of its apparent components. Arrange these in order of urgency or significance. Don't be diverted by trivia.
- Relate the foregoing where possible to the person or persons most directly concerned, involved or best informed and,
- Discuss the appropriate parts of the situation on a direct personal basis with each person.
- Then, and only then, determine your response to the crisis on a step-by-step basis and proceed to implement the response.

You may decide that a straightforward, candid confrontation with a key person or element of the crisis should be the initial response. Or you may decide that the response will have to come, at least partly, from higher up—in which case, the steps you have gone through will enable you to discuss the problem with your boss on an informed, unemotional basis. (Also avoid overly defensive postures.)

Remember, in the personality-clash type of crisis, it is usually best to let the other person have his say. If you try to shut him up before he has let off steam, the crisis may recur when his resentment comes to the surface again. On the other hand, if you let him give vent to his resentment and talk himself out, he will usually be much more receptive to what you have to say when it is his turn to do the listening. Many personality-clash crises are based on incomplete, inaccurate or biased information; your job is not only to ferret out the trouble and present the correct facts, but also to present this information at a time when the disputants are ready to listen, and in such a way that the information can be accepted without too much loss of face.

If you are worried about a deadline-type crisis, don't try to shield your group leaders from the seriousness of the problem—remember that participative management involves participating in the rough spots as well as the routine decisions. In other words, you can let your concern show so long as you do not imply that you have lost control of the situation and have no plan for dealing with it.

mechanisms of his company, and particularly those of the division of which his section is a part. You must now begin to develop the objectivity and skills that hopefully will help you to reward your subordinates in a manner that reflects their productivity and related factors.

You will have to cope with the difficulty of separating your personal feelings or friendships from other inputs in making salary administration decisions. Many otherwise fine managers are guilty of the serious flaw of permitting their emotional feelings about people to override the more objective considerations; they may find themselves over-rewarding those who think like they do, or who successfully curry personal favor with them. At the opposite extreme is the manager who gives the greatest rewards to those subordinates of whose loyalty or approval he is least sure, on the theory that this will win them over.

Whatever your eventual recommendations may be in the salary area, the final checkpoint is whether you feel comfortable about the basis for the decisions. If you feel that you have examined all the evidence as objectively and consistently as possible and your conscience is clear, you have done your job well.

Professional Development of Subordinates

The assessment of skills, strengths and weaknesses of your subordinates should begin on the very first day of your new assignment. Their individual career progress, as well as your section's productivity, can depend on what you do or don't do in this area.

It will be your responsibility to provide your men with appropriate opportunities for professional and personal growth. Initially, at this stage of management, it is probably most important that you attempt to determine whether the subordinate is in the proper job. If not, what other types of assignments would be more suitable? If he is in the right job, what are his potentials for further growth, and how can these be further developed and applied?

Be prepared to make specific recommendations and observations on all of these points—including appropriate training programs for your various subordinates. These would mostly be in the area of technical skills and would be closely designed either to prevent or repair technological obsolescence, or to develop specific professional knowledge required by the work objectives. We have already mentioned the need to be aware of trends that may affect your section's work objectives or methodology in the future—such awareness lends logic to long-range training goals.

Some of the most critical indications of your potential for high-level technical managerial responsibilities may well emerge from the concern and skills you display in "developing" your subordinates.

Perhaps the most difficult skill in this area is adjusting your own technical judgment and knowledge to that of your subordinates. If you insist that your own judgment and approaches prevail, how can you establish an environment in which your men can evaluate and further improve their technical competency so that they can be true professionals? Can you develop enough self-discipline to give them freedom to make their own decisions, and to accept accountability for these decisions after you, as the manager, have stated your own position?

The manager who feels that *because he is manager* he is necessarily superior to his subordinates in technical skills is not a big enough man to become an outstanding leader. This is not to say that the manager must always subordinate his own technical knowledge and judgments. Rather, he must develop the ability to adjust his actions to the particular situation and risks involved, recognizing when his own approach is merely different, and when it is clearly superior. In the latter case, a persuasive explanation of his viewpoint will do much more for the development and morale of his subordinates than mere pulling of rank.

Broadening of General Managerial Concept

Finally, as you master this level of managerial responsibilities, your viewpoints must begin to broaden out: In addition to appropriate concern for the needs and opportunities of your section, you must recognize when these have to be subordinated to the larger needs of the company.

This is a type of managerial maturity and discipline that many otherwise good managers fail to learn; but the outstanding manager must and does learn that the part can never be greater than the whole, and he proceeds accordingly. In this context, you must now search out opportunities and approaches that will enable you to support your supervisor in the development of larger opportunities for all his sections. You must learn how to recognize situations where your own group's assignments and responsibilities can be more supportive of those of other groups, and vice versa.

Because you may properly be expected to make suggestions in policy and procedural matters not only for your section but for the larger organization, try to learn more about your division, its organization, philosophies and policies. Also, some of your outside reading should now deal with the broader aspects of business administration, human relations, and community problems and affairs. ■

Custom-Building Your Management Career

Part IV: At the Crossroads

The time has now come to evaluate new inputs and reexamine four questions:
Where do you want to go?
When do you want to arrive?
How do you wish to travel?
What will it cost?

E. M. KIPP, Sun Oil Co.

Although this part of the series is pointed quite directly at the man who has proceeded through the novice and apprentice phases of his technical management career, I believe that even the recent graduate can derive some important insights from our discussion: it will help him become aware of the experience and values he will find useful if and when his own career reaches the "crossroads" decision.

At that point of decision, you—the section chief, assistant plant superintendent, or equivalent—must be ready to make a firm, realistic choice between technical management and some other type of career. Also, there are related decisions that must be made. Specifically, you must come to grips with each of the four major career-planning questions that were posed at the beginning of this series:

Where Do You Want to Go?

If final and firm commitment to a top-level management career is the answer, a tentative decision as to the ultimate target must be made. Is this to be unlimited in the sense of shooting for the presidency of the company? Or a vice-presidency? Or should you zero in on director of R&D or manager of engineering as your ultimate goal?

These questions, to be realistic, require that you also judge whether you wish to be associated with a large, complex corporation, a medium-sized one, or a relatively small firm. The working environments associated with various sizes and kinds of industrial organizations will differ markedly—for instance:

Small Companies—Generally speaking, a skill and liking for entrepreneurship may be the dominant personal asset here. Decisions can be made and implemented more quickly, because authority and responsibility are usually concentrated in a small number of key persons. Staff services may be lacking. Although consultants may take up some of this slack, resourcefulness and versatility are generally at a premium. If financial resources are limited, the technical manager feels maximum personal accountability for the success of new ventures. Top performance in rapidly growing small companies can lead to rapid advancement, but security may be lower than with the larger, well-established corporation.

Large Companies—Authority and responsibility may be spread among more people; the making and implementation of policy and work decisions may take longer (although some large companies are now trying to move faster in this area). In a stable organization, top performance or potential may not necessarily lead to instant advancement, but lateral transfers (e.g., to senior staff positions) can provide opportunities if the promotional channels for line management become clogged. High-level technical managers can be more specialized than in the small company.

Some effective soul-searching on matters such as the above will prove helpful in determining your answer to "Where do you want to go?"

When Do You Want to Arrive?

Many a career has gone astray for lack of a timetable of key checkpoints. There are only so many years in a career, and it is therefore essential that you develop target dates for arriving at each of the subsequent managerial levels.

This is probably best done by working backward

from the ultimate-goal position. If this is at a presidential level in a large company, the target date should normally be in the age bracket from 55 to 60 (or sooner if possible). The timetable can then be completed by providing training periods of from four to six years in each of the subordinate managerial levels below that of the presidency. (I believe most people would agree that the outstanding person can become proficient in a given managerial position within three years; the rest of the period would represent "peak productivity" years.)

Should you keep your own counsel in working out this timetable, or should you seek out your supervisor's opinion on what may or may not be realistic?

If you have a good record as an apprentice manager, and work for a reasonably people-oriented and forward-thinking supervisor, by all means discuss your goals and timetable with him.

I realize that you may feel uncomfortable or presumptuous in asking him whether he believes that you may be a potential candidate for, say, the presidency of the company; you may prefer to put the question in terms of how high he thinks you could go. Remember, the question should never be posed in a manner to suggest that you are asking for a promise of promotion or any other kind of commitment. Rather, make it clear that you are simply hoping to achieve two important communication objectives: first, to indicate your tentative career-planning goal; second, to find out whether your management regards this career objective as a reasonable one in terms of what you have demonstrated so far.

If you exercise tact, judgment, open-mindedness and sense of timing, this sort of interview can be advantageous to your company, your supervisor and yourself. It will help you plug new inputs into your timetable—e.g., the probabilities that managerial promotions will be open at the times you project that you will be ready for them.

Another factor that must be considered with candor and courage is the number and quality of other persons in competition. Hopefully, your supervisor will realize that it is false service to both his company and to you to paint an unrealistically rosy picture in order to keep you in his division. The more enlightened course is to be constructively candid—to give the competition its due and to appraise the effect of it on your promotional outlook under reasonable assumptions of corporate and personal growth.

How Do You Wish to Travel?

This pivotal question involves the personal values and philosophies that will largely determine your communications and interactions with people as you travel towards your ultimate goal.

The issues involved fall into three groups; the first can be considered a group of "stimulators" for the outstanding technical manager. Based on several studies reported in the literature, on personal communications and my own experience, these stimulators have been found most compelling in bringing about commitment:

Achievement—Above all else, the top performer values the opportunity to build and to get important things accomplished.

Recognition—Although he does not desire recognition to which he is not entitled, he does expect it will be given freely when earned.

Nature of the Job—Among these considerations, perhaps the most important is that of freedom to innovate, and to build the job to fit the man's primary strengths so as to challenge his full resources.

Responsibility or Accountability—The outstanding person, whether manager or engineer, responds to the challenge of accountability for what he does or does not do.

Salary—Aside from the spending-money aspects, salary is important as recognition for achievement and potential, and as a practical indication of supply and demand for your managerial talent.

Prerogatives and Status Symbols—These are valued primarily as physical or other obvious signs of a man's achievement and progress.

Opportunity to Shape Course of Events—This includes power or influence over financial, material and people resources, and over the selection, control and implementation of objectives and policies.

Respect—The actual or anticipated respect and loyalty of subordinates, colleagues and supervisors can act as a major stimulator.

When you are faced with the crossroads decision, make a conscientious effort to examine the foregoing factors in relation to your personal values and drives. Do these factors, viewed as a group, embody the primary incentives for your managerial aspirations, or does the possibility of extra money and prestige predominate?

Personal Attitudes and Philosophies

These attitudes comprise the next group of considerations that pertain to the "How Do You Wish to Travel?" question:

Technical vs. People Orientation—Managers with a compelling orientation to both technical and people problems are the most desirable, although by the same token in short supply. However, if your primary love and joy are associated with technical rather than people problems, you should seriously question whether the goal of high-level line management is the best one for you. For instance, a better alternative might be the type of high-level technical staff position that involves managerial decisions but little or no responsibility for supervision.

In attempting to make some judgments as to the nature and depth of your people-orientation commitments, consider such things as: Do you really like people? In general, do you trust or distrust them? Are you sensitive to your impact on others? Do you view people as a means to an end, with the end justifying the means—or will your personal fulfillment be complete only when it can, to whatever degree possible, be based upon the fulfillment of your subordinates, colleagues and superiors?

Delegation of Work Responsibilities—Excessive caution in delegating, or insistence on "do it exactly like I would do it," are traits that limit the effectiveness of many managers. Some who sincerely believe they delegate effectively are reluctant to learn that they may be the only ones having that opinion.

In your case, are you willing to really entrust a subordinate with an important segment of the work for which you as a manager have accepted full accountability? Do you enjoy giving credit and will you accept blame? Do you feel it incompatible with managerial prestige to admit you have been wrong?

Morals, Ethics and Courage—Despite widespread cynicism on this matter, you do need a sense of ethics in order to make defensible moves as a high-level manager. If responsibility for profitability as well as for technical success is going to lead you to an "anything goes" philosophy, this sort of destructive expediency will eventually make you wish you had stuck to technology. Of course, the same may hold true if your sense of doing unto others is so exalted that you view competitive battles with complete distaste, or if your philosophy is so antimaterialistic that you view any sort of cost-control pressure as the unreasonable demands of "petty accountants."

Within this framework, you need the courage to stand up for demands you believe to be reasonable. Sometimes, you will need the strength of character to subjugate your personal interests, and those of your subordinates, to the interests of the company; sometimes you may have to take the initiative in convincing the stockholders, board of directors, etc., in which direction your company's interests really lie.

Attitudes Toward Conformity—Noted psychologist Clark E. Moustakas states, "I believe that much of the human misery in the world today, the serious emotional problems and conflicts, result from an attempt to fit into conventional modes . . . from striving for goals of success, status, and power which provide no intrinsic value or satisfaction, and contribute to a meaningless existence."[1]

A preoccupation with conformity can lead to a meaningless existence in the corporate as well as the personal sense. I particularly call your attention to these rather well-documented observations that identify the significant difference between conformists and constructive nonconformists:

An effective nonconformist is neither unduly susceptible toward nor unduly driven from the pressures of the group. He displays an acute ability to sympathize with and understand people. He tends to be substantially freer from pretenses and affectations. When given a choice, he will choose high-risk opportunity against security.

These comments do not suggest or recommend that a nonconforming posture is always preferred but, rather, that the individual possessing the basic strengths and characteristics of the nonconformist will

How Will You Fare in the Race to the Top?

If you are quite honest with yourself and can answer the following questions **affirmatively,** this indicates a strong orientation to a high-level management career.[2] Furthermore, such a yes-answer profile can suggest to you what type of response to some of the other crossroads questions posed in these pages is most in keeping with top-management orientation.

1. Are you challenged by stress?
2. Do you enjoy being a constructive adventurer?
3. Are you a strongly motivated competitor who can bounce back from disappointments?
4. Are you relatively unconcerned with security?
5. Do you believe strongly in the constant need for change and forward movement?
6. Are you capable of disciplined concentration on pivotal problems of a complex nature?
7. Are you both comfortable and effective with either participative or authoritarian management patterns as the situation may demand?
8. Are you more interested in identifying and assessing positive rather than negative aspects?
9. Do you habitually tackle the most difficult problems first?
10. Do your have the ability to be effectively dramatic when a suitable occasion arises?
11. Are you vitally concerned with long-range questions—i.e., will you perform as a prophet when your role requires this?
12. Do you expect high standards of performance in matters such as deadlines, effective flow of communications, etc.?
13. Do you believe that the customer's profit objectives properly define your firm's objectives?
14. If so, would you then insist on spending some portion of your time in getting to know key customers and their needs even if you are in R&D or operations rather than marketing?
15. Do you believe that the best of us must continue to be students throughout our professional lives?
16. Can you relax at appropriate times?
17. Does the opportunity for making a major contribution motivate you more than any other single factor?
18. Do you feel you can work with people who may have an essentially cynical attitude, without becoming a nervous wreck (or without becoming a cynic yourself)?
19. Would you hire an important subordinate who might elect to wear a beard if he measured up in all other skills and resources?
20. Do you know how to support true innovation?

skillfully choose and implement this role when it is the intellectually honest approach.

Leadership Style—The young manager at the crossroads decision must identify his basic feelings and preferences as to leadership style. Perhaps the ideal is a natural preference for the participative, Theory Y, approach, combined with ability to recognize situations calling for the use of authority.

Performance Under Pressure—It is important to know oneself with regard to performance under strong and sometimes extreme pressures. Some indeed are challenged by this, and only perform at high levels when under considerable or even extreme pressure. There is, of course, a wide spectrum of ability to respond effectively under pressure.

Response to Adverse Criticism and Opposition—How do you handle and respond to negative criticism and strong opposition, particularly if the latter appears to you to be based upon emotional grounds or inaccurate understanding and information? The manager must be prepared to react constructively to these factors if he is to attain his managerial goals.

Managerial Methodology

To illustrate how your preference and skills in the area of managerial methodology play a part in the "How Do You Wish to Travel?" decision, here is an example: Presume you have decided on a final and complete commitment to a high-level technical (or broader) management career, and that you are about to be promoted to the directorship of an organization comprising many people and a substantial budget. How would you go about managing the enterprise? Presumably, you would develop some type of approach or plan to identify objectives, and to relate these to target dates and available resources:

- Would you develop this plan as a one-man operation or as a joint activity involving appropriate subordinates?
- Would you prefer a highly informal or formal approach?
- What is your version of "planning by objectives?" (This is a widely misunderstood topic and one that unfortunately has suffered from lack of understanding rather than lack of merit. In any event, every individual has his private version of what is meant by planning by objectives.)
- How high do you value forward-looking skills and activities? Is your evaluation based upon a short, intermediate, or long-time scale?
- Should the planning be limited only to immediate, obvious needs?
- Will your allocation of resources reflect a carefully considered balance of program in terms of short-, intermediate-, and long-range projects? Will these also reflect appropriate balance among low, intermediate and high-risk endeavors?

What are your attitudes toward boldness in innovation and creativity? Significant emphasis upon profitable, technical innovation demands changes of many kinds—including personal attitudes. Many organizations are disturbed by change. By and large, we are more prepared to deal with that which is familiar and rationally predicted or defined. Based on your career so far, how do you react to change?

From questions such as the above, the answer to "How Do You Wish to Travel?" can take shape.

What Will It Cost?

The final instalment of this series will take a closer look at the reward/cost relationship in higher management; for now, it is enough to say that the cost of climbing the management ladder can involve such factors as:

- Damage to mental and physical health—and strains in personal relationships—resulting from long hours, heavy travel schedules, or from inability to stop thinking about the job even at home.
- Frustrations because of inability to influence events far enough or fast enough, or because of need for painful decisions involving subordinates or colleagues.
- Feelings of isolation, irritability or self-doubt during trying periods.

What is the top price you are prepared to pay in striving for the high managerial job? This is a real soul-searcher, but must be dealt with. At what point will your loyalty and responsibility to wife and family supersede those to the company and vice versa? What of other outside interests, such as hobbies and leisure-time activities, when these are in conflict with the demands of the job? Some rare men are more able than others to retain the best of all worlds and to defer the final moment of truth.

Another possible "cost" element involves factors beyond your control. For instance, despite the skills and resources that you develop, it is possible that the career fulfillment targets will not be attained—perhaps because of competition from individuals of higher seniority, changes in the economic or corporate climate, etc. If your career runs into an undeserved roadblock or delay of this kind—one that you might not have encountered had you chosen some other career objective—will you be able to continue on a constructive path, or will you fall prey to despondency or cynicism?

So much for "What Will It Cost?," which is the last of the four career-building questions. Having addressed yourself to these questions with honesty and thoroughness, and having checked out key points with your wife, trusted friends and your boss, you can make your crossroads decision with considerable confidence. If a pattern of top-management orientation is apparent to you and to those whose counsel you have sought, then go, man, go! ■

References

1. As quoted by: Raudsepp, E., "Conformity's Role in Industry," *Hydrocarbon Processing*, Dec. 1967, p. 159.
2. Some of these questions are based on: *Harvard Bus. Rev.*, May-June 1967, p. 97-104; *Dun's Rev.*, July 1967, p. 40.

Custom-Building Your Management Career

Part V: Middle Management

If you have just been appointed manager of your company's R&D division, what will your first steps be? What actions will you take once you have established yourself? Here are specific guidelines on these matters—as well as help in preparing yourself for middle management in general.

E. M. KIPP, Sun Oil Co.

The middle manager's level of responsibility is equivalent to a director of R&D of a medium to large-sized company, or the technical director or manager of a division in a big and diversified corporation. The number of subordinates generally include from 50 to several hundred professionals; and a typical salary range for this managerial level in the chemical process industries would be \$28,000 to \$42,000. Most people who are now in middle management reached that level after 15 to 25 years of prior experience.

Characteristics of the Middle Manager

Position titles are not particularly informative, as these will vary widely from company to company and from industry to industry. A division manager in one organization may actually have a lesser job than a section chief in some other company. An operating plant superintendent in one company may actually have a more responsible position than that of a works manager in another industry. It is therefore better to describe the middle-manager level, in terms of the organization chart, as the level in the management structure that most generally will be two or three levels removed from that of the top man. This suggests that the middle manager will report to a vice-president, who may in turn report directly to the chief executive officer or to a senior vice-president. In very large or complex organizations, the middle manager may even drop down another level in the organization chart; whereas in much smaller organizations, the manager may report directly to the chief executive.

Rather than try to cover every situation, I have centered the discussion around the middle-manager level in an R&D or primarily technical activity. I believe the discussion will bring out most of the major considerations common to this level of managerial responsibility irrespective of whether we are speaking of an operating, production or design/construction activity. (The final part of this series, in which the job of vice-president of manufacturing will be examined, will help fill in what gaps there are in regard to the operations and production aspects of middle management).

A hoped-for extra benefit is to give engineers and technical managers outside of R&D a better insight into how that department operates, and to suggest areas of cooperation.

Adjusting to New Responsibilities

The middle manager's responsibilities and functions involve substantial commitments of corporate resources. Stewardship of these resources can have major impact on corporate growth and the generation of many millions of dollars in additional profits.

One of the most difficult problems of adjustment faced by a technical manager working his way up the ladder is that of learning how to drop or modify certain actions, responsibilities and interests that were pertinent to his prior responsibilities but may in fact now interfere with his performance in the more responsible position. Solution of this problem will be reflected in his effective delegation of various portions of his functions, and the staffing of subordinate positions with persons in whom a high level of confidence is justified. The middle manager must then resist the temptation to have a special "standard" for the man who inherited his old job.

Your Opening Moves

Let us say you have just been appointed director of R&D, having done a superior job as a senior section chief. What should some of your first moves be?

Your most immediate task is to develop effective communications with your boss, and to understand what he expects of the R&D department. Try to obtain some feel for your boss's responsibilities and philosophies, and the pressures under which he operates. Clarify the limitations of authority and responsibility under which you will operate, and the desired frequency and nature of communications between you and your boss:

- Does he want to be involved in the solving of R&D problems, or will he prefer that such problems be solved at the director's level to the greatest possible extent?
- In what kinds of discussions does your boss want to participate? What role does he wish to reserve for himself in the necessary two-way communications that must be established between R&D and the rest of the company?
- In communications with higher management, will you be expected to function as your boss's spokesman or "expert," or will your boss prefer to have you give him a full briefing, thus making your role a supportive one?

Clarification of these matters can sometimes be developed rather quickly and easily. In other cases, depending upon personalities, attainment of a good level of mutual understanding may require much longer. Sometimes it may never really be achieved, and in this instance everyone suffers.

The next task is to establish a more comprehensive knowledge of the company. Normally this is best done by understanding how the company is structured, and the primary reasons for this structuring. Establishing effective relationships with the heads of all significant company groups or divisions should be your constant objective; good personal contacts with these persons can mean the difference between a mediocre and an outstanding R&D operation. The prime goal should be to develop a thorough understanding of the functions and objectives of the other men's departments, and how the R&D department can assist them in more effectively attaining their objectives and solving their problems.

This is also a good time for you to candidly discuss your own concepts of your position, and the appropriate role of R&D. Mention the weaknesses and problems of R&D as well as its strengths and performance potential. In developing sound relationships with these other men, make it clear that their comments are welcome and will be considered even if they are critical and often of an uninformed nature. In this way, you will be developing an understanding and appreciation throughout the company of the R&D organization, while also strengthening both your own skills and those of the R&D group.

Unfortunately, many R&D directors seem to feel that R&D is their business only, to rule and administer. Too many R&D directors seem to fear the possibilities of destructive involvement by other men in the affairs of the department. Such attitudes are self-defeating; they reflect weakness, lack of direction. R&D is the entire company's business. If successful, R&D will *benefit* the entire company; if unsuccessful, it will *penalize* the entire company.

Having completed the foregoing communication tasks, you now have the inputs to structure the responsibilities and operations of your office, and to interpret the mission of your R&D department. This mission should be defined by a structured series of task objectives within the constraints of time, money, and corporate needs and objectives. Therefore, your first duty will be to develop these task objectives in the form of appropriate R&D program and budget proposals.

In your initial structuring of these proposals, you must relate them to the tentative or probable resources available to you (i.e., monies, buildings, staff, and such.) If the resources are not commensurate with the desired structuring of the R&D operations, you must then either revise the R&D plans, or take some other action to bring about a substantial balance. I will come back to this point a little later.

In administering the R&D program and budget, how can you make the most of the human resources available to you? Let us now look at the major personnel considerations, and illustrate the essential elements of effective administrative structures.

Line Organization

The line organization is intended to identify primary subordinate levels of managerial responsibility. It is important to keep the number of supervisory levels as small as may be possible in the light of several and sometimes conflicting considerations. A six-supervisory level arrangement such as the following is often quite acceptable:

1. Director of R&D.
2. Associate directors.
3. Assistant directors.
4. Division or department heads.
5. Section heads.
6. Group leaders.

If the organization is a fairly small one, say 300 to 500 employees, it is desirable to have only one or no subordinate levels between the director of research and the department or division heads.

It is normally much more effective to structure the departments or divisions on the basis of functions

How to Lead People Whose Work You Do Not Fully Understand

One of the controversial issues in the field of technical management is the relative importance of technical competency as one of the skills or resources of the middle manager. The question becomes particularly provocative and pertinent when we are speaking of a technological activity requiring contributions of several disciplines—such as chemical engineering, organic chemistry, metallurgy, physics and mathematics. Obviously, with the exception of a few geniuses, most middle managers cannot realistically aspire to being more knowledgeable in each of these disciplines than are individual professionals well trained in their respective disciplines.

Therefore, the question arises as to how the manager can proceed when he is confronted with problems or controversies of a highly specialized technical or scientific nature. There is no simple answer, but there are acceptable responses to this problem.

By this stage of his career, the middle manager should have developed a pretty reliable book on how he can best judge the comments and advice of one technical person versus another on his staff. For example, he will know that some of his technical experts are always optimistic and will generally overstate the case, with perhaps considerable oversimplification, whereas others are inclined to be very conservative, extremely meticulous in developing their arguments, and will be extremely fair-minded in considering views from equally competent fellow professionals.

Competent specialists should become able to rephrase highly technical concepts in a way that will get the main point across without the manager having to understand the technical subtleties.

A good middle manager will also be quite skilled in phrasing questions that will challenge his technical protagonists to explain how each one's position can be related to the company's benefit. He can ask what other technical alternatives have been considered. Also, if he is a good manager, he can enlist Individual A's constructive criticisms of Individual B's technology and vice versa in the presence of each. Having availed himself of such typical resources, the middle manager, not himself professionally competent in a given discipline, can make the appropriate decisions in a quite effective way.

The question is often asked how the middle manager can determine whether his specialists are current in their technological competencies—whether they are fully aware of recent techniques and know how to apply them. If the middle manager has followed the precepts so far dealt with in this series, particularly in regard to performance appraisal and assessment of the professional-development needs of subordinates, he should have little trouble in recognizing whether this is a problem. If he should have trouble, it is possible, under conditions acceptable to all concerned, to invite a neutral outside expert to assess the possible magnitude of this particular problem for the manager, and to suggest solutions in terms of special courses, sabbaticals, supplementary use of consultants, and similar steps.

rather than by scientific or professional skills and disciplines.

Another important and progressive policy in the operation of the line departments is the development and skilled use of task and/or project teams. In most instances, these serve a most useful function when a particular divisional or departmental project progresses to a point where people and material resources resident in the other departments in R&D or even elsewhere in the corporation are needed. When such a taskforce or group is established, special provisions must be made to accommodate whatever authority and responsibilities may be required for effective operation. If the director of research and his subordinate managers are of appropriate stature, this is not a major problem. However, many organizations do not possess the managerial leadership and skills for effective utilization of the taskforce or project-engineer concept within conventional line organizations.

The director of research must also provide for other corporate needs—i.e., supporting technical services to production or manufacturing, and to marketing and other corporate groups. For the most part, these needs should not be viewed as research and development, and the costs of such technical support is most effectively charged to and controlled by the corporate divisions requiring this type of service. Such functions must also provide for very short-term troubeshooting needs arising from the other corporate operations. Organization can be via separate groups reporting directly to the R&D director, or as a specified subordinate staff operation.

Staff and Service Functions

As director of research, you and your subordinate managers must determine and define the staff and service functions needed if your line organizations are to be most productive and efficient. Five staff operations will often suffice:

1. Technical economics.
2. Patents.

3. Marketing research or commercial development.

4. Administrative staff—i.e., budget control, accounting procedures, and the like.

5. Industrial relations, usually comprising recruitment, personnel administration, training, employee relations generally, etc.

As for supporting functions, these will most often comprise library and information services, grounds and buildings, analytical services, engineering, computer, and machine shop. In structuring and manning the indicated staff and service functions, your decisions will necessarily be guided by the availability of corresponding staff and services at the corporate level; by the size of the company and of the R&D organization; and by many other related considerations. If the company is a large one and does provide staff services of a comprehensive nature, an effective approach may be to establish satellite groups headed by persons who will report to R&D line management administratively, and to the corporate staff heads with regard to techniques, methodology, etc.

Professional and Managerial Training

Perhaps the most important responsibility you face as director is the constant development of your people's resources and capabilities. Because rapid growth in technology can quickly obsolete professional competency of the engineer and scientist, you must lose no time in determining how the professional training needs of your people can best be provided. Fortunately, there is an extensive literature and knowledge in this area; the various approaches to training needs are well established. The most difficult part of this operation is defining the needs for each individual, and relating these to his future opportunities.

A word about managerial training. Outstanding engineers or chemists do not necessarily make outstanding administrators or managers; people who not only have high technical competency but who either possess or can be trained in the diverse talents and skills of management are the ones who are in greatest demand and shortest supply. Identifying these persons within your organization is a most challenging opportunity. In your handling of the development of your people, imaginative and effective on-the-job training for managerial careers should rate very high on your order of priorities. The earlier parts of this series bear directly on how you can now recognize promising candidates for managerial training, and fulfill their expectations in regard to career guidance and other counseling.

Related to managerial training is the question of managerial philosophy. In my judgment, companies will require their R&D directors to be increasingly dedicated to the participative involvement philosophy of management. To effectively implement such a philosophy at all managerial levels, you and your key subordinates must recognize that individual persons vary widely in their desires and needs. For example, some prefer authoritarian supervision; they want to be told what to do and they feel most comfortable when their involvement is limited to carrying out specific orders. Therefore, the participative involvement technique must be adjusted to each individual.

It is important that the advantages and principles of participative involvement be understood and implemented through appropriate training programs.

Custom-Building Managerial Quality

As a new director of R&D, you have now completed the first cycle of your managerial activities. But however excellent, your performance so far will become meaningless unless you develop the skills necessary for continued improvement of your department.

The first consideration here involves the relative emphasis between Research, Development and Service. In reassessing the operation you have created, as defined by the first cycle, you can now take a retrospective look at whether your R&D department is a r&D, R&d, r&d&S department etc. You must decide whether this is as it should be, or whether to drive toward a different emphasis.

A further consideration involves reassessing your delegation of authority. Have you given your section chiefs and other subordinate managers appropriate

Balancing the R&D Program

Curiously, many R&D organizations do not properly emphasize balance of programs. By balance, I have two characteristics in mind:

The first is balance between basic and applied research. Every good research administrator has his own definition of basic research, applied research, technology, etc. An effective definition can be based upon the acceptable time span to be allocated to the achievement of a particular objective. If the objective is to be met over a period of, say, a year or longer, the research is more likely to include increasing elements directed to answering the question "why" (basic or fundamental research). If the time span is quite short, the emphasis will be on answering the question "how" (applied or development research). In any event, however he does it, the director must develop a balanced program in terms of short-term, medium-term, and long-term projects. The particular mix must necessarily reflect corporate objectives and resources.

The other element of balance is the factor of risk. Risk comprises technical and economic matters. Company managements will be more inclined to support a project with high technical risks if the profit potentials are substantial. Clearly, an R&D program cannot consist entirely of high-risk, low payoff; high-risk, high payoff; low-risk, low payoff; etc., projects. The director must seek to establish an optimum blend of these. Pragmatically, the effective director of R&D at the very least should try to ensure that there are enough low-risk, moderate-profit projects to sustain the effort for the more risky, but higher potential projects.

authority and operating elbow-room, even though you cannot pass the buck to these men in discharging your own responsibilities?

Try to clearly define those areas in which your subordinates will speak and act for you without first having to get your opinion. A subordinate manager unable to make decisions on the spot without in effect saying "I must first check with the boss" is not an asset to the organization. In this context, you cannot challenge your subordinate for having made the decision (if within previously agreed areas and limitations); you can only challenge him on matters of judgment, skill, timing, etc. A director who finds it impossible to clearly delegate appropriate segments of his job cannot be an outstanding performer.

Relating to Corporate Goals and Objectives

When you reappraise your program in terms of its match with corporate objectives and resources, this can provide you with important opportunities for improved skills and performance. You will recognize that clarification or review of corporate goals that depend on science and technology may be needed. Even more importantly, you now have a base line from which to create alternative research and development opportunities for consideration by higher management. A desirable type of top-level decision-making takes place when management is effectively challenged by the R&D organization, with an opportunity to select from an appetizing, high-quality menu of alternatives, the total cost of which exceeds currently approved R&D budgets.

You must continually reassess the implications of your division's progress in terms of the resources required to put the R&D output to profitable use. Many an otherwise fine R&D operation has failed because of inadequate attention to this requirement. An R&D operation may well be too rich for the corporate blood; it may also be too lean.

Longer-Term Program and Budget Planning

Construction of an annual budget is no longer adequate in meeting the problems and opportunities of the future. To the fullest extent possible, you and your staff should supplement your yearly program and budget recommendations with a minimum four-year projection (i.e., to form a five-year plan). In some cases, it is desirable and possible to prepare useful ten-year or longer projections. Long-term projections of programs and budgets are particularly appropriate to R&D operations where substantial lead-time is mandatory in the development of new products and processes.

One effective approach is to present budget projections in terms of expected or planned-for new products, processes, specific product and process improvements, etc. Thus, the projected budgets are related in new "business opportunities" terms to goals and objectives based on estimated or calculated profitabilities. The resources required to successfully attain these objectives are forecast. As projections extend beyond the immediately proposed yearly program and budget, they are increasingly speculative and qualitative.

A primary function of a five-year projection of the R&D program and budget is as a base line against which the director and his staff can respond more realistically to new developments. This type of feedback can quickly be reflected through the annual revisions of the five-year projections. There are other substantial advantages to this approach in relation to long-term corporate objectives: the earmarking of additional funds and monies that may be required if R&D's new investments opportunities do, in fact, materialize, etc.

Custom-Building Other Skills

At this middle level of technical management, the increasing breadth and complexity of the necessary skills is readily apparent. However important, the director's technical competency rapidly becomes only one of many other resources—of which the ones below are perhaps the most basic.

Your knowledge in these areas cannot and need not be as detailed as that of the professional specialist However, it is quite feasible for you to appreciate what these other functions are in terms of how they can reinforce the quality of your managerial decisions, and contribute to the work of the R&D department.

In other words, you can know in a general way, at least, what goes into the black box and what can come out of it, without understanding the intricate mechanisms operating within the black box.

Computer Services

Appropriately used, computers can increase by several magnitudes the work output of a person or group. Computer resources open entire new opportunities for the more extensive use of mathematical simulations for a wide variety of areas ranging from chemical-engineering pilot-plant operations to the better planning of an organic-chemistry synthesis project in a new area. Computer resources can lower the director's administrative costs in the area of salaries, information retrieval, budget control data, and many other such activities. On the other hand, computers also have serious limitations and pitfalls; these should be recognized and understood by the director.

Technical Economics

In the final analysis, support of a product commercialization or process improvement must be justified by the estimated profits, after allowing for the cost of the prior research and development on the project, plus the continued defensive research that may be needed.

An extensive and interesting literature has emerged

in the past few years and continues to gain momentum on new and improved techniques for evaluating the many different economic components of an R&D operation. The director needs to understand terms such as discounted cash flow, initial rate of return on investment, present worth, etc. He should understand the usefulness as well as the weaknesses of decision-aiding analyses such as: decision trees, Monte Carlo probabilities, PERT diagrams, terminal costing, linear programming, heuristic programming, single point estimates, uncertainty profiles, etc.; he should have enough of a feel for the R&D potential of these techniques to establish an informed relationship with appropriate staff experts.

Marketing Research, Commercial Development

To the extent that the director's department is charged with responsibility for developing new products or new markets for current products, it is mandatory that a marketing research or commercial development resource be available to him. These resources may be in the regular marketing or commercial development departments of the company. Sometimes they can be more effectively located in R&D.

I like to think of this type of group as one that is charged with exploratory and development research in the marketing area—somewhat analogous in its area to the operation of a pilot plant where final data are obtained before large funds are committed to the building of a plant. Here again, the director should make a constant effort to understand the various techniques that can be employed, the pros and cons of each, the limitations governing their use, and the costs of doing such research in terms of desired output.

Employee Relations

The values and expectations of our current graduates differ markedly from those of even 10 or 15 years ago, both in degree and kind. According to every indication, this situation will continue to change. The director of R&D cannot rely on memories of his own early years in deciding how to motivate the younger members of his team; he must remain attuned to what is happening today and what will happen tomorrow—this includes visiting the campus, acquiring a better understanding of minority-group problems, and being sensitive to ways of obtaining team effort without stifling individual initiative and recognition.

The director of research should also have more than a casual understanding of labor relations, laws governing activities of unions, prerogatives and restrictions upon union membership and upon management. In the context of this knowledge, he must establish effective approaches and policies for labor relations matters. It would be most helpful indeed (and many companies are able to provide this type of exposure) for him to participate as an auditor on occasion at bargaining sessions with union organizations.

Broad Social Concerns

The day is rapidly passing, if not already gone, when it is proper and sufficient for the director of R&D to concern himself solely with the impact of his operations upon corporate welfare, profits for the stockholders, and the like. It goes without question that in our private enterprise system, adequate corporate profits and returns to the stockholders are mandatory performance requirements. However, the responsibilities and opportunities of the R&D director as a corporate citizen of a community, state, and nation must also be dealt with boldly.

In the long run, if these social responsibilities are not met, society will react by changing the environment in which private enterprise must exist. Specifically, if the research director should be so fortunate as to have spawned a new scientific or technological advance of importance, it is no longer enough for him to translate this into profits for the company. He must now, and increasingly so in the future, analyze and consider what impacts this development may have on society. He cannot divorce himself from such responsibility, although others within his company may have more-direct responsibilities in the matter. If the outputs of industrial research are increasingly viewed by the people at large generally as destructive or restrictive of their opportunities for a fuller life, they will change the environment of our private enterprise system, however profitable the results of research may be to an individual company or group of companies.

Top-level research directors of today and particularly of the immediate future must become actively involved in civic and social matters at the community level as well as at higher levels. They should break bread occasionally with political and social leaders in various walks of life. Of course, such activities must be in proper perspective and in proper relation to the primary requirements of his position, but they must and will receive much more attention and involvement from the director than has been the case.

The Key Challenge

I envy Michael L. Haider's ability to summarize what I consider the key challenge to the director of R&D:

"The way in which a manager deals with his people can, quite literally, turn them off or turn them on. To hammer an organization into passive compliance requires little skill; to generate a climate in which people at every level feel free to suggest new ideas —to venture beyond the sure, the known, and the safe—is the mark of a true leader."*

The manager who can implement this philosophy in dealing not only with his subordinates, but also with his colleagues and counterparts in the company, is readying himself for top-level responsibility. ■

* *World Petroleum*, Nov., 1967.

Custom-Building Your Management Career

Part VI: The Top-Level Technical Manager

Want to be a vice-president of manufacturing or technology some day? Even if you do not, you may gain useful new insights into your company's managerial practices from this realistic picture of the "executive suite."

E. M. KIPP, Sun Oil Co.

Although this final article of the "Careers" series will deal with the lofty upper reaches of the management ladder—i.e., a vice-president (or manager of a major industrial complex) in a large company, or the president of a small company—the material is not primarily intended for those about to be promoted to this level. Rather, it is aimed at sharpening the thinking of all engineers or technical managers in regard to the aptness of a top-management career goal, and also in regard to the relationship between their present levels of responsibilities and top management.

There seems to be a great deal of misunderstanding regarding the top-level management position. Often, such a position is viewed in one of two extreme ways, as either a:

1. Comfortable job where the pay is high and where most of the work (and blame for bad decisions) can be passed down to lower levels, or

2. A thankless, man-killing job where you develop myopia to everything in life except profits and budgets, and where you are too tensed up to enjoy such few leisure hours as may come your way.

The "Soft Touch" Theory

Let us look at the first extreme. Can a vice-president lead a charmed life by hiring or promoting smart "assistants" who do all the heavy thinking for him? He can certainly try, but once the assistants discover that they are in effect making vice-presidential decisions, and are doing all the work for none of the glory, they will not want to stay assistants for very long. Such situations have a habit of resolving themselves to the vice-president's disadvantage.

And take the matter of security. I have heard it said that whereas engineers can easily get fired for a lapse in *engineering* judgment, vice-presidents can usually explain away any lapses in *vice-presidential* judgment, and can use all sorts of stratagems for shifting the blame to subordinates, labor unions, the government, the vagaries of the economy, etc.

It is of course true that relatively few Annual Reports contain such statements as "earnings were down 28% last year because we vice-presidents did a poor job of managing the company." Nevertheless, other-than-voluntary vice-presidential and even presidential changes frequently show up in Annual Reports, and in the business news media. At this time of writing, for instance, a story in the *Wall Street Journal* about a certain steel company starts out with the statement that the firm: "dismissed its chief executive amid indications that the company is actively seeking a merger partner to bail itself out of its current financial troubles." So it would certainly seem that vice-presidents—even chief executives—enjoy something far short of absolute security.

Of course, it would not be realistic to expect a vice-president to be fired the first time any one of his decisions turned sour. The ensuing discussion will make clear the highly subjective, ambiguous nature of many of the problems with which the vice-president must deal. In this setting, one can expect that a good vice-president will in the long run, turn out to be right on most of the big things, but it would be self-defeating for a company to pounce on him the first time he trips up on something along the way; this would encourage a very low level of risk taking.

The "Man Killer" Theory

Let's now look at the second extreme—i.e., that the top-level manager is a basically unhappy person who may wind up with little to show for his high-powered efforts except a rich widow and a bunch of maladjusted offspring who claim their father never had any time for them.

As pointed out in the "Crossroads" installment, top management does involve certain sacrifices—the job is lonely, time-consuming, and demands great self-discipline. These things almost no one enjoys; one can just make the best of them. But some other things that are often cited as drawbacks are not that one-sided. True, the job demands sustained periods of high concentration, but some people work best under such pressures. Also true, the tough decisions that have to be made may hold almost awesome implications for the company, community and for subordinates; but some people derive a sense of satisfaction from making such decisions even if it means losing a night's sleep or rearranging a weekend. They feel this is a small price to pay for the privilege of putting a direct personal stamp on the setting as well as implementing of corporate policy. In fact, they may be less likely to develop ulcers in making such decisions than if they worried about the impact of decisions someone else would be making.

Are you one of these people? If you still are not sure, I hope this final article will help supply you with new input for your eventual decision.

Vice-President of Technology or Manufacturing

The positions we will be discussing generally encompass an annual salary range of $40,000 to $60,000, usually supplemented by some form of incentive compensation. The majority of people who have reached such positions have had 20 to 30 years of prior experience in management and technology, although some people have made it in less.

The various types of top-level technical management positions are more alike than dissimilar in terms of the skills and attitudes required from the man filling the position. However, there are major increments

Deciding Whom to Promote, and How to Do It

Telling a candidate for promotion, "Congratulations. Starting next month, you're the new Chief Engineer" can be a delightful task.

However, a less delightful task may go with it—that of dealing with the disappointed candidates who may feel that they were equally qualified and that they now have nothing to look forward to for years. This sort of problem is fairly general at all management levels, though more difficult at the higher echelons. The toughest situation is that in which a junior man may have the special qualifications to jump over a number of his older superiors.

My suggestions regarding these problems may perhaps be more meaningful if presented in the broader context of manpower planning for the company. Here, a decision must be made on the depth of qualified personnel to be trained or otherwise made available for appointment to key managerial and technological posts, present or predicted. Should there be one outstanding candidate trained and ready for each position? . . . should there be two or three or even more? . . . or should the number of such persons vary with the level and kind of position?

Many companies, generally the larger and more complex ones, make it a strict concern to have at least three outstanding persons in reserve for each predicted key job need. Others are satisfied to have but one primarily qualified person identified. Still others—and these include large as well as small organizations—have no formal targets or programs. Rather, they improvise as each need or opportunity develops, on the theory that standout candidates are, so to speak, self-generated, and generally recognized by supervisors, peers and subordinates.

There are strong and divergent opinions about related problems. Some managers believe that the direct or implied identification of leading contenders may give such people an undesirable fair-haired boy image, thereby in degree of responsibilities and skills in bridging the gap between middle and top-level manager. Although this may be a minority opinion, I feel that the gap between these two levels is in some respects deeper than between any two of the lower levels.

I hope to illustrate the foregoing observations by dealing with two hypothetical, high-level managers: One, the vice-president in charge of research and engineering, the other, the vice-president in charge of manufacturing. Let's start with matters common to both executives, and then examine some questions primarily characteristic of each.

The hypothetical young engineer whom we introduced at the start of this series, and who is now years older and hopefully wiser, must make his final assessment of the four "career guide" questions: Where do you want to go? When do you want to arrive? How do you wish to travel? What will it cost?

His age and general health are particularly critical questions. The probability of a vacancy at this top managerial level is primary. If our manager is now about 55 or older, he may be well advised to round off his career at the middle-manager level unless higher management has unequivocally tabbed him for the vice-presidential or presidential level vacancy within the next two or three years at most. If he no longer aims toward this level of management, he must make the necessary emotional and other related adjustments. The hazards to be avoided are loss of involvement, and vitality of approach. Even though he may no longer be training for the higher responsibility, a "quality" manager realizes that continued improvement in his performance is important even though he advances no further in the organization chart. His job can grow even without a new title.

If, on the other hand, the choice is to drive to the apex of the managerial ladder, he must realistically be prepared to face such considerations as these:

As vice-president for technology in the company, his responsibilities may be quite broad, and may comprise R&D, engineering, market development, patents, technical economics, and semicommercial manufacturing operations. As vice-president in charge of manufacturing, he may in effect be managing a series of small manufacturing units or may be responsible for the manufacturing units of a number of operating divisions or corporate subsidiaries.

Furthermore, at this managerial level, be it in research or in manufacturing, responsibility for corporate affairs not strictly related to either research or manufacturing will more often than not be assigned. For example, corporate relationships of substance with official agencies, committees, or other activities of the federal, state or local governments may be a formal and important part of the job.

The vice-president may be the official company spokesman on major intra- and intercorporate committees. Those committees primarily oriented to technology, engineering and science will be more often assigned to the vice-president of research and engineering, while those with a greater orientation to public relations (i.e., between the company and the com-

---

encouraging intrigue, noncooperation, second-guessing, etc. Some executives also believe that identification of managers-to-be often reduces the motivation and commitment of other high-quality employees. Another fear is that key "trainees" could become frustrated to the point of quitting should unforeseen events delay or eliminate the advancements for which they are being prepared.

While these and similar considerations are indeed relevant and should never be overlooked, I feel that they are not controlling, and should not be the reason for turning down a policy of planned manpower development for critical management positions.

The basic premise underlying the whole approach to the custom-building of high-quality management careers is that a continuing performance evaluation and counseling program pays rich dividends. It can provide all of the necessary input data upon which to base effective manpower planning tables and forecasts; it is one of the most effective ways of assuring any company that it will indeed have the best possible men available as the need arises.

Even more significantly, this approach provides the flexibility of fitting the job to the strengths of the man rather than to force the man into compliance with the specifications of a position-description document.

The very nature of this approach, and the fact that it is applied to all employees over continuing periods of time, will minimize the objections to identifying leading contenders for higher careers by such things as training programs, job rotation assignments, or delegation of important segments of a superior's job, or even through appointment of a person to act for the boss during his absence. This, I believe, answers the problem of how a manager can deal effectively with the disappointments of those who are passed over. The disappointments will have been controlled because the individuals will have had, on a continuing basis, a realistic knowledge of their progress and performance, and a good basis upon which to judge their competitive positions.

Best of all, the superior will be well prepared to answer any remaining questions of a disappointed subordinate on the basis of extensive and well-documented specifics, rather than having to take shelter in a thicket of subjective irrelevancies.

Suppose such a supervisor were to say: "You have not been cast aside; as soon as you boost your skills in economics and in some of the other things we have been talking about, you will be a candidate for the next promotion. In the meantime, I'll be giving you more and more assignments that involve economic analysis so that you can apply your new training." If this supervisor has established his credibility in human relations—e.g., if he has given evidence of following up on what he says—then such a specific statement will usually be taken constructively, and could pave the way for a future promotion rather than for a transfer request.

munities in which the plants may be located) will be more appropriately assigned to the plant manager or vice-president in charge of manufacturing.

Either type of manager at this high level can serve the company effectively by seeking opportunities as a featured speaker at colleges and universities, at national or regional meetings of technical, trade, civic, and other associations. The common denominator is the great degree of responsibility carried by either of these men as symbols of the corporation. What they say and how they say it can be of great benefit or harm to the company.

As top-level managers, their voices are now substantially more decisive in the affairs of the corporation at the board and presidential levels of decision than heretofore. The middle manager's judgments and recommendations, although important, are nevertheless subject to approval at the vice-presidential level, whereas our top-level managers make the final decisions and propose courses of action directly to the board or president. The buckpassing train is now at the end of the line, or at least at the next to the last stop.

Both types of managers must not only evaluate the roles and performances of their departments within the company, but must also evaluate these within the larger and more complex reference of national and world progress in the sciences, technology, engineering, business and politics. Each vice-president must see to it that somewhere among his subordinate departments there are the organized skills to interpret these developments in terms of the company's future.

It is all too easy for a company to become so immersed in its immediate problems as to lose sight of these other matters, often suffering great damage thereby. Almost all of us can visualize the executive who will pick up the phone and call the director of research saying, "Hi, Joe, I have just read an article on new applications of psychic lunar vibrations to the problems of prospecting for moon dust. Have we thought about this and is your group doing anything on this?" Or, to be less facetious, the top-level manufacturing manager's attention may be called to some new development in online computer control that a competitor has applied, with the implied question, "Why haven't we been as smart in adapting these developments to our XYZ process?"

The point is that other executives—say, in marketing—either assume that the technical and manufacturing experts are current in their awareness or, finding that this is often not the case, they then assume that the technical and manufacturing people are not up-to-date and do not read the literature, other news-media, etc. In effect, the marketing executive then becomes part-time R&D director or plant manager.

Responsibilities to Subordinates

In essence, the top-level managers have one overriding responsibility: developing the kinds of departments the company must have for it to be profitably competitive and for its future to be reasonably assured. For his *subordinates*, the successful top-level executive, whether in technology or manufacturing, must be able to:

- Interpret the nature of the company, its opportunities and its problems, to each subordinate manager.
- Define the quantity and quality of available corporate resources, as well as the problems to be resolved if the more effective operation of his departments requires substantial adjustments in these resources.
- Clarify the questions to be dealt with by each subordinate department for improved performance.
- Confer with each of his department or division managers so as to establish clearly stated and mutually accepted work objectives. These must be developed within specific limitations of money, people, time and other resources. Provisions for effective management and implementation must be agreed upon.
- Provide effective leadership in planning and updating long-term projections of manpower requirements, both as to quantity and quality.
- Determine with his subordinates the extent to which managerial depth and technical talent are to be developed for each key position.
- Satisfy himself that each subordinate manager has developed and is implementing a sensible, quality program for developing and upgrading his people.

Responsibilities to Boss and Peers

For his *top management*, the vice-president of technology or manufacturing must exercise outstanding leadership in:

- Giving his fellow executives a realistic picture of his departments' roles, and how each department expects to contribute and is contributing to the building of a constantly stronger company.
- Periodically judging whether the size and scope of activities of each department are in optimum balance for the company's best interests. If a given department's budget and program is too lean or too rich from the overall corporate viewpoint, the vice-president must be prepared to analyze the matter effectively and defend his remedial recommendations realistically. He should specifically note and assess results that are expected if his recommendations are accepted.
- Remaining continually alert to signs of destructive strife, misunderstanding, or adverse competition among his departments and other corporate groups.
- Giving critical and continuing attention to communication matters. Intradepartmental lateral communications and upward communications are generally in greatest need of improvement. The universal need is not more communications but better quality and timing.

The vice-president has important homework to do

Advancement and the Young Engineer

Young men of ability and promise are being advanced more rapidly to levels of significant managerial leadership than has been true in the past.

Some of the reasons for this more rapid advancement are:

1. We are learning that younger men can be very effective in responsible managerial positions. What they lack in experience, they may make up for by being more adaptable, less inhibited, more decisive in coping with ambiguity, and more adept at responding to higher-risk situations.
2. Sabbatical leaves for longtime technical managers, longer vacations and earlier retirements are creating new opportunities to get started in management. Another factor is the gap in the group of 40-year-old employees that exists as a result of World War II impact on education.
3. New opportunities for young men of outstanding managerial talents are also being created by the need for industry to successfully cope with the almost revolutionary nature and shape of the sociological, political and economic trends in our environment.

Of course, the above does not mean that the management-minded young engineer should concentrate all his thoughts on the executive suite; one can worry too much about the future to the detriment of the present. You may in fact wonder whether orderly and controlled long-term career growth has perhaps been overemphasized in this series.

Actually, the issue in each case is to attain a good balance between career-objectives planning and doing justice to today's assignments. I have therefore tried to structure the material in the first five articles in terms of matters that will be immediately useful, and that will also provide the most critical elements upon which to successfully develop further progress in management.

before tackling the foregoing tasks. How well he tackles them can make the difference between crisis-prone departments that are barely able to keep things going, and efficiently versatile ones that can create and exploit new opportunities. This difference may be measured in millions of dollars a year.

Specific Considerations: Manufacturing

There are specific activities and responsibilities that must be dealt with by the vice-president for manufacturing (vpm) which, as will be seen, are quite specific to this area. The following listing is not intended to be a complete one by any means but, rather, representative enough to make the point.

Above all else, the manufacturing operations must be profitable, and this means that manufacturing costs must be below some ceiling. Although under extenuating circumstances, certain unprofitable manufacturing operations can be tolerated, these cases will distinctly be exceptions. The profitability of manufacturing operations requires an extremely intimate and effective degree of cooperation and lateral communications with the marketing or sales department:

1. The cost of selling the products to be manufactured must be known as accurately as possible, and must be related to volume of product and to cyclic variations in market demand.
2. The manufacturing costs for each product line must be known with accuracy. For a multiplicity of smaller-volume products, it may be only feasible to make a somewhat arbitrary distribution of total manufacturing costs to each product.
3. Any realistic manufacturing-cost data must be related through executive-policy agreement to the important question of product quality. Is premium quality to be a primary consideration? If not, what will the quality level of the product be in terms of specifications? In practice, the vpm must see to it that corporate policy decisions are made as to the particular point on the quality-vs.-cost curve to be occupied by each product. The decision may well be to manufacture a product at two or three quality levels and price values, so that the customer can have the prerogative of deciding how much premium he is willing to pay for "quality."

Capital Investment Costs

The vpm must accept the responsibility for basic decisions on capital investments. To what extent should manufacturing equipment such as chemical reactors, hydraulic presses or rolling mills reflect long or short projected useful lives? When is "overdesign" most economic, and when is it wasteful? When are investments in computer control and other forms of automation likely to be most productive? Factors such as maintainence and repair costs, downtime charges, replacement costs, and obsolescence must all be interrelated, and each factor given specific ratings.

The vpm must develop the best possible inputs from marketing and from research/engineering as to obsolescence risks for both the product and the manufacturing process employed. If both of them reflect high obsolescence risks because of consumer preference trends, advancing technologies, etc., this must be thoroughly defined and interpreted.

In effect, the vpm must operate his area on input data that will permit him to determine what kind of production tools and equipment are to be provided to the working force in making the products of the company. This determination is a prerequisite for decisions on the type of housing, plant locations, etc., needed for these manufacturing tools and their intended human operators. Flow of raw materials, semifinished and finished goods must be optimized so that there is minimum handling and transportation of materials at every step, and in such a way that each piece of equipment is utilized most profitably. Decisions must be made in consultation with the marketing department as to inventory for each product, as well as for raw materials or other semifinished goods required for manufacture of the products.

The vpm must make primary recommendations on the crucial question of production or manufacturing capacity to meet current and projected markets. Does he only build new plants when product demand exceeds supply? Or, is "getting a piece of the action" enough justification? How about lead time requirements? Should the starting materials be purchased, or can a case be made for extending the manufacture back to the basic building blocks?

Suppose that the R&D and engineering departments have reason to suspect that new developments in material sciences or process technology are imminent, and that these could mean important improvements in capital costs, or operating economies of process equipment? Clearly, if manufacturing and R&D are not effectively communicating, the company will suffer. The vpm must ensure that R&D is alert to any competitive technological trends and developments pertinent to present and projected manufacturing obligations.

Human Resources

Although the effective deployment of people is a challenge not unique to manufacturing, some aspects are certainly distinctive. In manufacturing, the team goal is meeting or exceeding specific production targets as quantitatively measured by volume of units at programmed (or lower) unit costs against a demanding production schedule. Therefore, the vpm needs people who derive substantial personal fulfillment from their involvement in this type of job even if it means getting out of bed at midnight to help track down a leak at the plant or to troubleshoot a balky piece of equipment.

The vpm must establish a framework within which his subordinate plant managers can structure their line organizations. Consider the matter of foremen, for instance. These first-line supervisors play a key role in coupling the workforce to operating line management; the difference between top-quality and mediocre maintenance foremen can have a tremendous impact on modern, single-train processing facilities.

The vpm must therefore assist the plant managers in establishing foreman-qualification philosophy. For instance, should foremen positions be filled by young engineers as part of their training in production? If so, what promotional opportunities does this leave for outstanding nonprofessional people—will they feel boxed in? If the policy is to be flexible, how many foremen positions call for creative engineering ability, and how many call primarily for first-hand familiarity with maintenance and operation of the hardware?

Having developed an effective line organization of plant managers and unit superintendents, the vpm must also check his subordinate organizations to satisfy himself that the appropriate supporting resources are available and structured for effective operations. Rather than "going by the book," he should only authorize those supporting organizations that are regarded by the relevant operating line managers as providing needed and clearly profitable support, and he should review this aspect periodically.

Representative supporting groups comprise arrangements for proper liaison and coordination with marketing, R&D, engineering, purchasing, and industrial relations. Staffs to handle purchasing, labor relations, cost accounting, personnel administration and recruitment will often be required in the larger manufacturing complexes; smaller manufacturing operations may consolidate a number of these functions in one man or one department.

Research and Engineering (R&E)

Let us now take a brief look at the managerial responsibilities that are particularly important to technological operations. (They may, of course, also be important in other functions such as marketing, manufacturing, etc., but they are particularly critical to R&D or R&E.)

If the company is in a business especially sensitive to technological and scientific progress, the vice-president's department will have a much larger role to play at the board and presidential level than that of simply providing appropriate technical support to marketing and manufacturing. Although an R&E department can function primarily as implementer of orders received from the top, it can also play an important leadership role in the formulation of such "orders."

For a company that is significantly vulnerable to science and technology, then, the vice-president must develop with the board and the chief executive officer a corporate policy and strategic concept that clearly defines the particular type of leadership role to be discharged by the R&E department.

It is particularly important that in doing this the initiative to be assumed by the R&E department should be reasonably well defined. The R&E organization can best implement this responsibility by formulating specific new business ventures, in terms of new products or product lines, or in terms of the additional new markets to be captured if the proposed new work is successful.

Many companies have—even unknowingly—denied themselves a far more effective R&E operation because of fuzziness about this cardinal question. The question is so fundamental, in fact, that it will control the operation of the entire department.

For purposes of these discussions, let us assume that R&E is charged with this larger, leadership role. The vice-president then should develop firm policies on such matters as:

- Are his immediate subordinates principally supposed to carry out his instructions, or are they also valued associates participating significantly with him in formulating policies and work objectives?
- How is over-all performance of R&E to be measured? Meaningful self-evaluation is particularly difficult in a department such as R&D, where evi-

dence of progress tends to be more subjective and less quantitative than in manufacturing or sales. Can realistic calculations be made of R&D's contributions to the company's profits? If not, what alternative approaches will be used upon which to judge fairly whether R&D is an asset or a burden? In what way and in what terms will the vice-president account for his personal stewardship of those important company resources entrusted to his care?

(There are several ways in which these types of performance assessments can be made, and the relative virtues of each should be thoroughly understood so that the most appropriate approaches from the viewpoints of all concerned will be used.)

On the personal level, is the vice-president going to be big enough to encourage and manage constructive conflict and dissent within R&E, and also in its dealings with other groups? To what extent will he be concerned about a prime occupational hazard of high executives—the protective "information filter-barriers" that inevitably, though not necessarily consciously, are built up by well-intentioned subordinates? How can he check out from time to time in a constructive way whether or not his subordinates are "telling it like it is" with regard to unpleasant information?

I believe that many of the frustrations characterizing all managerial levels stem from differences in the risktaking postures of people at various levels of line management under varying circumstances.

My personal risk-tolerance level, if my own money and resources are involved, may be substantially different from my risk-tolerance level when company monies and resources are at stake—often quite properly so. Clearly, a director of R&D feeding a high-risk-tolerance diet to a boss who can only digest a low-risk-tolerance diet is going to have problems! Therefore, an effective blend and interrelationship of risktaking attitudes that are supportive rather than disruptive must be established and recognized.

The risktaking attitudes of individuals can be rather accurately measured with reasonably little effort, yet this is almost never made a proper concern in the better management of technology! If this particular task is skillfully dealt with, the manager will be more readily able to determine how much and what kind of high-risk projects are suitable for his R&D department.

More specifically, he must determine the most appropriate balance of projects in the R&D and engineering areas. As pointed out in Part V of this series, R&D must reflect a commonsense yet fairly bold blend of high-risk/low-risk, short-term/long-term activities. It must also make appropriate provisions for the more mundane activities such as technical support of marketing and manufacturing and other service-oriented work.

The vice-president must obtain agreement on effective reference standards to help in the difficult task of selecting among many competing projects. Technical-economics knowhow can play an important role. Any such economic reference points should by all means be based upon sliding scales for the needed flexibility to adjust to other factors: Two competing projects may have essentially equivalent economic justifications, but one may still be much more attractive than another for reasons that are not strictly economic.

By the same token, with the collaboration of marketing and manufacturing departments, similar sets of guiding references must be constructed and agreed upon for proper control of a project as it escalates or as it may need to be escalated in approaching commercialization.

The vice-president must also assume the primary responsibility for assuring appropriate feedback from the board and presidential levels, particularly if the work on the project appears to justify increasingly important commitments of capital and other resources. There are few things that are more destructive of the morale of the researcher than to have become highly committed to a project only to have it develop belatedly that for nontechnical reasons, the project has never had a chance in the first place.

The Common Link

When all is said and done, the engineer who has successfully attained this top managerial post in technological management, whether it be in R&E, manufacturing, marketing, etc., has one overriding responsibility: The appropriate matching of the resources of his departments to the needs and opportunities of the company.

He will be successful to the extent that he does, in fact, understand the company's needs and can effectively interpret these to his subordinate departments in terms of challenges and opportunities to them.

He must be equally effective in providing other corporate groups with a realistic appreciation of the resources of his departments, so that these groups can take full advantage of them. The outcome? Effective coupling of needs and resources throughout the company. ■

Acknowledgment

I would like to thank S. L. Thompson and A. W. Pollock, both of Sun Oil Co., and Herbert Popper of CHEMICAL ENGINEERING for their useful suggestions and contributions.

2

Communications and Information Gathering

Because technical managers spend so much time obtaining, analyzing, and disseminating information, the difference between a good and bad manager frequently centers on communications – does he write informatively and efficiently, can he deliver a good talk or run a constructive meeting, does he know how to get the reliable information he needs, and so on. Of course, communications skills are not exactly wasted on the nonsupervisory engineer either.

Fast, Functional Writing

To start off, here is how a new writing-sequence concept—and a Report Highlights form of the type shown below—can help you . . .

Organize the Report for Fast Writing, Easy Reading

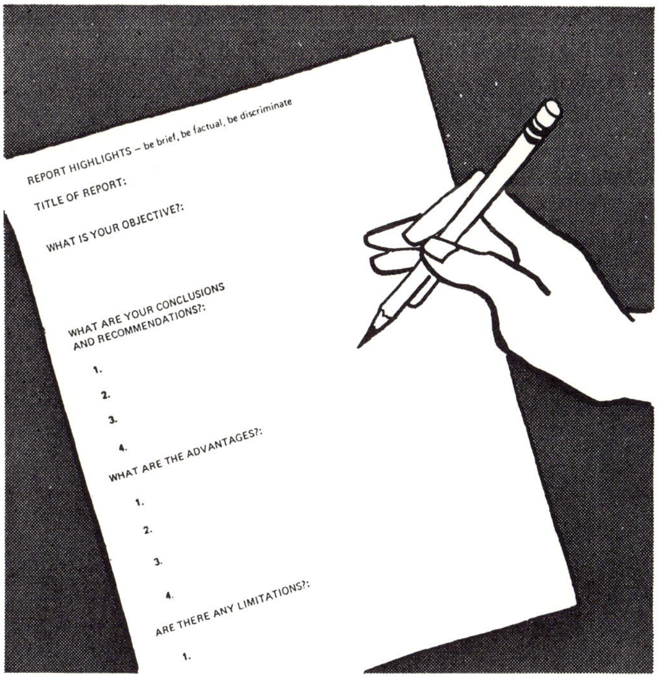

THOMAS P. JOHNSON, Writing Consultant

Not so long ago, the conversation reprinted below took place in the administration building of a large company. The participants: an engineer (E) and an iconoclast (I). The setting: alongside a coffee-dispensing machine, located in a closet exactly halfway between the engineering laboratory and the front office.

E: Have you finished reading my report?

I: Yes, I'm afraid I have.

E: Well, what's the verdict?

I: I'll give you the pleasant part first. You're a pretty good writer, it seems to me. I had no trouble understanding what you had to say, except for a few technical terms here and there. I'm glad you didn't imitate the typical business writer and drown your main points in a sea of pompous, abstract statements.

E: Thanks. Now let's get to what you really want to talk about. Where did I go wrong?

I: My friend, you did what too many other technical writers do. You produced an 80-page report that is a nightmare of disorganization. In other words, the outline you followed was a slap in the face of logic.

E: Explain.

I: Well, I read through your report the same way I would go through any other piece of writing—a novel, a newspaper story, a magazine article. I started at the beginning and read right through to the end.

E: Of course. What other way is there?

I: There's the old tried-and-tested "how to read a technical report" approach, which I deliberately chose to ignore here. You first turn to the *final* section—the one entitled Conclusions—so that you can get the main point of the report. Once you know what the author's pitch is, the course of action he recommends, you go back to the beginning and start plowing through the mass of historical detail that appears in Section One.

E: Wait a minute! I put my recommendations up front. You apparently didn't read my one-page summary right at the beginning of the report.

I: I certainly did. In fact, I read it three times—closely. So far as I can see, all it contains is a series of general statements. You state that the company is thinking about starting a new project, the one you're reporting on. Then you state that the results of your study are subject to the "advantages and limitations discussed in Sections IV, VI and VII"—whatever they are. You estimate "certain material and labor expenses," then refer the reader to several formidable charts in your report appendix. You hint that something wonderful will happen to your company—maybe —if the project is feasible. In short, you don't convince the reader, because you refuse to come to grips with your subject.

E: I'm not supposed to do that in the summary. After all, isn't the summary a general statement that indicates the scope of a report?

I: That's not quite correct. A summary should be more specific than that. Theoretically, it should be a clear, concise record not only of the report's scope but also of the conclusions and recommendations reached. But what actually happens is almost always far different. Given a blank page labeled "Summary," the typical report writer invariably produces the same sort of cautious, overly general prose you've dished up here.

E: By the way, I'd like to point out that I didn't begin writing my report with my eyes closed. To make sure that I'd be correct on procedures, I dug out my old college text and followed the outline it recommended. I don't suppose you're familiar with Professor Jones' "Technical Reports I Have Known"?

I: Yes, I know that textbook and twenty others just like it. They all say the same thing: "When you submit your report to higher management, you should preface it with a short summary. Then you should discuss your subject in the following order: (1) introduction to the subject, (2) discussion of the subject, and (3) conclusions and recommendations. Appropriate appendices round out the report."

E: Right. And what's wrong with that?

I: Let me answer you by asking a couple of questions. If you do recommend that your company begin the project, you're asking it to make a sizable investment. Now, what's the first thing your boss wants to know when he picks up your report?

E: That's easy. He wants to know whether we should initiate the project and, if we do, what it's going to cost the company.

I: Where in your report do you answer this question?

E: Well, I touch on it in the summary, of course, but I spell it out in detail in my conclusions.

I: On pages 78 through 80, to be exact. What you've done is adopt the technique of the mystery novelist, keeping your reader in suspense by withholding your most important information until the very end. Here's another question: Once your boss knows whether your recommendation is "yes" or "no," what does he want to know next? Think hard about this.

E: If you remember, I recommend that we do begin the project. Considering this, I suppose he wants to know why we ought to begin it, how I arrived at my conclusions, how the project will benefit the company—that sort of thing.

I: Precisely. He wants to be sold on your recommendation. He wants to know the advantages of the new project, together with the limitations. He wants concrete figures on things like cost savings, production increases, sales possibilities, and labor savings. All these things do appear in your report, but they aren't discussed until pages 39, 56, 63 and 72. Why not pull them all together and put them at the beginning of your report, preceded by your "yes" recommendation, so that these will be the first things he sees?

E (triumphantly): I'll tell you why! It's because I'm not allowed to. I can see the point you're trying to make, and I tend to agree with you. The only trouble is, our company policy is to follow the same academic outline that you object to. I couldn't deviate from it even if I wanted to.

I: That's an old engineer's tale, and it amounts to a sorry state of affairs. To put it bluntly, I think your company is wrong. If I didn't, I wouldn't be an iconoclast. There is a better way to organize a report, a much more logical way, and I'll be glad to show it to you.

E: Don't show me, show my boss.

Needed: A New Approach

In the above conversation, as well as in what follows, I am appealing to anyone who has to write technical reports, but especially to those who determine what the report format shall be. My contention is that today most reports fail to make the impact they should, simply because they are not logically organized. In the age of the computer, the report writer is still doggedly following precepts laid down in the 19th century. Usually he's doing so because he has been ordered to do it that way.

What is needed is a new approach to report organization, one that differs radically from the traditional form. Because it is senseless, however, to suggest change for the mere sake of change, I want to emphasize that the new approach should bring at least two benefits. It will help the report *reader* by directing him more quickly and more naturally to the main ideas in the report. And it will help the report *writer* not only to make his writing an easier task but also to sell his recommendations to the reader much more convincingly.

The conventional structure of technical reports, as prescribed by most textbooks and company style-manuals, is this:

A. Summary.
B. Text: 1. Introduction.
2. Discussion.
3. Conclusions and Recommendations.
C. Appendices.

Briefly, the new approach to organization suggests the following changes:

- The "Summary" is eliminated and is replaced by the section "Conclusions and Recommendations." The physical appearance of this new section is a form similar to the one reproduced in the cartoon on the previous page.
- The section "Introduction," which consists of

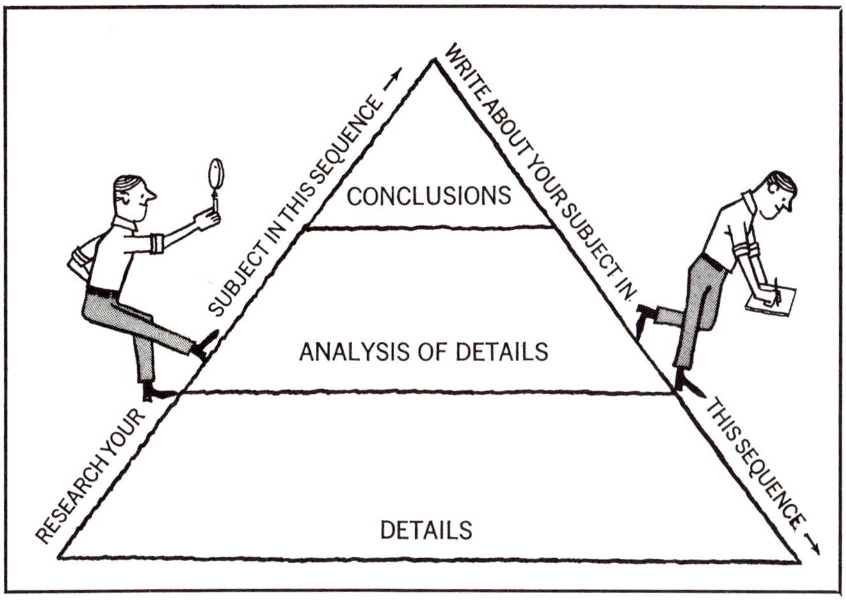

INVERTED-PYRAMID approach assures that the main ideas in the report—what the reader most wants to know—are at the beginning, not at the end.

background material and historical detail, is eliminated. (This material does reappear at various places in the new report structure, but not as a monolithic block.)

• The section "Discussion" is eliminated and replaced by detailed expansions of the conclusions and recommendations that appear on the "Report Highlights" form.

Where Reports Go Wrong

When a technical report is poorly organized, the writer is usually guilty of this fault: He has written his report in the same sequence he followed in researching his subject matter. Let's examine this statement in detail.

The research that precedes the actual writing of a report can be divided into three separate steps. First, the writer gathers all the details on the subject that are available to him. In this broad investigation, he collects background material, technical data, production details—anything that has any bearing on his subject. His second step is to analyze all this detail, separating it into logical groups that point the way to the conclusions he will reach. Finally, he arrives at those conclusions, which are accompanied by his recommendations.

The structure of this three-step research project is like a pyramid (see the diagram). And the most important part of this pyramid is the apex, where the main ideas (the *what* and *why*) of the author stand out clearly and boldly.

What happens when the researcher writes the report in the same sequence? The most obvious fault is that the reader does not have a full grasp of the *what* and *why* until the final pages. But there is another, equally insidious, fault: By following this academic approach—introduction, discussion, conclusion—the writer forces the reader to go through all the spadework and trivial detail he himself had to endure while researching his subject matter. Nothing is more frustrating to the reader at that point.

The pyramid structure mentioned above highlights what should be a key axiom of technical writing: *When you write a report, don't make the reader climb your pyramid; he may have trouble getting to the top.* Such an axiom carries this message to the report writer: When you actually start writing your report, flip-flop the sequence you followed in doing your research. Put your conclusions and recommendations at the beginning of the report. And put all those details down in the base of the pyramid, where they reinforce your main ideas, not precede and dominate them.

The Inverted Pyramid

This structure, which has been called the inverted-pyramid approach to report organization, is far and away the most logical for any technical report, because:

• At once, the reader learns what he wants to know. For instance, if several copies of a research report are distributed, some of the most important recipients may have neither the time nor the desire to examine every page; all they may want is a clear picture of the conclusions that emerged from the research—and the inverted-pyramid form can give this to them without their having to hunt for the right pages.

• Conclusions are usually concrete, rather than general. Thus the writer avoids a notorious trap when he begins with his conclusions. He is not forced to begin with a lengthy, general introduction before coming to grips with his subject matter. This is a major fault of the academic approach to organization, where a vague summary, followed by a vague introduction, creates all kinds of confusion. One has to get down to brass tacks at some point, so why not do it right at the beginning?

• For the writer, there is a great psychological advantage when conclusions come first. Once the reader grasps the writer's main points, and learns why the writer is promoting them, he is forced to consider the writer's point of view as he reads through the report. He may not entirely agree with the writer at the outset, but he can't ignore his main points, no matter how much he tries. In contrast, when conclusions come at the end of the report, the reader isn't influenced by them at all; in fact, he may have arrived at entirely different conclusions before he gets to the final pages.

• The report is much easier to write when conclusions appear at the beginning. This is perhaps the biggest benefit of all for the report writer. When conclusions are spelled out clearly at the outset, most of the traditional problems of organization tend to disappear. The remainder of the report simply becomes an expansion of the conclusions and recommendations. And quite frankly, the organization after this can be atrocious, yet the reader is not likely to object so vociferously. Why? Because the first part of the report gives him clues or guidelines that make it much easier to interpret the mass of detail that follows.

A Specific Proposal, and Its Benefits

As mentioned earlier, I strongly recommend that the traditional report summary be eliminated, and that it be replaced by a standard printed form. Such a form, filled in to represent a typical report, is reproduced on the opposite page, along with the conventional summary for this report. In the discussion that follows, it will be helpful to refer from time to time to both the filled-in form and the summary.

Why a form? Admittedly, few people really enjoy filling out forms. As I am writing these lines it is early April, only a week or so before that most infamous of all deadlines, for taxpayers. Yet let us be honest with ourselves: If anything could be worse than having to fill out an income-tax return, it would be having to furnish the same information with no form at all.

What the report form does is force the engineer to spell out his main ideas clearly and concretely. It gives him a set of guidelines to follow, which are far more helpful than a blank piece of paper. One of the most frequent complaints of engineers is that they find it very difficult to get started on the task of writing reports. The form makes this task a little less formidable.

Several improvements come when the printed form replaces the blank page:

• The section "What Are Your Conclusions and Recommendations?" presents the *what* of the report in detail. Notice that the conventional summary is much more vague. In fact, some of the conclusions do not even appear in the summary.

• The section "What Are the Advantages?" presents the *why* of the report in detail. This is extremely important, because it shows why the reader should bother to consider the writer's conclusions and recommendations. In the conventional summary, notice that most of these points are only hinted at; in other words, the reader has to do the interpreting if he is to get the full significance of the conclusions, because the writer has done little interpreting.

• The form is considerably more detailed than the summary, because the form forces the writer to be more specific. It makes him consider including five conclusions and recommendations, four advantages, and two limitations. (In the end, he may not actually have to fill in every blank space, but at least he has had to think about doing so. Thus, report writers will often include important details that they wouldn't think of putting into the traditional summary.)

• On the form, the writer must separate the various elements of his *what* and *why*. In other words, he must be more discriminating when he lists his main ideas. On the conventional summary, there is little of this discrimination, unless the writer is a far more skilled word craftsman than the typical report writer. As a result, the impact on the reader of these important ideas isn't nearly as strong as it should be.

• The form compels the writer to be more concise in his choice of words. Notice that the numbers on the form are arranged in a way that the writer is forced to state an idea in no more than three or four single-spaced typewritten lines. (See blank form on p. 43.)

• Finally, the form encourages the writer to introduce figures on costs, savings and the like. The writer can't simply say to the reader, "Buy my recommendations because they are bigger, better, and more wonderful," which he is likely to do in the traditional summary. Such a statement may help to sell soap or hair sprays, but it doesn't convince the reader of technical reports.

Selecting the Form

In a few enlightened companies today, forms like the one suggested here are producing impressive results. Of course, the forms in use vary considerably, because reports of individual companies or departments vary widely. But whatever the physical format, the best forms do compel report writers to arrange their main ideas in an orderly fashion. The chaos of the traditional summary is kept to a minimum.

Anyone who is responsible for selecting the format of his company's reports should seriously consider replacing the summary with a form. You are welcome to appropriate the form given here if you find it meets your needs. (An alternate form is shown on p. 48.) Usually, though, you will want to devise

Summary

This report examines the capacity of the bottling lines for Chemical Specialty XYZ, relative to projected demand over the next ten years. (The demand projections are based on Market Research Study No. 46 (1968), and are detailed in the Appendix.) It is concluded that the present theoretical capacity is adequate to meet demand until 1975, assuming certain additional warehousing facilities can be obtained for seasonal peaks. However, unless the present efficiency of the lines is increased so that they operate closer to theoretical capacity, it is likely that shortages will occur by 1971, unless a considerable amount of overtime work is engaged in. This report therefore examines a program of preventive maintenance, foreman and employee training, minor line modifications and other measures whereby the efficiency of the lines might be improved. Other benefits of this program are also examined. A discussion of costs and a proposed timetable are included on pp. 24-28. (It is anticipated that the costs of this program will be confined to operating expenses; no capital appropriation is being requested at this time.)

REPORT HIGHLIGHTS – be brief, be factual, be discriminate

TITLE OF REPORT: Chemical Specialty XYZ -- Bottling Capacity vs. Demand

WHAT IS YOUR OBJECTIVE?: To determine whether the bottling line for XYZ can meet demand over the next 10 years.

WHAT ARE YOUR CONCLUSIONS AND RECOMMENDATIONS?:

1. Capacity should be adequate until 1975, if we undertake programs to increase line efficiency from the present 75% to 90%.

2. A preventive maintenance program will account for 10% of the increase in efficiency.

3. Modification of the capping station (total cost, $10,000) will increase line efficiency another 5%.

4. Another 25,000 sq. ft. of warehousing space should be made available in 1971, to take care of seasonal peaks.

5. The new Tabor bottling process, which won't be on the market until 1974, remains the best candidate for future capital expansion.

WHAT ARE THE ADVANTAGES?:

1. No capital appropriation is necessary at this time. Program costs, an average $7,500 a year, will be absorbed by operating expenses.

2. On a per-unit basis, direct-labor costs will decrease in line with the 15% increase in line efficiency.

3. Increases in indirect costs will easily be absorbed by reduction of overtime premiums.

4. Adoption of a preventive maintenance program is long overdue. Here is an excellent opportunity to put some of our long-held theories into practice.

ARE THERE ANY LIMITATIONS?:

1. Overtime work may be necessary by 1971 if program proves inadequate.

2. Preventive maintenance must be rigorously enforced; to ensure this, a special training program must be set up for foremen.

HIGHLIGHTS PAGE (right) is only slightly longer than conventional summary (left)—yet is much more specific, and gives the reader a clearer grasp of the significant ideas in the report. The format can save writing time.

your own form. Here is some practical advice on this:

First you should make a survey of past company reports. This study should give you the answers to questions like these: How numerous are the *what's* and *why's* of past reports? What headings should be included? Should conclusions and recommendations be included as separate sections? When your survey is completed, you should have a pretty good idea of what form will work best for your own company.

Try to avoid oversimplifying the form. Make it as detailed as possible, so that your writers will have to consider all the things you feel are important.

Sometimes a form may have to be longer than the one-page form suggested here. You should try, however, to get everything on one page, adding at the bottom a phrase like "continue on reverse, if necessary." Incidentally, it's a good idea to reproduce each filled-in form and put the reproduction into a master file. Such a file would then become available to anyone in the company who wanted a clear, concise record of the *what* and *why* of any company report. If he then needed more information, he could always go to the report itself.

After you have devised the form, have copies printed and insist that every report writer fill one in. If anyone objects, complaining that "it won't work for this particular report," have him prepare the traditional summary and then show him how the form will work. Most writers will be grateful for the form, and there should be few objections. If complaints do become too numerous, don't despair. This is a good indication that the form itself isn't

REPORT HIGHLIGHTS – be brief, be factual, be discriminate

TITLE OF REPORT: Polymerization Study: Catalyst ABC vs. XYZ

OBJECT OF RESEARCH: To study the feasibility of replacing catalyst XYZ, costing $5/lb., with catalyst ABC, costing $1/lb., in the polymerization of phlegiston. Current net usage of catalysts XYZ is 10,000 lb./yr.

FINDINGS AND CONCLUSIONS:

1. Catalyst ABC can produce equivalent yields if temperature and pressure are maintained 15% higher than for catalyst XYZ. This would reduce potential savings from $40,000/yr. to $20,000/yr.

2. If temperature and pressure were kept at present levels, the loss in yield would more than offset the savings in catalyst cost.

3. Catalyst life, net quantities needed, and regeneration economics are roughly equivalent for the two catalysts.

NATURE AND SCOPE OF STUDY:

Economic evaluation based on 25 bench-scale tests under pressures of 500 to 1,000 psi. and temperatures of 200 to 300 F.

RECOMMENDATIONS FOR APPLICATION OR FURTHER RESEARCH:

1. Study temperature/pressure requirements for modified version of catalyst ABC, costing $1.50/lb.

2. Compare process-control problems by running pilot-plant test.

3. Make detailed end-product quality study to confirm interchangeability.

4. If outcome of above two items is favorable, run full-scale plant test in Eastern Div. plant prior to July 1 turnaround.

RESEARCH PROGRESS REPORTS could adopt this type of format for the Highlights Page.

properly designed, that it should be reworked and reworded to take care of the objections.

The Rest of the Report

Once the report writer has filled in the form, he has an excellent guide to what the organization of his report should be. In effect, the filled-in form becomes the outline for the report; everything that follows is a detailed expansion of the main ideas stated on the form.

To illustrate, consider once again the filled-in form on p. 47. Part One of the report would consist of a detailed explanation of the report's objective, together with an explanation of the most important conclusion (here presumably the first one written on the form). To support these, material from the "Advantages" and "Limitations" sections on the form should be introduced, along with any introductory background material that relates directly to the objective and conclusion.

Part Two of the report should be a detailed expansion of the second conclusion: Establish a preventive maintenance program. Again, this part of the report should be reinforced by advantages, limitations, background material, and the like. Part Three would give all the details on modifying the capping station. And so forth.

At the end of his report, the writer can sum up his case by including the traditional "Conclusions and Recommendations" section if he likes. But this is strictly optional, because this information is already available on the form.

Question: What happened to the conventional "Introduction" section that appears in most reports? Answer: As already noted, it no longer exists; that is, it is no longer a monolithic block designed either to bore readers to tears or to give them information they already know. Instead, it has been broken up into numerous little pieces, and each little piece has been put somewhere in the report where it will be more effective.

In the new approach to organization, background material is inserted to reinforce the main ideas, not to precede and dominate them. Definitions, historical data, and the like are thus subordinated to the main ideas. They are of course present, but they don't have a chance to obscure the conclusions and recommendations the writer is trying to sell to his readers. Every writer of technical reports should be required to memorize this key rule for effective organization: Always keep the *new* of your report (your conclusions and recommendations) uppermost in the mind of your reader; don't eliminate the *old*, but always use it to reinforce the new, not precede and dominate it.

Final Words to the Writer

Every aspect of business has its "don't rock the boat" champions. The format of technical reports is no exception. If you do consider this suggested approach to organization a logical one, but meet with violent resistance from higher authority, consider something further: Adopt the form anyway (secretly) and make the best use of it before you revert to the tired old academic approach to organization.

I'm suggesting that the next time you have to write a technical report, you should first fill out the form. This should bring three immediate benefits. First, it will help you to overcome the initial block about writing, which almost everyone experiences at times. Secondly, it will help you to organize your conclusions and recommendations more clearly and concisely. And finally it will help you to write a much better summary than you would have written otherwise. ■

To help you get more writing done in less time, via a secretary or dictating machine, here is practical advice on . . .

How to Be a Great Dictator

AUREN URIS, Research Institute of America

This article has been dictated. A dictated draft was completed in 1 hour and 15 minutes. Subsequently, it required 1 hour of editing by the author. Total time required to produce the manuscript, in the final form to be sent to CHEMICAL ENGINEERING: 2 hours, 15 minutes. Estimated time if dictation had not been used: one working day.

There's no doubt that the engineer or technical manager who can dictate not only observations, memos and letters but also reports and manuscripts can save himself hours of toil. For anyone who wants to make it in print, learning to dictate can be a major step in a time-saving "writing" skill.

Potential Dictators All

Anyone who can speak can dictate. And few people in industry would have any qualms about using a stenographer or a machine to dictate simple letters and memos. For example:

Pete Bramble, holding a written query he has just received from the front office, gets his boss's secretary to take a letter. He says, "All right, Gladys, here it is: 'To C. R. Castor. In reply to your question about the drums piled up in the hallway of Building B, comma, I'm happy to report that they've just been removed, period. Signed, Pete Bramble.'" He doesn't have to bother adding the date or departmental location. Gladys will take care of that. Then he goes on to dictate a couple of other short notes.

But suggest to Pete Bramble that he dictate a five-page report, and he'll think you're crazy, because everybody knows that lengthy communications, such as reports or an extensive written presentation, must be written out in longhand, or pecked out laboriously on a typewriter. But as often happens, what everybody knows isn't necessarily so. Practically everyone can develop a reasonable dictating skill sufficient to cut writing-time requirements by 50% or even more.

The potential to be gained in time is immediately evident when you consider: the man who sets out to provide a report in longhand can probably produce the finished product at the rate of about 30 words a minute. If he dictated to a capable stenographer, he would be able to triple that speed. And if he used a machine, he could get down his thoughts almost as quickly as he could find the words for them.

The advantages of being able to dictate are so extreme that it's worth the while of every professional to at least give this technique a trial.

Obviously some people will be more successful than others. Interestingly enough, however, it doesn't necessarily follow that the fluent speaker is the one who makes the better dictator. In some cases, the man who is less fluent and even stumbles over his words may actually be dictating "harder" copy: If his thinking is orderly, his dictation may be more concise than that of the man who speaks easily but whose thoughts meander.

There's no mystery in how to go about mastering the dictating skill. These points can help you:

What to Dictate

While almost anything that is to be written can be dictated, there are some types of writing that effect greater time savings by dictation than others. Reports, for example, are particularly worthwhile for dictation since they tend to be lengthy; undictated they represent large investments of time.

Letters and memos, of course, are also prime communications to be handled by dictation. As a matter of fact, a trained stenographer can often compose an entire letter when her boss simply dictates one or two key sentences and leaves it up to the secretary to fill in the usual formalities of address, greeting, closing and so on.

Unsuitable for dictation are names or other proper nouns, numerical tabulations, and highly mathematical copy (i.e., where more than half of the copy is comprised of equations).

Girl or Machine?

As already indicated, the dictator may have a choice between a stenographer and dictating machine. Recent years have seen dictating equipment refined. Machines are lighter, operate more easily. Some companies have dictation systems where one dictates into a mouthpiece or telephone: a recording tape or other medium, in a centralized location, records the dictation, which is subsequently handled by a typing pool.

Regardless of particular mechanism, machine dictation and dictation to stenographer have advantages and disadvantages. The chart on the opposite page compares the two.

Getting Started

"My trouble in dictating," says one technical manager, "is to just get off the ground."

It's a common problem. But there are several preparatory moves that can help beat the hangup:

Gather your information, background material, and so on, in advance. Don't start your dictation and be forced to interrupt yourself because you are missing a piece of reference material. Have available all the backup material you will need—if possible, in the order in which you will use it.

Make a rough outline. An outline has the virtue of giving a track to run on, a framework on which to build your final product. And, except for lengthy or complex projects, you needn't get involved in the kind of elaborate outlining that is sometimes taught in school. At a minimum, a number of words or phrases that give you a logical sequence of ideas will do the trick. For example, here is the outline that was used in dictating this article.

Introduction: Benefits

Everyone can do it to some degree

1. What to dictate
2. Girl or machine?
3. Getting started (after the preparation steps)
4. Common hangups and how to overcome them
5. Improving your copy
6. Reviewing your weak and strong points and improving your technique

Conclusion: Byproduct benefits

Other preparations such as physical arrangements for convenient working—a comfortable chair, a table or desk for notes and background materials, a comfortable chair for your stenographer, or a convenient location for your dictating apparatus, round out the preparation steps.

Common Hangups and How to Overcome Them

Discussions with individuals who have attempted to dictate material other than letters or memos reveal some common problems:

• "Writer's block." Some people say that when faced by the need to start dictating, they freeze. This is similar to the paralysis that occasionally stops even the professional writer in his tracks. To counter this problem, check back and make sure that all your preparations have been made. In some cases, the feeling of not being ready to dictate may be the obstacle. Then, start! Throw hesitation to the winds. Don't worry about finding the "right" word, phrase or sentence. Plunge in! In other words, an effective way to end a starting problem is simply to *start.* Another way is to begin with a brief or "easy" piece of dictation. This bit of pump-priming usually gets you going.

• "I can't get down to brass-tacks. I ramble and digress." The solution here lies both in having and *following* an outline. If you find you're straying from it, go over it to make sure it covers your subject adequately. Then stick to it, forcing yourself to stop when you seem to be digressing.

• "It's disconcerting to have a stenographer waiting for me to grope for an idea." Try using a machine; or, change stenographers; or make some disarming or humorous comment, letting your stenographer know that you don't expect to talk smoothly and incessantly. If you have not been doing much dictating, it is no disgrace to admit it, and to imply that your stenographer will have plenty of time to catch her breath while you gather your thoughts or grope for the phrase you want.

• "I never liked gadgets, and talking to a lifeless machine is disconcerting." Try a stenographer. Or, the particular make or model of your dictating machine may be a poor choice for you. In today's market, one can find a wide variety of dictating equipment, involving everything from tape to plastic sleeves. If you are having trouble with a machine that you have tried, a trip to an office supply company will make it possible to test various types of equipment. You should be able to come up with one that overcomes the problem.

• "I lose too much time searching for a word or phrase or even an idea." This is a common and major hangup for many people. To overcome it, bull it through—put down any word or phrase that will get you over the hurdle, knowing that you will be able to improve the word or phrase or idea later on.

The important thing is to work for a *rough draft,* because once you have your copy in black and white, you have something to work with. The professional writer will often push on to complete his draft regardless of how many word gaps or idea gaps there may be in his copy. He knows that these points can be rectified later on. "Finish the draft" is the slogan of the goal-oriented professional.

Improving Your Copy

The greatest misconception about dictation is that you say it once, and that finishes the job. It's true that *neither* the dictated nor written copy is "right" the first time. Every potential writer will tell you that the *rewriting* is an essential part of the operation.

So consider the typed pages you get either from your stenographer or from the typist who has worked from your machine-recorded material as merely a draft, a preliminary version. What you should strive for is to turn your first rewrite into what professional writers call "final copy." In some cases, you may have to do two rewrites; that is, revise your copy twice before you have a final draft.

It's in the editing or rewriting that you can eliminate any oral habits that are objectionable. The repetition of the words "you know," for example, and such stilted phrases as "to be sure," which in verbal communication go unnoticed, may become objectionable when written. Count on your editing to eliminate and tighten up.

Improving Your Technique

After you have tried dictating once or twice, review your technique. Use the above discussion of hangups as a kind of checklist to spot your weak points. Then, try to improve your technique via practice. Even someone very much used to dictating letters cannot expect to undertake longer communications without some preliminary difficulties.

If first attempts are discouraging, try again, dictating material on which you have a good grasp. Don't give up too quickly, because the rewards of success can be substantial. Repeated attempts can help you score a breakthrough.

Further, although we have mostly emphasized timesaving, some dictators claim that the *quality* of their writing improves when they dictate: "I find I can think more clearly when I talk than when I type or write out my ideas in longhand," says one manager. "My correspondence sounds less stilted and more readable when I dictate it," says another.

You may be the greatest dictator of all times—but you can't know just how great until you try.

Stenographers vs. Machines: How They Stack Up

STENOGRAPHER

Advantages	Disadvantages
Useful rapport can exist between steno and dictator.	Usually works 8-hr. day or less, and may resist overtime.
Can make corrections and insertions on the spot, and supply quick readbacks.	May tire after an hour or so.
Can be helpful in giving instant feedback to dictation (in terms of reaction to wording, ideas, etc.). Can ask questions immediately if something is not clear.	If steno is absent from work and has not completed transcription of her notes, chances are no substitutes can read her shorthand. This means a delay until steno returns to work.

DICTATING MACHINE

Advantages	Disadvantages
Doesn't complain, has no personal habits or traits that can irritate you.	Can lend themselves to corrections, but some people find machines difficult in this respect.
Available 24 hours a day, and is tireless.	Cannot help supply a missing word or phrase.
Some people are less selfconscious when they dictate to a machine. There is no need to apologize for pauses or interruptions.	If transcriber has difficulty in discerning some words or phrases dictated, the dictator may not always be available to be asked.
Conserves secretarial time, and can be taken along on a business trip without precipitating office gossip.	Transcriber might feel it's a waste of time to listen to the entire recording before typing from it, preferring to take it down in shorthand to facilitate transcription of corrections, changes, and punctuation.

Dictation and the Engineer

A selective approach to dictation, with special attention to technical material, may be necessary to reap full benefits and avoid pitfalls—particularly if the secretarial situation is less than ideal.

How do you spell $C^*(j\omega) = \frac{1}{(j\omega)^2} \int_0^\infty \sum_{k=1}^{\nu} a_k u_1 (t - t_k) e^{-j\omega t} dt$?

JAMES R. FAIR, Monsanto Co.

In principle, I fully agree with the previous author's premise that most of the writing needs of engineers as well as of nontechnical people can be satisfied by dictating to a girl or machine. And the time- and effort-saving advantages of dictation can be tremendous.

In practice, however, many engineers face some special hurdles, especially in regard to technical material. My purpose here is to focus attention on how some of these hurdles can be surmounted.

Secretarial Resources and the Engineer

In order to gain some perspective, let's first consider two idealized situations involving the engineer-dictator. The first involves a prosperous technical consultant named Brown, who must prepare a 16-page typewritten report for a client. He gathers reference documents and data, prepares a detailed outline, and buzzes for his private secretary, Mrs. Green. After arranging for another secretary to take all calls, she settles herself in front of his desk, notebook in hand, and prepares to take down his report in shorthand.

Brown starts by dictating title, distribution, and table of contents. He then goes through the entire report, speaking carefully and distinctly to Mrs. Green. When he comes to a figure, he notes its place in sequence and dictates its title. When he comes to a table, he dictates title, column headings, and hands her the penciled figures to be inserted. He is careful about style, does not overuse certain words or expressions, and modulates his speaking so that Mrs. Green will have no difficulty in handling punctuation and paragraphing.

Mrs. Green has worked closely with Brown for many years. She is college-trained and does a bit of writing on her own. Thus, she takes the dictated material, types it up in final form, proofs it with another girl, and hands it to Brown. He, in turn, glances through the report, nods approvingly, and says "Send it out!"

The second idealized situation: Project Manager Gray dashes to the airport to return home from an all-day conference with an engineer-contractor. Aboard the plane, he finds that the seat next to him is not occupied, so he spreads out his papers, takes his lightweight dictating machine from his briefcase, and prepares to dictate his trip report. Airborne, he declines the drink offer from the stewardess, starts with a fresh tape in the machine, gives his secretary guidance on the format and distribution of the report (e.g., telling her to type it up in final rather than in draft form) and then launches into an account of the conference while the details are still fresh in his mind.

Next morning, Gray hands the tape to his secretary at 8:30 a.m. Two hours later, the trip report has been distributed to key people in the building, and at 1:00 p.m. the document is being used by several of the recipients in a project review meeting. While the secretary was typing the report, Gray was concerning himself with other needs for the meeting.

Now, just to make the story complete, let's consider an unidealized situation—a fairly realistic one, I believe, in a number of firms. Engineer Black has to write to several companies for information on catalyst supports. He has been told by his boss that he should develop dictating ability, so he decides to give it a try. After thinking about what he wants to say, and perhaps actually scribbling out some parts of the letters, he attempts to locate a secretary.

The trouble is, Black shares the secretary, Miss Blue, with 9 other engineers. At the moment, she's on an errand for one of his colleagues. When she finally returns, he calls her in and starts dictating those letters. After several fumbles, mumbles, and

When Black finally gets the typed letters (let's not forget that Miss Blue has her own priority system), he finds that they look pretty bad. Sentence structure is poor, capitalization is wrong, and some technical terms are misspelled. He's not sure whether the letters really convey his desired meaning, and he doubts whether they would represent properly the image that he and his company want to present. He tries to edit the letters, but finally decides to do them over in longhand. In the end, he is disillusioned and Miss Blue is upset. "This business of dictating is for the birds," claims Black.

Developing the Skills for Efficient Dictation

The above three situations, which cover some typical communication needs of the engineer or technical manager, illustrate that conditions could be right—or wrong—for dictating effectiveness. If conditions are wrong, this does not mean that the engineer should stay away from dictation but rather that he should use it selectively at the start, and gradually expand the amount of dictation as his proficiency increases and as obstacles are overcome.

Materials to be communicated fall into these categories:

1. Notes for personal files, etc.
2. Office forms—e.g., for disbursements, supplies.
3. In-house memoranda.
4. Correspondence with suppliers, customers, etc.
5. Formal technical reports and papers.

It is easy to see that this listing follows a gradation of demand for dictating skill. To put it another way, even the novice can skip the rough-draft stage in dictating, for example, recollections of a meeting he attended if these recollections are intended for his personal files or as an informal note to a colleague. Conversely, even the accomplished dictator may want to go through the rough-draft stage for key portions (e.g., the summary, conclusions, recommendations) of a formal technical report, or for the first few pages of a manuscript for publication.

On a percentage basis, the biggest savings in dictating come about for work where you can skip the draft stage. Dictating memos and letters, and then just signing the finished product, can be several times as fast as writing them out longhand, as the previous author has pointed out. But if the material has to go through a draft stage, the total time spent in outlining, dictating and then later editing the draft may not always be very much less than typing and writing out the draft yourself, making corrections or erasures as you go along, and then giving it to your secretary for the final typing.

In my opinion, the engineer's goal should be to avoid the rough draft except in those cases where wording and style are quite critical. The basic approach should be "try to do it right the first time."

Dictating Equipment

The dictating system comprises the **dictating machine** and a compatible **transcribing machine.**

Dictating machines may be characterized by their portability and by the type of recording medium. A machine just for office use will weigh about 12 lb., cost about $400, and include features that emphasize clarity of recording and flexibility of operation. A machine emphasizing portability will weigh about 2 lb., cost $225-400, use battery power, and have limited flexibility of operation (for example, it will not serve well for recording conferences). And there are "hybrid" machines incorporating some features of both the fixed and portable types. All types can be adapted for recording telephone conservations.

Another method for characterizing dictating machines is the type of recording medium: disc, belt or tape. Discs give high recording clarity, accept rapid dictation, and are readily mailable. The grooves show the extent of dictation, but are not erasable for reuse. Belts give generally good clarity, are easily mailable, and can be erased if they are of the magnetic type. Tapes give fair clarity, are somewhat cumbersome to mail and, being magnetic, are erasable.

Belts and discs are easily scanned (by moving the needle), but tapes require relatively slow winding or rewinding. The reusability advantage of magnetic tapes and belts is offset by the difficulty of locating points where correction or playback is needed.

All-in-all, there are many choices. See your dealer. A sampling of types and characteristics appears below:

Brand * Battery-powered	Type	Erasable?	Record. Time, Min.	Speed, Words/ Min.	Weight, Lb.	Cost, $: Recorder	Transcriber	Manufacturer
Executary	Belt	Yes	28	70	13	430	405	IBM
Voicewriter	Disc	No	30	85	12	425	405	McGraw-Edison
Envoy	Tape	Yes	45	60	9	200	405	McGraw-Edison
Escort*	Disc	No	30	70	8	375	410	McGraw-Edison
Time-Master	Belt	No	15	70	7	435	410	Dictaphone
Travel-Master*	Belt	No	15	70	5	355	410	Dictaphone
Compact*	Tape	Yes	120	60	3.5	200	405	McGraw-Edison
Dictamite*	Tape	Yes	60	~60	1.9	215	275	Dictaphone
Executary 224*	Belt	Yes	28	70	1.8	425	405	IBM
Stenorette Versatile*	Tape	Yes	45	~60	1.1	70	130	DeJur-Amsco

In line with this, the dictator should force himself to:

- Develop a detailed outline.
- Picture just how the type will look on the page.
- Use direct readback to cut out repetition or clumsiness.

I realize that there is much to be said for the adage "to write is to rewrite." Up to a point, the revisions are well worth the effort. I also realize that some people are very slow writers but fast dictators and fast editors—for them, dictating has advantages even if there are two drafts before the final version. But in general, I feel that *indiscriminate* use of the rough draft tends to be a crutch—it promotes sloppiness. Here is how you can use rough drafts more judiciously:

1. Start out by dictating notes for your own files; also routine paperwork and in-house memoranda. Dic-

Dictating Technical Material

Below are a few paragraphs from the manuscript for a report that CHEMICAL ENGINEERING edited and published last year ("Design of Modern Distillation Columns," in the Apr. 22, 1969 issue). I chose these paragraphs not because they are particularly well written, but to illustrate what's involved in dictating technical expressions.

At right is a reconstruction of the way these paragraphs were dictated. The italics indicate nontranscribed material; my secretary distinguished between this and the material to be transcribed by noting my intonation and gestures.

I had made up a fairly detailed outline before starting to dictate. The material was transcribed directly, without a draft. About 25% of the 70 double-spaced transcribed pages in the manuscript needed corrections of some kind, but very few of these involved redoing the entire page.

PREDICTIVE MODELS FOR EFFICIENCY

Work sponsored by the AIChE Research Committee led to the development of the "AIChE Method" for predicting the mass transfer efficiency on a crossflow-type (bubble cap, valve, sieve)* distillation tray. The model is described in the AIChE Bubble Tray Design Manual, (1) and explained very lucidly by Smith (54). It is based on the two-film theory, and is directed toward the prediction of point efficiencies, with most of the experimental foundation work having been done in small columns where liquid mixing on the trays was essentially complete.

According to the film model, the point efficiency can be defined as:

$$E_{OG} = 1 - e^{-N_{OG}} \quad (34)$$

where N_{OG} (the number of overall gas phase transfer units) is equal to an overall conductance, the reciprocal of which is the sum of the individual phase resistances:

$$\frac{1}{N_{OG}} = \frac{1}{N_G} + \frac{\lambda}{N_L} \quad (35)$$

The phase transfer units are obtained from the general expressions:

$$N_G = k_G a' t_G \quad (36)$$

$$N_L = k_L a' t_L \quad (37)$$

It is immediately apparent that the effective interfacial area term a' is difficult to evaluate. Hughmark (32) has correlated it empirically with gas rate, weir height, liquid viscosity, liquid density, vapor density and surface tension.

*Tray or plate devices that force the frothy liquid to flow across each plate.

Now, Beverly, we completed the section on Plate Efficiencies. Today, let's start with the subheading PREDICTIVE MODELS FOR EFFICIENCY. *Capitalize and underscore all that, please.*

Work sponsored by the AIChE Research Committee led to the development of the *quote* AIChE Method *close quote—and capitalize method* for predicting the mass transfer efficiency on a crossflow *hyphen* type *parenthesis* bubble cap *comma* valve *comma* sieve *close parenthesis, asterisk* distillation tray *period. (Now, Beverly, I would like to tie this footnote into that asterisk:* Tray or plate devices that force the frothy liquid to flow across each plate *period, and that should end the footnote. Now, back to the paragraph.)* The model is described in the AIChE Bubble Tray Design Manual *comma, parens reference* 1 and explained very lucidly by Smith *parens reference* 54 *period.* It is based on the two film *(hyphenated)* theory *comma* and is directed toward the prediction of point efficiencies *comma* with most of the experimental foundation work having been done in small columns where liquid mixing on the trays was essentially complete *period, paragraph.*

According to the film model *comma* the point efficiency can be defined as *colon (now, Beverly, leave space for equation* 34) where capital N *sub* OG *parenthesis* the number of overall gas phase transfer units *close parenthesis* is equal to an overall conductance *(underscore) comma* the reciprocal of which is the sum of the individual phase resistances *colon; space for equation 35; paragraph.*

The phase transfer units are obtained from the general expressions *colon (space for equations 36 and 37)* It is immediately apparent that the effective interfacial area term a′ (*Beverly, that should be a lower case "a" with a superscript prime mark)* is difficult to evaluate *period.* Hughmark *H-u-g-h-m-a-r-k parens reference* 32 has correlated it empirically with gas rate *comma* weir height *comma* liquid viscosity *comma* liquid density *comma* vapor density and surface tension *period.*

tate these items carefully without asking for a draft. When you see the final version, ask yourself whether the writing is concise, whether the meaning will be clear to the intended audience, and whether the material seems to represent the work of a reasonably literate person. If some of the answers are no, you may correct the material if you feel necessary, but your main goal should be to avoid these shortcomings in future dictation. So make some mental or actual notes.

2. As you improve, start dictating most outside correspondence, and the noncritical portions of technical reports (e.g., the appendices, description of procedures, etc.), keeping in mind the pointers on handling technical and mathematical material discussed elsewhere in this article. Don't ask for a rough draft. When you get the typed material, some pages may have to be corrected or retyped, but if you have dictated carefully, many of them will be acceptable as is. (Keep in mind that dictated material will rarely reach perfection in the mind of the dictator; but carefully worded longhand may not always do this either.) Again, take note of pointers that apply to future dictating work, so that the percentage of pages that have to be corrected or retyped goes down the more you dictate.

3. As for particularly sensitive correspondence, and the critical portions of reports, you have your choice of dictating into draft form from a rough outline, trying to dictate into final form from a detailed outline, writing the draft in longhand or (if you are a reasonably fast typist) typing the draft. Personal preference can be the legitimate deciding factor: You may find you can do the best job of putting complex, crucial thoughts down on paper when you have a pencil in your hand and can see every sentence take shape. Or you may find that you can uncoil most easily by dictating these thoughts rapidly without worrying about style, and then carefully editing the draft. In other words, after you have developed the skills to dictate most of your material efficiently, use whatever seems to work best for really critical portions of your writing needs.

Engineers and the Office System

In recommending that engineers use dictation for more and more of their writing, I realize that the office system is not always ideal for engineer-to-secretary dictation.

Most engineers share a secretary with many others. So it follows that the secretary cannot handle all of "her men" on a demand dictation basis. Directly or indirectly, the men are encouraged to use longhand or a machine. There are cases where a secretarial pool makes possible dictation on demand, but under such circumstances the engineer may have a different girl almost each time he calls for assistance.

In most office systems, machine dictation does provide distinct advantages over laborious longhand, especially for noncritical work. Also, some companies are able and willing to supply their technical departments with additional stenographic help if the engineers can prove that this would boost their own productivity, and that the additional help would be used efficiently—in other words, that a girl earning a stenographer's or secretary's salary will not be used primarily to type up handwritten material. (The latter is not only inefficient but can cause discontent and high turnover among stenographers.)

Most secretaries working in a technical organization can learn to handle such esoteric terms as phthalic anhydride and contact catalysis quite readily. But they cannot be expected to clean up sloppy dictation or writing. And they probably will not understand the individualistic desires of the dictator regarding punctuation.

In most cases, the best approach is to dictate punctuation marks, paragraph designations, and unusual capitalization (see box, opposite page). Further, the dictator must be alert to words with alternate spellings. It may take some practice to do all this without losing one's train of thought, but it can be done.

Technical Material and Other Problems

When we discussed Consultant Brown, we touched on the special problem of dictating tabular material. There are similar problems with much of the material that an engineer is likely to be writing.

First, there are the mathematical expressions. They can be dictated directly if the secretary has a degree in mathematics, but one doesn't find such secretaries very often. The best plan is to write out the expressions and designate them so that the secretary will add them in sequence at their proper place in the dictated text. One might say "blank for equation 16," or words to that effect. If a rough draft is planned, the dictator can write in the equations himself.

Other special problems involve symbols, Greek letters, foreign words and phrases, and so on. The important watchword is understanding of the secretary and her capabilities. A scratch pad on the desk between dictator and secretary can be very helpful.

As for dictating equipment, p. 53 provides a fair sampling of what is available. In all cases, care must be taken to erase old dictation; to identify current dictation with subject, date, and instructions as to whether it is to be in rough draft or final form. The problem of advance notice of correction is especially difficult on some machines, and this must be recognized.

To sum up, engineers should not expect instant success in emulating Consultant Brown's "great dictatorship." In his case, the propensities of both man and system to dictation effectiveness are high. Such a combination is not often found, however.

I feel that the man, not the system, usually determines the degree of resistance. Almost every engineer has *some* secretarial support, and by exploiting it efficiently, he can lay the groundwork for more support.

The economic potential of effective dictation remains high. It's just another case where man/woman and man/machine interfaces need more exploitation. ■

Six Guidelines for Fast, Functional Writing

Here is how you can get more writing done in less time, without becoming sloppy—in fact, some of these toil-saving suggestions actually improve the informativeness of reports, letters and memos.

HERBERT POPPER, Chemical Engineering

Some types of improvement in the quality of one's writing can only be achieved at the expense of quantity—for instance, by spending more time on the editing of drafts. However, regardless of whether you use longhand or some form of dictation, there are several areas where quantity and quality go hand in hand, and where it can actually take less time to produce an effective piece of communication than an ineffective one. The six guidelines that follow take aim at that area.

Because writing problems vary in kind and degree from person to person, you may find that some of these suggestions either don't apply to you or appear self-evident. However, "fast, functional" writing skills are so important to most engineers and technical managers that even if you only find one or two of the suggestions really useful, this should more than repay you for your reading time.

1. Know What Your Audience Expects

A vice-president of operations was making his annual tour of the company's outlying facilities. At one plant, he chatted with the assistant manager about the upcoming labor negotiations with the local union.

"I wish you'd send me a fairly detailed report of those negotiations," asked the vice president. Flattered by this request, the assistant plant manager resolved to do an outstanding job of reporting. By the time negotiations were over, he had more than 50 pages of handwritten notes, which he decided to amplify with background information on some of the points the union or the management team had raised.

It took him two weeks after the close of negotiations to organize this material into a rough handwritten draft. Because of a secretarial bottleneck, it took him two more weeks to get this draft typed and edited, and then another two weeks to get the massive document typed up in final form. Nevertheless, as he signed the letter of transmittal in which he apologized for imperfections in the typing and commented on the shortage of good secretarial help, he felt rather proud of himself: The report read like a courtroom drama, and had a great degree of polish.

Unfortunately, before the report could even reach headquarters, the vice-president had made an acid phone call to the plant manager. Apparently, what the vice-president had really wanted was brief, day-by-day reports that he could review while negotiations were still in progress, and that he could pass on to another plant that started negotiations a week later. When he finally got the assistant plant manager's monumental opus, a lot of the information was old hat, the key points having been already passed along by the plant manager in a lengthy, long-distance phone call just before the close of negotiations.

Next year, the plant manager decided to handle the job of reporting himself; at the close of each day's session, he dictated his notes into a recorder—his secretary would type them up on the next day, so that he could ink in some comments and send them out as soon as he finished that day's negotiating. Since he didn't use a draft, his daily reports were not particularly polished; the inked-in comments and corrections looked rather informal—and yet these reports were acclaimed by the vice-president.

Moral of the story: Make sure you know what the audience wants. Don't be afraid to ask. There are times when speed, rather than quality or quantity, may be of the essence. In the example above, which is based on an actual situation, the assistant plant manager could have saved himself hours of toil by finding out what sort of details the vice-president did or did not want to know about, and whether quick dictation without a draft was acceptable.

Failure to find out or to really understand the audience's expectations is perhaps the single biggest timewaster in technical writing. An example of such inefficiency involves the engineer who uses exactly the same writing approach regardless of whether his report is primarily intended for the departmental archives, his immediate supervisor, or a financial executive on the appropriations committee.

If the report is just intended for a supervisor, for instance, there is nothing wrong with using technical jargon, and for keeping background information very brief. But if the report is to go to a financial executive —particularly one who lacks any sort of technical background—the engineer must avoid technical jargon, and explain some things that he would not need to explain to his boss, while leaving out technical details. In such a report, it is particularly important to make the first page or two tell the bulk of the story, in terms that are meaningful to the administrator.

If you are writing a dual-function report—say, one intended for both a financial executive and your boss—by all means get squared away with the latter on how the two approaches can be reconciled (e.g., by eventually giving your boss additional sections or informal notes that are omitted from the administrator's copy, by coming up with a preliminary version of the first two pages that you and your boss can review jointly before finalization, etc.).

When writing for publication, rather than for internal use, finding out the audience's expectations can be an equally great timesaver. Many magazines have booklets that discuss their "expectations" in regard to such things as writing style, quality of drawings, compatibility with readers' interests[1]—and are further prepared to conserve the author's time by reviewing outlines in some detail so as to minimize the need for revision of the final manuscript.

2. Mobilize Your Subconscious

Legend has it that on the day the opera "The Magic Flute" was scheduled to get its first performance, someone reminded Mozart that there still wasn't any overture, whereupon the composer calmly sat down and dashed off the magnificent overture just in time to give the score to the orchestra before the curtain went up. Actually, I think if this legend is true, the chances are that Mozart must have "precomposed" the overture in his mind days before he set it down on paper, so that leaving the actual writing until just before the deadline may not have been as risky as it seemed to others at the time.

Unfortunately, many of us tend to display a Mozart-like confidence in being able to meet a last-minute deadline regardless of whether we have done any precomposing or not.

If you have a writing project with a far-off deadline, and you don't feel like starting the actual writing right away, then don't. But I would still suggest that you draw up a rough outline as soon as you can, see how much information you already have available and how much more you will have to dig out, decide when you should start this digging out so that the information will be available at the proper time, and discuss the outline for the project with your boss or with a colleague. That way, you will not only get a feel for the emphasis needed (as discussed in the previous section), but you will be giving your subconscious mind a chance to come up with ideas while the project is incubating.

To come up with useful ideas, your subconscious should have some sort of framework—that way, perhaps some of the framework will be filled in for you while you shave in the morning. But if you wait till the last minute, even to prepare an outline or think about an approach, all your subconscious can do is to provide you with a feeling of anxiety.

Another constructive use for the prewriting period is to do some preselling. People hate shocks—which is what may happen when your recommendations based on a full year's work are coldly dropped on your de-

Got a minute to review the outline for that "brief engineering report" you asked me to prepare last year?

partment head's desk. It is fairer to him, and kinder to you and your writing time, to use the "let's let him in on part of the secret" technique. So, give your boss some inkling of what your report is going to recommend. Don't try to do a complete job of selling, but do give him a chance to get some exposure to your ideas. From his initial comments, you may also be able to obtain pointers on how to best link your ideas to his, and to the current objectives of the department. This can save a great deal of writing time in the long run, and will often let you get the constructive involvement of the reader much sooner than if you use the "This may come as a shock, but . . ." approach.

3. Build Up and Exploit Your Momentum

"I sit down with a fountain pen and paper and the story pours out. However lousy a section is, I let it go. I write on to the end. Then the subconscious mind has done what it can . . . The rest is simple effort . . . going over a chapter time and time again until, though you know it isn't right, it is the best you can do." That's how W. Somerset Maugham said he got his thoughts down on paper. If you have trouble getting started and building up momentum on a writing project, then the Maugham approach of putting down your thoughts in any old undisciplined way and eventually going back to polish them up, has a lot to recommend it. Note that this is also the basic approach to combating writer's block that Auren Uris suggested in his "How to Be a Great Dictator" section.

What do you do if you still don't succeed in unplugging your thoughts, regardless of whether you are using pen, pencil, stenographer or dictating machine? Here are a few additional pointers:

- Go over your outline (or Report Highlights Sheet, as suggested in an earlier section) and pretend a good friend asked you to explain each point to him. If he were to ask questions such as "Why don't we just stick with the old process?" or "How does this gimmick work?," the chances are you wouldn't be at a loss for words—you would just say it like it is. So, pretend your friend is asking you questions that correspond to your outline, and write it like it is.
- Pretend that this is an examination, and that you have five minutes to get something down on paper for each thought on your outline—i.e., that if after five minutes you have nothing down on paper on the first point, you have nothing that you can be graded on, and you would thus get zero. (A couple of years ago, I took an aptitude test in which I was given a cast of two or three characters and had to shape them into a plot for a short story within ten minutes; then stop and shape another cast of characters into a new plot; then do this a few more times. Not considering myself a really fast writer, I was amazed at how much I could get down in ten minutes when the pressure was on, and when I knew I would be graded on content rather than style. It ocurred to me that if I could pretend that I was in a similar situation when starting on a writing job, it would get me going—and it works!)
- Pick the best time to get started on a tough writing project. For most people, this may be the very beginning of the day, before things have come up to distract them. Others work best right after lunch. Still others find that the best time to unlock their thoughts is after dinner at home, and that once they get started that way, they can continue the project at work without much difficulty. If you haven't already done so, try to find a time pattern in the daily periods during which you feel sharpest and loosest; then exploit these periods to get started on tough projects.

Of course, getting started does not necessarily mean writing a brilliant beginning; it means getting started on the "meat" of the project. Very often a meaty fact-filled beginning is the most brilliant one anyway. But if you do feel the urge to think up something particularly striking or original, you may be better off doing this after most of the project is down on paper, rather than letting it be an initial stumbling block.

Once you have built up your momentum, how do you keep it?

First of all, try not to stop. Let someone take your phone messages. If you see a chance to finish the project by working late, do so. Resist the temptation to take a break until you start slowing down. (Personally, I find that sipping coffee or puffing on a cigar while doing difficult writing is preferable to running the risk of getting sidetracked by taking an unnecessary break.) Putting it into industrial engineering terminology, make sure you have a long enough production run to justify the setup time.

If you can't finish the project by working late, then it is usually better to leave it in the middle of a paragraph than to try and finish the particular section or thought. Finishing the thought the next day will be relatively easy, and this will serve to prime the pump for the next thought. I used to make the mistake of staying up until whatever hour was necessary to get a major section finished, only to find that it would take me all the next morning to stop resting on my laurels and to uncork my thoughts for the next section. Conversely, when I don't try so hard to come to a logical stopping place, I can usually continue the section without any difficulty the next morning and then launch right into the next section. (I can't take full credit for this approach; someone named Hemingway recommended it in a book called "A Movable Feast.")

4. Watch Out for Time-Wasting Verbosity

Not everybody has the uncoiling problems dealt with in the previous section. In fact, some people who can uncoil most of the time at the drop of a pencil encounter the opposite type of problem: How to avoid the verbosity that wastes their own time as well as that of the reader, so that they can make the most of their fluency as writers.

The first section in the 1966 "Efficient, Effective Writing" report[2] gave good advice on ways of recognizing and avoiding destructive verbosity. Although that section was aimed at making writing more informative, most of the suggestions can also lead to faster writing. For instance, the writer can usually save time by cutting down on passive verbs, abstract nouns, and prepositional phrases. (Obviously, it would have taken me longer to write the previous sentence had I said "Time can be saved by the writer if proper consideration is given to a reduction in the use of passive verbs, etc.")

Here are a few additional suggestions:

- Don't be legalistic in a nonlegal piece of writing. Lawyers have to be extremely careful to cover every possible contingency—hence they may feel justified in using strings of words that may differ only very slightly in meaning (e.g., null, void, of no legal force, etc.). But engineers are seldom justified in coming up with a phrase like "The *development, establishment,* and *implementation* of process-control *philosophies* and *policies* must take place at an *early* or *incipient* stage in a project." There just isn't enough difference between the italicized words to warrant using more than one of each; if you say "philosophies," 99 out of 100 readers will assume that this includes "policies."
- Don't waste time on excess hedging. For instance, when you say "Based on bench-scale experiments run at ten different temperatures, 300 F. produces the best yield," you have clearly indicated on what you base your conclusion, and there is no need to add hedge phrases such as ". . . 300 F. may tend to produce the best yields, assuming that these bench-scale results are freely applicable to commercial conditions . . ." Putting it another way, readers generally realize that almost any technical, business or philosophical statement that can be made is subject to limitations and qualifications; if you insist on pointing out limitations that are either obvious or unimportant, you are wasting time (and probably producing dull writing). Particularly wasteful and deadly is the "triple hedge"—e.g., ". . . may, under some circumstances, tend to . . ."
- Use more tables, illustrations and in-text listings to cut down on verbosity. Earlier in this report, Johnson pointed out how a "1, 2, 3" approach to the Report Highlights page can avoid the verbose generalities that sometimes go into report abstracts; the same sort of item-by-item approach can result in tighter writing of all kinds, including the body of the report. (At the risk of appearing to disregard my own "hedging" caveat, I should point out that itemization can be carried to dull extremes, particularly once the triple sub-indentation stage is reached. But if you haven't been doing much itemizing, the technique can save you a lot of words in reducing generalities to specifics.)
- Consider leaving out some details altogether. Resist the temptation to tell the reader everything you know about the subject—just tell him what you think he needs to know. The latter is quite different from just telling him what you think he would like to hear—or from using a dual standard whereby you exclude all unfavorable details while including all those that are favorable. The memo or report that burdens the reader with every conceivable detail has not only taken the writer much more time than necessary, but suggests that the writer was unsure of himself because he left it to the reader to decide which of the details were significant. (If you are afraid that a short report won't adequately reflect the work that went into a particular study, you can always indicate the *type* of additional data that can be supplied on request.)

Memo to department heads: It has been established that time has been wasted and misunderstandings may have been given rise to because use of the passive, impersonal construction has been overdone in technical communications. It is hereby suggested that use of the active, personal construction be not necessarily frowned on by engineering supervisors. . .

Of course, the classic example of detail elimination involves the man who got a letter from his landlord asking if he intended to vacate his apartment. The answer consisted of "Dear Sir: I remain, Yours Truly, Henry Smith."

Unfortunately, it's not always that easy to combine succinctness with politeness in handling one's correspondence, particularly when one is dictating it. Things that can waste the dictator's (and the reader's) time include:

- Marathon sentences. An occasional long sentence can supply useful variety, and may be needed in order to relate ideas to each other, but strings of runaway sentences impede efficient communication.
- Ditto for marathon paragraphs.
- Shotgun or overkill attacks on the topic, whereby the dictator, not being sure he has said what he really wanted to say, goes on to re-attack the topic in several other, equally roundabout, ways.
- Tendencies to throw in cliches and meaningless phrases just to keep the dictation process going. It is less wasteful to keep your secretary waiting for a minute while you find the right phrase than to dictate a phrase that adds nothing.

5. Save Time via the "Example" Technique

Most of us find the going slow when we write on an abstract, philosophic level for any length of time. True, the amount of such writing can be minimized by translating abstract concepts into dollars and cents,

other manifestations of inexperience, Black gets the job done—and Miss Blue swishes out of his office.
but this cannot always be done. For instance, if you are trying to change the attitudes of foremen towards some community or labor-relations problem, neither payout time nor discounted cash flow is going to be of much help to you. But this does not mean that your communications with the foremen must be entirely on an abstract philosophic, level.

Good philosophy is hard to write. There are all sorts of pitfalls: failure to define terms, or to relate them to the reader's frame of mind, failure to avoid either oversimplification or obtuseness, etc. But fortunately, you can write about an abstract topic—e.g., a desired change in attitude—without staying on the abstract level for very long. The idea here is to translate the abstract into the specific by using analogies, miniature case histories (either actual or hypothetical), projections of what might happen if the status quo were maintained, etc.

Of course, this "example" technique is not limited to communications that deal with corporate philosophy but should also be used when generalities and abstract concepts crop up in technical writing. Here, a one-sentence "for instance" can often take the place of a much longer explanation; a liberal sprinkling of such sentences can keep the discussion on the ground, and can result in an easier-to-write yet more-informative communication than one in which all abstract technical concepts are explained or promoted in only theoretical terms.

6. Know Thyself, and Thy Writing

The difficulty in many articles dealing with improving your writing is their generality; they presuppose that all writers have the same problems. This report has tried to be more specific—for instance, by showing how *less* discipline can be a timesaver for the engineer who has trouble uncoiling, whereas *more* discipline can be an eventual timesaver for someone whose problem is verbosity or poor organization rather than lack of writing fluency.

Where do you fall within these two extremes? Should your prime emphasis be on boosting your writing output or on boosting its quality? If you have difficulty answering that question, you can get help quite readily.

For example, the 1966 "Efficient, Effective Writing" report[2] supplied some simple tests whereby you can evaluate the informative quality of your writing. If your score is low, perhaps your prime goal should be to make your writing more informative. At the beginning, this may reduce your page output because you will be spending more time on editing and rewriting. However, if you try to learn from your editing, you will find that output will eventually increase along with quality. This means spending a few minutes reviewing an edited draft to see what type of corrections are prevalent, and to think about ways of minimizing the need for these types of corrections in subsequent writing.

Discuss your writing with your boss. Some bosses are reluctant to initiate such a discussion because they have found that it takes almost as much tact to constructively criticize a subordinate's writing as it would to constructively criticize his wife or family. But if *you* initiate such a discussion (I mean about your writing, not your wife), the chances are that overly defensive postures can be avoided, and that you and your boss will both gain a better insight into problem areas.

Ask your boss whether he thinks you are spending too much or too little time on writing. For instance, based on samples, would he settle for draftless dictation on some reports in order to give you more time for engineering?

If one of your writing problems is that interruptions are always breaking your train of thought, ask your boss about ways of getting more privacy, or about occasionally working at home to make headway on difficult writing chores.

In order to discuss other aspects of your writing, you may want to go over a sort of writing-inventory checklist, such as the one published in the *Harvard Business Review* article "What Do You Mean I Can't Write?"[3] This can be a good starting point in getting your supervisor's views on such matters as whether you sometimes present too many opinions and not enough supporting data (or vice versa), whether you tend to under- or overestimate his familiarity with your work, etc.

A concluding thought: The engineer who progresses in his company tends to do more writing every year. Eventually, he may also have to supervise the writing of subordinates; many of their reports and memos will go out over his signature and will reflect on the caliber of the work done under his supervision. Thus the engineer who becomes adept at fast, functional writing is in an enviable position—he can save a significant and increasing amount of time, establish his credentials as an efficient communicator, and eventually help his subordinates solve their own communication problems. We hope that the suggestions in various sections of this report will help point the way, and will let you apply the "work smarter, not harder" principle to your communicating. ■

Acknowledgement

Some of the ideas in this final section stem from various members of CHEMICAL ENGINEERING's editorial staff, and also from Manny Meyers of Picatinny Arsenal, and Peter J. Rankin of Basford Inc.

References

1. See, for instance, "How to Write for CHEMICAL ENGINEERING," a booklet available from CHEMICAL ENGINEERING, 330 W. 42 St., New York City.
2. Johnson, Thomas P., How Well Do You Inform?, *Chem. Eng.*, Mar. 14, 1966 *(see next page)*.
3. Fielden, J., *Harvard Bus. Rev.*, May-June 1964, p. 147.

How Well Do You Inform?

Here are some provocative tests that reveal how well you are meeting the four criteria of informative writing--criteria that apply to all types of reports and memoranda.

THOMAS P. JOHNSON, *Writing Consultant*

Probably no good engineer has ever been fired simply because he was a poor report writer. Far too often, however, good engineers have missed out on promotions and other rewards because their poorly written reports did not do justice to their technical achievements.

Of course, older engineers find that writing skill is vital in the transition to management responsibilities. But the ability to write may be even more important to the young engineer: Until he becomes a participant at committee meetings, executive lunches and other conferences, he may have to rely almost exclusively on the written report to present his work and his ideas to higher management.

The four simple tests that I will describe are not the solution to all technical writing problems. But they will help the report writer answer the elusive question: "How well am I informing my readers?" Nothing else in engineering writing is nearly as important as being informative.

Informative writing can be defined as being easy to understand; it need not necessarily be "readable" (i.e., easy to read). In recent years, many books and scores of articles on the subject of readibility have appeared. Written primarily to counteract the excesses of ponderous, cautious, "academic" writing, these publications seem to offer some rather simple solutions to writing problems—don't use big words, and don't use long sentences. Unfortunately, although short words and sentences are desirable if they can be used without oversimplification, they are not a panacea for writing problems, and do not necessarily lead to informative reports.

For instance, in a tabloid newspaper or popular magazine, the announcement of a new lubricant will be presented in very readable, perhaps even exciting, prose. However, to obtain the information he needs on the new lubricant, the engineer must turn to the manufacturer's brochure or to a business magazine, where the language may contain such "unreadable" multisyllable words as "coefficient," "viscosity," and "pressure differential." If the chemical engineer becomes confused or unhappy with this language, it will be because of the way these words are used, and not because of the words themselves.

Actually, informative writing involves much more than the lengths of words, sentences and paragraphs. While it does not involve "style" in the sense of grace or elegance, it does involve attention to sentence construction and choice of words. Also, the writer must know how to present these details logically.

These requirements will become clearer as we proceed with the four tests. Essentially, the tests should show the engineer not only what presentations are usually best, but also why they are the best.

(A word of warning: The tests require some knowledge of grammar. Because time is unkind to memory, grammatical terms will be redefined where appropriate. If the reader is still hazy after finishing this article, he should run, not walk, to the nearest grammar text. A few minutes of intensive study should be enough to clarify things.)

1—Handling Details

In technical reports, where detail must play a far more significant role than in most other forms of writing, the goal of the engineer must be too present pertinent details—major and minor—in an informative way.

Basically, there are two ways to present details in a sentence; let us first examine the bad way:

> The unlikelihood of meeting orders from the majority of its new customers is of concern to the company, due to tardiness in the installation of its new manufacturing line.

Here is a good example of what might be called *catalogical writing,* in which the writer has cataloged or equated all his details, so that minor details become as important as major ones. In other words, there is no discrimination among details, so that the sentence is more difficult to understand than it should be.

Catalogical writing of this sort abounds in technical reports; in fact, more poor writing can be traced to the cataloging of details than to any other source. Happily, there is an infallible clue to the cataloging of details: *There is an inordinate use of the prepositional phrase.* By way of proof, let us repeat the example, this time with prepositional phrases—there are eight of them—in parentheses:

> The unlikelihood (of meeting orders) (from the majority) (of its new customers) is (of concern) (to the company), due (to tardiness) (in the installation) (of its new manufacturing line).

In sentences, prepositional phrases act as adjectives or adverbs, that is, they add qualifying detail. (Adjective: The writer *of technical reports* is intelligent. Here the phrase modifies the noun "writer." Adverb: The book is lying *on the table.* Here the phrase modified the verb "is lying.")

When the qualifying details in a sentence are put into prepositional phrases, they are all treated in precisely the same way, technically speaking. But some details obviously are more important than others; hence the cataloging of details clearly is a detriment to informing, because the main details simply do not have the impact on the reader they should have. To repeat, the cataloging of details is the most serious error in most poorly written reports.

How can the technical writer eliminate as many prepositional phrases as possible? The following rewrite of the example says essentially the same thing as the original:

> The company does not know whether it can meet most new-customer orders, because its new manufacturing line is not yet installed.

Here is a good example of *analytical writing*, which is the chief asset of good informative writing. If the clue to catalogical writing is overuse of the prepositional phrase, the clue to good analytical writing is *a judicious use of the dependent clause.* In the rewritten sentence, there are two dependent clauses: "whether it can meet most new-customer orders" and "because its new manufacturing line is not yet installed." There are no prepositional phrases.

For at least three good reasons, dependent clauses help make sentences more informative:

- They analyze detail, highlighting what is important. The very words that introduce clauses strengthen this analysis, because they tell the *when, if, because, who* and so forth of the clause. These are strong words that help the reader interpret what is being described.
- They force the writer to break up the equating of detail that is characteristic of a series of prepositional phrases strung together.
- They make writing sound much more natural. Anyone who doubts this, should reread the examples in the section aloud.

In analytical writing, dependent clauses act as adjectives or adverbs in a sentence; in addition, they may act as nouns. Here are some examples:

I heard *what you said* (noun, object of the verb "heard").

The book *(that) I read* was very interesting (adjective, modifies the noun "book").

He sings *when he takes a shower* (adverb, modifies the verb "sings").

With a little practice, anyone can learn to recognize dependent clauses quickly. These three points need to be remembered: (1) A dependent clause is not a complete statement; (2) It must have an introductory conjunction or relative pronoun *(if, when, which, that, because, since* and the like), although the relative pronoun *that* can often be omitted, as in the second example above. (3) To qualify as a clause, it must have both a subject and a verb, such as *you said, I read,* and *he takes* in the above statements.

Here is another example of the difference between catalogical and analytical writing; prepositional phrases are in parentheses, dependent clauses are italicized:

> CATALOGICAL: The need (in the company) (for great increases) (in space), caused (by a steady growth) (in research work), has necessitated the addition (of a new wing) (to the laboratory).
>
> ANALYTICAL: *Because its research work has been growing steadily,* the company is adding a new wing *that will greatly increase its laboratory space.*

Once again, the rewritten sentence eliminates every prepositional phrase. In most catalogical sentences, however, it is difficult if not impossible to eliminate every prepositional phrase, nor should a writer strive to do this. Yet it should be possible to reduce the number of these phrases by at least one-half. The report writer's tool for doing this is, of course, the dependent clause.

Here is a longer example of catalogical writing; the subject matter has to do with two-phase flow systems:

> Variance of the mechanism for momentum transfer with respect to the flow pattern is to be expected, but this is not too well understood. Uncertainty even surrounds the correctness of reliance on visual description of flow patterns as a means of identifying the regions involved in possible changes in the

Why taxpayers get gray

The following excerpt is from the booklet **"New York State Income Tax Forms and Instructions for Filing,"** published by the New York State Income Tax Bureau. The excerpt does not inform as well as it might, chiefly because of the way it presents details. To determine why, readers are invited to subject the excerpt to Test 1.

> The return for the period before the change of residence must include all items of income, gain, loss or deduction accrued to the taxpayer up to the time of his change of residence. This includes any amounts not otherwise includable in the return because of an election to report income on an installment basis.
>
> Stated another way, the return for the period prior to the change of residence must be made on the accrual basis whether or not that is the taxpayer's established method of reporting. However, in the case of a taxpayer changing from nonresident to resident status, these accruals need not be made with respect to items derived from or connected with New York sources.

> mechanisms for momentum transfer. Consideration, in addition, of entrance effects, as well as the possibility of transient or unsteady-state behavior being of importance, makes the relative efficiency of existing correlations seem even more remarkable. Knowledge of the flow pattern and all the physical and geometric properties of the system enables one to predict, in the best correlations, pressure drop within an accuracy of about 25%.

This four-sentence paragraph contains a large number of prepositional phrases: 24, to be exact. There are no dependent clauses. This is a dull, uninformative catalog of details, which the reader must interpret if he is to extract concrete meanings. After wading through several paragraphs of this sort, most engineers would be ready to head for the hills.

A more analytical approach would have benefitted the reader immensely. Actually, the original version of this passage was more analytical; the above excerpt represents this writer's catalogical version of a well-written paragraph that appeared in a recent issue of CHEMICAL ENGINEERING (Dec. 6, 1965, p. 139). Here is the original paragraph, with its analytical approach restored:

> It is to be expected that the mechanism for momentum transfer might vary with the flow pattern, but this is not too well understood. It is not even clear whether visual description of flow patterns is sufficient to identify the regions where the mechanisms for momentum transfer might change. When, in addition, entrance effects are considered, as well as the possibility that transient or unsteady-state behavior is of importance, it is remarkable that the existing correlations work as well as they do. If one knows the flow pattern and all the physical and geometric properties of the system, the best correlations will predict the pressure drop within an accuracy of about 25%.

The statistics here are much more encouraging: The number of prepositional phrases has dropped from 24 to 12, and there are now eight dependent clauses, listed below by introductory word, subject, and verb only:

that mechanism might vary
whether description is
where mechanisms might change
when effects are considered
that behavior is
that correlations work
as they do
if one knows

Because each passage is quoted out of context, and because some technical terms might not be familiar to everyone, the passages cannot inform completely. But few persons would deny that the analytical version is much more informative than the catalogical one; it is also much more natural.

The report writer who has begun to suspect that his writing is more catalogical than it ought to be should take the following simple test:

Test 1: Take a short passage from one of your reports (one typewritten, double-spaced page is long enough). In the passage, blot out every prepositional phrase with a red pencil and every dependent clause with a green pencil. Blot out the whole phrase or clause, not just key words. If, when you have finished, there is more red than green on the page, you can be reasonably certain that you are not handling details as informatively as you might.

2—Handling Ideas

If details within a sentence must be presented carefully, it is equally important to concentrate on presenting ideas—sentences themselves—in an informative way. As with details, sentence construction can be catalogical or analytical. Learning to connect ideas or sentences analytically is a major step toward more informative writing.

In any paragraph, some ideas obviously are more important than others. But the writer who catalogs ideas refuses to recognize this fact of life. As a result, his technical reports are apt to follow the pattern of the following paragraph (in each sentence, the main subject and verb are in italics):

> This expanding drum *brake* for motors *has* shoes made of electrical steel laminations, mounted on a common anchor pin. *They/serve* as armatures for the brake release solenoid. The *drum/is* aluminum. The brake *lining/is bonded* directly to laminated shoes. The release *solenoid/is connected* across electric motor terminals. *It/is energized* when the motor starts, holding the shoes away from the drum. *Interrupting* current to the solenoid automatically *applies* shoes to the drum, with brake pressure applied by a helical spring between the shoes. Manual *release/is* by a cam, applying pressure below the pivot point of the shoes. The drum *brake/is* available for mounting on NEMA C frames or independently housed and foot mounted. *It/is* available with 2 to 20 foot-pound ratings.

A quick examination of this ten-sentence paragraph reveals an interesting fact: Almost every sentence begins with the main subject and verb. What we have here is a tedious catalog of ideas, each technically constructed in precisely the same way. By limiting himself to this approach—subject-verb, subject-verb, subject-verb—the writer has made informing virtually impossible. In this paragraph, the main ideas simply don't get through clearly to the reader, because minor ideas are given the same construction. Too, it is extremely difficult to tie ideas together when this pattern predominates; this sort of writing doesn't "track."

In informative writing, sentence construction is much more varied. Consider the following excerpt from p. 168 of the Dec. 6, 1965, issue of CHEMICAL ENGINEERING (once again, the main subject and verb only of each sentence are italicized):

> Operating and maintenance *supervisors* of an oxygen generating plant *were* quick to credit polyelectrolytes with increasing plant-operating efficiency, reducing downtime, and eliminating manual mud removal.
>
> The *plant*, located on the site of a steel mill, *supplies* oxygen for steel making. Although it has a closed cooling system and makes up with clean water, a sticky *mud/was settling* in its pipes, heat exchangers and compressor jackets because atmospheric dust from the mill was entering the water through the cooling tower.

> The *mud* greatly *reduced* heat-exchanger efficiency, *caused* power consumption to go up and operating efficiency to go down. Because the operating temperatures of the oxygen compressor ran high, the *compressor/had* to be shut down every two months and manually cleaned to restore temperatures to design levels. Besides the expense of cleaning, *oxygen/ was vented* when the compressor was down.

In this passage, arrangement of main subject and verb is much more varied, as it should be. Some of the main subjects and verbs are preceded by dependent clauses, others by phrases. The writing does "track," and, incidentally, is much more natural than the catalog of ideas previously quoted.

Test 2: In a passage from your writing, underline the main subject and main verb in each sentence. (In this test, ignore the subjects and verbs of dependent clauses.) If the main subject and verb appear at the beginning of more than half the sentences in the passage, you have good reason to believe that you are not presenting ideas informatively.

The percentage (one-half) given in this test is of course arbitrary and perhaps a bit high, but it does give you a goal to strive for. If it does nothing else, it should provoke you into making a few thoughtful changes, such as putting a few connective phrases or clauses between ideas.

3—Abstract vs. Concrete Nouns

We can define the ideal engineering report as one that every reader will interpret exactly the same way.

Very often, engineering reports are hard to interpret not because they are too technical but because they contain too many ordinary, everyday words of a certain kind: abstract nouns. Abstract nouns are by definition ambiguous, and the more often they crop up, the less informative the writing.

Admittedly, abstraction appears in all forms of writing, including technical reports. Words like *profits, industry, economics,* and *manufacturing* are as abstract in their own way as words like *truth, love,* and *justice*. Whenever a writer has to make a general statement (and he often does), abstract nouns are likely to creep in. The following statement, for example, is open to several interpretations, chiefly because of the abstract nouns (italicized) that are present:

> The *question* of where the company will locate its new plant depends upon such *factors* as the employment *situation*, local tax *conditions*, and transportation *facilities*.

Here is one concrete interpretation of this statement: "The company wants to build its new plant near a major rail center, where local taxes are low and skilled labor is plentiful." This is, however, only one interpretation—not necessarily the correct one. No one except the actual writer of the statement can hope to know precisely what is meant.

In many technical reports, there are sentences that could be expressed more concretely. Generally, the main subjects of these sentences are abstract nouns. Here are some examples:

> The condition of the centrifuge is such that it rerequires replacement.
>
> The problem of increasing production was resolved when the company purchased an additional machine.
>
> The use of a computer will help a company to process its payroll faster.

Corrective surgery on sentences of this sort, to eliminate abstract nouns like *condition, problem,* and *use,* is not difficult. Usually, all the writer need do is look immediately behind the abstract noun, where he will

Leave abstraction to the philosophers

The excerpt below, taken from John Dewey's "The Quest for Certainty" (1929), is a good example of the abstraction that is typical of philosophical writing. Readers who subject the passage to Test 3 will discover that 17 of the abstract nouns and awkward constructions listed in the text appear in the excerpt.

There is one common character of all scientific operations which it is necessary to note. *They are such as disclose relationships.* A simple case is the operation by which length is defined by one object placed end upon end upon another object so many times. This type of operation, repeated under conditions themselves defined by specified operations, not merely fixes the relation of two things to each other called their *length*, but defines a generalized concept of length. This conception in connection with other operations, such as those which define mass and time, become instruments by means of which a multitude of relations between bodies can be established. Thus the conceptions which define units of measurement of space, time and motion become the intellectual instrumentalities by which all sorts of things with no qualitative similarity with one another can be compared and brought within the same system. To the original gross experience of things there is superadded another type of experience, the product of deliberate art, of which *relations* rather than qualities are the significant subject-matter. These connections are as much experienced as are the qualitatively diverse and irreducible objects of original natural experiences.

Avoid these pallid, passive verbs

The passive verbs in the following list are especially weak and should be avoided whenever possible by one of the three techniques described on the next page.

is accomplished	is facilitated
is attained	is featured
is based	is involved
is considered	is made
is designated	is observed
is effected	is permitted
is employed	is provided
is enabled	is used
is enjoyed	is utilized

find a prepositional phrase with a concrete noun as its object. Promoting this noun to main subject produces a more concrete sentence:

The centrifuge needs to be replaced.

Production increased when the company purchased an additional machine.

A computer will help a company to process its payroll faster.

Sometimes the concrete or real subject of a sentence does not come immediately after the abstract main subject. But with a little practice, any writer can learn to spot the real subject of a sentence quickly.

Abstract nouns are bad for informative writing in another respect: They often attract awkward construtions that have no place in good writing. If the writer replaces an abstract noun with a concrete one, he often will find that the awkward construction disappears automatically. Here are some examples (abstract nouns and awkward constructions are italicized):

ABSTRACT:

The *nature* of helium is *such that* it is a gas at room temperature.

The *reason* production stopped was *due to the fact that* the workers went out on strike.

The *use* of the new compounds *in connection with* the cleaning of boiler tubes will result in the elimination of scale.

CONCRETE:

Helium is a gas at room temperature.

Production stopped because the workers went out on strike.

The new compounds will eliminate scale from boiler tubes.

The following abstract nouns should be eliminated, if possible, from technical reports:

ability	effort	order	relation(ship)
activity	employment	persuasion	respect
basis	extent	policy	responsibility
case	facility	position	result
character	factor	possibility	situation
circumstance	instance	practice	standpoint
concept(ion)	intent	problem	substance
concern	interest	prospect	system
condition	manner	purpose	type
connection	measure	quality	use
course	method	question	utilization
degree	nature	reason	view
effect	necessity	reference	

Some of the nouns in this list can, of course, have concrete meanings at times. But there is a great deal of difference between the concrete "The instrument *case* is in *use*" and the abstract "In the *case* of the missing instrument, the *use* of our insurance policy may minimize the loss." The objection here is to the abstract usage of these nouns, which results in needless verbosity.

Here is a list of awkward phrases that crop up with great regularity in technical writing. These should also be eliminated:

according as to whether	in the course of
as far as. . . .goes	in the form of
as regards	in the interests of
associated with	in the light of
as to	in view of
by means of	on the basis of
due to the fact that	on the order of
for the purpose of	on the part of
for the sake of	relative to
from the standpoint of	such that
in a manner of	to the extent that
in connection with	with reference to
in order to	with regard to
in relation to	with respect to
in the case of	with a view to

The above lists of abstract nouns and awkward constructions suggest a third test for the report writer:

Test 3—In the passage from your writing that you are analyzing, draw a circle around every abstract noun and awkward construction that appears. If more than six circles appear on the page, your writing probably is more abstract than it should be.

4—Passive vs. Active Verbs

In informative writing, active verbs usually outnumber passive verbs by a large ratio. When too many passive verbs creep into writing, prose tends to become unnatural, less direct, and far less forceful than it should be.

Passive verbs are easier to recognize than to avoid. Any verb that consists of some form of the auxiliary *to be*, followed by the past participle of the verb, is passive. In sentences, the subject of an active verb initiates the action described by the verb ("Jones *drives* his car to work."); the subject of a passive verb, on the other hand, receives the action described by the verb ("The car *is driven* to work by Jones.").

The following passage, in which every verb but one is passive, is typical of the unnatural writing that appears in many reports (verbs are italicized):

> This handbook *has been written* to provide an authoritative single source of information on all facets of temperature control in these systems. It *is issued* for use by all who *are concerned* with design, development and utilization of components and/or complete systems. This revision, as *will be* the case when future revisions *are considered* to be appropriate or necessary, *is based* on experience gained in usage of the handboook over a given period of time during which certain advancements *have been made* in the state-of-the-art. This book *is* not *intended* to supersede any regulations, contracts, or documents referenced therein.

The contention here is that at least half the passive verbs that appear in technical writing can and should be replaced by active verbs. When this happens, writing almost always becomes more natural than the passage just quoted, and more informative. Our argument is not with the passive verb itself—no one can hope to avoid it completely—but with its consistent overuse.

There are, of course, many ways to rewrite sentences so that passive verbs become active. The three following techniques, however, accomplish this goal perhaps more easily than others:

1. In a sentence, look immediately behind the passive verb, where you will often find an infinitive that can be converted into an active verb that replaces the passive verb. Examples:

PASSIVE:	Coal is used to heat the plant.
ACTIVE:	Coal heats the plant.
PASSIVE:	Tests were made to verify the hypothesis.
ACTIVE:	Tests verified the hypothesis.
PASSIVE:	In the process, a catalyst is provided to speed the reaction.
ACTIVE:	In the process, a catalyst speeds the reaction.

2. If an infinitive does not follow a weak passive verb, try converting a noun that appears in the sentence into an active verb. Examples:

PASSIVE:	Sulphuric acid was employed in the removal of surface impurities from the specimen.
ACTIVE:	Sulphuric acid removed surface impurities from the specimen.
PASSIVE:	Completion of the process is effected by a rust-preventive coating.
ACTIVE:	A rust-preventive coating completes the process.
PASSIVE:	Correction of production errors was obtained through laboratory tests.
ACTIVE:	The laboratory tests corrected production errors.

3. Switch from an impersonal style, which is usually passive, to a personal style, which is usually active. Examples:

PASSIVE:	Several laboratory tests were conducted on the specimens.
ACTIVE:	We conducted several laboratory tests on the specimens.
PASSIVE:	After installation of the new equipment, production was doubled.
ACTIVE:	After installing the new equipment, the company doubled production.
PASSIVE:	Procedures were initiated to reduce plant accidents.
ACTIVE:	They initiated procedures to reduce plant accidents.

Some report writers do not favor a personal style. The tone of technical writing, these people feel, is objective and impersonal. There is, of course, nothing wrong with an impersonal style, except that too much of impersonal writing is passive, abstract and catalogical. Also, there are times when the reader will want to know who did what. Note, for instance, that the statement "The reactor that was ordered was inspected" is not only more awkward but also less informative than "the materials engineer inspected the reactor that Dept. X had ordered."

Although impersonal writing can be informative, most professional writers prefer to be personal, simply because it makes the task of informing easier. In personal writing, people do something, but in impersonal writing something is done to things. The writer who is consistently impersonal tends to become consistently unnatural in his use of language.

In seeking to reduce the number of passive verbs in his writing, the report writer should try first of all to eliminate the really weak passive verbs that appear in the box on page 65. These verbs almost never carry their own weight in sentences. When they appear, sentence construction is likely to become awkward and overly indirect.

Here is a fourth test that will show the report writer whether he is informing as well as he might:

Test 4—In a representative passage from your writing, underline the first 20 verbs. Count two points for each passive verb that appears in the list on page 150, one point for all other passive verbs. If your total score is more than 10, you have evidence that your writing is more passive than it should be.

What the Tests Show

From the examples that appear here, the reader should see that the four tests are intimately related. Almost always, catalogical writing is passive, abstract and impersonal. Analytical writing, on the other hand, is usually concrete, active and personal. Very often, making corrections in one area will automatically improve writing in other areas. Thus, the writer who consciously avoids the cataloging of details is likely to present ideas informatively, actively and concretely. ■

Putting Technical Illustrations to Work

Illustrations can be made to function like any other discrete tool, when the engineer learns to use their content to serve the over-all purpose of his technical report.

Relative Ease in Solving the Four Problems for the Presentation of Tables, Figures and Equations – Table I

Type of Illustration	Problem			
	To be correct	To be accurate	To emphasize	To be easy to read
Tables	Easy to do	Relatively simple	Usually difficult and time consuming	Easy to do
Figures	Extensive thought required	Care needs to be taken	Easy	Quite a few simple principles
Equations	Usually a lot of work	Usually easy	Easy	A few simple concepts

DONALD R. WOODS, *McMaster University*

The engineer who is going to succeed in the environment of contemporary technology must be able to formulate his technical thinking — his ideas, data and know-how — within the disciplines imposed by the conventional technical report. An essential feature of such technical reports is the illustrations.

However, the choice and design of illustrations for technical reports is rarely treated as a discrete tool which can be mastered as part of the requirements of the profession. Too often, when the design of an illustration is separated from the rest of the thought content, engineers will regard that design as something outside the scope of their interest. Yet technical illustrations are more easily mastered when thus separated. The engineer needs only to identify what the illustration is to do, as well as what it is to say—to identify its primary purpose correctness, accuracy, emphasis, or ease of comprehension.

These four purposes of a technical illustration can usually be accomplished to a greater or lesser degree through one of three general types of illustration—tables, figures, and equations,—depending on the contents of the report, as indicated in Table I. Since the thought-contents are only sometimes suited to the author's purpose, the main effort required in the design of any type of illustration is overcoming its weaknesses. Thus the main effort required for tables is that of incorporating the table into the text for emphasis; for figures it is that of making them accurate and not misleading; and for equations it is that of making them correct and easy to comprehend.

Tables

Tables are perhaps the most useful method of presenting data. They give exact numerical values and are not limited by the number of variables. The development of the digital computer has added a new dimension to the use of tables; data are fed to, and received from, typical computers in tabular rather than graphical form. However, tables do present discontinuous information rather than continuous information, which makes them less suitable to demonstrating those relationships necessary for making a point.

Correctness

It is easy to tabulate correctly. Simply label the rows and columns with name, symbol and units; assign a number and title of the table; and for record tables, include the original data as well as the calculated or corrected results.

Accuracy

It is also relatively simple to make tables that are not misleading: (1) keep the number of significant figures within the uncertainty of the data, (2) report values less than 1.0 with a zero before the decimal, as 0.25; (3) use a consistent set of units; (4) indicate whether the entries are calculated or measured, and (5) avoid expressing very small and very large numbers as powers of 10, which presents confusion as to whether the entries have been multiplied by this factor.

Emphasis

It is frequently difficult to make a table prove a point. In this regard, tables may be classified according to two types: integral and record. Integral tables appear in the body of the report; all data are not necessarily included; and the table is used as supporting evidence for the discussion of the work. Record tables should appear in the appendix and include all data. Whereas integral tables demand a lot of attention, record tables are relatively easy to arrange.

The first step in devising integral tables is to tabulate all of the measured data plus the pertinent calculated information in a record table, which will be appended to the report. This information is then interpreted; trends are noted; a decision is made on the points that the data can be used to support; appropriate data are selected for these points and these data are included in the integral table. This may mean that the record table is divided into small integral tables (or graphs) to prove several points. Also, the same data may appear in more than one integral table, since direct reference should not be made in the body of the report to data reported only in the appendix record table.

Following this approach, tables often illustrate better than a graph, when the percentage error is too small to be shown on a curve. And they may be used to emphasize points by grouping similar items together. Consider, for example, two tables from the body of a report:

How decreased cone angles increase calculated collection numbers in centrifugal-type cyclones – Table ABC

Inside Body Diameter Cm.	Cone Angle Degrees	Collection Number %
60 ± 1%	30 ± 2%	81 ± 2%
	20 ± 2%	90 ± 1.5%
20 ± 0.5%	10 ± 1%	84 ± 3%
	5 ± 1%	93 ± 2%

Effect of increased Reynolds numbers on the flow behavior inside pipes, for water and methanol at 25C – Table XYZ

Fluid	Observed Flow Pattern	Calculated Reynolds Number
Water	Smooth	830 ± 10%
	Wavy	1780 ± 8%
	Continuous-break	1810 ± 10%
	Diffuse	2100 ± 9%
	Diffuse	2800 ± 7%
Methanol	Smooth	300 ± 12%
	Smooth	700 ± 11%
	Diffuse	2160 ± 10%
	Diffuse	2380 ± 8%

Ease of Comprehension

Follow four rules: (1) put the principal comparison in columns, rather than rows, (2) try to eliminate ditto marks by grouping similar items together, (3) avoid vertical rulings, and (4) try to locate the title caption above the table.

To summarize, tables are most useful in presenting multivariable, discrete information; and they are easily made correct and accurate. They can also be used to prove a point, as integral tables in the body of a report.

Figures

The word "figure" is a general term that includes drawings, sketches, photographs, maps, graphs, histograms, bar charts, pie diagrams, flow diagrams and balance sheets. Some general characteristics shared by all these various types of figures are worth noting. Figures . . .

- work best with specific yet simple titles that tell what is interesting.
- need to be related to the text by a reference number.
- are more effective when placed near the text that refers to them, but never before that text.
- are more conveniently attached as a foldout to the last page of the report when they are referred to repeatedly.

Graphs

The primary function of graphs is to show trends in a set of data; they usually do this more effectively than tables. However, it is often difficult to construct graphs for systems with more than three variables; and unless the search for a trend is an important aspect of the report, it is often best to first try a table for complex systems. Even if the table is not entered directly into the body of the report, its time of preparation is not wasted, for the table then serves as a reference in the appendix.

Correctness: Certain conventions should be observed with respect to graphs. The dependent variable is plotted on the ordinant (y-axis) with the independent variable along the abscissa (x-axis). Both the coordinate axes are clearly shown and carefully labelled with the name of the variable, its symbol, and the units. For reports, the graph is completely identified by the name of its originator and date drawn. (This is particularly important when personnel are shifted from one project to another, since the information may need clarification, and one can then seek out the originator for details.) All the data points are clearly identified; and if there is more than one set of points a code key is used to identify the sets of points. All pertinent data points representing experimental results are included, except perhaps when a calibration line is presented. Finally, no data should appear on the graph, unless they are tabulated in the appendix or otherwise stored for reference.

Accuracy: Graphs are easily misleading. The same variable should not be included on both axes. Dimensionless groups should not be used for only two or three changing variables, unless a note specifies that this has been done. Both ordinate and abscissa are dimensionless, when a dimensionless group is used. In choosing dimensionless groups, the following order of preference is suggested.[5]

1. Those derived from fundamental equations such as Reynolds, Euler, Cauchy, Froude, Weber, Peclet, and Strouhal numbers.
2. Those used as dimensionless methods of expressing experimental data, such as friction factor, drag coefficient, dimensionless, torque coefficient, catalyst effectiveness, dissipation coefficient, Nusselt, Power and Sherwood numbers.
3. Dimensionless combinations of physical properties, such as specific heat ratio, Prandtl and Schmidt numbers.
4. Simple dimensionless ratios, such as particle-to-tube diameters, lengths to distance ratios, density difference to liquid density.
5. Groups derived from combination of other dimensionless groups, such as the Stanton number.

In selecting the important dimensionless groups, the choice among various methods should proceed in the following order of preference.

1. Normalization of fundamental equations, including the equations of change.[1, 2, 3, 4]
2. Similitude of ratios of forces used to indicate the important phenomena.[3, 4]
3. Dimensional analysis.

Different plotting papers can yield misleading results. Linear-linear, logarithmic-linear, logarithmic-logarithmic, isometric, triangular, polar coordinate, linear-probability, logarithmic-probability, Cox, Poisson—these are some of the plotting grids available on graph paper. For most applications, linear-linear, logarithmic-linear (semilog) and logarithmic-logarithmic (log-log) graph plots are used. Fig. 2 shows how lines that are straight in one type of plot become curved when replotted in another. Also, uniform error bands on linear-linear plots are no longer uniform for semilog or log-log plots. If a plot is selected so that the line is as near straight as possible, and the slope near unity, equal accuracy is more probable in all regions of the graph.

To select a plotting paper that will not mislead, first plot the data on linear-linear paper. Any relationship that appears among the variables may then be used as a guide to the choice of a more appropriate plot. All too often log-log plots are made first to smooth out variations, and it has been suggested that log-log paper should only be used to find the constants for equations of the form $y = ax^b$, and to present all data clearly on one graph when the data range over several orders of magnitude.[5]

Scattered or "rogue" points that are far removed from a proposed theory or correlation must not be suppressed. Sometimes there is a temptation to discard these data. Sometimes the graph misleads by placing a coordinate axis, title, or an information box so close to the "rogue" data that they are not easily noticed (Fig. 3). Rogue points are very important because they can warn of a correlation that does not accurately describe the data and of other variables that may be significant. Fig. 4a presents a reasonable correspondence between data and theory, with the exception of one rogue point; it contrasts sharply with what a detailed literature search revealed for Fig. 4b.

Also, the correlating line should not extend the range of the data through implication. Nor should trends of the data be exaggerated. The smallest divisions of the plot should approximate the experimental errors.

The zero point tends to aid accurate interpretation

Effects of plotting on curves—Fig. 2

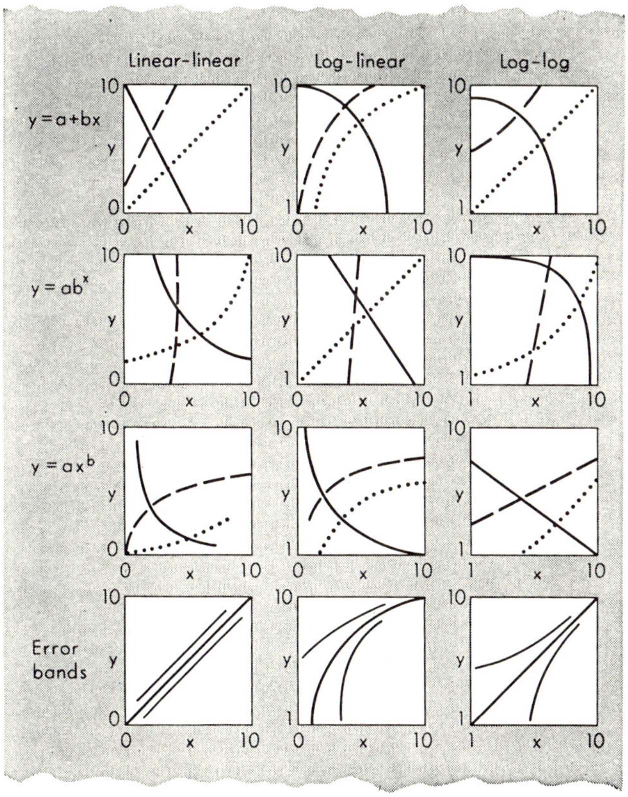

How concealed points mislead—Fig. 3

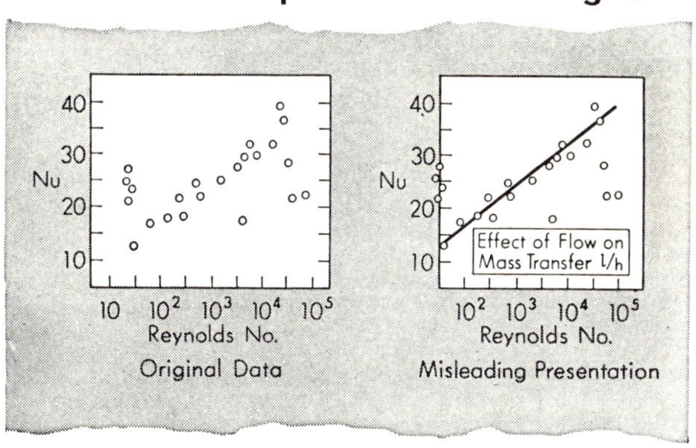

How one rogue point can reveal deviant phenomena—Fig. 4

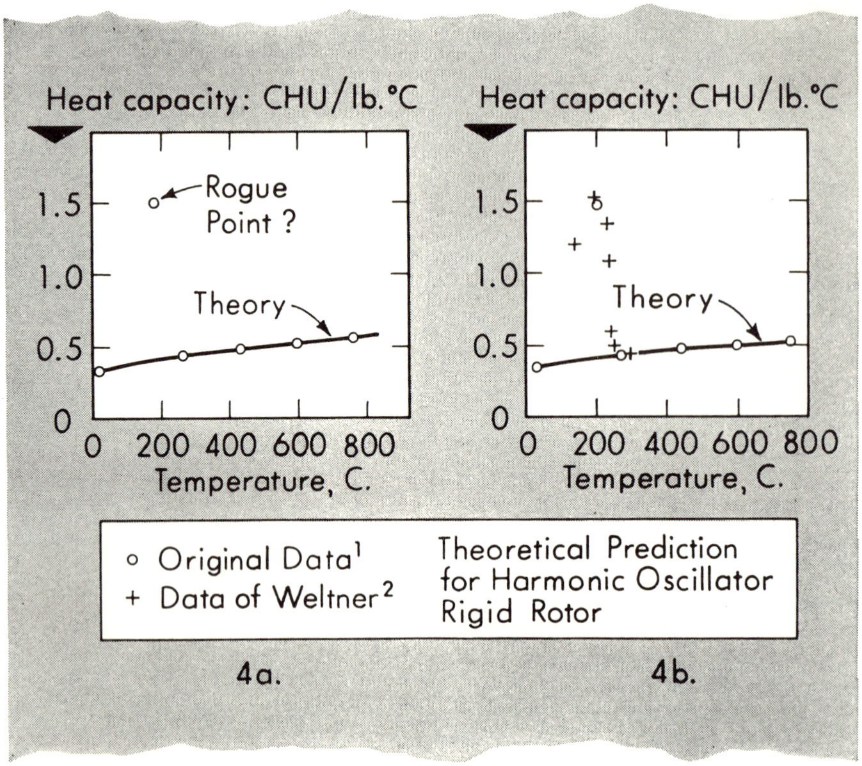

and should be shown where practical. Scales should be easily subdivided for accurate interpretation and reading. All the data should refer to the same phenomenon, and when there must be exceptions, care should be taken to point out the differences. Powers of 10 should not be used in the label for an axis. Any factor should be directly associated with a number and not with the units. Finally, the calculated, theoretical, and measured data should be clearly distinguished.

Emphasis: A graph is used for a reader to quickly visualize the relationship among the variables, and it is thus very effective in proving a point. This function of the graph should not be interfered with by too many lines, particularly when some of these lines might duplicate data of an integral table. Data that do not belong to the set should be omitted. Also, when nonarithmetical plotting is used to obtain a straight line, the text should point out the effects of the particular plot.

Ease of Comprehension: Several techniques can be used to improve the appearance of graphs. Heavy lines should be used for the data, moderate lines for the axes, and the minimum weight for the grids. Grid lines should not contribute to a cluttered appearance. Data and parametric lines should be labelled. Symbols defined in the key should reflect in their order either an increase in one variable or the order of the parametric lines. Data points should be indicated by conventional symbols, such as ○, ▽, △, □, +, ⊙, and ×.

Multicolor figures should be avoided, unless there is a technique for adequately reproducing the colors. The figure should read upright on the page, or from the right-hand side. The graph should be surrounded by a box that excludes explanatory details, scales and captions. And there should be plenty of space.

Bar charts

A bar chart is a simple and useful tool for making one-dimensional comparisons. Its main function is the comparison of discontinuous variables—different compounds, different instruments, years, months, etc. The most important problem encountered with bar charts occurs in the choice of a scale that will adequately show the differences without exaggerating the results. Three common devices that misrepresent in this fashion are (Fig. 5): partial scales that suppress the zero, percentage scales with 100% based on one item in the comparison, and right-and-left sections.

The advantages of the bar chart are best illustrated in Fig. 6, which effectively shows the change in the relative characteristics of meters in going from low to high external resistance. Sometimes an attempt is made to increase the visual communication by replacing the bars with symbols, such as little soldiers, dollar-signs, or bags of flour (Fig. 7).

Pie Diagrams

These are useful for showing the relative percentages of a whole (Fig. 8). However, they can also misrepresent, if the pie is converted into a cylinder and part of the wedge removed, since the wedge does not then represent the correct proportion of the whole.

In summary, graphs, bar charts and pie diagrams

Three bar charts that mislead—Fig. 5

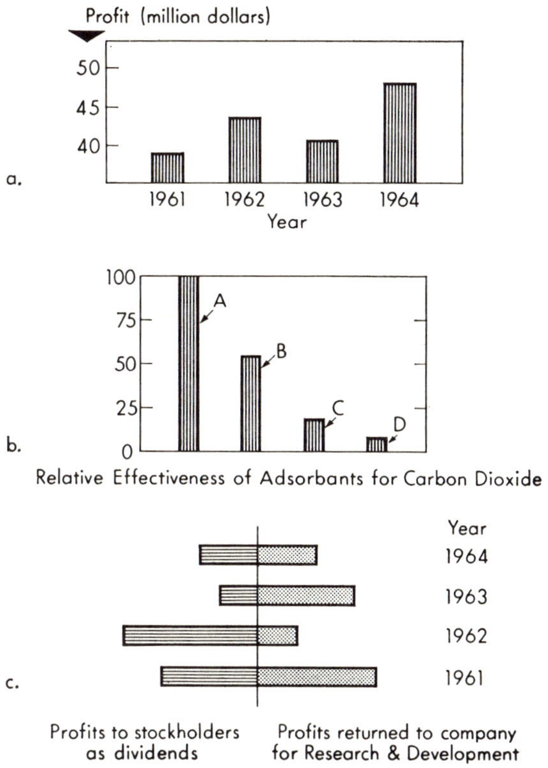

Correct use of a bar chart—Fig. 6

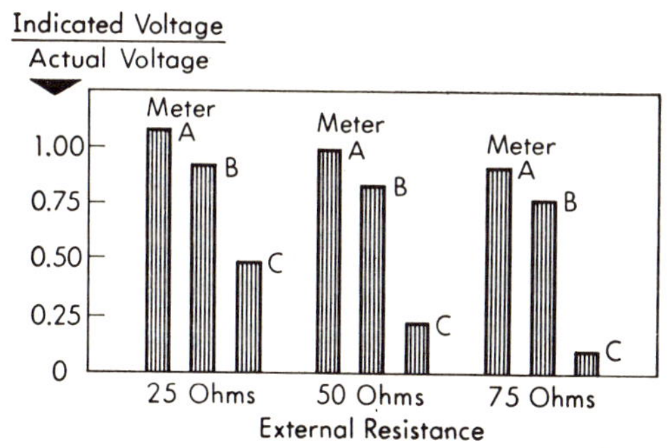

How symbols give emphasis—Fig. 7

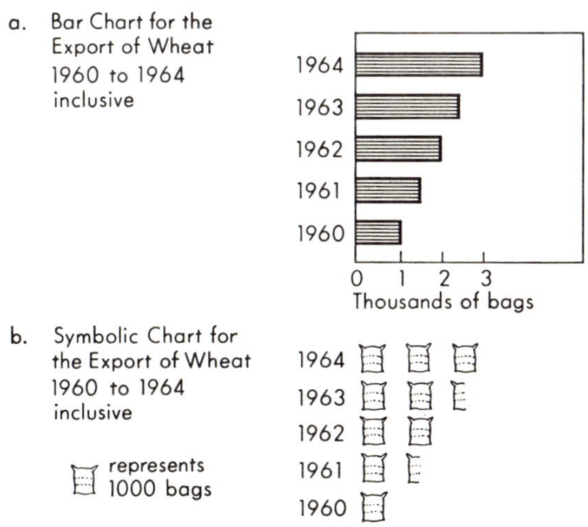

How a pie diagram can mislead—Fig. 8

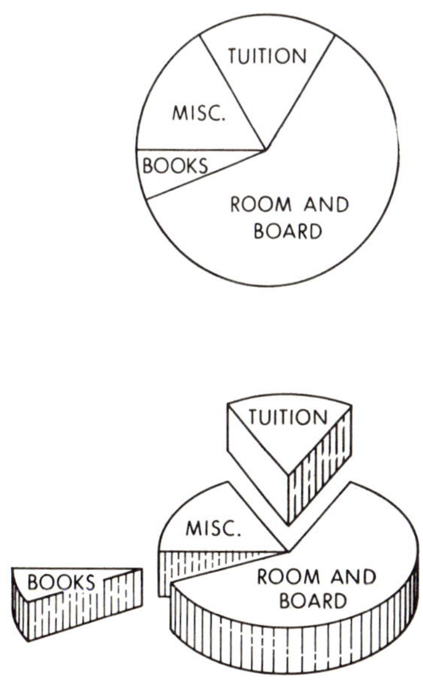

are powerful tools for making a point, but care must be taken to prevent these figures from misleading.

Also there is that class of illustrations called flow diagrams, which will not be discussed here fully. Two classes of flow diagrams are commonly used: Schematic and detailed engineering flow diagrams. Schematic flow diagrams illustrate the general system and its components, without comprehensive details. Detailed engineering flow diagrams show everything necessary for applying the general system to a specific installation. These differences in the amount of detail indicate how and where the flow diagrams are used in communication. Schematic flow diagrams are used in the body of reports like annual reports, company descriptions, etc. Detailed engineering flow diagrams appear with field construction instructions, exhibits to contracts, etc.; they lie intermediate in function between illustration and engineering design drawings; and some companies use detailed engineering flow diagrams to show relative elevations of critical equipment.

Photographs

Although photographs are inconvenient to reproduce in a technical paper, there are times when they become invaluable for lending reality to abstract concepts. Some guiding principles: Unnecessary background should be removed. Some means of identifying scale should be included in the photograph to give the reader an appreciation of size. Use identification labels where possible. Equipment should be photographed from the point of view of its operator or tender. The captions should aid in bringing out the point for emphasis. Interesting features might be identified by arrows or other pointers superimposed on the photograph. And finally, the source of the photograph should be given.

Equations

Experimental results and conclusions are best illustrated by use of graphs or tables. When further calculations using the data are required, however, it is convenient and often necessary to have mathematical equations that fit the data. As far as technical writing is concerned, such equations are valuable for neatly summarizing large amounts of tabulated or plotted data. Sometimes the mathematical form of the equation may suggest useful application or extension of the data.

Correctness: The primary problem in the applications of equations is that of obtaining correct representation. A three-step process is normally used for deriving an equation. First, an expression is evolved, which appears to describe the data—an expression chosen from theory, or found empirically from trial-and-error plotting, or developed from a finite-difference study. Once the equation has been found, values are assigned to its constants. Five common techniques for assigning constants are the slope-intercept method from a straight-line plot, the selected points method, the method of averages, the method of least squares, and the polynominal fitting method. Finally, the equation is checked for a fit between correlations and the data. Methods of checking are: (1) minimizing residual differences, (2) the Gauss residual, and (3) statistical techniques.

Accuracy: To prevent an equation from being misleading, it should be accompanied by a statement of percentage accuracy and range of application. Final equations are classified as being theoretical, approximate solution, numerical solution, semi-empirical correlation and empirical correlation.

Emphasis: Equations alone do not prove a point without some text to complete them. When the point is simply that the data can be represented by the equation, then standard deviations should be offered as evidence; the equation by itself is not evidence.

Ease of Comprehension: Ensure that the assumptions, initial conditions and boundary conditions are clearly stated for every working equation. Definitions of the symbols should be clearly set forth. Chemical equations or complicated equations describing a phenomenon should be aided by words that identify terms, as for example, the equation of motion in fluid dynamics:

$$\frac{\partial \rho \mathbf{v}}{\partial t} = \underbrace{-\,[\nabla \cdot \rho \mathbf{v}\mathbf{v}]}_{\text{Internal Forces}} \underbrace{-\,[\nabla \cdot \tau] - \nabla p}_{\text{Surface Forces}} \underbrace{+\,\rho g}_{\text{Body Forces}}$$

Finally, the appearance is improved if the report does not end a paragraph or section with an equation. More text should be added, as here. ■

References

1. Bird, R. B., W. E. Stewart and E. N. Lightfoot, "Transport Phenomena," Wiley, New York (1967).
2. Bird, R. B., W. E. Stewart and E. N. Lightfoot, *Chem. Eng. Prog.* Symposium No. 58, Series Vol. 61 (1965).
3. Johnstone, R. E., and M. W. Thring, "Pilot Plants, Models and Scale-up Methods in Chemical Engineering," McGraw-Hill, Toronto (1957).
4. Kline, S. J., "Similitude and Approximation Theory," McGraw-Hill, Toronto (1965).
5. Rowe, P. N., "The Correlation of Engineering Data," *The Chemical Engineer*, No. 166, p CE 69 (1963).

Meet the Author

Donald R. Woods is an Assistant Professor at McMaster University, Ontario, where he has taught courses in technical communications for several years. Before coming to McMaster, he received his Ph.D. and M.Sc. (1961 and 1958) at the University of Wisconsin. He received his B.Sc. at Queen's University, Kingston; and he is a licensed professional engineer in the Province of Ontario.

Your boss' initial reaction to your next suggestion may well be negative, no matter how brilliant the proposal . . .

Good Ideas Need Selling!

HERBERT POPPER
Assistant Editor

Once there was a chief engineer whose boss was known for his speed in making decisions, and for his reluctance to reverse a decision once made.

The chief engineer, whom we'll call Jones, sent his boss a memo one day, outlining a project that he felt to be important and wished to undertake. The memo was promptly returned with a red *No* above the boss' initials. Still convinced of the soundness of the proposal, the chief engineer called on his boss and outlined the idea in more detail. The boss listened carefully and asked some pertinent questions.

"Jones," he said finally, "what you propose makes a lot of sense. It's a good idea. It's really a shame that we can't go ahead with it—because I already said *No!*"

Actually, after Jones and the boss had a good laugh, the "No" decision was reversed.

Using this little story as a starting point, we're led to some important principles for getting ideas accepted.

First of all, let's say that Jones' basically sound idea was not an immediate hit for some or all of these objections:

1. Jones didn't bring out all the advantages of the project. As a result, his boss didn't think the benefits would justify the investment.
2. A number of possible disadvantages occurred to the boss, and he therefore felt that Jones hadn't thought the idea through.
3. The boss just wasn't ready to accept the idea on the basis of a short, impersonal memo.

While other objections might have played a part in vetoing Jones' idea, the three I have picked seem to come up time and time again when engineers submit "difficult," but sound, recommendations for approval. Let's therefore look at these obstacles individually, and see what can be done.

Emphasize All Advantages

When trying to put across an idea to someone, this is a good rule to keep in mind: What do I have or what can I do that will help him, not me?

Let us assume that Jones' idea was to convert a small batch operation to a continuous process, at a capital cost of $100,000. In his initial memo, he stated that one operator, earning $7,000 a year including fringe benefits, could be reassigned. If the boss took this at face value, he might conclude that a pretax yearly saving of $7,000 on a $100,000 investment is not sufficiently attractive and disapprove the expenditure.

But what if the proposal had other advantages that weren't brought out? For instance, the change to a continuous operation might also:

- Improve the quality and uniformity of the product.
- Eliminate a bottleneck that sometimes affected other departments or customers.
- Provide extra capacity for future needs.
- Reduce maintenance costs by replacing troublesome equipment.

These factors might all affect the decision, and unless Jones brings them out, it's probable that his boss will miss some of them. Once the boss knows all the advantages to expect, he'll be interested in hearing more on how the project should be carried out.

Sometimes the disadvantages of maintaining the status quo have to be emphasized. Perhaps if Jones didn't get his new continuous system, the old process would become more and more costly to operate because of increased overtime premiums, skyrocketing maintenance costs, increases in the hourly wage rate.

If any of these could be translated into actual statistics, so much the better. For instance, Jones might want to cite and extrapolate the old unit's history of maintenance costs, the number of breakdowns per year, or the overtime premiums at various projected capacity levels.

To sum it up: don't leave any doubts about the benefits of your proposal.

Meeting the Objections

So far we have spoken only of advantages that are inherent in Jones' idea. But what about any dis-

The sales side of engineering . . .

"An engineer with any degree of management responsibility must recognize that he is a salesman. When he decides on a course of action, he must sell it.

"Sometimes, he sells it downward. This is fairly easy if he has the respect of his people. Much more difficult is the task of selling it upward, usually because an expenditure or policy change is involved.

"When rejections occur, it is difficult not to become discouraged and irritated with the 'blind, stupid, unreasoning shortsightedness of the higher-ups.' While this may be a good emotional release, it does not solve the problem that caused the recommendation to be made in the first place.

"Since the customer is always right, our engineer-manager must accept the fact that his sales pitch did not prove effective. He must accept the responsibility for this failure, explore his pitch for its weaknesses, strengthen and reorganize it, and go on doing this and resubmitting it until it is either accepted, or he begins to see that he was actually wrong in the first place."

George F. Dappert
Manager, Process Engineering Dept.
General Aniline & Film Corp.
Linden, N. J.

advantages? For instance, if he mentions that the continuous equipment, by being new and better designed, will require fewer hours of maintenance, shouldn't he also mention that this reduced maintenance presupposes greater skill and, perhaps, special training on the part of the maintenance men?

Yes, he should. But first, he ought to go over the disadvantages carefully in his own mind. When the time comes to resubmit his proposal, he can then put these disadvantages in their proper perspective.

For instance, he can point out that while the maintenance men may need additional training to handle the continuous-process instrumentation, the training could be quite short since the instrumentation wouldn't be very different from that in the rest of the plant.

Jones can also make changes in the project to remove or minimize a particular disadvantage.

Finally, by considering the disadvantages, he shows his boss that he has at least thought the project through.

Weighing the Pros and Cons

Here's a simple system that can be helpful in evaluating the merits of an idea:

1. Divide a sheet of paper into three vertical columns.
2. Label the left-hand column "Pro," and in it briefly list all the advantages of your proposal that you can think of.
3. Label the middle column "Con," and in it list all disadvantages, including cost factors, disruptive effects, training problems, demands on engineering manpower, and so on.
4. Use the right-hand column for commenting on the disadvantages. For instance, if "demands on engineering manpower" is listed as a disadvantage, you might be able to suggest a solution, such as "Cheaper to get a consultant" or "Wait until Project X is finished."
5. After you have done all this as objectively as possible, put the sheet aside for a day or two. (If you don't have the time, even a couple of hours is better than nothing.) Then, add any new pros, cons, or comments that have occurred to you.
6. Weigh the pros against the cons. Keep the "cons" and comments in mind when deciding whether there isn't a slightly different but better way of doing it.

When you have finished, if the pros are most significant on balance, go ahead and write up or present your idea.

Selling It the Second Time

If, as George F. Dappert says elsewhere in this article, you should keep resubmitting a proposal until it is either accepted or you begin to see that it was wrong to begin with, how can you do this without straining your relationship with the boss?

Here are some suggestions:

1. Listen carefully to your boss' objections, then see if you can't tailor your idea to overcome some of them. As long as he sees you have an open mind, and are not just listening to be polite, he will probably be willing to discuss the matter quite fully. In the process, he may often sell himself on some of the desirable features of the project. In any case, the more you know about his objections, the better your chances of overcoming some of them.
2. Every time you resubmit the idea, have some new material. Don't repeat the old advantages verbatim; try to shed new light on them, or find new ones. Sometimes, current happenings in the company or in the industry can have a bearing on your proposal, and can give you an excellent reason for resubmitting it.
3. If it becomes clear that you can't get immediate approval, try to leave the door open for future action by requesting time to get additional data or to make additional observations.

There are some ideas that take a long time to put across.

For instance, a certain chemical company executive took about two years to sell his superiors on the necessity of a work simplification program. When asked why it took him so long, he said that he had to let his superiors become convinced that other alternatives—

The small-dose approach . . .

"Getting management to accept engineering recommendations is a sales problem. Large projects have to be sold in small doses. People hate shocks, which is what often happens when a full year's work is coldly dropped on someone's desk. It is better to use the bring-them-along technique. When you have lunch with your boss, or meet him in the plant, tell him what you are investigating. Keep him posted on progress.

"I am constantly amazed at how a major change that our department has been advocating for a number of years is suddenly adopted. We used to think that special circumstances were at work; but the more I see of these cases, the more I am convinced that they illustrate the sell-in-small-doses principle. Long exposure to the idea behind a major change apparently brought about acceptance."

Robert H. Maskel, Manager
Industrial Engineering-Operations Research
Kimberly-Clark Corp.
Neenah, Wis.

like getting better foremen, or doing more training—weren't the answer. He did not want to use the bulldozer approach because he needed not only approval, but active support. He figured the best way to get this was "to nudge the pendulum a little bit at every swing, instead of trying to set it in motion with one abrupt movement."

Of course, not all recommendations require this "pendulum" approach. Often, just a talk with your boss can be a great timesaver, either taking the place of a report or preparing the way for a short and simple written description of your idea.

However, you should be well-prepared for this talk, having reviewed your proposal as carefully as you would have in preparation for writing a report.

Which Are the Hardest Ideas to Sell?

It's best at the outset to recognize which proposals are going to be easy to sell and which are going to require a major effort.

Money isn't always the key. Production engineers in a cash-rich chemical company, for instance, may have little difficulty getting funds for process improvements if the project promises to show a certain minimum rate of return. These same engineers, however, may have considerable difficulty selling ideas that involve a change in personnel policy.

In other companies, where many projects compete vigorously for capital appropriations, the reverse may be true.

The key here is timeliness. If your boss is under fire for high maintenance costs, he will obviously be more receptive to an idea that will take some men off the payroll rather than one that requires additional maintenance help. This will be true even if the proposal may reduce the backlog of work orders and improve the plant's appearance.

The really difficult ideas are the ones that are not particularly timely at first glance and require a change in somebody's thinking.

These, like the previously mentioned work simplification project, usually can't be sold in one or even two sittings. It's not fair to expect your boss to fully accept such a concept—with all its risks and implications—within the time it takes him to read your report or listen to your talk. So plan your approach accordingly. Robert H. Maskel has some helpful advice on this elsewhere in the article.

When Did You Sell Your Last Idea?

Keep in mind that it is usually better to have 80% of ten ideas adopted than 100% of one idea. If you've become discouraged because some of your early recommendations didn't see the light of day, then try again with other proposals.

And don't be disheartened if you lose some preliminary skirmishes. If you still have faith in the recommendation—faith based on logic rather than wounded pride—you are doing both yourself and your company a disservice if you don't marshal some new arguments and try again. ■

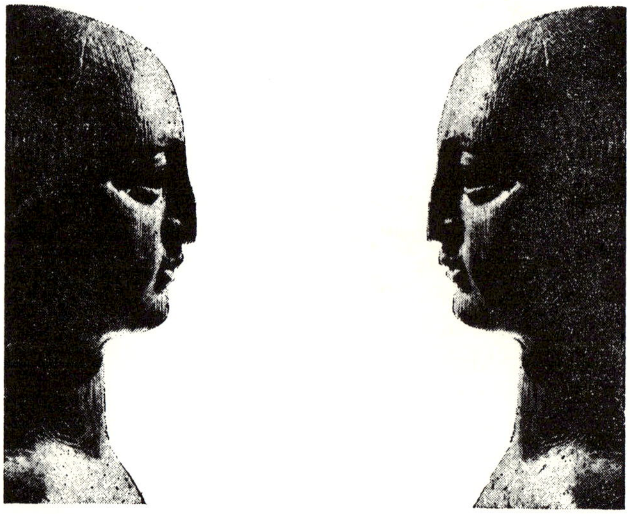

Improving Oral Communication

Although we spend a large part of our lives in talking and listening, we don't do either very well. This series will tell how to do better.

ROBERT HAAKENSON, *Smith Kline & French Laboratories*

A few years ago, alumni of Purdue University were polled regarding what had been most conspicuously lacking in their education. The responding alumni, predominantly engineers, declared overwhelmingly: "communication."

The engineers had discovered that technological proficiency is not enough. The professional person must also be able to explain what he does, and why, and how; and to understand what others are saying to him.

For reasons that will be given in a moment, we will concentrate on the spoken word in this series and entitle it, "Improving Oral Communication." Further, we will concentrate on formal settings—oral reports, public speaking—but will not minimize conversation, interview, committee meeting, conference, etc. Essentially what is effective in platform presentation is also effective in the less formal modes. The challenge is to find and apply the universal principles.

Embarking on "Improving Oral Communication," then, let us attempt to answer briefly these questions:

- How much of it do we do?
- How important is it?
- How good are we?
- How do we improve?

In this installment, we will discuss the first three questions and offer a beginning on the fourth. But the full discussion of the fourth will be the subject of several ensuing articles.

How Much Oral Communication?

In 1926 a young man named Paul Rankin offered a dissertation in support of this Ph.D. candidacy at Ohio State University. His study produced surprising results. He concluded that the typical person spends 70% of his waking hours engaged in some form of verbal communication. This suggests that more than 11 of the normal 16 waking hours are spent in exchanging words.

Rankin went further. He analyzed the 11 verbal communication hours and came up with this apportionment:

9% writing
16% reading
30% talking
45% listening

It is interesting that we spend 75% on the spoken medium (talking and listening) and 25% on the written (writing and reading). Our formal schooling, incidentally, exactly reverses the commitment of time.

Other studies paralleling Rankin's have been conducted at various times and in various places since, yielding similar results.

The December 1966 *Communications News* proclaims boldly that the computer rapidly is dominating communication—men talking to computers, computers to men, computers to computers, and computers to machines. Yet the article freely admits that spoken communication has grown steadily, and will continue to increase.

Thus we conclude that we spend a lot of time talking and listening—substantially more than a quarter of our lives.

How Important Is It?

There are many ways of responding to this question. One is to consider the time, energy and resources we pour into the endeavor. Loren Reid of the University of Missouri wrote:

> **A General Electric official reported that management spends half a billion dollars annually in communication to employees. Vance Packard estimated that the 100 largest corporations spend 50 million dollars each year on public relations. *Fortune* in 1952 reported that American business spent 100 million dollars a year selling free enterprise to Americans . . . The financial editor of the New York *Herald Tribune* gave 168 lectures in two years. The president of the U.S. Chamber of Commerce makes 100 speeches annually; Kenneth MacFarland of General Motors makes 200. The Dale Carnegie company teaches 50,000 students a year from General Motors, RCA, Coca Cola, and other corporations and businesses.**

There are now more than 3,600 Toastmaster clubs in 50 countries and territories throughout the free world. *Toastmasters International* describes itself as a "nonprofit, nonpartisan, nonsectarian educational organization which has helped more than one million men through its program of self-expression and self-improvement."

A club is described as "an organized group providing its members with opportunities to improve their abilities to speak in public, conduct meetings and develop executive abilities. In congenial fellowship, ambitious men help each other through practice, mutual constructive criticism and the assumption of responsibilities within the organization."

Another way of considering the importance of spoken communication is our intent. We talk and listen primarily to exchange messages: to achieve understanding. This is basic to any sort of efficiency in getting on with the world's work.

Less obvious, but powerfully important is morale. Because personality is brought directly to bear, spoken communication has special potential in elevating human relationships. One authority suggests that the finest thing one human can do for another is to listen—fully and empathically.

Edward Annis was a fine but unsung surgeon of Miami, Fla., until the American Medical Association discovered his speaking talents at a time when the Medicare issue was flaming. Dr. Annis vaulted into the presidency of the AMA and became its forceful number one spokesman.

Again, listening comes into prominence. Pfizer Laboratories entered into a contract with Basic Systems, Inc., to develop programmed instruction in listening. When the course was written the first employees to use it were the sales representatives.

Dr. Robert T. Oliver, Professor of Speech at the Pennsylvania State University, addressing the Annual Convention of Toastmasters International, New York City, Aug. 20, 1965, stated:

"Winston Churchill with a rifle in his hand, crouched behind an earthen rampart along the Dover coast, might have repelled two or three Nazi invaders. But the same Winston Churchill, speaking with his matchless oratory, was able to marshal the global resources and inspire the will to victory that toppled Hitler's empire and preserved the democratic civilization of the Anglo-American world."

Let us not minimize the individual's selfish reasons for polishing his oral skills. A chemical engineer experienced slow but steady progress during several years of employment with a firm. Suddenly he received a handsome promotion with expanded responsibility. Befuddled, he analyzed his meteoric rise. He traced it back to an important oral presentation he had made to the firm's top management.

In his talk to the Toastmasters convention, Robert Oliver cited experiences of some great Americans:

". . . Henry Ward Beecher, America's greatest preacher, traced his intellectual awakening to the speech class taught by John Lovell in Mount Pleasant Academy, in Amherst, where for the first time he encountered a kind of teaching that was less concerned with the input of information into his mind than with the outflow of influence from his whole personality.

Henry Clay attested that whatever he was and whatever he achieved he owed to his early and constant training in the art of public speaking. Andrew Johnson could not even read and write at the time of his marriage, but he educated himself by hiring a boy to read aloud to him the great orations of Burke, Fox, Erskine, and Pitt while he worked busily at his sewing in his tailor shop. Lucy Stone left her farm home to enter Oberlin College where, as a mere girl, she was not allowed to give speeches but prepared herself for leadership as an eloquent advocate of women's rights by sitting as a mute auditor in the young men's public speaking class.

Woodrow Wilson wrote his first published essay on oratory and organized a debating society because he was convinced that skill in speech was the basic requirement for intellectual development and personal leadership."

In summary, then, we can say there are many compelling reasons for improving oral communication: for getting on with the world's work—to enhance understanding, to dignify human relationships and elevate morale, and to strengthen persuasion, both for convincing and motivating; for the individual—to possess the powerful tools of good speaking and listening and to enjoy the rich pleasure of confidence in one's self-expression.

How Good Are We?

Both as talkers and listeners we are, in a word, poor. After conducting a series of studies, Ralph Nichols of the University of Minnesota concluded that in the typical audience situation, not more than 25% will be able to report accurately even the speaker's central idea. This reflects badly on both speakers and listeners.

We are insensitive—to ourselves and to those with whom we are supposedly communicating. Our messages are poorly conceived in topics, amplification and support, and organization. Our words, inflections, gestures and audio-visual aids are either inadequate or overdone. We fail to exploit the power of felicitous language to clarify and motivate. We are not wise in the choice of media (e.g., speech, discussion, interview, telephone call) and in timing. We fail to elicit feedback and use it to refine ensuing transmissions.

Loren Reid told of a two-part Civil Service examination used to create a manpower pool of young men and women to fill important junior posts in the federal government. Though the need for personnel is great, only one of five passes the written portion. Then, alas, of every five who survive the written test, only one passes the oral.

How Do We Improve?

If as a result of what we have said up to this point, we have a new appreciation for oral communication—its reach, its importance, and our general ineptness—we have already begun to improve.

Above all, we will wish to conceive of (and cast our) communications in the form of dialogue: Mr. A and Mr. B striving for a meeting of minds in an ongoing series of verbal exchanges.

One theorist conceived of communication as "helical" in its formation and suggested the "Slinky" toy as a communication model. One tip end of the coil might represent an initial transmission. A point diametrically opposite would represent the receiver, whose "feedback" response would be represented by the "return" semicircle. But the response does not return to the precise point of origin, but to a point just "above" on the coil, showing the process nature of communication.

R. W. Revans, Director of the European Association of Management Training Centers, published in the Aug. 26, 1965, issue of *New Society*, the report of a study by Manchester University:

> **Ten courageous mathematics teachers in ten different secondary modern schools, in and around Manchester, volunteered each to give five lessons of 40 minutes upon a subject of their common choice. They met together several times and decided upon the use of logarithm tables; as far as it is possible to standardize the presentation of this subject among 13-year-old B-stream pupils, they agreed among themselves to do so. Each of the lessons was then photographed at three second intervals by two cameras, one facing the class, one the teacher; each lesson was also tape recorded.**
>
> **Each class had about 35 pupils; since each class was photographed 800 times per lesson and since 50 lessons were observed, the total number of child photographs was well over one million. We used 16 mm. film with a wide-angle lens, and were able to work under normal classroom lighting; the children got familiar with the apparatus and, in our subsequent scanning of the films, we found it very rare indeed, not once in a thousand times, to observe a child distracted by the camera.**

The students' achievement in learning the logarithm tables was measured and the pictures analyzed painstakingly, searching for important clues. The major conclusion reached was that the most signicant factor in the students' learning achievement was entering into dialogue with the teacher. Revans concluded:

> **Is this result not perhaps an illustration of a more general law, that institutions in which subordinates feel free to ask questions of their superiors are those that adapt best? . . . May it not be that one of the more important inalienable rights of small children is to be both seen and heard? Is it not intelligibility rather than happiness that most of us are after in this life? What, after all, is education other than learning how to ask questions? . . .**

In future installments, we will forge ahead—improving our oral communication by discussing these topics:

A. Oral reports and public speaking
   1. Composition: content (topics, supports) and organization.
   2. Audio-visual aids.
   3. Language.
   4. Delivery.
   5. Question and answer.

B. Group leadership and participation.

C. Listening.

Summing Up

In summation, how much talking and listening do we do? A lot; typically more than eight hours per day. How important is it? Critically important: for understanding of messages, for morale, in persuasion and in individual development. How good are we? Most of us are poor talkers and listeners. And even the good ones can improve. How do we improve? Lots of ways.

We begin by attempting to communicate in the form of ongoing dialogue. We will aggressively seek feedback from our receivers, and use this response to improve our succeeding transmissions. Then, in the months ahead, we will develop some principles on speech content, organization, audio-visual aids, language, delivery, handling the question and answer period, discussion methods and listening. ■

Meet the Author

Robert Haakenson is Community Relations Manager of Smith Kline & French Laboratories, 1500 Spring Garden St., Philadelphia, Pa. 19101. In 1959 he launched the company's nationwide Speaker's Bureau which now has 500 members. He has taught at the University of Iowa (where he received his Ph.D. in Speech), Temple University and the Universities of Missouri and Minnesota. He is past president of the Pennsylvania Speech Assn. and a member of the National Council of the National Society for the Study of Communication as well as numerous other organizations.

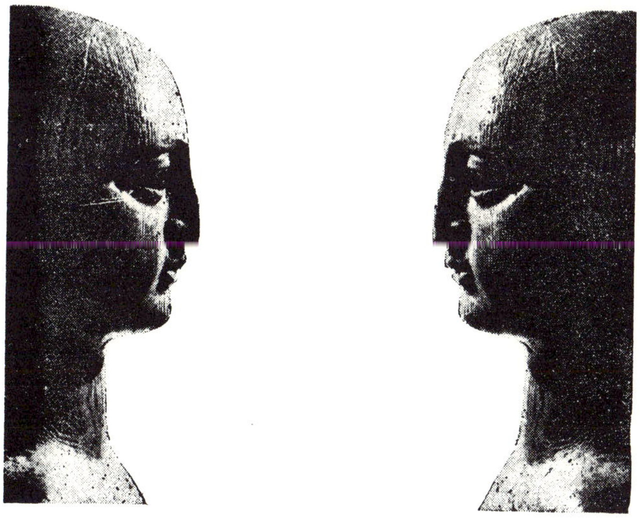

Organizing Your Talk

Good organization makes your talk more effective and easy for the listener to follow, plus making it easier both to write and deliver.

ROBERT HAAKENSON, *Smith Kline & French Laboratories*

"First I tell 'em what I'm gonna tell 'em; then I tell 'em; and then I tell 'em what I told 'em." This is the reported success formula of a backwoods speaker.

"A strong proposition based on three compelling reasons." This is the reported success formula of interscholastic debaters.

Put the two formulas together and we have the basis for a functional pattern of speech and report organization.

Simple, hard-hitting organization is the basis for versatility, eloquence and effectiveness in communication. We will speak of:

- The basic plan (schema).
- Patterns of analysis.
- Introductions and conclusions.
- Sample outline.

Good organization is helpful first to the listener. With a clearcut matrix, he has a frame of reference in which to fit the particulars as they come to him. But do not minimize the value to the speaker. If he is working from a simple, well-conceived outline, he is freed from the memory burden of intricate sequences, and is able to concentrate on winning from his listeners the desired response.

Schema

Actually, we may think of any well-organized communication in this simple schema:

The long first line represents the *central idea*. The shorter lines represent the *main points* that undergird the central idea. There should be not fewer than two nor more than five main headings.

The central idea may come instantly and automatically to the speaker, or it may present a very difficult challenge. If we are asked to speak in behalf of joining the civic association, or to explain the new data processing installation, the central idea fairly well declares itself. If, however, we are asked to speak on civic reform, or to advance ways in which corporate procedures can be streamlined, the selection of this central idea may be a long, even agonizing, procedure. If the subject area is emotionally explosive—civil rights, partisan politics, moral standards—the assignment may be even more trying.

When decision on the central idea is difficult, there is no easy way out. The speaker simply must commit the necessary time and thought. One consideration might help: We know that as any experience, including listening to speeches or reports, fades into history, it narrows to a sort of unified single impression. If we can decide, when the tumult and the shouting, the sound and the fury have died away, that *this* is what we would like to persist in the minds of the listeners, we have a good clue to the central idea. What, above all else, should be retained? If we can phrase this in a simple declarative sentence, we probably have a good statement of the central idea.

Patterns of Analysis

Having selected the central idea, the main headings ought to come more easily. There are five basic patterns of analysis, with certain stock designs in each category.

We say "not fewer than two main headings and not more than five" because if there are fewer than two we have not subdivided. If there are more than five, we have not discovered natural groupings and have lost the major benefit: ease of remembering by both speaker and listener.

Many speakers are fond of 10-point expositions. Who can remember 10 points? Five is a sort of upper limit. Also, as will be evident in examples to follow, natural groupings are found in two, three, four and five. Interestingly, the most popular number of main headings is three. Some persons attribute this to the

SPENDING TIME on organizing your talk pays off later.

religious trinity; others feel that it derives from the concept of continuum: two extremes and a middle ground.

Here, then, are the five major so-called "patterns of analysis," with illustrative "stock designs" for each:

1. *Chronological or Time Pattern*—There is only one stock design here: past, present, future. It is probably the most used stock design. This is understandable. As we approach any topic we ask: How did it start? Where are we now? Where do we go from here? All of this is fine, except that when it comes time to report our analyses, there are probably much more significant relationships.

2. *Spatial, Topographical or Geographical*—Following are some stock designs in the spatial relationship:

Federal	Metropolitan	Left
State	Suburban	Center
Local	Rural	Right
Top	Near	Inside
Middle	Mid-distant	Outside
Bottom	Far	

3. *Topical or Distributive*—Here the basic relationship is neither time or space but is substantive, e.g.:

Who	Theory	Background	Criteria:
What	Practice	Problem	Quality
Where		Methodology	Price
When		Results	Service
How		Implications	Beauty
Why (sometimes)			

A very useful device both for analysis of topics and synthesis of outlines is the so-called "decachotomy" (a ten-way breakdown):

Political
Social
Economic
Scientific
Educational
Religious/Moral
Military
Psychological
Cultural/Esthetic
Philosophical

Using the "decachotomy," the speaker can determine which facets of any subject are the most important to the talk. Indeed, when he organizes his speech or report, he may use the self-same considerations as his main headings.

4. *Logical Pattern of Analysis*—There are basically two stock designs here: problem-solution and cause-effect. We commented earlier that past-present-future is the most-used stock design and probably overworked. Instead, the most-used should be problem-solution. This is because we are a purposeful people. We communicate to meet needs, solve problems and, in general, advance the human condition. Further, the stock design should include "damage":

Problem
Damage (consequences)
Solution

If, for example, we deal with the problem of air pollution, demonstrating undeniably that the air is being polluted, and demanding that laws be passed and enforced, the member of the audience might say to himself, "Sounds reasonable. I agree." But all of this is detached, at arm's length.

If, on the other hand, the speaker demonstrates that so long as the problem of air pollution exists, there is direct damage, dire consequences, to these *immediate listeners,* their families and heirs, then the member of the audience becomes involved, ready to respond actively, no longer arm's length, but shoulder-to-shoulder, ready to move.

Remember, we tend to rely most heavily on past-present-future, but because Americans, above all others, are progress-oriented people, we should be employing problem-damage-solution. We wish to identify problems and needs, and to move forward purposefully to resolve them.

5. *Psychological*—The psychological pattern of analysis takes into account the psychological set of the listener as a member of the audience. One of the most famous of the stock designs is Richard Borden's "Ho-hum" formula:

Ho hum
Why bring that up?
For instance?
So what?

First the speaker must catch the listeners' attention, then demonstrate that his topic has relevance, then demonstrate its validity, and finally carry it over into the action phase.

A combination of psychological and logical (problem-solution) designs is the "motivated sequence" of professor Alan Monroe of Purdue University:

Attention step
Need step
Satisfaction step
Visualization step
Action step

The famous format of advertising is "AIDA":

Attention
Interest
Desire
Action

Thus we have five basic patterns of analysis, with illustrative "stock designs" in each. These, obviously, are merely suggested ones. The speaker or writer of the oral report may create his own main headings and can use these stock designs as patterns or suggestions.

Introductions and Conclusions

In one of Plato's dialogues, Socrates comments to the effect that a speech should be like a living being with head, body and feet, and all parts complementary to the whole. Ever since, it has been fashionable (and properly so) for a talk to have an introduction, body and conclusion.

The body of the talk is the fulfillment of the schema, i.e., the central idea and two to five main headings. In a long or complicated talk or report, the speaker may develop subheadings and even sub-subheadings. In most presentations, however, the simplicity of the central idea and the two to five main headings is advantageous.

What about the introduction and conclusion?

In the introduction, our speaker will want to "tell 'em what he's going to tell 'em." He does this with a preview. He reveals his pattern or outline by announcing his central idea and main headings. If he wishes to build suspense, he will do this in an abbreviated or perhaps somewhat mysterious fashion. At the very least, however, he will provide good clues to his central idea and main headings.

In his conclusion, our speaker will want to "tell 'em what he told 'em." He does this in the summary. After a preview in the introduction and development in the body, the speaker leaves nothing to chance, and recapitulates his schema emphatically.

What else remains?

Sample Outline

I. Introduction
- A. Icebreaker
  - 1. Story of mass production of pencillin in World War II
- B. Preview
  - 1. Challenging criteria in chemical production
    - a. Quality
    - b. Economy
    - c. Safety

II. Body
- A. Criteria in chemical production
  - 1. Quality
    - a. Translating chemical process into chemical production
    - b. Solving the challenge of faultless production of massive quantities
  - 2. Economy
    - a. Chemical engineer strives for lowest possible unit cost consistent with quality and safety
    - b. Efficiency is critical where throughput is enormous and profit margin slim
  - 3. Safety
    - a. Production workers
    - b. Community
    - c. Subsequent handlers and consumers

III. Conclusion
- A. Summary
  - 1. Chemical engineer must fulfill demanding standards in chemical production
    - a. Pinpoint-accuracy in massive quantities is first requisite
    - b. Production economy may determine market life
    - c. Safety concern extends beyond production workers to ultimate consumers.
- B. Haymaker
  - 1. Genius of American chemical engineering, which produced enough penicillin to treat all Allied casualties in Normandy landings, is working for us round the clock.

SAMPLE OUTLINE for a talk on the drug industry follows the suggestions given in the outline. Note the introduction with its icebreaker and preview, the body with its three main topics, and the conclusion with its summary and haymaker.

Our speaker will begin artistically with an "icebreaker." Its purpose is to call attention to, and rouse interest in, the speaker and his topic, and, secondly, to create a pleasant communicative rapport between speaker and audience.

The best type of icebreaker is the case-history type of narrative that illustrates the central idea, or at least leads into it intriguingly. Audio-visual aids are often useful in breaking the ice and in previewing. Audience participation, e.g., show of hands, guessing answers, also is good. Sensational statements, "shockers," should be used with caution. If jokes are relevant and well told, they are fine. Spontaneous spur-of-the-moment humor is better. Definitions may be used. The speaker may quote scripture, maxims or poetry. References to the audience, occasion, speaker, or all three, are fine.

Thus, to call attention to, and arouse interest in, the speaker and his topic; and to create a pleasant rapport between the speaker and audience, any one of a variety of icebreakers is available. If time is plentiful, the speaker may use several, e.g., reference to audience and occasion, humor, participation ("Who will tell us what material this container is made of?"), a case history.

The speaker then offers his preview ("tells 'em what he's going to tell 'em"), e.g., "I want to share with you some of the miracles of chemistry in health. We can get a pretty good idea in brief time if we cover: First, discovering the molecule's health-giving properties; second, synthesizing the molecule; and, third, quality-control manufacturing in quantity. Or, in brief: one, discovery; two, synthesis; and three, manufacture." Now both speaker and listener have a plan of attack, an orderly approach.

Occasionally, some concept or term is so basic to the entire talk that it should be defined in the introduction. If the speaker is to report, for example, on "Information Sharing in Decentralized R & D Locations," and he plans to make extensive use of the concept of "telecommunications," obviously he should define the term early, probably immediately after the preview and just before getting into the discussion proper.

At the other end of the talk is the conclusion. Here, the speaker first summarizes ("tells 'em what he told 'em"). We know that we forget 50% of what we hear in one 24-hour period during the ensuing 24 hours.

In our speaking and reporting, we wish to do everything we can to improve retention. One aid is restatement. Therefore, the speaker recapitulates central idea and main headings. If there has been true development in the body of the speech, and if the speaker has any artistry whatever, there will be steady growth from preview, through development, to summary. The final summary can be hard-hitting and emphatic because of this growth. It will not seem to the listener to be mere repetition. Listeners will accept more redundancy than the speaker might at first think possible.

Finally, the speaker will wish to end climactically. The orators of ancient Greece and Rome called this peroration. Here the speaker made his final impassioned plea for his cause. Today we are less flamboyant but the principle stands. The "haymaker" may be almost any one of the items suggested for "icebreakers": Narrative, instance, quotation, participation, humor, audio-visual aid, etc.

A particularly pleasing touch, though not essential, is for the "haymaker" to refer back to the "icebreaker," bringing the talk 360 degrees, full circle. If the "icebreaker" is a case-history-type narrative, for example, and the "haymaker" brings the narrative right to the moment, there is a sort of esthetic completeness that is artistically satisfying. Obviously, the "haymaker" above all should give impact to the central idea.

In the "telecommunications" example just suggested, for instance, the "icebreaker" might be an interesting narrative of the amount and variety of information on a given topic a computer can produce in a matter of seconds.

The "haymaker" might be the posing of an R & D problem confronting this immediate group that might be susceptible to the information potential of the data processing installation. The talk has come full circle, the central idea is reinforced, and a challenge is put to the listeners.

Summary

We have said, then, that clear-cut organization is the basis of versatile eloquence. It enables the listener to proceed methodically and frees the speaker to concentrate on winning the desired audience response. Effective organization usually multiplies the speaker's efficiency in his preparation. If he can quickly decide on his central idea and main headings, he then can go about completing the talk and creating audio-visual aids methodically.

The basis of functional organization of any communication is the *schema:* Central idea based on two to five main headings. The main headings will reflect one of five major patterns of analysis: Chronological, spatial, topical, logical, or psychological. Well-known "stock designs" (past, present, future; federal, state, local) may be used themselves, or may suggest main headings the speaker creates himself.

The speaker opens artistically with an introduction that "breaks the ice" and then previews. He concludes artistically and emphatically by summarizing and by landing a "haymaker."

Henceforth, we will combine the genius of the backwoods speaker who "tells 'em what he's gonna tell 'em, etc." and the inter-scholastic debater who advances his proposition relentlessly on the basis of "three compelling reasons." ■

The Content of Your Talk

Once you've chosen your topic, you are faced with the problem of just what items to include in your talk, and what to omit. Here are some hints for solving the problem.

ROBERT HAAKENSON, *Smith Kline & French Laboratories*

Late in the 19th century, Dr. Russell Conwell gave a talk at a reunion of some of his old Civil War comrades. Entitling it "Acres of Diamonds," Conwell considered it a good effort, but not outstanding. He described it as "a mere accidental address." To his surprise, it was requested again and again.

The upshot: Over the next many years, Dr. Conwell delivered "Acres of Diamonds" more than 6,000 times, for revenues of $7 million. Most of the money, incidentally, was plowed into the education of needy young men and the founding of Temple University in Philadelphia.

Perhaps the outstanding feature of the talk is that it develops a highly unified theme by narrating one illustrative instance after another. The theme is simply that opportunity lies everywhere about us if we will but scratch the surface.

The supporting instances range the breadth of human experience. They begin with the legendary Ali Hafed spending his life in search of the fabled "acre of diamonds" and—on returning home, old and exhausted—dying, just before the treasure was discovered on his own land. Additional instances, actual and more current, include the discovery of gold at Sutter's Mill, California; the discovery of oil at Titusville, Pennsylvania; the millinery venture that launched the fortune of John Jacob Astor; and dramatic success stories in trout farming, toys, and hatpin design.

This famous talk illustrates several principles of speech content, notably topics and supports. Let us discuss:

- Analysis of audience and occasion.
- Delimitation of topics.
- Amplifying and supporting particulars.

Audience Analysis

As he begins to prepare, the first thing the potential speaker should do is focus on his target. And as he continues his preparation, he should keep refining his aim. For example, while gathering materials, he will continue to analyze the audience and the occasion. His principal source of information will be his sponsor —the program chairman, president, or other officer— but he will not stop there. He will contact prospective members of the audience and former speakers, and consult published descriptions. He may be able to attend intervening meetings of the group. His objective is to get in his mind's eye as comprehensive and accurate as possible a picture of his forthcoming speaking engagement:

Who will be present? How keen is their interest in his topic? How much do they know about it? What is their concentration span?

What are the circumstances of the occasion? Is it a serious meeting, or largely social? What precedes? What follows? Will a long program have fatigued the listeners? Will a lengthy agenda worry them about preserving vitality for what is to follow?

Delimiting Topics

Often a recommendation is made to: "Narrow the topic; don't try to cover so much."

This advice is doubtless well intended and often good. But it is not infallible. Simply narrowing a topic endlessly may not necessarily achieve the desired result: e.g., chemistry, synthetic materials, plastic, plastic toys, plastic flying saucers, and so on. As the topic is narrowed, to be sure, the speaker is permitted to delve more deeply in the specified length of time. But what is the most meaningful for the audience? Perhaps exposition on the broader theme of "synthetic materials" may be best for this audience at this particular time.

In delimiting the central idea, therefore, the speaker will realize that he can interpret it broadly and provide limited coverage. Or he can shrink the central idea and plunge in depth. Or he can choose a middle course.

At the outset, then, the speaker must make the basic decision: Is it more useful for these listeners at this time to say more about less, or less about more?

Amplifying and Supporting Particulars

Amplifying and supporting details fill out the speech outline. They supply clarification for understanding, support for belief, stimuli for action, and interest to attract and sustain attention.

Following are some of the chief types of such specifics:

Definition.
Description.
Comparison/contrast.
Figure of speech.
Example or illustration: real or hypothetical.
Instance.
Case history.
Date/event.
Enumeration/listing.
Narrative/anecdote: true or fictitious, serious or humorous.
Figure/statistic.
Quotation: witness testimony, authority opinion, maxim, scripture, poetry.
Humor: quip, pun, repartee.
Audio-visual aid.

Case History

The case history is a first-rate type of amplifying particular. It attracts attention, holds interest, and persists in the listener's memory. Properly selected, it clarifies, convinces and motivates.

Members of the Smith Kline & French Laboratories Speakers Bureau have addressed more than 14,000 audiences in the past 8 years. One talk, "Prescription for Tomorrow," in its various revisions, accounts for more than 6,000 of the talks. The speakers consistently report that the case histories explain the popularity. Here is an example:

"Most of us have had prescriptions that call for taking medicine every three or four hours. We all sometimes forget or think it a great inconvenience, especially when we have to get up at 4:00 a.m.

"A team of our scientists searched for a method of solving this problem. They thought it would be possible to make one dose that would provide continuous, uniform medication for 10 to 12 hours. They reasoned that this could be accomplished by dividing a single does into a number of very small parts that would disintegrate at different times. But this took some doing. Month after month went by without finding the proper combination of size and ingredients.

"Finally one of our scientists, while shopping in a grocery, spotted those brightly colored candy beads used to decorate cookies and cake icings. These were the size particles he was looking for. An idea clicked in his mind, and candy beads like these, *coated with intricate patterns of medication,* became the basis of an entirely new pharmaceutical form.

"We now have 18 products in this special form called 'Spansule' capsule, indicating that the medication is released over a long *span* of time. In each capsule are more than 600 tiny pellets, each of which will release its medication at the proper moment. That simple project —from candy beads to finished product—simple though it seems today, took seven years or 35,000 man hours to complete.

"'Thorazine,' that pioneer tranquilizer, is available in 'Spanule' capsule form, and in several strengths. Think of the advantages when this medication can be given only once, or at most twice, a day and bring continuous tranquility to a once-disturbed mental patient."

Source

There are four fruitful activities for gathering amplifying and supporting particulars:

- Inventory: taking stock of what one already knows.
- Talking.
- Reading.
- Observation.

A frequent tendency is to depend chiefly on reading to gather materials. This is regrettable because it overlooks many vital data, and, secondly, misses a great deal of the fun of research. The communicator should read, to be sure, in a wide variety of published sources and in great quantity. But he should also give himself credit for already possessing vital and useful information. And he should certainly interview others—formally and informally. This provides not only firsthand specifics, but is a kind of audience analysis, yielding insight into the way the ultimate audience may react. Finally, in all topics, wherever possible, nothing is more valuable than direct, firsthand observation. What, after all, is more convincing than, "I was there. I saw it with my own eyes."

The speaker should gather an abundance of particulars. A good rule of thumb is that he should begin his outlining with not less than four times the amount of information he can possibly include in the talk. And not only should we have such copious quantities, but they should be of varied types and from varied sources.

Criteria

Now the outlining—and perhaps drafting—of a manuscript can begin. The speaker applies these demanding criteria:

- Relevance. Irrespective of any other merit, the particular is worthless if it is not germane. It must amplify or support the point under consideration.
- Accuracy. Truth must be served.
- Human interest. Richard C. Bates, M.D., writing in the February, 1967, *Hospital Physician,* on "How to Keep Medical Audiences Awake," suggests, ". . . unadorned traffic statistics will put an audience to sleep; . . . details of specific accidents in the community will do the opposite."
- Quantity. Are there adequate details to clarify the point and make it convincing, but not so many as to labor it?
- Proportion. Are the specifics nicely arrayed in terms of type, length and source? One good suggestion is to "balance the specific with the general," i.e., illustrate the point by citing an actual instance, then demonstrate that the instance is typical, representative of the general situation, by citing statistics or quoting authority opinion.

Successful meeting of many of these criteria is illustrated in a recent lecture. Dr. Hubert Alyea is professor of chemistry at Princeton University and a celebrated lecturer who addresses audiences all over the world in all sorts of circumstances. He has presented one talk on the atom 2,800 times, and another, "Unlikely Accidents: Great Discoveries by Prepared Minds," 500 times.

On this recent occasion, he was introducing a new talk, "Curiosity and Satisfaction." Its theme is the romance of science, and it is intended for after-dinner audiences who may be groggy from food and drink and sweltering in a banquet hall that is too hot, too dark, too smoky, too noisy. Dr. Alyea faced up to the challenge with an admixture of amplifying and supporting particulars worthy of any rhetorical alchemist. He knows well the time-honored factors of attention: surprise, shock, humor, narrative, emotion, conflict, novelty, the familiar in a new setting, personal data.

Dr. Alyea led off with surprise, reversing the ordinary situation by handing his speech outline to the audience. He quickly revealed that he would intersperse his commentary with demonstrations. He admitted that these laboratory pyrotechnics would be more for recapturing flagging interest than for teaching science.

His commentary consisted of one narrative after another, much in the fashion of Dr. Conwell's "Acres of Diamonds." But Alyea went Conwell one better: Almost every one of Alyea's instances involved him personally, providing interest and identification.

An example: Alyea told of Albert Einstein's visiting him in his research room at the Kaiser Wilhelm Institute in Berlin. There, Alyea had a Rube Goldberg device for the investigation of certain optical phenomena. The device consisted of a pipe closed at one end and, at the other, two lenses unequally lighted. Experimentation included twisting the lenses to measure light ray interaction. Curious, Mr. Einstein approached the device and peered into it. Furious, he complained, "Can't see! Can't see!"

Professor Alyea noted that he was peering into the closed pipe, explained that he was at the wrong end and directed him to the lenses. Einstein was elated, "Wonderful! *Wunderbar!*"

Alyea reported that in later years, with both men resident at Princeton, Einstein frequently attended his lectures. During his explanation of a point, Alyea would watch Einstein. If he found a frown, even immobility, he would try another explanation, and another, keeping it up until he saw Einstein's head begin to bob up and down and a smile to spread across his face. Then Alyea knew he had made it clear.

At this point, it may seem that the Alyea lecture was exclusively gimmicks—human interest stories and laboratory sleight of hand. But along with this, the professor had imbedded a very solid message in his discourse. His point regarding Einstein, for example: a true scientist does not hesitate to admit his ignorance.

The Larger "Message"

The larger "message": Alyea was very caught up in the project of bringing chemistry education to the disadvantaged people of the developing nations. He described and demonstrated the ingenious steps taken to reduce costs of a workable audio-visual projector from $400 to 25¢. He described and demonstrated the "unicell," in an inexpensive Plexiglas container that alone does the work of 200 audio-visual devices formerly required.

There are millions who need science education but their circumstances permit only penny expenditures. The selfsame science they need to learn may nevertheless make possible the achievement of their educational goal.

Alyea states that laboratory costs for a chemistry student in a U.S. university total around $500 a year. With devices like those described above, he has brought this down to $5. He points out that this is significant not only to the students of the impoverished, developing nations, but also to students of overcrowded schools everywhere. The miniaturization makes it possible for each student to have his laboratory on the folding arm of an auditorium chair. This project, called TOPS (for Tested Overhead Projected Series) is described in an article, "Microchemistry Projected," in the June, 1967, issue of *Journal of Chemistry Education.*

Documentation

A specific piece of information has a way of being about as good as the source from which it comes. The matter of divulging sources of information we call "documentation." This is a study in itself. There are rules: legal, moral and stylistic. We will suggest here some rules that reflect—at least informally—those regulations:

- Reveal the source when credit is due: The ideas belong to another; the information or the data are the fruits of another man's toil. If the material is copyrighted, of course the source must be acknowledged, the limitations of quote and paraphrase respected, and —for extended passages—permission obtained.
- Reveal as much information about the source as necessary to give the listener a clear idea: Who said it? When? Under what circumstances? Is it a direct quote? Paraphrase? Where is it published? Use language precisely, to distinguish clearly what is quoted or paraphrased.
- Accentuate the positive. Wisely handled documentation adds interest and credence; awkwardly handled, it invites inattention and reduces credibility.

Summary

Speech content, then, consists of the topics we select for our talks and reports, and the amplifying and supporting particulars that give body to those themes.

The first step in preparation of speech content is careful analysis of the audience and occasion: Establishing as comprehensive and accurate a picture as possible of the persons who will attend and the circumstances that will prevail.

We delimit topics effectively when we define that breadth of scope that will permit desired depth of treatment in the time available.

We marshal amplifying and supporting particulars effectively when we know all the types available, use all methods of gathering them, analyze audiences and occasions perceptively, and employ functional criteria in selecting details for inclusion: relevance, accuracy, human interest, quantity and proportion.

We employ documentation (revelation of sources).

Russell Conwell gave a talk 6,000 times and netted $7 million. Hubert Alyea launched a new talk that exalted the first audience, and doubtless will do the same for hundreds more. Each of these talks illustrates critical considerations in speech content: wise delimitation of topics, and moving use of amplifying and supporting particulars. ■

Delivering Your Talk—I

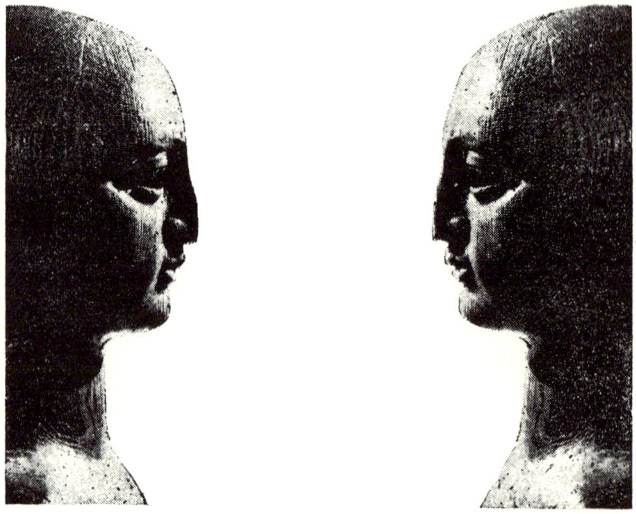

Even the best thought-out talk can be ruined by poor delivery. Here's how to make the audience sit up and take notice.

ROBERT HAAKENSON, *Smith Kline & French Laboratories*

Which is more important: what the speaker says, or how he says it?

This battle has raged for years but is needless. Obviously, the speaker should have something well worth the saying, then must say it well enough to gain a fair hearing.

It is true that many persons with fine ideas go unnoticed because of colorless presentations. Other persons, gifted in expression, often attract more attention than their thoughts truly deserve.

Our objective here will be to perfect our skills in delivery and then to use them in behalf of worthy causes and sound ideas. In order to do this, we will consider four points:

1. The challenge.
2. Goals.
3. Options.
4. Success formula.

We will use the words "delivery" and "presentation" interchangeably. Further, we will differentiate "delivery" and "presentation" (how the speaker says it) from "content" (what the speaker says) and "organization" (how he structures the talk). "Content" and "organization" have been covered in previous articles.

The Challenge

Talks are given under widely varied circumstances. They differ in numbers of persons attending, length of presentation, formality of occasion, setting, etc. A typical community audience numbers 40 persons. A typical setting is a meeting room that is little more than adequate: noisy, modestly to poorly outfitted, fairly to poorly ventilated and illuminated.

The speaker would like to believe that when he opens his mouth to expound, a hush will fall over the group and a magic spell will prevail. This is rarely the case.

There are always certain factors present that compete with the speaker as he strives to stir his listeners. These factors may be broadly classified as internal and external.

Internal competitive factors are within the intended listener. Each member of the audience brings his own collection of concerns: ill health, physical or intellectual fatigue, preoccupation, worry, overindulgence in beverage or food. Wrap them up and they become a great bolus, a sizable barrier separating the speaker from the desired listener response.

External competitive factors: even in the newest, most ingeniously designed meeting rooms, there are sources of distraction: excessive noise, traffic, other extraneous activity, excessive heat or cold, humidity or aridity, smoke. Air-circulating equipment often is amazingly noisy.

Stop for a moment, wherever you are, ask everyone to remain silent, and listen. You will be surprised and distressed to hear the roar. This is just one of the many external competitive factors that get between the speaker and the listener.

The foregoing description may seem to be overdrawn and morbid. There are some settings, true, that approach the ideal. But the safest thing for the speaker to do is to assume the worst. He should, in fact, prepare himself to survive a "Roman Circus."

Another precaution is to assume there are present greater competitive factors than meet the eye—especially, the internal. The human attention or concentration span is notoriously limited. Further, the things that unfold in the report or talk often will remind listeners of old worries, or stimulate thought in peripheral or unrelated areas.

The challenge, then: Win the desired response while operating in the arena of the "Roman Circus."

Goals

One good definition of communication is "an endeavor by one individual through a system of symbols to win a desired response from another." We can state these goals in a progression of aspirations, i.e., for the speaker to be:

- Heard.
- Understood.
- Believed.
- Inspiring.

One should be heard. Many individuals are concerned —and properly so—about "acoustic pollution." For years, the noise level has been rising, especially in the big-city complexes. In the 15,000 talks by members of the Smith Kline & French Laboratories Speakers Bureau, the most pervasive single problem has been that of being heard. Even where public address systems are available, speakers find it difficult to rise above the clatter of dishes, the uproar from the adjoining room or the din of traffic outside. In short, being heard is no minor aspiration.

Secondly, the speaker strives to be understood. Wishing not to demean his audience, and wishing to present himself as an authority in his field, the speaker is naturally motivated to provide challenging ideas and information. But even with sophisticated audiences, it is easy for him to get beyond his audience conceptually and in vocabulary. Today's increasing specialization complicates matters. Despite our very best intentions, it is easy to fail to be understood.

In much, perhaps most, of our speaking there is an element of controversy. In some cases, the audience response is openly hostile. In others, it is simply one of resistance to change. Therefore, much of our speaking is dedicated to achieving change (*we* think of it as "progress"). Thus we strive to influence attitudes and belief. We strive to change convictions and that is a challenge.

Finally, we strive for the ultimate objective: action by our listeners. Possible actions include: to carry forward the message, to initiate a project (or cancel or modify it), to sign a petition, to vote in a certain way, to campaign, to become a candidate, to enroll as a volunteer, to make a financial contribution, to join an organization, to endorse a proposal, to approve a budget.

Options—Four Modes of Delivery

There are four modes of delivery for presenting our reports and speeches: memorization, reading, extempore and impromptu. Each has certain advantages and disadvantages. The truly "compleat speaker" is effective in any of the four, and knows when to use each.

Memorization has the advantage of detailed preparation, exact timing, and freedom from any memory aid (manuscript, note cards, visual aids). It has the great disadvantages that (a) it is hard work, (b) few speakers recite well, and (c) memory is fickle—the chain is as strong as the "weakest link"; if it snaps, the whole presentation has an air of fraudulence.

Reading from manuscript has many of the advantages of memorization plus the greater security of the manuscript—almost anyone can read aloud so long as there is proper lighting. The major drawback of reading is that few people are communicative in this mode. The manuscript becomes a barrier between speaker and audience. This barrier, however, is not insurmountable.

In extempore speaking, the speaker is fully prepared, except that he stops short of putting his ideas and information into final words. This is done on the platform. Thus this mode has the advantages of full preparation (including rehearsal), adaptability or flexibility, and spontaneity. The disadvantages are that:

- The word supply, which was abundant in rehearsal, may fail before the audience.
- Sometimes the inspiration of the moment stimulates the speaker to say things he didn't intend to and later regrets.
- It is easy to lose track of time—particularly, to talk too long.

The speaker may use memory aids—notes, an outline, visual aids, or even a manuscript as an outline. It is particularly effective to speak extemporaneously, while free from any notes or other aids.

Impromptu speaking is completely off-the-cuff or ad lib, and is therefore different from the three preceding modes in that it is not prepared. It is splendid when it is the only option available, but unforgivable otherwise. Occasionally a speaker will procrastinate, failing to prepare and forcing himself to speak impromptu. The inspiration of the moment may bring resounding success and he will resolve never again to undergo the agony of preparation. Such success without preparation is rare and, therefore, a snare and delusion. The few successes in such impromptu speaking are outweighed by thousands of dismal failures.

There is a current trend toward reading from manuscript: speeches must be approved by employers or sponsoring organizations—sometimes, indeed, by representatives of the group to be addressed. Timing must be exact; occasionally, visuals must be coordinated with a projectionist. Often the text of the talk must be released in advance to the news media and other publications. There is increasing attention to this secondary (reading) audience, and properly so. For these reasons, today's effective speaker must learn to talk communicatively from manuscript.

Success Formula

To overcome the competitiveness of the internal and external environmental factors that make up the challenge; to fulfill the objectives of being heard, understood, believed and inspiring so the desired audience response is won; and to be successful in all four modes of delivery—memorization, reading, extempore and impromptu—a success formula must be found. In a single word it is "impressiveness"—impressive content, impressive organization and impressive delivery. Content and organization have been discussed elsewhere. Impressive delivery can be described by considering the speaker's duty at the lectern step-by-step.

Getting under way

(1) Focal or initial pause

(2) Salutation

(3) Smooth transition into the discussion proper

Under way*

(1) Authority
(2) Involvement
(3) Communicativeness
(4) Vocal expressiveness
(5) Physical expressiveness

Climactic finale*

(1) Hard-hitting windup
(2) Terminal pause

Getting Under Way

Focal or Initial Pause—When the amateur speaker hears his name called by the program chairman making the introduction, his immediate inclination is to begin talking—even before he has risen from his chair. He is self-conscious and over-eager. The veteran speaker, on the other hand, will proceed to the lectern deliberately, even slowly. With a pleasant facial expression, hopefully a smile, he will look out over his audience.

At a convention in New York City, Bishop Fulton Sheen was introduced to an audience of more than 2,000 persons meeting in the grand ballroom of a major hotel. He was given an ovation. Then, as the audience settled down to hear him, he looked out over the assembly, smiling serenely. Even after the last chair scrape had quieted, he continued this initial pause. It may have been as long as 30 seconds. Because he was seemingly comfortable in silence, his audience was comfortable. It gave them the opportunity to ready themselves to listen and to anticipate what he might say. It gave him the opportunity to feel the stimuli of this anticipation.

If a speaker begins to jabber immediately, his words are competitive with the visual stimuli he is transmitting. Listeners have to satisfy their visual curiosity. If words are emanating simultaneously, they are in conflict.

Thus the speaker serves both the audience and himself by indulging himself lavishly in this silence. It is the initial pause to permit the listeners to focus their attention on the speaker and specifically on what he is going to say. The veteran speaker will use this time to sense the audience anticipation, to compose himself, to get his notes, manuscript, visual aids, microphone, drinking water in order; or perhaps, he will simply look out over the group making visual contact or enjoying the pleasant scene before him.

The self-conscious amateur who finds this hard to do can occupy himself with certain business, e.g., running through a little checkoff list: Jacket buttoned? Manuscript in order? Microphone at right height? First words (salutation) clearly in mind?

* These subjects will be covered in the second part of this article.

Salutation

The salutation should be full, spontaneous and as free from cliches as possible. First, the speaker will want to acknowledge the nice introduction he has been given. He should state his thanks spontaneously, using the title and name of the person who introduced him. Then he will wish to acknowledge the others present by name and/or title, and most especially the name of the organization that is the basic audience.

Sargent Shriver addressed a large banquet gathering in Philadelphia. After his focal pause, he said, "Thank you, Mr. Chairman, Judge Reiml, for your glowing comments and lively introduction. President and Mrs. Gladfelter . . ." At this point, Mr. Shriver was faced with a head table that stretched the width of the ballroom. How could he keep the "pecking order" straight in a lineup of this sort? He resolved it by starting at his extreme left, graciously acknowledging each individual, with an occasional impromptu comment for some, and working his way naturally and easily across the width of the head table.

When we are introduced socially around a living room, we try to respond, repeating the name of each individual to whom we have been introduced. Public speaking protocol follows basically the same principle. The program chairman has introduced us to this group assembled and we respond by acknowledging the presence of the program chairman, the dignitaries, the organization and guests.

The salutation should be natural, spontaneous and cordial. That protocol be respected is far less important than that there is honest awareness of persons present and their identities.

If the situation is informal and time is limited, the salutation will be trimmed accordingly. At the very least, it should be an acknowledgement of the program chairman and his introduction and most important, the identity of the group itself.

Smooth Transition Into Discussion Proper

Next our speaker should have "parry and thrust" ad libs. Unless time is pressing, he should make references to the occasion, to the audience, to himself, and to his topic. His purpose is to seize attention, arouse interest, and weld common bonds in a nice, easy, communicative rapport. Humor is wonderful if appropriate and deftly put, but unnecessary and downright bad if heavy-handed.

This ad-lib parry and thrust should provide a smooth transition into the text of the talk. The speaker should know the opening paragraphs of the manuscript "cold" so he can slide smoothly into the discussion proper.

Delivery of the talk, once underway, and the handling of the climactic finale will be covered in Part II of this article. ■

Delivering Your Talk—II

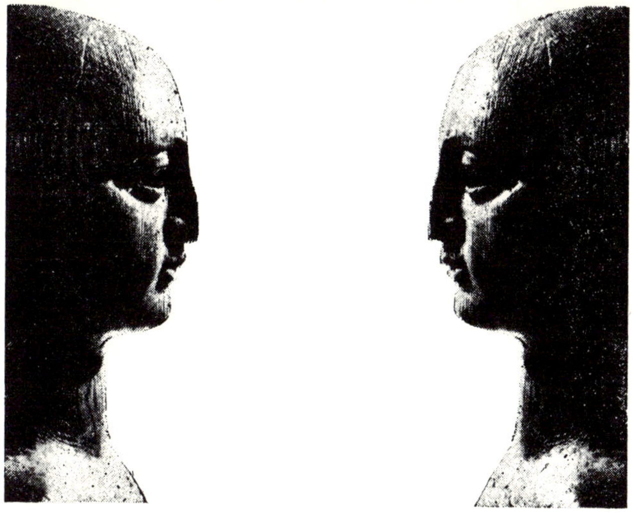

Once your talk has started, the way that you handle yourself has a marked influence on the audience. Here's how to do it right.

ROBERT HAAKENSON, *Smith Kline & French Laboratories*

In the last article, we discussed how to get a talk under way. The speaker's next responsibility is to present the talk itself with impact. To do so, he should fulfill the five essentials of impressive delivery:

- Authority.
- Involvement.
- Communicativeness.
- Vocal expressiveness.
- Physical expressiveness.

Authority

Here we speak not of content expertise, but of the speaker's frame of mind or psychological set.

Members of audiences empathize with speakers. If the speaker radiates assurance. the audience reposes its confidence in him.

Perhaps the best phrase to express what we mean is that of the military: "situation well in hand." In speech terms, this means that the speaker need not be absolutely free from stage fright or apprehension but that he is present, determined to communicate. He reveals that he has control of fears and has adequate confidence to get on with the job of communicating.

A couple of alphabet-soup expressions might be helpful:

PTP: "Prime the Pump"—Thousand of speakers report that they lack authority in the opening moments of a talk, but once under way gain confidence and assurance. The challenge then is to bolster performance during those first minutes.

Speakers who have tried it report that they can successfully "prime the pump." By exhibiting a little courage or bravado that they may not actually possess, the desired effect can be achieved. It might be described as making an extra effort, or, in theater parlance, assuming the posture. One might say that the speaker can "psych" himself into an attitude of confidence that will strengthen his opening and give him continuing authority.

If, at the beginning, the speaker reveals that he is in fact taking charge, and that he is at least that confident, he has requisite authority. If he can also radiate his comfort with this responsibility—i.e., natural confidence, self-assurance, poise — effective communication will be enhanced.

KBF: "Keep a Bold Front"—There are really two dimensions here. One is at the outset of the talk. Speakers often open with apology. They didn't know until 11:00 last night that they were to speak. Their health is deserting them. (It is true, of course, that apprehension about a speech can bring on or aggravate symptoms.)

The speaker opens his talk by saying, "I didn't know until 11:00 last night that I was to be your speaker. As you can hear, I am having an attack of laryngitis but, if you will bear with me . . ."

Having said this, the speaker has experienced catharsis. Verbalizing his feelings of inadequacy provides a kind of purge.

The member of the audience does not feel the same relief. Instead he thinks, or senses subconsciously, "Here is a man who did not know until late last evening he was to be the speaker; here is a speaker who by his own admission is not in tip-top condition; and yet he has the gall to take up a precious half-hour of my life." Thus we see what a masterpiece of destructive psychology the speaker's alibi is.

What is often demonic is that the speaker, no matter how sorely tried by tardy invitations, poor health, or other circumstances, secretly expects to do all right. But with his alibi he has protected himself. Should he flop, then the blame is on the dilatory program chairman or the doctor responsible for his poor condition. On the other hand, if he does all right, as he expects to do, he will look all the better because he has labored under such trying circumstances.

The second dimension of "Keeping a Bold Front" is in the handling of mishaps under way. The tendency of the amateur speaker is to magnify his mistakes by calling attention to them. True, such verbalization probably reduced his tension, but it also spotlights inconsequential errors such as an incorrectly pronounced word, a transposition of phrases, a lost place. If the error is insignificant, the speaker should KBF and sail right on, not even bothering to correct it. If the error muddies understanding or disserves truth, then, of course, it must be corrected. In a quiet, workmanlike way, the speaker goes back, rights the situation, and resumes his presentation.

No matter how trying or unfair the circumstances, we should not prejudge our talk, or the audience's reaction. We should sail in and do the best we can. And, happily, the chances are very good that we will do all right.

Involvement

Involvement is the speaker's sensing deeply the meaning and emotion that his words are intended to convey. Involvement slips away very easily when the speaker speaks from memorization or reads from manuscript. And when it does, there is no panacea. If the speaker realizes that he is merely parroting words, he can only draw himself up short, pause, and work his way back to the meaning and emotion.

Several things will help. One is *pause,* frequent use of pause—either for emphasis (holding eye-contact) or simply for speaker and audience to reflect over what is being said.

More alphabet-soup may be helpful:

TTT: "Think the Thought"—If the speaker is involved, he will himself be thinking the thoughts that he is attempting to share. If he is thinking the thought, he is almost certain to reveal his involvement to his listeners. TTT is helpful to the speaker under way. He asks himself, "Am I 'thinking the thought'?" If not, he must do something about it: Pause, and deliberately work his heart back into his work.

LKF: "Little Known Fact"—This is a special kind of ad lib. The "Little Known Fact" is something known to the speaker alone or perhaps also to one or a few of the audience. The little known fact demonstrates that the speaker has a personal connection with the subject, thus a substantive involvement. His spontaneous expression of this fact communicates involvement in delivery.

The reason the ad lib reinforces involvement is that while the speaker is talking, free association may trigger recall of some pertinent item, possibly something in the day's news or something said by his luncheon partner just preceding the talk. The speaker is so deeply immersed in what he is saying that these relevant experiences are spontaneously brought to mind.

Many speakers will deliberately mark the manuscript at places where they can ad lib if circumstances permit.

Communicativeness

Think of yourself in your last act of "friendly persuasion." You were endeavoring to persuade some friend to some belief or course of action for your friend's best interest as you saw it. Here is what you were doing, probably subconsciously:

You were bringing into play every argument, every contention, every emotional appeal, every shred of evidence, every piece of description that would make your case persuasive. In short, you brought into your campaign everything in the way of *content.*

Secondly, you structured your campaign. You might have said, for example, "Okay, we're agreed up to here, are we not? All right, then." And you would launch off into "no man's land." You would penetrate as deeply as you could, until you thought you could penetrate no deeper, being rebuffed. Then you would fall back and regroup, consolidating your gains. After a summary, you would launch off on a new tack into "no man's land."

Thus you were structuring intuitively. You were using *organization* for maximal persuasiveness.

Thirdly, your personality ran its full gamut, from overpowering forcefulness to humble supplication. All facets of presentation varied dynamically, persuasively: voice, articulation, bodily expression, language. Your vocal inflections varied from thundering to whispering. Certain words were enunciated with compelling precision. The words and phrases themselves were your most persuasive. You circled about your friend and may have towered over him or even drooped before him. Your face was alive with expression. Your eyes did not leave his. In short, your personality and presentation ran the gamut, intuitively, without deliberate or self-conscious intent. Thus, the maximum of your potential in *delivery* was exercised.

Fourthly, you were so caught up in the act of communicating, striving to elicit the desired response from your friend, that you were totally oblivious of self. Your total attention was riveted on response, noting when you had made headway, when you had to amplify, to reinforce, or to offer more supporting detail.

Vocal Expressiveness

It is said that Madame Eleanora Duse, a great Italian actress, once brought an audience to tears merely by her moving recitation of the alphabet in her native tongue. Properly used, the voice by itself can be powerfully persuasive. Most of us use only a fraction of the potential.

The greatest danger is that we will not be heard. Next, our voices will be dolefully monotonous. Thirdly, there is the danger of artificiality, forced inflection changes.

There are four variables with which we work:

- Pitch—soprano, alto, tenor, bass.
- Loudness—ranging from whisper to shout.
- Rate—rhythm or tempo, how fast and how slow and how much pause.
- Quality—timbre.

Over the first three variables, the speaker has a great deal of control.

The fourth variable, timbre or quality, is harder to change. This factor is best defined by negative characteristics: strident, hoarse, husky, harsh, breathy, excessively nasal, etc. Favorable attributes include resonance and mellifluousness. Timbre is in fairly large part hereditary, and improvement therefore may require clinical attention, or perhaps cannot be improved at all. But proper use of the first three variables can compensate for poor quality, and may sometimes actually overcome timbre problems.

The best assurance, however, is that each of us, even though his voice may be quite ordinary, has enough control over pitch, loudness and rate, to make it effective.

To overcome monotony, the simple counterpart is variety. To overcome artificiality, the inflections must be generated genuinely, spontaneously. The speaker freely lets his voice give expression to his true feelings. To overcome speaking too quietly (or too loudly, rapidly, slowly, shrilly or rumblingly), the speaker merely avoids the extremes.

Remember that your voice is spontaneously expressive in your "friendly persuasion." A good rule is: "Bring to the lectern your conversational manner with close friends, only put a little more energy into it."

Since the most frequent problem is that of inaudibility, a wise procedure is for the speaker to recruit someone to serve as monitor. This should be arranged before the speech begins; but, if necessary, the speaker may ask in his introductory comments, "Can you hear me? May I ask the gentleman in the middle of the back row—you, sir—to give me a signal if I should become too quiet or too loud?" Always the gentleman in the middle of the back row is happy to oblige. The speaker then must see to it that he watches for any signal, and responds promptly and fully.

Microphones and public address systems add whole new collections of problems. The speaker can do several things: (a) observe what works for the speakers who precede him; (b) recruit the monitor described above; (c) throw himself on the mercy of his audience: "Can you hear me? What do I need to do?"

There is another dimension of vocal expressiveness: *articulation.* This consists of accurate pronunciation, the right syllables in the right order, and crisp, distinct enunciation.

There are occasional problems of mispronunciation, e.g., "nuculer" for "nuclear", "appraised" for "apprised", "revelant" for "relevant," but for the most part speakers do well. A wise practice is to always look up the pronunciation if you have the slightest doubt.

In the second aspect, *enunciation,* however, most of us are mediocre at best. We are typically lip-lazy, slovenly. Actually, on the public platform, the speaker should exaggerate the crispness and precision of his enunciation.

Physical expressiveness

Meaningful bodily action includes movement about the platform, stance, gesture, facial expression and eye contact. The best general rule for physical expressiveness is: "Plentiful, meaningful and spontaneous". When we think of such pantomimists as Charles Chaplin and Harpo Marx, we realize how eloquently communicative bodily action can be.

The speaker should move about the platform as freely as the setting allows. Perhaps the worst circumstance, and a frequent one, is when the speaker is locked between program chairman and president, with wall behind him and lectern in front of him. Yet, with no more than a two-foot square in which to move, the speaker can shift from side to side, step backward or forward, or lean purposefully across the lectern, thus enhancing his effectiveness.

Such bodily expression is helpful to the audience because it freshens the visual stimulus—making attention easier, interpreting meaning, and providing emphasis. Such movement helps the speaker because it relaxes his tensions, helps him use up additional energies his emotional stimulation has created, and gives him greater total expressivity. Movement about the platform can be overdone, of course, as is often illustrated in the caricature of the professor who paces back and forth in front of his class, sentry-duty style.

The second opportunity for the speaker to use effective physical expression is in stance or posture. The basic rule is "erect, yet comfortable." The speaker should stand tall, leaning slightly forward while balancing his weight on the balls of both feet. He should not clutch the lectern, but might rest his hands lightly on the edges of it.

A good basic position is to rest the wrists lightly on the hips. With elbows thus bent, the speaker's arms are cocked, ready immediately to respond to the impulse to gesture. If, however, his hands are clutching the lectern, the paralytic grip must be relaxed before the impulse can be followed, and then it might be too late. Also, good gestures must have some lead time.

With the speaker's arms across his mid-section, and his elbows bent, the fingers of one hand might lightly touch the fingers of the other; but one hand should not grasp the other for the reason just given. The speaker may prefer to bend only one arm which, of course, is perfectly all right. Some speakers will bend the arms so sharply that the hands are in front of the chest, at the lapels. This initially might seem awkward to the speaker, but, suprisingly natural to the audience; and in this position it is almost impossible *not* to gesture.

The emphasis on taking a stance that will facilitate gestures indicates the importance of the latter. Gesticulation at its best is a visual aid. If the speaker draws an imaginary diagonal, this is an effective visual aid to depict an upward or downward trend. Tallying on the fingers visualizes enumeration. Relationships of size, placement, growth, diminution, speed, direction, and many other concepts can be effectively visualized by good gesture.

Gesture, like proper stance and movement about the platform, helps give rise to total expressiveness.

The fourth factor of physical expressiveness is facial expression. The general rule is "alert and friendly." Do you recall the wire service photo of General Eisenhower when he was given the news of General Douglas MacAuthur's recall from his Pacific Command? Here was a mobile facial expression that communicated worlds.

The speaker's facial expression should be essentially friendly. Most speakers are friendly; but many, perhaps most, appear to be unfriendly. This is because they are apprehensive, and fear shows up less as fear than it does as unfriendliness: brow furrowed, eyes narrowed, corners of eyes and mouth pulled down, etc.

The problem is simply how to bring to the surface the friendliness the speaker feels. The answer: the speaker must smile. Yes, the best way for the speaker to bring his friendliness to the surface is to smile. Although a smile is not infallible, it is a universal sign. A way of brightening the facial expression without smiling is simply to raise the eyebrows. The speaker may also employ: (a) humor; (b) melodic vocal inflections; and (c) jovial cordiality with the other people who are present. A lively, pleasant interaction suggests a friendly speaker.

The fifth and final factor of physical expressiveness is eye contact. A general rule is that the speaker should maintain eye contact 90% of the time—eye contact with all his listeners in an irregular, roving sweep.

Genuine eye contact is not achieved by a blank-eyed stare in the direction of the audience. The speaker must not only see his listeners, but he must note their reactions and adapt to them. If they appear bored, he must spark his presentation; if confused, he must amplify; if disbelieving, he must offer support; if unmoved, he must offer inspiration.

Closely related to physical expressiveness is the problem of what to do with the manuscript itself. There are two alternatives: (a) when possible, keep the manuscript inconspicuous. It is merely a memory device and the speaker who is fully prepared will refer to it only in infrequent and swift glances. "True art conceals art"; communicative reading conceals reading. (b) When references to the manuscript are unavoidably conspicuous, e.g., the speaker has not had opportunity to steep himself in it, or there is no lectern available, the speaker should display the manuscript boldly. It is reported that Winston Churchill said the speaker should virtually flaunt the manuscript, figuratively beating the listeners about the head with it.

Either way, the important consideration is that the speaker emerge predominant and the manuscript only as an aid.

A verse by an unknown poet sheds light on the problem:

Of all the kinds of lecturers
The one I most detest
Is he who finishes a page
And places it behind the rest.

I much prefer the kind who takes
The pages as he finishes
And puts them in a mounting pile
As pile the first diminishes.

But I like best the lecturer
Who gets them in confusion
And halfway through he lets escape
The phrase, "And, in conclusion . . ."

Climactic finale

Hard-hitting windup—The speaker should conclude as smoothly as he began. He should know the last several paragraphs of his talk "cold." This should free him completely from both the manuscript and the lectern.

His "haymaker" (peroration) should be a truly climactic finale, positive, resounding, inspiring. He may wish to come around in front of the lectern, closer to his listeners.

Terminal pause (holding eye contact)—When his final word is said, he should begin a terminal pause, holding eye contact with his audience. He should challenge himself to make it a matter of artistry that he can end thus climactically without limping off on the weak crutch of "thank you." Employing this phrase to end a talk is such a cliche that the artistic speaker should vow never to use it.

If the speaker can't think of anything else to do during the terminal pause, while he is waiting for his audience to applaud, he can count silently to himself: "1000-1, 1000-2, 1000-3." If he gets to "1000-8" and the audience still has not begun to applaud, he may contemplate abandoning his speaking career. The terminal pause obviously is the counterpart of the focal pause. At the outset we emphasize the smile; at the windup we emphasize eye contact.

Summary

Assuming that our speaker has ideas and information well worth his audience's hearing, he will set about providing proper expression. He begins by realistically facing up to the challenge—an abundance of internal and competitive factors that may add up to a Roman Circus. He "goes for broke" in his aspirations: to be heard, understood, believed, and inspiring to the point that the listeners take the desired action. He develops expertise in all four modes of delivery and selects that which is most appropriate for the occasion: memorization, reading, extempore or impromptu. Ultimately, he delivers the speech impressively with "situation-well-in-hand authority;" "think the thought involvement;" "drive for the desired response communicativeness;" melodic, crisp vocal expressiveness; and plentiful, spontaneous, meaningful physical expressiveness.

Cicero described the ideal orator as "the good man speaking well." ■

Audio-Visual Aids Bolster Your Talk

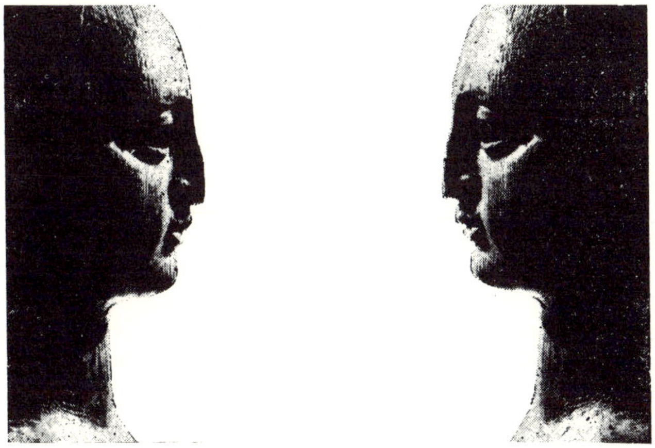

Nothing but talk can be dull. Visual aids can make points clearer and add interest, too.

ROBERT HAAKENSON, *Smith Kline & French Laboratories*

A speaker was talking about molecular manipulation in chemical research. He explained that tiny modifications can mean extreme changes in the chemical's effect. The point was reasonably clear and the generalization accepted by the audience.

On another occasion, a speaker advanced the same point. He, however, displayed a molecular model of the benzene ring, the so-called "wedding ring of organic chemistry." He described the properties of the industrial solvent benzene, represented by the molecular structure. Then he added a group of atoms to the model, explaining he now had benzoic acid, a food preservative. Finally, he added a third group of atoms, explaining that the new compound has pain-killing qualities: *aspirin.*

The description by the second speaker obviously was clear and vastly more memorable because the visual aid was employed. Additionally, this particular story illustrates many principles of effective audio-visual (a-v) aids usage, as we shall soon discuss.

It would be wrong to imply that a-v aids are good *per se.* Properly used, they strengthen the impact of the talk; improperly used, they detract.

To develop the principles of effective a-v aids usage, let us discuss:

1. Rationale for their use.
2. Types available.
3. Criteria for selection.
4. Techniques.

A basic consideration in effective a-v aids usage is determining and stopping short of that point where the aids begin to dominate, reducing the speech to mere demonstration, and the speaker to the role of narrator. Proper use keeps the items as *aids,* i.e., they bolster the presentation but do not take over.

Rationale

It is reported that 85% of what we retain in our minds is gained from visual stimuli. What we gain through listening, therefore, is only a portion of the remaining 15%. This means that the oral-aural medium is a fickle one indeed. On the other hand, it is widely reported that the medium of communication that has the greatest impact is the spoken word, especially that described as "eyeball to eyeball." The reconciliation of these apparently contradictory statements may be found in R. Benschofter's article, "In-Service Training Aids" published in *Proceedings of Nebraska's Inservice Training Conference, Oct.* 21-22, 1964, Omaha, Neb., Psychiatric Institute:

Methods of Instruction	*Recall 3 Hr. later, %*	*Recall 3 days later, %*
Telling (when used alone)	70	10
Showing (when used alone)	72	20
When a blend of telling and showing is used	85	65

A-v aids enhance the communication process in various ways. First, members of the audience are served. Their attention is attracted, interest aroused, understanding benefited and motivation stirred by good aids.

Secondly, the speaker—especially the amateur—is served. Preparation of aids tends to sharpen preparation over-all. Attractive aids give the speaker confidence. Handling the aids provides opportunity for purposeful "stage business"—something to do with one's hands, good reason for making direct contact with the audience.

Thirdly, good aids serve the sponsor of the meeting by fostering more-satisfactory treatment of complex topics in less time.

Types

A-v aids are almost limitless in type. The box contains a classification that is illustrative, not exhaustive.

This list cannot be exhaustive because virtually anything to be found in our universe may become an aid. Think of the variety of items that have been "Exhibit A" or "B" or "C" in courtroom trials.

There is a broad classification that may be useful: (1) items as they exist in their natural state, e.g., an ear of corn; (2) items modified for a-v usage, e.g., an ear of corn halved and specially mounted to show (in section) ideal proportions at full maturation; and (3) items specifically created, e.g., stop-action motion-picture visualizing the maturation of an ear of corn. Wise use of a-v aids, then, requires awareness of the limitless variety of types available.

Criteria

Eyes opened to the array of possibilities, we must now have some guidelines for judicious selection. Criteria may be summarized as "RSVP"—*Relevance, Subordination, Visibility* and *Portability.*

Relevance—A speaker who was to describe the new technology in oil refining made a special effort to include visual illustration. Splendid. But he fell victim to the lure of "interesting scenes," the temptation of irrelevance. He showed slides of company officers, foreign offices, oil depicted in art classics, oil in cartoon humor—some amusing, the majority interesting, but too many *irrelevant.* His central idea and main points were obscured by the scrapbook display.

The speaker must resist this temptation to include visuals simply because they are attractive, novel or "cute gimmicks." Aids should be relevant. They should underscore the central idea and main points. They should help the speaker lead the listeners in the right direction. One good way to assure relevance is to use one good aid for each main point.

Further, among the two to five aids thus employed (one for each main point), it is a good idea to use different types. In a speech to reveal the impact of synthetic textiles in our lives, the speaker illustrated (1) *comfort* by balancing the weight of several garments against one made from natural fibers, and using an electric fan to show air penetration, (2) *convenience* by demonstrating ease of cleaning, (3) *cost* by showing large swatches of synthetic and natural fiber fabrics on which were painted expenditure-per-hour-of-service figures, (4) *variety* by a brief motion-picture clip of range of textile applications, colors and patterns, and (5) *industry growth* by a graph portraying growth in dollar volume.

Once in a great while, it is okay to toss in an a-v "gimmick" to interject variety, recapture interest or relax the audience with a little humor.

The real test of the aid is that it amplifies and supports the point under consideration. For example, a Colorado psychiatrist projected a humorous cartoon to introduce each of five major types of disorientation he was describing.

Subordination—Characteristic of the scientific meeting is the speaker stepping to the lectern, saying, "House lights off, please; first slide, please"; then droning on in the darkness through 45 slides in 45 minutes; and concluding, "Last slide, please; house lights on, please; thank you very much."

The member of the audience has only the scantiest notion of even the speaker's looks. The talk was a *spoken* aid to a visual presentation; in short: a demonstration. The speaker was no more than a voice. This is not speaking. Most of the all-important impact of the speaker's *person* is lost. A good rule to follow is: *The speech must be able to stand by itself, yet be a better speech because the aids are used.*

If a person is present as the speaker, or the maker of the oral report, his personality must merge with the thought-content in order for his theme to make its unique impact.

Visible (audible)—A-v aids should be large enough for all to see, or loud enough for all to hear. To be visible to an average-size audience of 40, an aid about the size of a full newspaper page is satisfactory if illustration and text are proportionately large and simple. For big audiences in big auditoriums, aids may have to be the size of a large motion-picture screen.

To make sure audio aids will be heard by all, the speaker should check acoustics and amplification system before he begins his talk. Since audio aids are predominantly electronic, the nontechnical person should consult with a specialist.

Visuals should be properly displayed. Ideally they will be in sight only when in use. Otherwise, they may be a distraction, competing with the speaker for audience attention. If a large visual has to be mounted in advance, it can be covered by paper or cloth except when in use. Handouts should be distributed *following* the talk, perhaps simply made available at the door.

A notable exception to the out-of-sight-except-when-in-use rule is the organizational visual: a visual presentation of the talk's structure, i.e., key words or phrases that signify the central idea and main points. This may be in view throughout the talk. Speakers of the Smith Kline & French Laboratories

I. Visual aids

A. 3-dimensional
   1. Persons
   2. Objects
   3. Models/mockups

B. 2-dimensional
   1. Photos
   2. Drawings: schematic, cartoon
   3. Maps
   4. Posters: graphs, charts, enumerations
   5. Devices: blackboards, flannelboards

C. Projected
   1. Slides/filmstrips
   2. Motion pictures
   3. Overhead projector ('Vuegraph')
   4. Opaque projector

II. Audio aids

A. Human voice: imitations, songs
B. Sound instruments: musical, electronic
C. Phonograph records
D. Tape recordings
E. Radio

III. Audio-visual aids

A. Sound motion pictures
B. Video Tape recordings
C. Television

Bureau use this in all talks. In a talk on chemotherapy in the treatment of the mentally ill, "The Silent Revolution," the card lists "Bedlam, Banishment, Breakthrough, Bridge." As the speaker develops his theme, he refers to the visual at appropriate points along the way.

Reasoning that a speaker's face will be in view for all hearers, we can generalize that visuals should be displayed at head height, and near the speaker's head. A poster-card visual, for example, fits nicely into the angle formed by head and shoulder. With an aid that is easily handled, the speaker can avoid using a long pointer, or turning his back on his audience. The speaker himself should look at the aid one-fourth to one-third of the time to focus audience attention on it; the rest of the time, he should look at the audience. Use of a miniature pointer, such as a pen or pencil, adds a touch of precision.

A major part of visibility or audibility is simplicity. Let us say we were going to make an organizational visual for our present discussion. We could present the outline verbatim:

Improved Oral Communication
Through Good A-V Usage

Reasons for using a-v aids

Types of a-v aids

Criteria for selection

Techniques of good usage

Using letters of reasonable size, our card or screen would have to be enormous.

Instead, we would abbreviate, perhaps as follows:

A-V Aids Usage

Rationale

Types

Criteria

Techniques

A-V Aids

Why

What

Which

How

It is not bad, indeed it is good, to abbreviate, to be cryptic. This enhances the spoken word that will explain and fulfill the visual impressions. The same principles of abbreviation and simplification apply to projected visuals and audio aids. Keith W. Sehnert, M.D., in his article, "Your Slides: A Visual Aid or Annoyance?" (published in the September-October, 1966, issue of *The Journal of New Drugs*) includes the following items as five of his "Ten Commandments for Speakers":

- Thou shalt not use slides with blue or black background.
- Thou must be able to read 2 x 2 in. slides without a magnifier . . .
- Thou shalt not use more than three comparative results per slide, or have multiple-line graphs.
- Thou shalt make graphs tell a story at a glance.
- Thou shalt add color or pictures to slides whenever possible.

Portable—A-v aids should be portable. The advantages of being able to carry them any place, any time, and without special transportation, are obvious. Free from bulky briefcases, portfolios, clumsy easels, or other display equipment, the speaker avoids much inconvenience. What may be lost in size and impressiveness is gained in mobility.

Dr. Hubert Alyea, professor of chemistry at Princeton University, developed a remarkable projector by silvering a 25¢ auto-headlight lamp and connecting it with a car battery. He also developed a companion Plexiglas cell with which chemical reactions can be projected.

The overhead Vuegraph projector is very versatile. Its present design is compact and lightweight—easily portable. Its light source is so intense that clear images may be attained on an ordinary wall while room illumination is normal. Facsimile transparencies may be made speedily on a photocopier machine. And the speaker, with a grease pencil, can modify the visual as he speaks—adding, deleting, underscoring. ■

Techniques for Using Audio-Visual Aids

Audio-visual aids are so important a part of your talk that it's essential to use them correctly. Here are details on working out your own visuals, providing the needed action and variety, and avoiding practices that detract rather than help.

ROBERT HAAKENSON, *Smith Kline & French Laboratories*

Audio-visual aids have become a veritable art form. Educators were using them increasingly when World War II began. Then the Army and Navy, faced with the job of teaching millions of persons new tasks in a short time, relied on every known a-v aid. The lowly comic book became an educational institution. Before invading North Africa, for instance, the military supplemented its lectures with a comic book travelog featuring Donald Duck.

Today, educational institutions, industry, advertising agencies and the military have full-blown a-v departments and conduct far-reaching research. Television stations use aids for newscasts, commercials and speeches.

Speeches are becoming shorter and visual aids more numerous. The typical address is now closer to a half-hour than the hour or more of 50 years ago. Time limits imposed by radio and television have influenced this, as has the pace of modern living. A-v aids help the speaker say more in less time.

Principles of a-v aids, and corresponding techniques, have evolved. Here are some:

- "Roll your own."
- Variety.
- Action.
- Projected visuals.
- Audio devices.

"Roll your own"

The speaker is wise to prepare his own visuals, at least to participate aggressively in their design. Here are some of the reasons: (1) he can select aids to reinforce exactly those points he chooses; (2) aids can be kept free from distracting details; (3) size and complexity of aids can be made appropriate to the occasion; (4) the speaker will be intimately familiar with the aids and able to use them better; and (5) the speaker will be strengthening his over-all mastery of the talk.

Artists' show cards or poster boards make excellent visual aids. Or, using a large art tablet on a sturdy easel, the speaker can sketch lightly with pencil before the talk, then while speaking "create" his sketch with bold strokes of crayon or felt-tip pen. Large blackboards with colored chalk offer good possibilities.

Crayons, marking pencils, india ink and other materials are now available in most colors. The basic visual may be black on white, but color should be used generously to provide interest and emphasis. Dark green on a light green background is said to be the best color combination for legibility.

Variety

Where several a-v aids are to be used, different kinds will be particularly suited to making different points; variety heightens interest.

A medical speaker described the human heart with the aid of a skeleton, a ceramic heart, a schematic drawing, and an oscilloscope portrayal of heartbeats. In a music appreciation expert's talk designed to inform the audience about musical cadence and tempo, the speaker used enlarged time-signatures from musical scores and two audio aids: a metronome and the humming of well-known tunes.

A speaker comparing the success of two competitive products used: (1) samples, (2) a tabular presentation of the makeup of the two, (3) a line graph of sales trends, (4) statistical tables comparing sales and profits, and (5) writing, on a large paper pad, possibie reasons for the differing success of the two products being compared.

Action

Changing the a-v aid while in use is desirable. "Creating" the drawing from the near-invisible sketch, described above, is an example. Or a speaker might start with an item prepared for some totally different purpose—for example, a highway map—and adapt it to his purpose, such as by designating an area scheduled for industrial development. In presenting statistics, one prepared figure, line or column can be boldly circled or otherwise marked for special attention. The same can be done with a list.

Blackboards and flannelboards almost always involve action; stripcharts do also, by the speaker's peeling the cover off specific portions at appropriate times during his talk.

Projected Visuals

Projected visuals are both the speaker's joy and despair: slides, filmstrips, transparencies, motion pictures, videotapes. In varying degree, they are expensive, weighty, bulky and "demanding," i.e., they need a darkened room, projection screen, power supply, operator, etc. But they can provide vivid illustration, distinct images even for vast audiences, rich artistry, lively animation, etc.

Thus we must weigh the advantages carefully. Here are some tips to supplement those of Dr. Sehnert, quoted in the previous article:

Projected visuals must be kept simple just as any other visuals. Because the screen is big, speakers are inclined to cram their slides with data. The result is a screen full of barely legible symbols. *Abbreviate, simplify*. Cut to the bone and leave the rest for the audience to imagine, or the speaker to verbalize.

Cluster the slides and filmstrip sequences so the house lights need be darkened only once or twice (maximum three times). Only under rarest circumstances should lights be down for the entire talk.

Special Projectors

The opaque projector is elephantine in its bulk and weight, but it is swift, powerful and economic. Any printed material may be fed into its breech; it will accommodate more than a full-page in width and the thickness of an encyclopedia. Indeed, encyclopedia illustrations are often projected. A toggle permits the operator to manipulate an arrow-shaped beam of light to single out items in the projected image.

The so-called "overhead" projector overcomes many of the objections to other devices. Some models are costlier, bigger and heavier than a slide projector, yet most are compact and portable. Transparencies can be made as quickly and easily as ordinary photocopies. The light source is so powerful that a blank wall is a satisfactory screen and the image is sharp even in normal room illumination.

Because the transparency rests on an easily accessible deck, the speaker can stand right up front, facing the audience, near the image on the screen, and add, delete and alter to his heart's content. He can provide on-the-spot artistry with his grease pencil or felt-tipped pen, supplying the "action" recommended earlier. The "buildup" technique of one transparency overlay superimposed over another, and another, and another, is very useful in portraying complicated relationships.

Integrate With the Image

The speaker should try to integrate himself with the projected image. Ideally he will stand right alongside it and physically point out items of interest. The worst is when the speaker is totally remote, perhaps the width of the stage or the length of the room away. Then he is no more than a narrator. The least he can do here is to employ a flashlight, with its beam masked into an arrow shape, so he can do some visual designating.

The principle is unity of stimulus. If voice and visual are physically separated, they tend to be divisive and competitive. If they are together, the audience can focus its attention. If the speaker must be his own slide-projector operator, he should obtain a remote control cord and switch.

These problems seldom occur when using the overhead projector. It is quite near its projected image, and the sense of unity of stimulus is achieved further by the speaker marking up the transparency as he progresses.

Audio Devices

There should be no long stretches of tape or phonograph recording. It is very difficult for the speaker to integrate himself into an audio aid unless he is supplying a share of the dialog "live," with responses on the transcription. During "pure" listening to such transcription, the audience ideally should close its eyes, because any visual stimulus is unrelated and distracting. Short audio excerpts, however, offer novelty and limitless possibilities: interviews, speeches, discussions, sounds of all types—nature, music, warfare, industry, electronics, languages. A speaker who was describing the communications program of a great U.S. corporation included in his presentation tape-recorded excerpts of interviews with employees' families, and a sample of a "canned" message-of-the-day available to executives by picking up their phones and dialing a number.

Finally, audio aids should, if possible, be pretested in the auditorium in which the speech will be delivered. Tested in an ordinary room or office, a tape recorder or phonograph might prove sufficiently loud, with lots of gain remaining on the volume control. Later, in the meeting room, however, when the volume is turned up for the larger room, distortion might be introduced because the recorder or phonograph speaker simply is not large enough for fidelity at high volume.

Under these circumstances, the speaker will have to play his tape or phonograph record using the auditorium's public address system, or connect his own machine into it.

Summary

Today's public speaker recognizes the value of audio-visual aids. Audiences tend to be less patient and more fickle than they were in generations past. A-v aids can help capture their attention, sustain their interest, clarify their understanding, bolster their conviction, and motivate their action.

Types of aids available—persons, objects, 2- and 3-dimensional representations, projection and other devices, audio-aids—are limited only by the speaker's imagination. By making certain that his aids are relevant, subordinated, visible (or audible) and portable; and by making certain that they are well-conceived in terms of the speaker's "rolling his own," injecting variety, incorporating action, and making intelligent use of projected and audio aids, the speaker can enhance his content, organization and delivery.

Confucius is reported to have said that a picture is worth 1,000 words. Who can estimate the worth of the right pictures *combined* with the right words? ■

The Q&A Period Can Make or Break Your Talk

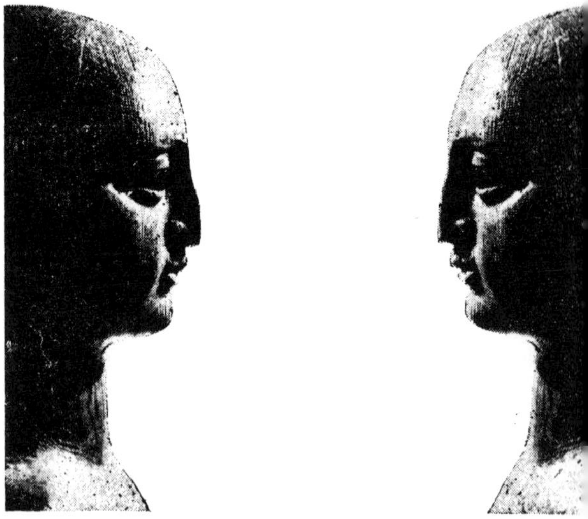

The question and answer period lets you emphasize main points and clear up problems.

ROBERT HAAKENSON, *Smith Kline & French Laboratories*

Do you recall the University of Manchester experiment in teaching logarithms to junior high school students? From test scores and thousands of classroom photos, the experimenters concluded that the students who learned the most were those who entered most freely into the classroom dialog.

R. W. Revans, who reported the research, commented:

> **Is this result not perhaps an illustration of a more general law, that institutions in which subordinates feel free to ask questions of their superiors are those who adapt best? . . . May it not be that one of the more important inalienable rights of small children is to be both seen and heard? Is it not intelligibility rather than happiness that most of us are after in this life? What, after all, is education other than learning how to ask questions?**

In platform presentations—public speaking and oral reports—our best opportunity for such dialog traditionally is the question and answer period (Q&A).

Some speakers, it is true, are able to field questions as they go along. Properly handled, this is ideal. The speaker receives continuing indicators of audience response: interest, understanding, acceptance. But he must be skilled. Otherwise such audience participation may squander his time and lead him astray.

If the speaker has this skill, or if time is plentiful, he should at least experiment with handling questions as he goes along.

Traditionally, however, the Q & A session is saved until the conclusion of the formal presentation. Whichever form it takes, let us say:

- We *want* the Q & A.
- We will *obtain* the Q & A.
- We will *exploit* the Q & A.
- We will *follow up* the Q & A.

Although the Q & A is often translated as the question-and-answer "period" or "session," it is good to think of it as Q & A *discussion.* A good Q & A promotes lively and widespread participation.

Some people foolishly think that to question a speaker elevates them in the eyes of others. Some wish to show off their knowledge. Whatever the motivation, the speaker should accept brief statements, for they provide him an opportunity to expand his own message by refuting or amplifying the rejoinder. "I will be happy to respond to your comments as well as your questions, but please keep them brief."

Therefore, in the pages which follow, when the word "question" appears, please add silently "or comment."

We Want the Q&A

More than one veteran member of the Smith Kline & French Laboratories Speakers Bureau has said, "Keep the speeches brief. Make them just long enough to stimulate a Q&A. If we do any persuading, it's apt to be in the Q&A. This is where communication really happens."

Here are some benefits of the Q&A:

- People can ask about those aspects of the topic that interest them most.
- The speaker reinforces his message through recall and elaboration of main points.
- The speaker gets to demonstrate that he knows his subject and can handle himself on his feet. (How many speakers fumble in formal speaking, but flourish when fielding questions!)
- Speakers and their sponsors glean valuable feedback on audience reaction to the speech subject.

Therefore, we want the Q&A. In a moment, we will take up the subject of getting the Q&A.

First, however, let's put the shoe on the other foot. Let's put ourselves in the audience. Here, too, we should want the Q&A. As questioners, alas, we typically range from mediocre to poor. We don't raise questions promptly. When we do, we grope and mumble. (Usually it is good practice for the questioner to stand.) We phrase questions poorly. We don't listen to the answers.

The worst fault is wasting the opportunity to raise questions. As members of the audience, we should want the Q & A for many of the same reasons as the speaker. Also, if we have grievances on the topic, here is a fine forum. We can enjoy the emotional hygiene of a little public ventilating. We can give the speaker a boost, introducing topics he skipped, or re-opening issues where his support was weak. And we can even drop some bombs for the speaker to lug back to his sponsor.

Thus, the speaker and audience should be equally eager for the Q & A.

We Will Obtain the Q&A

It is important that the Q & A be planned, and not be a hit-or-miss, afterthought venture. If there will be barely enough time for comfortable delivery of the speech itself, no Q & A should be scheduled. If five minutes or more can be made available, however (perhaps by shortening the talk), it is worth planning the Q & A. This involves (a) letting others know, and (b) getting yourself ready.

Letting Others Know

A good place to start is the moment the speaking engagement is nailed down. As the topic, date, time, place are confirmed, simply ask, "Will there be a Q&A?"

The answer is usually, "If you wish."

Then your response should be, "Yes, indeed!"

Next there are advance announcements of the speech. Ask the program chairman to include mention of the Q&A. Audiences like to know they will have the privilege of asking questions, even if they do not exercise it later.

Before the actual program begins, the speaker should arrange the details of handling the Q&A with the chairman. The chairman, when introducing the speaker, should state clearly that the Q&A will follow the talk. Completion of such an arrangement should not put the speaker at rest, however. Chairmen have human frailties and in the excitement of the introduction may fail to mention the Q&A. If the speaker notes the omission, he will make the announcement himself. For example:

> **At the outset, I thank you for providing time for a question and answer period. I admit I will be disappointed if you don't have something for me. I welcome comments as well as questions. Please feel free to jot down items as I speak.**

This makes a good icebreaker comment, of course, even if the program chairman has *not* forgotten, e.g., "I am happy you allowed time for Q&A discussion. I second Chairman Hulbert's suggestion that you have some questions and comments ready."

The speaker's noting whether the chairman has brought up the Q&A is just one of the many reasons why a speaker should be a very good listener, totally attentive to what is going on about him throughout the duration of a speaking engagement. This is not easy to do when one is preoccupied with his own speech, but it is important. First, it is simple courtesy to one's hosts. Secondly, the environment will provide valuable clues for audience analysis and adaptation in introductory comments, ad libs during the talk, and illustrations in the Q&A.

Getting Yourself Ready

Piling up the ammunition is half the battle in a well-planned Q&A. The speaker will know, from earlier experience with the speech topic, from what is prominent in the news, and by the nature of his audience, the kind of questions he is likely to encounter. With such prospective questions in mind, he should do a little reflecting, reading, talking and observing to have his answers sharp. He may wish to bring with him to the lectern some critical supportive materials. He will be wise to organize them in a ring binder or some other indexed, easily consulted form.

Most questions raised during a Q&A are surprisingly easy. Facts obvious to the speaker because of his specialized knowledge often mystify and intrigue audiences. These simple questions virtually answer themselves and require little discussion here. But, inevitably, there are other types: complex questions that challenge a speaker's intellect, hostile questions that challenge his tact, partisan questions that challenge his persuasive skill. We will have in mind these tougher ones, particularly as we move on to exploiting the Q&A.

We Will Exploit the Q&A

First, we must get the questioning started. The transition between speech and Q&A should be made smoothly and without delay. After concluding the talk, the speaker acknowledges the applause gratefully and modestly, pauses briefly, yields the lectern to the chairman to open the Q&A, or opens it himself.

A speaker should be prepared to wait several seconds for the first question. If, after a good long wait, silence still prevails, the speaker has a decision to make. First he should remember he controls the mood: the audience will be only as embarrassed as he shows himself to be.

In this situation, the speaker asks himself, "Is there really enough interest to go ahead with a Q&A?" If members of the audience are looking at their watches or slinking toward the door, there is little to be gained by further wooing. The speaker wraps things up gracefully and calls it a day.

On the other hand—and this is frequently the case —the audience is interested but inhibited. No impatience is evident, but neither is there a question. There are several alternatives. One is to wait, comfortably, for what may seem to be an interminable time. Given that time, some member of the audience will get his nerve up, or get his question phrased.

In another approach, the speaker may comment:

Almost invariably, people will come to me after a talk to ask questions. I am pleased to respond and hope that this will happen today. What often disappoints me, however, is that the questions would have interested the entire audience. So, if you are planning a question or comment, let's hear it now while all can share.

Perhaps the surest alternative, and a justifiable one, is to use a "plant"—someone primed in advance to raise the first question. The plant, preferably in the audience rather than at the head table, should hesitate momentarily to see if questioning will begin spontaneously.

If it does, he need not seek recognition unless he is genuinely interested in raising his question. The justification for this seemingly dubious practice of using a plant is that the Q & A often will flourish if someone will break the ice.

The speaker should recognize questioners with precision: "The gentleman halfway back on my left." "The lady at the far table." The sharply pointed index finger is efficient, but since artist James Montgomery Flagg created the Uncle Sam "I want you!" recruiting poster, it may be considered threatening. The questioner can be designated accurately with a gentler gesture or perhaps by nodding the head and looking directly at the person.

In conducting the Q & A, as in spoken communication generally, use names whenever possible. It demonstrates the speaker's personal interest in the members of the audience. And it offers greater precision than even the intimidating index finger.

It is the speaker's responsibility to make certain that a question has been heard and understood by all before he begins his reply. He should ask the questioner to repeat it only if he got no part of it. If he got any part, then he can say, "Let me see if I understand. Is what you're saying . . . ?"

If the speaker has any doubt about the audience hearing and understanding—and he should remember that a questioner in the audience is directing his voice *away* from all those behind him—he, the speaker, should restate the question, simplifying it if necessary. Thirdly, it gives him a little extra time in which to collect his thoughts and frame a reply. Amazingly, the very restating of the question often gives the speaker the thought he wishes to begin his answer. If the question obviously is heard and understood by all, however, the speaker should begin his answer promptly. Blatant delaying tactics will offend the audience.

Analyze the Question

In the few seconds that pass between the speaker's first hearing a question, repeating or rephrasing it, and replying, he must get in some fast thinking. What kind of question has been asked? Is it "loaded" or otherwise hostile? Does it have more than one part? Is it really a question in the first place?

Spotting the Loaded Question

The loaded question is one based on false premises or assumptions. It may be the innocent result of ignorance. Or the questioner may be baiting the speaker. In either case, tact is the order of the moment.

For example, an agricultural chemicals speaker was asked, "Since your pesticides are killing off all the birds and fish, why don't they control insects?"

The speaker replied by first pointing out, firmly but pleasantly, that he could not accept the premise that pesticides are killing off all the birds and fish. He said that except for a few highly publicized exceptional instances, birds, fish and other wildlife are far better off *because of* biochemicals. Then he went on to the remainder of the question (on the effectiveness of insecticides), dignifying it by giving his best possible answer. Note two things the speaker did. First, he killed the "load" so that the audience would not be left with a false impression. Then, he went on to speak forthrightly on a controversial subject.

Divide Complex Questions

Many questions raise two or more points. When the speaker detects this and divides such a question into its component parts, he helps both himself and the listeners.

An SK&F speaker was asked, for example, "All I want to know is: Are those delayed action capsules manufactured so I get even results, no jolts or blanks, so I really get some benefit?" The speaker responded —properly: "Are you not asking two questions—first, does the medicine cure, and secondly, is it reliably manufactured?"

Dismiss Irrevelant Questions

Occasionally a question is clearly out of place. It may have nothing to do with the subject, or be unfairly personal. Such questions should be quickly dismissed: "I believe that is another subject. Next question, please."

Or, "I'm sorry that I can't respond to that question as it is outside my field of competence."

Or, "That is a very involved, personal question. If you wish to discuss it with me after the meeting, however, I will be happy to do so."

Or, in the case of outright or implied insult, simply: "No comment. Next question, please." ∎

Oral Communication: Answering Audience Questions

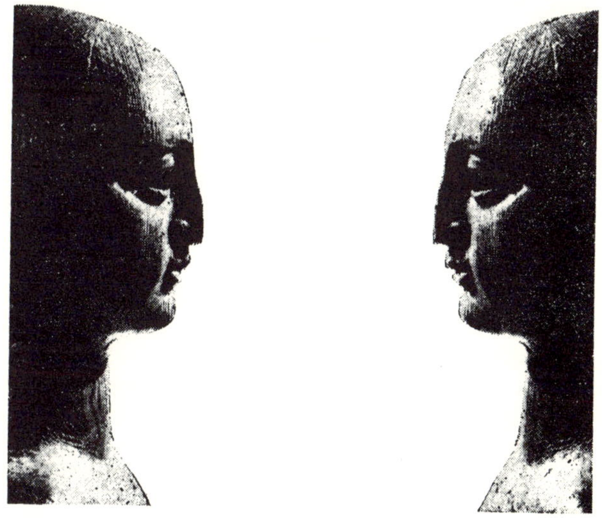

For some speakers, the question period can be a trying time. Here are some hints on how to keep things moving, how to handle hecklers and sticky questions, and how to end smartly.

ROBERT HAAKENSON, *Smith Kline & French Laboratories*

After you have given a talk, you are faced with questions from the audience. Many will be easy to answer, but others pose difficulties. The speaker should welcome—not dread—stickler, heckler questions.

A British political scientist noted that prime minister candidates' formal campaign presentations were dull as dishwater but that give-and-take with hecklers not only aroused interest but bolstered the candidates' performances. He suggested in all seriousness that a heckler should be a regular member of the campaign retinue.

A large measure of the speaker's over-all success will rest in the way he handles "hot" questions and comments. In all cases, his answers should be straightforward. There is a tendency to evade controversy. Don't. Meet it head on.

There is a danger of going on the defensive unnecessarily. The speaker or his sponsor is accused of wrongdoing, or otherwise impugned. Even if the charge is justifiable, the speaker should "accentuate the positive": "I know the instance of which you speak. I must insist that that unhappy case—and I assure you we are doing everything in our power to prevent its recurrence—was a rare exception. The over-all record is one of competence and dedication."

Answer forthrightly. Take a partisan stand. One warning: don't look for trouble. Do not conclude that because a question seems hostile and the questioner a heckler, that this is necessarily the case. Assume no worse than neutrality until you have good reason.

On the canvas of contemporary persuasion the advocate must make bold strokes with a broad brush. Forceful assertiveness is expected of him.

Factual Support

In the final analysis, that speaker will prevail in the Q & A period who has his "meat and potatoes." If he has his facts, common sense will probably keep his reasoning from going too far wrong, and the facts will impress his listeners. Thus there is no substitute for the speaker's knowing his subject.

For amplification he will want an abundance of definitions, descriptions, figures, comparisons, anecdotes, figures of speech, visual aids. For support he will want case histories and other specific instances, statistics, statements by experts and firsthand observations.

In some instances it is helpful to document the specifics, i.e., to cite their sources. This is called for when the information is hard to believe or when the source itself will add credibility: ("This is directly from the commissioner himself"). Documentation should include the name of the speaker or writer quoted, the place where the words were spoken or published, the date and page number, right down to chapter and verse.

The speaker who can toss instances and statistics right off the top of his head is to be envied. Most of us, alas, can keep only limited information on the tip of the tongue. For this reason, many experienced speakers refer to printed reference materials during the Q & A.

Ideally the speaker will review his Q & A reference material just before the meeting. He then can give the gist of it from his head, while he inconspicuously turns to the actual supporting data.

What must be avoided is delaying an answer altogether while searching for a page. The speaker *must not* say, "I have an answer for that here somewhere; just a minute," and then, after a deadly pause, proceed to draw the entire response from the printed material. If he cannot locate his place while making the initial reply, he should complete his preliminary response and say, "I have some specific data that support my answer; one moment, please."

Avoid Cliches

Traditionally the Q & A is shot through with cliches: "Did everyone hear the question?" "Will you repeat the question?" "That's a good question." "I'm glad you asked that question." "That's the $64 question." "As I said in my speech. . . ." Each of these has a place in isolated, and infrequent cases. But they become meaningless and insulting when parrotted over and over again.

In controversial questioning, one should avoid, "Does that answer your question?" Give your answer; do the best you can; move on: "Next question, please." If the answer did not in fact satisfy the questioner, leave it to him to renew his inquiry.

There is, however, one expression which is pardonable: "I don't know." Into every speaker's life comes that awful moment when, to put it bluntly, he's stumped. When the speaker doesn't know, even when he should know, he should say aloud what he is saying to himself: "I don't know." These can be the three most eloquent words in our language. The speaker who uses them need not fear losing the respect of his audience. If anything, listeners will admire him for his candor.

When the speaker doesn't know it is often a good idea for him to offer related thoughts he has on the subject. Another solution, especially if the speaker really should know, is to say: "I'm going to find out and I'd like to share the answer with you. Please jot your address on a scrap of paper."

Help From Allies

"There's a temptation for a speaker to offer his views on anything that comes up in a Q & A. Don't succumb. Stay within your area of expertise. Don't attempt to answer questions that are in the province of others. On the other hand, don't ignore the possibility of calling for help from brothers-in-arms in the audience. Phrase the invitation as tactfully as possible: For example, "I notice that Professor John Doe is with us this evening. I know him to be an expert in the field about which you are inquiring. Dr. Doe, do you wish to comment?"

Or even more cautiously, "This is a ticklish matter requiring some thought. It isn't fair, therefore, to put anyone on the spot. I notice that Professor John Doe is in the audience, however, and I know he is experienced in the area. Dr. Doe, would it be fair to ask you if you have comments?"

Make it easy for Professor Doe to wave you off if he wants to decline. He may be pleased at your attention but might also have several reasons for declining.

Share the Wealth

If you have sat in an audience during a Q & A, waving your hand in vain for recognition, you support fair play by the speaker. It is his responsibility, as moderator, to give everyone a chance. He should respect: (1) *time* and (2) *geography.*

The speaker is fair in the distribution of time when he keeps answers brief—two minutes at the absolute maximum, most less than a minute in length.

The speaker is fair in terms of geography when he acknowledges questions from all parts of the room: left, center, right, front, back, head table, balcony. When an area is ignored that portion of the audience begins to feel left out. When several hands are raised simultaneously, the skilled speaker will set up a "batting order": "You, ma'am; then you, sir; and you, sir"; and carry it out faithfully.

Allow any given questioner no more than one follow-up in succession. If the Q & A degenerates into a wrangle between the speaker and one member of the audience, the whole situation is apt to fall apart. If the questioner persists after two turns, the speaker can say, "I'm very pleased by your interest, but in all fairness we must hear from others."

As a matter of general principle, when the speaker has rough going, he properly should invoke the common good. He deals with any nasty situation not to save his skin, but because *the responsibility rests on*

him to protect the best interests of all assembled. This is justification, for instance, for squelching a heckler or a filibusterer.

Keep Things Moving

There is a rhythm to a good Q & A: questions and answers volley back and forth in a brisk, staccato fashion. The speaker does not prolong an answer, even though he might have more to say, because he is confident that if amplification is desired, another question will elicit it. Brief questions, brief to-the-point replies, with many of the audience actively participating—this is the essence of a vigorous and productive Q & A.

"Quit while you are ahead." It is much better to end with questions unanswered, even if it means stopping short of the allotted time, than to respond to an interested few, or stand waiting, when most of the listeners have had their fill.

Conclude Smartly

Anticipate the windup of the Q & A. Keep an eye on the clock, or ask the program chairman to signal you when time has almost run out. Then announce, "We have time for one more question. Whose will it be?"

Be prepared with some appropriate closing remarks. The speaker might acknowledge the audience's interest in the subject, comment on his own enjoyment at responding to their questions and comments, and offer to meet anyone interested in further discussion following adjournment. It is effective to end with a *summary statement* that wraps up the essential message. For example:

> **As you could see, your interest in this vital subject of citizen participation in community government has been exhilarating for me. Government begins at the grass roots level. The happy reward for active participation is threefold: (1) improved government, (2) pride in civic service, and (3) old-fashioned sociability and fun. I hope you will discuss these ideas with your acquaintances.**

Keep the Q & A moving briskly; stop while you are ahead; anticipate the end; and wrap it up with a summary statement.

We Will Follow Up the Q & A

An informal Q & A frequently develops spontaneously after adjournment. If time and circumstances permit, make yourself available for this type of discussion and be cordial and polite to everyone who stays to talk.

Frequently the problem arises that many wish to file by and simply express thanks for the talk. To hold them up while discussing another's point of concern is awkward. Try to sort out the "discussants" and the "leavetakers." A comment such as the following may work: "Let me quickly shake hands with those who have to run, and then we can give uninterrupted attention to the points you wish to discuss."

This postadjournment discussion offers many advantages. For one thing, it may lead to another speaking engagement. Then, too, this is a good opportunity to talk at greater length on controversial topics raised during the Q & A. Skeptical questioners often can be more easily persuaded through informal, personal give-and-take.

With the main program completed, there is a natural tendency to feel relaxed and casual. The evening is over, you've done well, and you feel good about it. This is fine, providing discretion is not tossed to the winds. Keep in mind the same guidelines that helped make the regular Q & A so successful: be fair, avoid arguments, stay within your area of competence.

If you've promised members of the audience additional information, be doubly sure you have names and addresses. If you found your expertise thin in spots, scribble yourself a reminder to shore it up. If you agreed to communicate certain items to appropriate authorities, follow through. Promise only what you can deliver, and deliver as promptly as possible.

Summary

We want the Q & A session whenever we can get it because we know what it does for audience understanding, message reinforcement, the speaker's personal impact, and feedback for the speaker's sponsor.

We get the Q & A successfully launched when we let others know it is going to occur and we get ourselves, as speakers, ready.

We exploit the Q & A when we assure that questions and comments are clearly understood by all, when answers are forthright and factually supported, and free from cliches.

Q & A "field strategy" is properly exercised when we utilize help from allies in the audience, share speaking time among all interested persons and keep the session moving to a smart conclusion.

We round out the Q & A ultimately when we follow through on promises we made—to obtain information, initiate inquiries, pass the word along, etc.; and we report feedback to our sponsors.

In the Q & A session in the Rocky Mountain states, a distinguished-looking member of the audience arose and said, "All right, sir, that's a very fine talk you have given us on this remarkable health progress. Adding 10 years to the life span may be just fine, but have you thought about me? I am a mortician." Laughter erupted.

The speaker was ready as soon the laughter subsided: "Please understand, sir; we don't prevent death; we merely postpone it."

The Q & A will not prevent failure. But it certainly can help. ■

If the last meeting you ran ended up in chaos, perhaps you can use this advice on . . .

How To Run Better Meetings

It's no secret that many people think meetings are a waste of time. They probably feel this way because many meetings *are* a waste of time.

But this need not be the case if the meeting organizer follows a few basic rules. It's a shame to allow poor technique to build up anti-meeting sentiments, because in many instances meetings are the most efficient way to deliver information, resolve conflicts, solicit ideas or solve problems.

Your first step, as organizer, is to decide what type of meeting it will be, for this will determine your approach. The majority of all meetings fall into three categories—tell, sell or solve.

The Tell Type—Purpose of this meeting is to pass information along to a group, such as explaining the company's new life insurance plan.

This type of meeting is the easiest of the three—about the only time there will be discussion is when someone asks you to clear up a point that he doesn't understand.

The Sell Type—This meeting is supposed to win the group's acceptance of the leader's proposal. Example: selling a new cost-reduction compaign.

A sell-type meeting usually divides equally into two parts: presentation by the leader and discussion by the group. Discussion is essential because people tend to believe that something is true, valuable or necessary if they can arrive at the same conclusion after general discussion, particularly with their associates.

The Solve Type—The goal here is to solicit help from a group in solving a particular problem—probably the most difficult type of meeting to run. For example, a production superintendent may be seeking ideas from his subordinates on ways to cut down on the amount of raw-material inventory.

The leader's role is to present the problem, then to keep things hot after the group gets warmed up. The group should do 80-90% of the talking in order to ensure that everyone's special knowledge is utilized.

Once you have determined the type of meeting you are going to conduct, there are three things to prepare: yourself, the room, the group.

A good way to start preparing yourself is to review notes or minutes of the last meeting (if there was one). Refresh your memory on exactly what happened and how things stood at the end of the session.

Next, know precisely what your meeting is to accomplish, then determine the best way of attaining your objective. Suppose the problem is to reduce waste coming from the A unit. Set a realistic cost-reduction goal and a reasonable time for its attainment. Then figure out the simplest, most convincing way to present the problem to your people. This means having at your fingertips background information, facts and figures, pertinent questions to stimulate discussion, provocative statements.

Prepare a complete agenda and plan to make the meeting as brisk as possible. A 30-min. meeting is ideal; an hour-long session is OK—but never let a meeting run over an hour and a half.

A final must is to check the list of persons you intend to invite to make sure no one is left out who can contribute to the solution—and to make equally sure no one is invited who has no business there!

Physical Arrangements

Advance checking of the meeting room can prevent chaos later on. On your checklist should be such things as ventilation, heat, lighting and chairs. If your meeting involves discussion, always arrange the chairs in an informal circle rather than in schoolroom rows, which suggest a lecture rather than free exchange.

Among the other items to check:

- Plenty of ash trays, paper and pencils.
- Name cards.
- Visual aids (e.g., charts, slides).
- Blackboard with an eraser and plenty of chalk.

Another important item is the time of the meeting. Remember that late-afternoon hours are tricky because a man's thoughts are likely to be devoted to worrying about missing his car pool. Meetings scheduled too close to noon put your words of wisdom in competition with lunch.

A successful meeting leader gives his people plenty of advance notice—not only because they may have other commitments but because he wants them to have enough time to do some thinking about the subject of the meeting.

Be sure to emphasize *promptness.* One company recently found that about three out of ten meeting goers were from two to thirty minutes late. As a result, leaders must repeat what the slowpokes missed, hold up the starting time of the meeting, or be stymied because three out of ten are only partially informed.

Whenever possible, the best policy is to ignore latecomers. Make no attempt to bring them up to date. This is one of the best ways to cure chronic tardiness.

Make a Fast Start

Whether you call it a "warm-up" or "motivating the participants," you've got to capture the group's interest in the first few minutes or they start to doodle. This process is especially important for the solve-type meeting where you want the group to come forth with constructive suggestions.

The burden rests with the leader; he should play his part to the hilt just as an actor does. Forget all your personal problems and give the group the kind of performance that will send them away reacting the way you want them to.

A relaxed, informal style is the most productive. Greet people by first name as they come in, make small talk as they settle down. As soon as everyone is ready to begin, be sure to welcome the group as a whole; this clearly establishes your role as host.

Plunge right into the purpose and objectives of the meeting. State goals concisely; preferably emphasize them by writing on a blackboard. Express objectives in terms of the group's self-interest. If each man realizes that a satisfactory solution is important to him, he will be more apt to take an interest

ANY OF THESE AT YOUR LAST MEETING?

Proper care and feeding of these creatures can prevent a meeting from degenerating into a zoo-house nightmare.

Loud-Mouth

Don't squelch him right off the bat. His volubility may come in handy later —and he really may have something to contribute. If he just wants to hog the spotlight, however, try tactfully ignoring him. If, as a last resort, you have to squelch him, do it firmly and make it stick.

Tough Guy

He loves to argue and has a talent for making other people furious. Don't lose your temper; try to find something in his argument with which you agree, then move on. If his statements are clearly inaccurate, toss his remarks to the group and let them handle the situation. If he gets completely out of line, talk to him privately. If you can't win his cooperation, exclude him from future meetings.

Eager Beaver

Before you can finish a question, he is springing forth with an answer—without taking enough time to think it through. Because he is sincere, your only resort is to sidestep his eager proposals, thank him, then ask for answers from the others.

Fuzzy Thinker

His conversation follows so many paths that he gets lost in irrelevant issues. When he stops to catch his breath, restate the subject and try to get him back on the track.

Antagonists

Clashing personalities can destroy a meeting by splitting it into factions. Try to separate duelists by playing up points of agreement, playing down disagreements. Sometimes throwing the question to a third party can produce a compromise that will neutralize bad blood. If the clash gets personal, tell both to leave personalities out.

Buzzer

He is forever whispering in his neighbor's ear. There's no point in giving the offender dirty looks, or stopping until the dead silence reprimands him. Best way to end the side show is to call one of the talkers by name and ask him a direct question.

Silent Type

If he's merely bored or indifferent, fire a direct question at him. If he's shy, direct a question to the man next to him, then ask the timid soul to tell his neighbor what he thinks of it. When asking his opinion, phrase the question as though it were a two-man conversation rather than a case of airing views to the whole group.

in what's said and to participate actively in the discussion.

Remember also that conducting a good meeting is a little like "show biz." Use dramatic effects—like a good film or recording, colorful diagrams, or some other audio-visual aid. Try a shock treatment—like greeting your listeners with "adjustment costs on Product X are 300% higher than two years ago. We have exactly 30 days to cut these costs down to par. I need *your* ideas on this subject."

Especially for solve-type meetings, such a beginning makes three facts crystal-clear:

- The meeting is a *group effort.*
- Everyone is expected to take part because he has a personal stake in the problem.
- The leader's role is simply to explain the objective and keep the discussion moving.

Pitching Questions

Once you have outlined the objective, throw the ball to the group. If someone doesn't pick it up, call on an articulate person and ask: "What do you think, Joe?"

If Joe lets you down, try another member of the group. If after a couple of tries the question approach doesn't work, suggest a solution that you know in advance no one will agree with. Someone is bound to challenge it—and the discussion is on. Remember that the purpose of the meeting is not to air *your* opinions; if someone asks you a question, toss it right back to the group.

Use "what, why and how" questions. If someone says, "I agree with Charlie," ask him why. Ask debatable questions, avoiding those that can be answered with just "yes" or "no." Ask three or four people the same question. To assure full answers, clamp down on interruptions, like saying, "Just a minute, Ray! Don has the floor." Don't hesitate to rephrase anybody's answer in the interest of brevity and clarity. But be sure to keep the same thought.

While many of your questions will come spontaneously from the discussion, don't push your luck. Think up some good questions *before* the meeting and fire them into the discussion when things slow down.

Disagreements are healthy; if there isn't a single dissenting voice in your meeting, the chances are that some of the people are too disinterested to care or else the group is loaded with yes-men. Don't attempt to gloss over disagreements but rather try to define exactly what the objection is and, if necessary, explain why the majority decision must prevail. In this way, the dissenters will still be encouraged to contribute on other topics since they feel their objections have been neither ignored nor steam-rollered.

When one topic is exhausted, make a 30-second summary and move on. Summaries act as a score-card on progress and let everyone know exactly where you are in the meeting. By moving through your agenda point by point, you break the subject up into blocks of information that can be easily digested by everyone.

Critical Time: the Wrap-up

The most important part of the meeting is the end. After all of the decisions, solutions, and conclusions have been discussed, it is still up to the leader to solidify the results.

Whatever the decisions were, be sure to repeat them and hammer them home to the group. Otherwise, you'll be saying, "Thank you for coming, gentlemen," to a lot of men who are uncertain as to what was decided and who accepted what responsibility.

Don't hesitate to spotlight some of the rugged aspects of the meeting: tough assignments, unpopular decisions, possible conflicts. The leader who shies away from these hard facts reveals that the meeting probably was a failure and the follow-up will be weak.

Group decisions are like passing the plate in church: some put in more than others. Individuals who have contributed the most to your meeting deserve credit—out loud. Joe will be even more helpful at your next meeting if he hears you say: "Joe Jones kicked in with a fine suggestion for handling the XYZ problem."

On the other hand, if no major decision was reached, don't let the group leave with a sense of failure or irritation. This takes a bit of doing and you sometimes must play it by ear. Some obvious devices are: emphasize what *was* achieved; express appreciation for valuable suggestions; announce need for another meeting to dig deeper into the problem; nail down definite follow-up assignments—"Joe, you're going to dig up such-and-such information; Bill, you're to have figures on XYZ by next Monday."

If you fail to follow through on your promises to the group they will be less enthusiastic about contributing the next time a problem crops up.

If you have promised your people a brief summary of the meeting in writing, get it into their hands at once. This is doubly important if you depend on certain individuals to act on your report.

If you have saddled some of your group with operating details, do some personal follow-up. Make certain that they deliver what they were supposed to.

If a report is to be made to top management, get it written as soon as possible. Outline the objectives of the meeting, the decisions reached, and the action required as a result of those decisions. Here again, give credit where credit is due.

Following through on meeting decisions will show everyone that meetings really *can* accomplish something and will make your people more eager to participate next time. ■

About This Article

Material in this article is taken from a booklet "Let's Hold Better Meetings" published by Goodyear Tire & Rubber Co. as a guide for its employees.

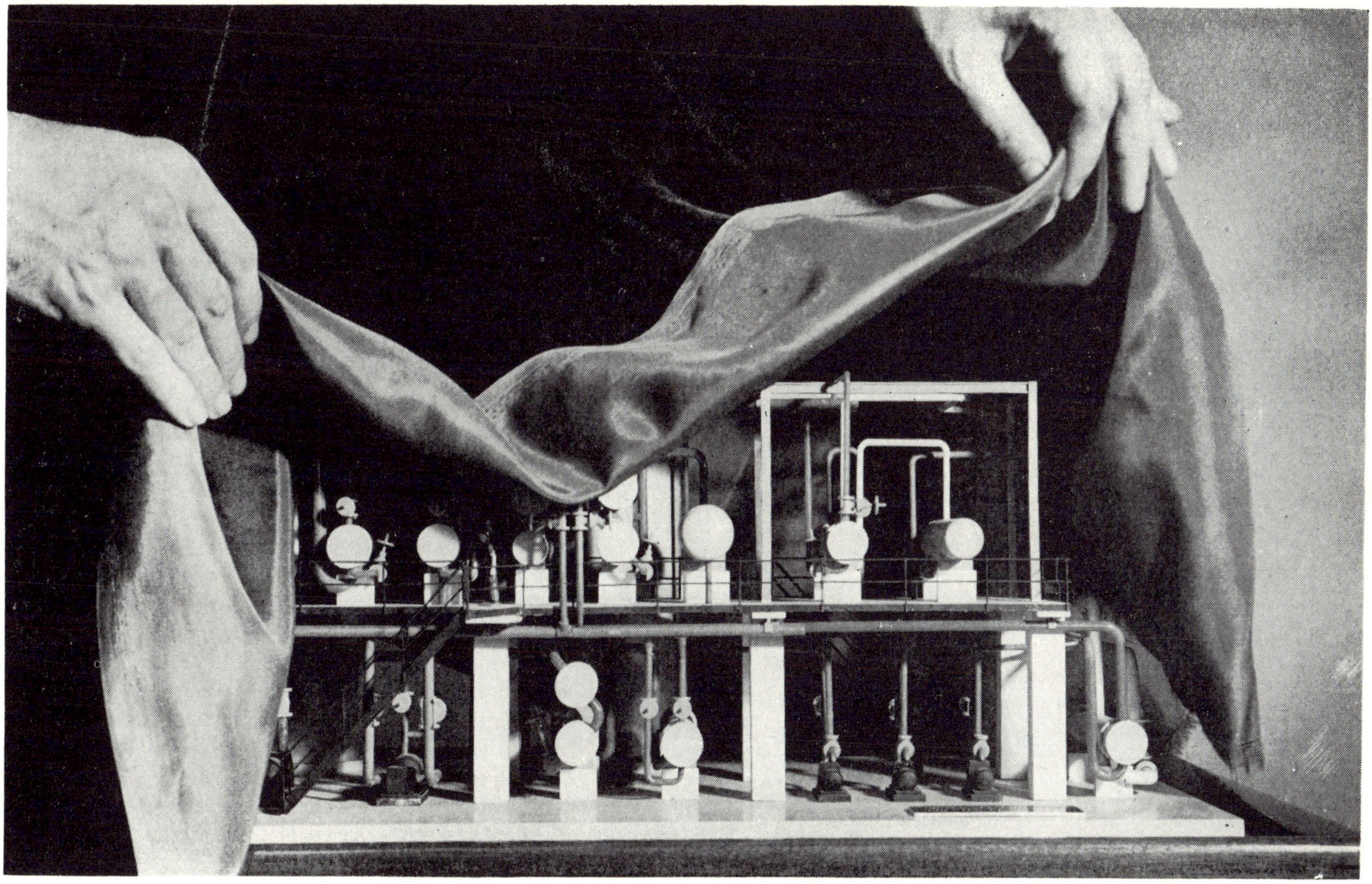

Uncovering Your Competitor's Costs

The gathering of competitive cost data suggests cloak-and-dagger activities to some. But there are many legitimate ways to gather this type of vital information.

About this Article
This article is based on a panel discussion presented at the sixth annual convention of the American Assn. of Cost Engineers. Among participants on the panel were the four co-authors of this article: Ralph T. Cline, James C. Cowan, Gustav Enyedy, Jr., and Thomas E. Hedge.
The authors point out that this material is based on their personal observations and is not intended as their own recommendations nor as a description of the methods of their current employers, except where specifically stated.

One of the most important aspects of doing business is knowing what your competitor is doing—and how much it costs him to do it. The gathering of this information is one of the most challenging jobs in the whole field of cost estimation.

This article will look at some of the ways to compile competitive cost information and will try to draw a line between the ethical and shady methods—recognizing that some unethical methods may be completely legal, and that some seemingly ethical methods may be illegal. For instance, chatting openly with a competitor at a convention may seem completely ethical, but under certain conditions the antitrust investigators would consider it quite illegal. On the other hand, hiring a key employee from a competitor may be completely legal—but could be quite questionable from an ethical point of view, depending on the circumstances.

The first section of the article will deal with the value of published materials. The following sections will treat personal contacts—both formal and informal—and the deductive method.

Tapping Public Sources

JAMES C. COWAN, *Pittsburgh Plate Glass Co.*

Much information can be obtained about competitors by reading the available literature. Certainly the first place to look is competitive price lists. How does one obtain these lists? I have heard that some companies make a practice of sending their latest price lists to competition, particularly when competitors in some lines are customers in others. If you can't get price lists via direct request, almost every salesman has a friendly "in" with a customer who will obtain this information for you.

Specification lists, prospectuses and brochures certainly should closely follow the price list as valuable sources of information. To the informed engineer, these publications can be very valuable. The means of obtaining such materials are about the same as for price lists.

Annual reports should also be high on the list. If a company is very large, these reports may be broken down by division or product. By careful study, you may be able to find out whether the item in which you are interested is a profitable one for the competitor.

If you have looked over the list of government publications recently, you know that for a quarter or less you can find a government-printed report on almost anything, probably including the item about which you are interested.

Under the same heading, we could place patents. You may not wish to pay for use of the patent, but at least this will tell you how someone else is doing it.

Following government reports in value are public court proceedings. Some will be reported in the newspapers, but don't overlook the court records themselves, especially in antitrust and price-fixing suits.

Utility companies often are a good source of information. What kind of power is being used and how much? In almost the same class are the transportation companies. What are they bringing into a competitor's plant and in what quantities? Also, freight rates may throw some light on the subject.

Maps, photos and pictures should not be overlooked. They may be obtained as pictures in magazines, from government agencies or, perhaps, through your own camera as you drive by. In one instance, a company discovered the exact process a competitor was using by means of a photograph that appeared in the competitor's own trade literature. Company A already knew the type of process and could deduce flowsheets and costs, but could not pick the precise route being used by Company B. But the published photograph ended speculation on this point.

Wage rates certainly form a large part of our own company's costs as well as the competitor's, and what better source for this information than the trade unions?

Establishing exact costs for a given locality can be helpful, though often this is like gilding the lily since there are so many other unknowns in the equation. However, for an ammonia plant in area X, you might find the local wage rates, pollution requirements (since such investments are considerable for this type of plant), cost information from local raw materials suppliers, and miscellaneous local rates from the Chamber of Commerce.

A widely used—but probably unethical—method is talking to a local subcontractor to see if he is willing to say a little more than he should about the last job he did, in hopes of getting more work.

Other things are a matter of public record, such as tax assessments in some cases. These things would give an idea of the investment involved, though you would have to be very skillful in interpreting the data. Some people have been known to fairly accurately back-calculate the investment in plant from the tax assessment. This requires a good knowledge of the relationship between assessed value and installed cost.

Sometimes the one bit of information needed to complete a cost picture is right under our noses and we overlook it. For instance, quotation files on business not obtained can be checked out and may possibly yield the competitor's quotes.

The moral question of information gathering is the most difficult one to assess. What is ethical to one person in one situation may not be to another person in different situation. Remember, also, that legal and moral are not the same.

And finally, be particularly careful about any direct contact with a competitor. If you discuss prices, costs, sales or production rates, you may be accused of price fixing. It may appear that you are, even though this is not the case. The safest course is never to raise these questions at all.

The Value of Formal Contacts

THOMAS E. HEDGE, *Diamond Alkali Co.*

In obtaining information, the difference between formal and informal contacts can be thought of as trying to get answers from a stranger rather than from a friend.

One of the problems of formal contacts as a source of information is that you must be more particular about legal and ethical implications than you would with informal or friendly contacts. The formal contact does not always have you or your company's best interests at heart.

You are risking more with a formal contact since you may stand to tarnish what is known as your company's "image" more than in an informal conversation. Further, you may give away more information than you gain. A formal contact might see through your subtlety and, if he is clever and so-inclined, can get you to reveal more than you ought and then not give you the desired information. For all these reasons, the formal contact requires the utmost in finesse.

Probably among the most productive of these types of information sources are the salesmen and manufacturer's representatives who in their enthusiasm to sell you something may reveal more than they intend about your competitor's operations. Equipment suppliers might occasionally drop a valuable piece of information in the same way. Your customers and suppliers might also have dealings with your competitors and your many routine contacts with them could produce some useful information.

Also high on the list of potential information sources would be any group that deals in selling services for a fee. Included here might be consultants, research organizations, or engineering firms. It is unlikely that any of these in a formal contact would intentionally reveal any information about a competitor's process or costs—their continued reputation depends on protecting former clients' confidence. However, these groups can offer specialized know-how and broad industrial contacts that can be useful in making deductions about a competitor's process.

For example, assume your competitor has recently developed a new process for a product that you produce also. From meager information in the patent and trade literature, you know the process is based on a fluid-bed catalytic reaction. Within your own company, you might have very little experience in the design, investment and operating costs of such a unit. Here, a consultant with experience in fluid beds could help you translate and interpret patents and other information into probable flowsheets. Cost figures based on past experience might also be available. This information, along with an estimate of the raw-material cost, could be combined to form an acceptable appraisal of the process economics.

The market research consultant can also be valuable as a result of his broad contacts in industry, which often are not available to the personnel of your company. Consultants seem to have a pretty good grapevine system, and as a result they hear who's building what, and who's thinking of building what. Much of this information is swapped at trade or society meetings and it is well to cover these functions with a discreet and attentive representative.

If the process that you wish to evaluate is one that is available for licensing, your task is going to be much easier. Engineering firms that license these processes often have published, or will make available, very complete data on construction and operating costs. You have only to adjust these costs to the particular situation, using wage, utility, and raw-material costs obtained from other sources, to arrive at a fairly accurate estimate of what it costs your competitor to operate the process at a certain location.

The overseas formal contact can be a fruitful source of information. There is an axiom that industry people are more close-mouthed with a potential domestic customer than with a European counterpart. So, if you need data on a domestic process, make use of your overseas contacts.

BEWARE OF CONVERSATIONS WITH COMPETITORS!

Current developments in the antitrust field indicate that the legal hazards in the practice of engineering are greater than have been previously realized.

Recent court decisions suggest that you must never talk with a competitor of your client or your employer about prices to be charged, quantities of goods or services to be produced or sold, territory in which sales are to be made, or the boycott of a particular firm.

You must also not talk with the customer of your client or employer about the prices to be charged by the purchasers on the resale of articles sold to him by your firm. By price, I mean everything that affects price. This involves freight, credit terms, quality differentials, discounts and the rest.

If you ignore these danger signs, not only loss of job but also a possible jail sentence—with all its dire consequences—could be involved. You need only read about what happened to certain men at General Electric to see what can come of talking to competitors about prices and related topics.

You don't need to ask a lawyer's advice on what you can or can't do in this area. It's clear as crystal that you cannot talk about these subjects with your competitors.

From a speech by John M. Crimmins, vice president, general counsel and secretary of Koppers Co., delivered at the Los Angeles meeting of AIChE in February.

With so much of this competitive intelligence directed at product and process innovations, the research man becomes a particularly vulnerable target. If he's got something good he sometimes is so enthused that he's got to tell someone about it. If you are involved with research and development work, you've probably been approached at technical meetings by sincere sounding fellows who ask plenty of tactful questions.

In my work, I am frequently called upon to estimate the economics of competitors' new or existing processes. In most cases, the information available in the literature combined with engineering deduction is sufficient to develop a process flowsheet and proceed with the necessary cost estimates—once an estimate of the market size has been provided. However, in some cases the additional information that can be derived through formal contacts proves indispensable.

In one instance, a new product came upon the market that was in direct competition with one of our own newer products. I was given the assignment of comparing the economics of this competing product with our material. The only information we had at the time was the structural formula of the compound.

Some of the chemists in our organic synthesis group were asked to suggest the reaction steps by which this compound might be produced. Six or seven possible routes were proposed that offered a broad choice of starting materials. One of the routes appeared to be the most direct, involving the fewest number of synthesis steps. However, it required an expensive starting material and a somewhat exotic catalyst.

So we asked our purchasing people to check into the cost and availability of this particular catalyst. A potential supplier inadvertently let slip that another of his customers was using large quantities of this material at a particular location. The customer in question turned out to be our competitor.

Then, on a routine call on an entirely different matter, one of our product development men picked up the information that this competitor was making his new product at the very same location. Putting these facts together helped confirm which of the synthesis routes was being used.

Because the basic chemistry of the reaction steps was well-known, we were able to make an estimate of the product cost. We concluded that this was a high-cost product that would not survive in a competitive market. This could well have been the reason that the competitor withdrew the product from the market a year or so later.

This incident points up a truism of intelligence gathering: very seldom will one source give you all the information you need. It is a matter of combining isolated facts to get the necessary foundation on which to base a judgment. Some companies even make a practice of filing and cross-referencing all such data on competitors' processes or costs as they come to hand.

Make Use of Informal Contacts

RALPH T. CLINE, *Sohio Chemical Co.*

The first and most obvious method of informal contacts would be the use of personal relationships, such as casual conversation over cocktails with friends and acquaintances—particularly those employed in competitive or related industries. Clearly, you must exercise great discretion at this point lest your friendship come to an abrupt end.

I'm sure most of you have made use of trade shows and conventions to initiate contacts with individuals from other companies. I believe that most people attend for this very purpose, and not to hear formal papers. Especially valuable can be contacts with equipment suppliers, manufacturers, and engineering and construction firms. After all, these people may have worked with your competitors and, if they aren't bound by secrecy agreements, may be at liberty to disclose useful process and equipment information.

A firm with a strong market research staff can use its many and varied industry contacts to great advantage. Frequently an important bit of information is gained by market research field-contacts as an informal supplement to the main purpose of the call.

Typical of the groups contacted by market research personnel would be: purchasing agents, sales staffs, your company's customers, your competitor's customers (who might incidentally be your own customers some day), development people and, of course, the market research personnel of other companies.

What sort of information can one hope to obtain from these sources? Purchasing agents, for instance, generally have a good grasp of current and near-term commodity pricing developments, particularly in relation to spot price movements. Often, spot price weakening may come to the attention of the purchasing agent first, especially if his company is a volume buyer. For instance, benzene has undergone a rather severe series of price cuts caused largely by the actions of coal tar producers. In many cases, the initiation of an isolated spot price reduction by a coal tar producer presaged a general round of benzene price cuts by the entire industry.

Early storm warnings of this type must be analyzed carefully, of course, as they could impinge upon your company's market-entry decision. As with so many of the techniques of intelligence gathering, how good such interpretations prove to be depends largely on how familiar you are with the industry and process under consideration. If you are completely unfamiliar with both, these methods can lead to misleading appraisals and poor decisions. On the other hand, there probably are many companies that could tabulate rather well-known cost figures and estimate competitor's costs quite accurately when considering entry into such a popular commodity as ammonia. In such cases, the prime question might well be the competitor's marketing position.

Market research contacts with sales department personnel frequently reveal them to be more willing to disclose information regarding a competitor's marketing activity than their own. Perhaps one might obtain information regarding a competitor's product movement in relation to a marketing area, volume and customers. Information of this type might be used in determining the logistics of marketing your product and to further appraise the sales and freight expenses of your competitor.

Your own company's customers and your competitor's customers may provide a valuable extension of marketing information developed from sales contacts. Your competitor's customer (or soon-to-be competitor's customer) under certain circumstances may divulge information about his supplier, particularly if he is interested in market entry by a new supplier in the hope of creating price softening.

Contacts with competitors' customers can also detect

the existence of reciprocal trade relations. It may be extremely difficult for you to sell to certain customers because of the existence of tie-in sales between certain companies. Also, frequent field contacts with the trade can serve as an indicator of new or improved products being offered for field trial, which could create an upsetting influence on your product entry.

Assuming that the study of your competition requires a reasonable knowledge of process economics, contacts with development and licensing personnel of a company with a process for sale may result in an offer of information not only on their own process, but on competitive processes for the same product. This, however, frequently requires signing a secrecy and know-how agreement and paying for the information.

Lastly, don't overlook information exchanges between corporate market research groups as a means of cross-checking much of the information sources already mentioned. If you elect to employ informal methods of this type to elicit information, be prepared also to discreetly trade information or else you may find that formerly cooperative witnesses will soon invoke the Fifth Amendment.

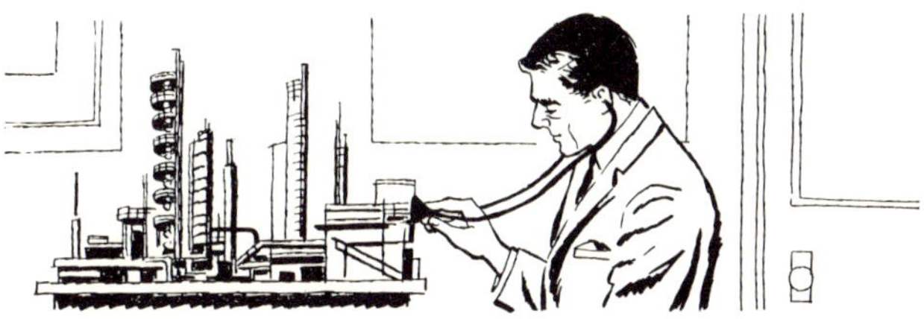

Applying the Deductive Method

GUSTAV ENYEDY, JR., *Diamond Alkali Co.*

To illustrate what I mean by the deductive method of industrial intelligence gathering, let me draw an analogy with the way a doctor gathers information from his patient. Through skillful questions, the doctor assembles all the relevant data concerning the patient's ills. Putting these bits of information together and drawing conclusions from them often allows the doctor to arrive at an accurate diagnosis without having to resort to physical entry into the patient.

Similarly, despite all of the methods for gathering industrial information, a company can still find itself without the data about its competitors that it seeks. This may be because the information is simply impossible to get honestly. In many other cases, the seeking company does not have the time, money or manpower to spend on a comprehensive cost-gathering program.

Thus, the deductive method. With only scraps of information as a start, it is possible to piece together process flowsheets and material and energy balances. The calculated material flow rates can then be used to estimate manufacturing costs and to size equipment, which will give plant investment costs.

This may appear obvious, especially to those in the chemical process industries who do it all the time. Whether we want to establish a selling price of a new material and its relationship to a competitive product, or whether we hope to grab a larger share of the market by cutting our price, we must know what is the bottom or break-even price of the competitive material. This we can estimate, and must do so accurately or the results can be disastrous.

Mathematics is the essential tool for such work. Even strictly empirical correlations can be useful. As an example of this approach, I would like to mention the Learning Curve that can be used to predict the consumption and price of a given commodity. The cumulative average price of the product plotted against the cumulative production on log-log paper is generally a straight line. This correlation results from the well-known observation that as a new material becomes accepted in the market, its selling price will tend to decrease with the increase in production. A disadvantage of this correlation (which is strictly empirical) is that some production figures must be known or estimated.

Perhaps the area where the largest chance of error exists in these techniques is in evaluating a competitor's overhead. Different firms use different yardsticks for estimating such costs. Some take a percentage of labor costs or payroll; some a percentage of capital investment; some a combination of these. A bit easier is an estimate of the manpower used in a particular type of plant or process and the calculation of labor costs for the type of labor needed to run it. You can estimate these labor requirements quite closely from the plant manager down to the plant guards, and this is likely to give more accurate results than taking a percentage of some other factor.

Another technique that is useful on occasion is that of similarity. Sometimes a competitor's cost can be figured with amazing accuracy by studying a similar, and more familiar, process. Let me give two examples:

1. Several companies manufacture caustic potash (KOH) and you are interested in becoming a producer. Although you do not make KOH at present, you do make caustic soda (NaOH), a chemically similar product. The evaluation of your competitor's capital investment and manufacturing costs in this case would be extremely simple. You wouldn't have to spend a cent on any outside source of intelligence. Since the processing steps and equipment required are identical, you could transfer your own cost information directly. The only difference you would have to allow for would be raw-material costs. Other variations would have only a minor effect on the accuracy of the resulting cost data.

2. You and your competitor both make the same product, but he has switched to a new process, substantially different from yours. You know this because he has advertised it with a two-page spread in the business press. You are still in a pretty good position to make a realistic estimate of his costs since you know (or should know) the peculiarities of the steps in the manufacture of the product in your own process. Much of this information can be used directly to work

out the complete details of the competitive process.

I was associated with a project that used this technique to its fullest potential. Our firm makes soda ash by the ammonia-soda process. A competitor developed a method for making soda ash from natural mineral deposits, and we were interested in estimating his costs and appraising what impact his production might have on the market. By using some of the methods described earlier, we had some idea of the processing steps. Details, however, had to come from application of the principles of physical chemistry, as this process turned out to be a complex crystallization operation.

Solubility data for the ternary system of sodium carbonate, sodium bicarbonate and water had to be assembled. An extensive literature search for this information resulted in a large but conflicting amount of data. We had to make correlation studies and check results against our own data from a kindred process. We needed data on solution heat capacities, densities, vapor pressures, and heat of crystallization of the salt. What was not available in the literature was determined in the laboratory. With these physical-chemical data, a continuous operation was traced on the phase diagram. Maximum crystal yield was chosen as the process goal. The yields at various dissolver and crystallizer temperatures were then compared with cooling water requirements (a major cost variable). The theoretical optimum for a given set of conditions was then established.

It was still necessary to refine the results to fit the actual conditions that would be encountered during operations. Corrections were made for such processing steps as the incomplete dissolution of the salts present in the ore, and loss of solution during purification and filtration, and the imperfect separation of mother liquor from the crystals discharging from the centrifuge. As it turned out, this process was pieced together in great detail. It is true that the optimization was possible only through the use of an electronic computer. However, by operating a mathematical model of the process we ended up knowing probably as much about the operation as the user.

With operating details established, it was relatively easy to estimate the manufacturing costs for any desired level of production. The operating details also made it possible to size equipment, and thus to evaluate the capital investment requirements and the annual depreciation charges. The other cost figures necessary in any evaluation, such as land, mining rights, taxes, overhead, research and development, administrative burden, etc., were established from standard corporate procedures. We found that the two costs that had the greatest range were mining and transportation.

Now we had a detailed probable cost breakdown of our competitor's process that represented his optimum process as well as one that was accurate. Hence, if we wanted to enter the market with the same raw material as a starting point, we would have to better our competitor's costs. We would also have to develop a process that was free from patent infringement, otherwise there would be an additional cost for licensing. Having our research and development objectives thus very precisely defined helped us to complete the study in a relatively short time and at a minimum cost.

I want to emphasize that when we took the competitor's product, process, market, volume and costs in the example above, we optimized his operation and found his *best possible costs.* Then, and only then, did we evaluate our own new process and product against this to determine whether or not to go ahead with development.

Besides the obvious advantages in running a sound business, the deductive method has a further attractive attribute: there is no question about the ethical aspects of this method. We do not even have to leave our office to accomplish amazing results. ■

ACKNOWLEDGMENT

The editors wish to acknowledge the extensive assistance of Arthur Zimmerman of McGraw-Hill's Cleveland Bureau in preparing the manuscript for this article.

Meet the Authors

GUSTAV ENYEDY, JR., *is a chemical engineer engaged in process evaluation and design at Diamond Alkali Co.'s Painesville, Ohio, research center. He has a B.S. and M.S. (1955) from Case Institute of Technology and is a member of AIChE and the American Assn. of Cost Engineers. He also teaches chemical engineering and economics as a part-time lecturer at Fenn College, Cleveland.*

THOMAS E. HEDGE *has a B.S. in chemical engineering from Case Institute of Technology and an M.S. (1958) from Ohio State University. He is employed in the process development section of the Research Dept. of Diamond Alkali Co. Hedge is a member of AIChE and American Assn. of Cost Engineers.*

JAMES C. COWAN *is a chemical engineer currently engaged in cost estimation for the Development Dept. of Pittsburgh Plate Glass' Chemical Div. at Barberton, Ohio. Prior to his association with PPG, he worked as a production supervisor at Hercules Powder Co. Cowan is a member of the AIChE and American Assn. of Cost Engineers.*

RALPH T. CLINE *is an economic analyst for the Sohio Chemical Co., Cleveland, a subsidiary of The Standard Oil Co. (Ohio). He was previously affiliated with the Atlantic Refining Co. and Socony Mobil Oil Co. Cline received a B.S. in chemical engineering from Syracuse University and an M.S. from Cornell. He is a member of AIChE and the American Assn. of Cost Engineers.*

Sharpen Your Interviewing Skill

If called upon to interview a job applicant, will you be able to bring out the information needed for an intelligent hiring decision?

JOSEPH W. GILMARTIN, *Chemical Construction Corp.*

The "hire" or "don't hire" decision that is handed down on applicants for engineering jobs is generally based on interviews conducted by the company's engineering department. Yet despite the crucial importance of the resulting decisions, few engineers have ever had any formal guidance in interviewing techniques.

An attempt will be made here to show how engineers can make the most efficient use of the interviewing time and improve their skills as interviewers.

While it is important to recognize that interviewing is a complex professional job, any engineer even without formal training can nevertheless conduct a useful interview if he is emotionally mature, objective in his appraisal of others, has a good knowledge of the job opening, and is a good listener.

In the normal progression of the employment process, the initial interview is conducted by a member of the personnel department staff. In the case of highly technical personnel, this interview is limited to details of employment such as previous experience, education, training, family background, and so on. The comprehensiveness of this interview will depend on the nature of the job to be filled, the special skills or personality traits to be appraised, and the interviewer's technical training and knowledge of the specific job.

The personnel department interviewer, in the case of highly trained technical personnel, then refers those applicants who have high potential to the line department. Ideally, the engineering department head, chief engineer, or supervisor should be the one to conduct the departmental interview. He is in the best position to give the applicant first-hand details about the work, working conditions, and potential associates. Since the performance of the department will be judged by the performance of its personnel, it is of the utmost importance that the head of the department have a deciding voice in the selection of his personnel.

Scheduling the Interview

While the schedule of a busy engineering department chief is subject to last-minute changes and delays, it is nevertheless important that he make a determined effort to give adequate time to each interview. He should make certain there will be a maximum of privacy and freedom from distraction, telephone calls, and interruptions by third parties, all of which can cause an applicant to feel self-conscious and reluctant to speak of personal matters.

At Chemico, we make it a practice to have most interviews conducted in private rooms in the personnel office. This not only assures a high degree of privacy but has the additional advantage of making it unnecessary for an applicant to parade through a departmental work area. While it may be important for him to be acquainted with his potential work area, it is often preferable to defer this visit until later in the employment process unless he himself requests it. Many applicants who are presently employed feel like fugitives from justice because they fear recognition and possible recrimination if word gets back to their present employer. This is particularly true in a specialized industry such as engineering-construction where many engineers make the circuit of perhaps four or five of the leading companies during their careers and have friends in all of them.

Use this checklist to organize the interview

The important points for the interviewer to determine are: (1) Is the job right for this particular applicant? and (2) Is the applicant right for this particular job? Thus, the interviewer must elicit the answers to such questions as the following:

- ✓ **What is the applicant's objective?**
- ✓ **What recognition does he seek?**
- ✓ **What participation does he seek?**
- ✓ **What freedom of self-expression does he require?**
- ✓ **With what type colleague does he wish to associate?**
- ✓ **What opportunities for advancement does he desire?**
- ✓ **What degree of job security does he seek?**
- ✓ **What opportunity to continue his education does he desire?**

In evaluating factual information to determine whether the applicant is right for the particular job, the interview will revolve about more precise information such as the following:

- ✓ **What companies did he work for?**
- ✓ **How comparable were his previous companies in organization, size, etc.?**
- ✓ **What exactly did he do in his past jobs?**
- ✓ **What is similar between what he has done and this job's requirements?**
- ✓ **What are the true reasons for his separations from previous jobs?**
- ✓ **Why is he seeking this present connection?**
- ✓ **What are his expectations regarding salary and salary adjustments?**
- ✓ **What have been his reactions to supervisors he has had in previous jobs?**
- ✓ **What has been his experience in getting along with fellow workers in other jobs?**
- ✓ **What is his home situation? Marital status? Family roots? Willingness to relocate? Willingness to travel? Transportation problems? Hobbies or special interests?**

Twofold Aim of the Interviewer

The aim of the interviewer is twofold: first, to learn what he can of the applicant's qualifications; second, to inform the applicant about the job, the company, the working conditions, the opportunities that might exist for recognition, participation, promotion, and similar details. The interview is therefore a 50-50 proposition and the interviewer is both a buyer of personnel and a seller of opportunity.

The interviewer's questions serve to start the applicant talking and to draw out factual information—material that the applicant may not have already volunteered. When artfully directed by the interviewer, they serve to steer the conversation and coordinate what otherwise might tend to be a disjointed picture.

Ability in questioning is the mark of a good interviewer. Mere familiarity with questions is no assurance of being able to use them. Effectiveness can be improved, however, by close attention to two important factors: clarity and suggestiveness. For clarity, the interviewer's approach should be adjusted to the language level of the applicant. The questions are of little value unless their precise meaning has been understood. Simplicity should be the rule.

Also, in evaluating responses to questions, it is very important to know the precise meaning of what has been said. For example, in a discussion of previous experience, the applicant might indicate that he had been "responsible" for certain functions, or was "familiar with" certain activities. There can be a difference between being "responsible for" a function and being qualified to actually perform that work.

Things to Avoid

At this point, it might be well to mention certain danger areas that can be detrimental to a good interview. The most common are:

1. *Unrealistic standards.* All new employees are not destined to become chiefs. Look for exceptional merit only in those qualities that are most important to the job, conceding the need to live with some imperfections. Keep in mind that on simple jobs, turnover is greatest among men with the highest intelligence, and on complex jobs, it is greatest among those with the least intelligence.

2. *Personal bias.* There are few among us who do not have likes, dislikes, opinions and beliefs that influence our thinking and behavior toward others. Make every attempt to disregard them. There is seldom any relationship between the personal characteristics that you may have noted and the applicant's ability to perform the job.

3. *Amateur analysis.* The professional interviewers, particularly psychologists, have been schooled in such areas as motivation, the influence of experience on present behavior and adjustment, the personality characteristics revealed by various modes of expression, and the symptoms by which emotional immaturity and maladjustment, mental illness and other deviations from normalcy may be detected.

The engineer does not have this background of training and experience in the field of psychology and should not attempt to make psychological diagnoses. There can be no denying, however, that impressions will be received as an interview progresses that will in the final analysis be a significant part of the total judgment of the applicant.

4. *Mechanical approach.* A mechanical approach to an interview on the part of the interviewer is likely to beget a series of mechanical responses on the part of an applicant. At best, this approach will reveal little more than that disclosed by the application form. The interview should be adapted to the individual and the interviewer should make every attempt to introduce

freshness, interest, and originality into each meeting.

Observing Attitudes

From the first moment the interviewer sees the applicant, he should be listening to what the applicant has to say and how he says it. He should be observing his attitudes with respect to every topic of conversation throughout the interview. Since these observations are important to the final judgment, the interviewer must be constantly on the alert to differentiate the real from the superficial in what he observes, keeping in mind that the applicant may be nervous and may say things that he doesn't really believe.

In terms of attitude, one of the rarest and most welcome characteristics that interviewers can possibly find in an applicant today is that sincere feeling on his part of "What can I give to this job?" instead of the unfortunately more predominant attitude of "What can this job give to me?"

References supplied by the applicant are seldom of any value. Bad references are rare. Most companies, when giving a reference in writing, steer away from unfavorable comments. If a personal reference is involved, the applicant is usually certain when he lists the name that he will be given a favorable recommendation. Of greater value are verbal reports received through telephone calls to the applicant's former supervisor. However, the reliability and the objectivity of the reference should be carefully weighed.

Taking of Notes

There have always been differences of opinion on taking notes during an interview. In support of note taking are the arguments that it is difficult to remember everything; that facts and observations can be easily distorted or confused when trying to recall them at a later time. This is particularly true when several applicants have been interviewed in succession.

In opposition to taking notes are the arguments that it is distracting to the applicant; that he is liable to stop talking while the interviewer is writing; that some applicants object to having their conversation recorded; and that some applicants' conversation tends to become disjointed and hesitant.

Whether or not to take notes depends a lot on the interviewer, his skill in jotting essential facts down unobtrusively, and the reaction of the applicant in terms of his observed responses. If the interviewer feels that there is an over-all disadvantage of taking notes during the interview, then it would be best to wait until immediately afterwards. However, some interviewers have found that a mere jotting down of a word here and there during the discussions is unnoticeable and adequate for subsequent accurate elaboration.

Closing the Interview

If it is felt that sufficient information has been obtained upon which to base a decision, or a definite basis for disqualification has been established, the interviewer should tactfully but firmly steer toward a closing.

Generally the communication of the final decision is deferred until after notes are compared with the personnel department and, in the case of an affirmative decision, the salary to be offered has been decided. The following practices should be observed once a decision has been reached:

1. If it has been decided not to hire the applicant, he should be told in a tactful manner.
2. If an offer is being made to the applicant, be sure to summarize all of the details of the job, the working conditions, the fringe benefits, etc.
3. Do not make any promises unless you have the complete authority to carry them out.
4. If an offer is made and accepted, outline the next steps to be taken; i.e., if a physical examination is required, arrange for it, and do not leave details up in the air so that the applicant isn't sure of what happens next.
5. If the applicant declines an offer, find out why. Possibly it is because of some misunderstanding. Assuming that there has been no misunderstanding, the reason why the applicant has turned down the job should be analyzed. Possibly the knowledge can be of constructive value, and many times the differences can be reconciled.

No matter what the decision is, it is important that the applicant leave with a good feeling toward the company and its people. Nothing travels faster than reputation—if that reputation is one of fairness and courtesy toward a prospective applicant, it will stand the company in good stead, both as a place to work, and as a company to deal with. Many of today's unsuccessful applicants might well be tomorrow's customers. ■

Meet the Author

JOSEPH W. GILMARTIN *is director of personnel and office services at Chemical Construction Corp. After graduation from New York University in June 1941, Gilmartin served in the Navy until September 1945—his last post being that of executive officer on an attack transport during the Pacific landings. After the war, he was employed by the Great Atlantic & Pacific Tea Co. in several managerial, personnel and training capacities, then joined Chemical Construction Corp. in February of 1957.*

You Can Read Faster—and Learn More

WILLIAM KULICK,* *Illinois Institute of Technology*

An engineer came to us recently because a mass of unread printed matter was piling up on his desk. He knew that there was information in that pile that would help him with his current project—if he could just find it.

After about 17 hours in a reading program, he knew enough about good reading habits to plow through the journal jungle, obtaining the significant information and not wasting time on material that was useless to him. His only regret at the end of the program was that he hadn't undertaken it 8 years earlier; 16 years of schooling had failed to teach him a few, simple learning techniques.

Nor is his case unique. Many engineers and businessmen struggle with their reading, hoping to get what they can out of manuals, policy statements and professional publications. They read too slowly, substitute hard, time-consuming work for efficient, goal-directed reading and study. One of their most frequent complaints is that they cannot keep up with the mountain of printed matter that is routed to their desks. It might be your problem, too.

Suggestions made here for improving your own reading skills are no substitute for formal, systematic remediation of severe reading handicaps. But if you are a reasonably competent reader who wants to improve your reading efficiency, the specific recommendations will let you embark on such a program on your own.

A good reader seeks ideas that will be meaningful to him in some way and will satisfy his needs. Many of these concepts might be used, if we had time to "discover" them and determine how they might apply to our work.

As he practices his reading skills, the reader develops a regular, stable pattern of behavior. It begins with the approach to the reading process, and it often ends with a change in attitude, thinking or behavior. The effective reader, of course, tries to improve his grasp of the various skills in the reading process.

Basically, the aim of the good reader is to recognize the meaning of a passage, identify important related ideas and note the crucial details that make the writing more meaningful. Finally, he attempts to integrate this new knowledge into his thinking to develop his intellectual powers. Without this important step, the reading pattern is incomplete. But from a careful selection and absorption of key ideas, the productive reader can go on to synthesize new ideas.

Preparation for Good Reading

Certain characteristics identify the poor reader. He looks at individual words. He takes each word separately, trying to get meaning from it without recognizing that ideas require groups of words for their expression. He places equal stress on each word, erroneously feeling that he will miss something if he skips a word or drops insignificant details. Accepting statements uncritically, he does not evaluate, does not try to compare this newest experience with his store of memories.

He also tends to re-read erratically, so that words don't flow smoothly to form sensible ideas. Reading becomes a hopeless jumble of nonsense because ideas

* WILLIAM KULICK is director of counseling services, Institute for Psychological Services, at Illinois Tech in Chicago. Besides industrial experience after taking his Ph.D., he has counseled at Northwestern University. He is a member of the American Psychological Assn.

aren't perceived in a sensible order. For this non-reading "reader," however, there will be no jarring effect on his conscious thinking. He continues along, blithely believing that he's extracting some meaning from the jumble.

Sometimes he is hampered by inner vocalization, a hangover from early training in word recognition and oral reading. Almost anyone who reads 250 words/min. or less does some of this vocalization in his silent reading. Its manifestations are lip movement, throat movement and internal auditory perception.

Busy sounding out words, this reader responds to the pattern of the words rather than their meaning. The slow sounding of every word is a needless step in the silent reading process, however. With training, the individual can learn to react to visual perceptions directly. But without it, the slow reader finds his mind rushing ahead and wandering—supplying ideas and thoughts that may significantly alter the author's meaning.

All these poor reading habits tire the reader because of the tension and anxiety they produce. Understandably slow to start and quick to stop his reading, he finds nevertheless that he does not know what to look for, nor does he understand the significance of whatever he finds. His problem of poor reading, which may be yours too, can be helped by some of the suggestions that follow.

Reading as a Stage Process

The first stage in reading is to quickly explore the passage at hand. It's necessary to evaluate what you have and determine whether it meets your need. This seems obvious, but many people approach reading in an uncritical manner, hoping somehow that it will have pertinence.

Many engineers read an entire article, afraid to eliminate what proves pointless after a short, critical inspection. Here, a clear concept of what you want from your job, and what your job demands of you, becomes a useful standard to guide further reading. Such a personal standard makes it easier to discriminate, sharpens your judgment and adds precision to your work, study and reading.

After the initial exploration, skim through the article and note easily discovered ideas. A sampling of reading on each page sometimes yields enough information so that you need go no further. In any case, skimming leads to more-efficient thorough reading. And skimming reinforces the initial search to enable you to form a general plan of attack on thorough reading, and to determine what you want from the article.

Only in the third stage do we reach the point that most people think of as reading. The reader looks at each line, proceeds from beginning to end in a regular, systematic fashion. This is not to say that he should read word-by-word. But the eye registers images only when it stops moving. It should fix on phrases, however, not words. The good reader may pause for two or three fixations per line.

Span of perception—length—of these phrases should be extended whenever possible to picture certain ideas. Sounding-out of words, in hope that ideas will magically come to mind, should be minimized. At this stage, you seek answers to pertinent questions about the reading and try to discover what the author has said about a point. But in looking for the larger meaning of a passage, don't interpose your own ideas too quickly—find out what the author had in mind.

The fourth stage, after reading the passage has been completed, is to take stock of what you have read. Determine what you understand of the author's ideas. A crucial step, it is ignored by many individuals. This doesn't mean that you should blindly accept all that has been presented. But you can't accept, reject or modify ideas unless you clearly understand the author's point. Don't be so concerned with your own ideas that you don't recognize his.

In the fifth, final stage there is integration of all that was read with your past experience. Here, you try to go beyond the intent of the author to apply what you have learned. For some, this leads to the discovery of new ideas; others discover new ways to apply old ideas. The final creative act in reading, this difficult-to-accomplish integration requires originality. It isn't predictable, and certainly there is no way to train one to do this successfully, but it is resistance to this final, creative act that ultimately leads to reading problems.

Some Self-Help Hints

If you want to take some steps to improve your reading skills, you may do so informally with some profit. In planning your own program, set aside a regular time, perhaps in the evening, to read and study.

Keep a record of what you read, and how quickly you read. Time one to three pages each evening. As you master good reading habits, you will find this record shows your improved skill. Study of your progress each session makes you aware of things that lead to fast or slow reading. Then you can adjust your speed to better meet your needs.

While reading, it's not necessary to push too hard, or worry about your speed. Your major task is to discover significant ideas of the author. Proceed through the material calmly; strain causes distraction, and then you lose the ideas you seek.

As you read, ask yourself questions that are as specific as you can make them. These should be worded to find answers about the meaning of the passage, the main associated ideas and the crucial details. You might outline particularly important passages so that you will have an objective response to evaluate. A comparison of what you outline with what the author has said makes it possible to clarify your thinking and increase accuracy of your perception.

Don't forget the final step in any reading that you do: consider how the material applies to you, and how you can modify your thinking—can explore ideas—as an outgrowth of your experience and your reading. The exhilarating experience of this last, creative act in reading will gradually lead you to want to read faster—and learn more. ■

Where To Find Business Information

Keeping up with the business side of his industry is a prerequisite for the engineer who wants to get ahead. Here is an extensive discussion of worthwhile sources for all kinds of commercial information.

CONRAD BERENSON, *The City University of New York*

If he aspires to move ahead on the management route, the young engineer must find out a little more about the commercial aspects of his industry.

There are many ways to do this, of course, such as attending the commercial symposiums that are put on by some of the technical societies, or by taking a university business course designed specially for the technical man. But the engineer should also know where to find and how to use the wealth of business information that is available in published form.

It is the purpose of this article to point out some of the major sources of business information that are available in the literature. The sources we shall discuss are mostly periodical publications (but excluding the better-known trade magazines) rather than books. This is so simply because business conditions change too quickly for books to be of primary value to the engineer. Another limitation is that the sources refer to what is known in business circles as secondary data, that is, data collected by others.

Sources of Data

Here, then, are some of the major sources of business information to be found in the periodical literature:

Business Periodicals Index—More than 100 magazines devoted to business are indexed in each issue. The index comes out several times yearly, and is cumulated yearly. Coverage includes business policies, management, accounting, marketing, economics, finance, etc.

Although information is cross-indexed, the searcher needs a little persistence to find all the references bearing on his field of interest. References are in the form of title of article, author, journal, volume number, issue number, page and date. In my opinion, this index is the best single source of general business references. Published by H. W. Wilson Co., New York.

Public Affairs Information Service Bulletin—The index discussed previously does not cover government pamphlets and reports to any significant extent. This lack is made up by the bulletin covered here. Published monthly, with an annual cumulation, it provides name of paper, author, etc., as above. Its emphasis is on political science but it deals with many areas of interest to chemical engineers—for example, federal laws affecting stream and air pollution, government attitudes toward R & D spending, the economics of different geographical locations, and the supply and demand for technical workers. Published by Public Affairs Information Service Inc., New York.

Applied Science and Technology Index—Similar in frequency and type of data offered to the preceding two, this index provides references to trade, technical, and technical society and association journals not given by the others. Thus, while the *Business Periodicals Index* will refer to some of the general business magazines of the chemical industry (*Chemical Week* and *Chemical and Engineering News*, for example), the engineer interested in the commercial and technical problems of the thermoplastics segment, say, or the paint and varnish industry would find references in the *Applied Science and Technology Index* to such specific publications as *Modern Plastics* and *Paint and Varnish Production*. This index is also published by H. W. Wilson Co., New York.

Chemical Market Abstracts—Published monthly, with an annual cumulated index, this publication does for the chemical process industries what the *Business Periodicals Index* does for business in general. It is a "must" source for any economic or business study of the CPI, and it is published by Foster D. Snell Inc., New York.

The Engineering Index—Published annually, this index offers references not only to technical articles in publications such as CHEMICAL ENGINEERING, but also to articles covering the commercial aspects of technological industries. It differs from the indexes already discussed in its technological orientation and in that it gives a brief summary of the article, as well as all the data provided by the others. Up-to-the-minute data are available on file cards to supplement the yearly appearance of the index. Published, with the assistance of the National Science Foundation, by Engineering Index Inc., New York.

Monthly Catalog of U. S. Government Publications—This index presents comprehensive coverage of the vast number of reports issued each month by the different agencies of the government. An annual cumulated index is also available. Information presented includes catalog number, price, issuing agency, title, and whether or not the publication is available at one of the depository libraries established by the federal government and located throughout the U. S.

Once again, as with all the preceding indexes, ingenuity and persistence are required to find all the relevant material. Published by the U. S. Government Printing Office, Washington.

The New York Times Index—The coverage of this newspaper is so complete that its index presents a gold mine of information important to the chemical, or any other, engineer. This index is published twice monthly and is cumulated annually. The data in it are classified alphabetically and chronologically by person, subject or organization name. For each article, the title, date, page and column are given. Published by The New York Times Co., New York.

A Bonanza of Directories

Directories are available by the thousand, covering such a diversity of subjects that any attempt to do more than indicate their existence would be foolhardy in such a brief article. There are directories that are specialized according to field, for example, *Chemical Week Buyers' Guide* and *Electronics Buyers' Guide*; there are directories of company directors in various fields (*Poors' Register*); there are directories citing the advertising policies and other relevant data for magazines (*Standard Rate and Data Service*); there are even directories of directories.

A directory of particular interest to the chemical process industries is:

Chemical Statistics Directory—In mid 1964, the U. S. Dept. of Commerce brought out the third edition (the second having appeared in 1949) of this invaluable reference. It provides a guide to U. S. Government statistics relating to the chemical industry and lists specific reference works, with page numbers, for thousands of chemicals.

It is safe to say that if you want information about a vendor, an executive in a competing company, a trade name, transportation rates, export and import regulations, or any of hundreds of similar data classifications, there is almost certainly at least one directory to help you.

Other Sources

Business Services—There have recently been introduced a variety of specialized reports, newsletters, loose-leaf services, confidential letters, etc. Depending upon his needs, an aspiring or already arrived executive might find one of these services useful. For example, a project leader whose firm uses basic commodities as a raw material might decide which of several alternative process routes to follow, depending on the raw-material price and availability trends revealed in a news letter.

There is, of course, a directory to these business services. It is *The Directory of Business and Financial Services* and it is published by the Special Libraries Assn., New York.

Bibliographies—A popular technique for dealing with today's avalanche of literature is the annotated bibliography. This is a published list of article (and/or book) references that an organization (a trade or professional association) or specialist (scholar or consultant) has culled from the mass of available references. Such lists usually give title, author, journal, volume and issue number, page and year of publication. Many bibliographies also describe the approach and the distinctive features of the work discussed. There are, naturally, directories of bibliographies.

Finding Out About Another Company

Sometimes the situation arises in which you would like to have business information about another company. Perhaps you are evaluating a job offer, or looking for a new vendor, or are just interested in the workings of a competitor. There are several readily available and useful sources of information about CPI companies:

Annual Reports—Most of the larger firms in the CPI are publically owned to some degree, and their stock is traded in regulated markets. This means that these companies are obliged to file comprehensive annual reports and, in most cases, abbreviated quarterly statements.

These reports reveal such information as sales, profits, dividends, taxes, balance sheets, profit and loss statements. Also included are lists of plant locations and their products, new products, discontinued products, expectations for the future, discussions of the company's performance, its problems and opportunities, lists of officers and directors, and so forth—in short, a broad perspective of the company in which you are interested.

Proxy Statements—Proxy statements sent to stockholders are available from libraries and from the issuing company itself. Among other facts, these statements must provide the names of the directors and the money paid to them, the three highest paid officers whose direct remuneration exceeded $30,000 in the previous year, a list of nominees for the board of directors showing the amount of stock held by each, and the direct remuneration paid to all directors and officers during the last fiscal year (without naming them).

The Wall Street Journal Index—Because *The Wall Street Journal*, the nation's largest daily financial newspaper, is so comprehensive in its coverage of the business world, its index is likely to contain a good deal of valuable information on the firms in which you are interested; for example, profits, losses, executive changes, plant expansions, anti-trust actions, international activities and the like. The index is published annually.

Dun & Bradstreet Million Dollar Directory—This is an annual with monthly supplements. It lists, for all firms with a net worth in excess of $1 million, the firm name and address, annual sales, number of employees, names of executives, type of business, names of divisions, etc. Published by Dun & Bradstreet, New York.

Thomas' Register of American Manufacturers—Also an annual, this directory provides the following data about manufacturers: products, address, names of principal officers, estimate of capital invested, subsidiaries, trade names, trade marks, etc. Published by Thomas Publishing Co., New York.

Poor's Register of Directors and Executives—Published annually with three supplements a year, this register lists executives and directors of corporations in the U. S. and Canada. The firms are grouped according to the type of business in which they engage. Data include names, titles and addresses. Published by Standard & Poor's Corp., New York.

Moody's Manuals of Investments—Separate annual volumes of exceeedingly comprehensive scope cover the fields of real estate, public utilities, transportation, government, and general industry (most applicable for those interested in the chemical process industries).

The data provided include a brief history, a description of plants and products, names and titles of officers, directors, a five-year comparison of the income account and balance sheet for large companies, a two-year comparison for smaller ones, complete dividend records for 15 years, stock quotations for 10 years, and a complete description of the company's security issues. The information in these manuals can even be revealing about one's own company. Published by Moody's Investors Service Inc., New York.

Standard Corporation Records—This loose-leaf publication, which is revised constantly, contains information on many thousands of U. S. and Canadian corporations. Data include balance sheets, earnings, market prices, security transactions and issues, and dividends. Published by Standard & Poor's Corp., New York.

Standard & Poor's also publishes more than a dozen other services designed for particular situations.

Reports of Stock Brokerage Firms—Most brokerage firms these days have research departments, some of which produce comprehensive and well-reasoned business analyses of particular firms and industries. These reports are ordinarily given free by the individual brokerage firms on request.

Where to Find All This Material

There are four basic places to look for all the printed matter we have been discussing.

Public Libraries—This may seem obvious, but remember that public libraries include more than just the municipal lending and reference facilities in your city. In the world of business, some of the most important "public" libraries are those operated by the U. S. Dept. of Commerce. There are 42 of these, located in field offices in major cities across the U. S.

The federal government also supports a network of special libraries in its various bureaus and agencies in Washington; for example, the libraries of the Federal Reserve Board, Dept. of Labor, Federal Trade Commission, Securities and Exchange Commission. Most states also maintain libraries and often will furnish information by mail or telephone if extensive research is not required.

Private Libraries—This category includes libraries in universities and business firms. A number of leading universities maintain first-rate libraries for their faculty and students. It is usually possible to visit these facilities or to arrange an inter-library loan through your own firm's library or your local public one.

Many companies also operate libraries with excellent collections of both business and technical material, which other companies or public libraries can often arrange to borrow.

Trade and Professional Association Libraries—These libraries usually contain excellent collections of material relevant to the members' fields of interest. The facilities are generally available to members and to others interested in the field.

Trade Publications—Many trade publications maintain outstanding libraries of business information relevant to their readers' interests. This information is usually made available to anyone who requests it, and the publication will frequently go to a great deal of time and expense to answer a query from anyone with a legitimate interest.

Additional Reading

For further reading on information sources, the following texts are recommended:

Coman, Edwin T., "Sources of Business Information," Revised Edition, University of California Press, Los Angeles, 1964.

Johnson, H. W., and McFarland, S. W., "How to Use a Business Library," Southwestern Publishing Co., Cincinnati, 1958.

Manley, Marian C., "Business Information, How to Find and Use It," Harper & Bros., New York, 1955.

Schmeckebier, L., and Eastin, R. B., "Government Publications and Their Use," Brookings Institution, Washington, 1961. ∎

Meet the Author

Conrad Berenson is associate professor of marketing at the Baruch School of Business and Public Administration of The City College of New York, and executive director of the research foundation of City University of New York.

Where and How to Obtain Computer Programs

Here is a directory of companies and government agencies that provide computer programs for free, for purchase, or on a rental basis.

SIDNEY KRAVITZ and MANNY MEYERS, Picatinny Arsenal

The link between the computer and the working engineer is in the type of computer programs that are available.

Computers cannot run without programs and someone has to write them to meet particular needs. However, before expending the time and money to develop a computer program for your specifications, it may be that such a program already exists. Even a computer program not precisely tailored to your goals can still yield valuable suggestions and prevent unnecessary research and expense.

There are organizations and sources available to the chemical engineer to help in searching for the proper program—particularly if your firm is not yet geared for an in-depth, large-scale computer operation. Various options are also possible because computer programs are available for free, on a rental basis, or for purchase.

However, if you don't want to start with a substantial handicap, you must be aware of two things when asking for a program: source language and documentation.

Matching the Language to the Computer

Source Language. When you receive a program, it should either be on tabulating cards, magnetic tape or paper tape. If you receive a printed listing (printout) of the program rather than the cards or tapes, the burden will be on you to have tabulating cards punched.

In any event, the program you receive should be in a source language: FORTRAN, COBOL or ALGOL. Any of these source languages is compatible with computers of any manufacturer. However, should you receive your program in assembly language, then it is important that the assembly language be compatible with the computer available to you. (Assembly language is the interior language with which each computer translates program requests into machine language. In this respect, computers are almost like individuals; and different models, even of the same manufacturer, use different assembly languages. Therefore, a program is useless to the requester if the same computer model series is not available.)

Documentation. When you request a program, make sure that documentation will be provided. If a computer program has not been documented, it does not exist. The documentation will tell you what the program can do, how to prepare the input data cards, and what output data cards the program can generate; and it should also tell you something about the equations or analytical methods used. If the charge for the use of a program is significant, the wise procedure would be to order the documentation first and then to make the decision whether to spend money on it.

Who Has the Programs

The following is a partial listing of some of the primary sources featuring programming solicitation and sales: associations and their publications, including some federal government-sponsored programs that are available; interchange organizations; and a listing of periodicals devoted to the data-processing field.

• *The International Computer Programs Quarterly* allows any subscriber to list a required program in the programs-wanted section without cost. Those offering programs for rental or sale may also list them without charge. In addition, a roster of contract-programming firms and an index of timesharing installations are given. The quarterly is published by International Computer Programs Inc., 2511 East 46 St., Indianapolis, Ind. 46205, for $60 per year or $15 for a single issue.

For example, on page 76 of the April 1969 issue, a multi-column distillation and flow simulator program is offered for sale by Simulation Sciences, Inc., of Rosemead, Calif. This program simulates single towers or complete separation systems consisting of any combination of towers, absorbers, strippers, coolers, reactors and flashdrums. Product purities, recoveries and properties, and vapor-liquid flow at any tray may be specified by the user. There are four versions of this program, selling for $20,000 to $33,500.

Don't Overlook Magazines and Journals

A long list of periodicals devoted to computers, programming, and data processing is given in *Ulrich's International Periodicals Directory,* edited by Marietta Chicorel and published by R. R. Bowker Co., 1180 Ave. of the Americas, New York, N. Y. 10036 for $15. A partial list of periodicals follows:

Latest Developments and Equipment

Computerworld, a weekly published at 60 Austin St., Newton, Mass., $9 per year.

Data Processing Digest, a monthly published at 1140 S. Robertson Blvd., Los Angeles, Calif. 90035, $36/yr.

Data Products News, a monthly published at 545 Fifth Ave., New York, N. Y. 10017, $7.50 per year.

Data Systems News, a monthly published by United Business Publications, a subsidiary of Media Horizons Inc., 200 Madison Ave., New York, N. Y. 10016—free to qualified persons.

EDP Europa Report, a bimonthly published by IDC Europa Ltd., 59 Grays Inn Rd., London WC 1, England, $65 per year.

Software Age, a monthly published by Press Tech, Inc., 2211 Fordem Ave., Madison, Wis. 53701—free to qualified persons.

Software Digest, a biweekly published by EDP News Services, Inc., 514 10th St., NW, Washington, D. C. 20004, $45 per year.

Data Processing, Analysis and Evaluation

Computers and Automation, a monthly published by Berkeley Enterprises, Inc., 815 Washington St., Newtonville, Mass. 20160, $15 per year.

Data Processing Magazine, a monthly published by North American Publishing Co., 134 N. Thirteenth St., Philadelphia, Pa. 19107, $8.50 per year.

Computer Decisions, a monthly published by Hayden Publishing Co., 850 Third Ave., New York, N. Y. 10022—free to qualified persons.

Computer Design, a monthly published by Computer Design Publishing Corp., P. O. Box A, Winchester, Mass. 01890—free to qualified persons.

Computer Group News, a bimonthly published by the Computer Group of the Institute of Electrical and Electronic Engineers, 16400 Ventura Blvd., Suite 321, Encino, Calif. 91316, $6 a year to members.

Datamation, a monthly published at 35 Mason St., Greenwich, Conn. 06830—free to qualified persons.

Modern Data Systems, a monthly published by Delta Publications, Inc., 3 Lockland Ave., Framingham, Mass. 01707—free to qualified persons.

Technical Details of Programming

Computer Aided Design, a quarterly published by Heywood-Temple Industrial Publications, Ltd., 33-39 Bowling Green Lane, London EC 1, England, $18.50 per year.

The Computer Journal, a quarterly published by the British Computer Soc., 23 Dorset Sq., London NW 1, England.

To reiterate, all these sources may not exactly supply the chemical engineering program that will accomplish your particular purpose, but they surely will give you extensive leads in the right direction.

International Computer Programs sponsors one-day executive seminars on the single subject "Computer Software, How to Sell It, How to Buy It." Topics covered in detail include the preparation of a program to make it marketable; how to buy a program with consideration given to the evaluation of the seller and to the evaluation of yourself as the buyer; and legal aspects, including copyright and patent protection, accounting, taxes, and contract. The seminars cost $115 per person. The April 1969 quarterly also presents a valuable discussion on how to buy a package. The article states that the decision to buy a program should be based on an economic analysis that includes an evaluation of the costs of the program, of conversion to your system, of making changes to adapt it to your specific needs, and of operating it.

• *Computer and Information Systems* (formerly *Information Processing Journal*) is another useful journal devoted to theory, fabrication and application of computers. Sources include periodicals, U.S. Government reports, conference proceedings, books, dissertations and patents. (Section CX7, for example, deals with chemistry, chemical engineering, and metallurgy.) It is published by Cambridge Communications Corp., 1612 K St. NW, Washington, D.C. 20006, and costs $90 per year.

• The Assn. for Computing Machinery (ACM) at 1133 Ave. of the Americas, New York, N.Y., offers two useful publications for those searching for computer programs. This association publishes the *Communications of the ACM* (CACM) which is free to ACM members (annual dues $25) and others for $25/yr. The association also publishes *Collected Algorithms,* which costs $15/yr. to ACM members and subscribers of CACM, and $25/yr. to others.

• The National Aeronautics and Space Administration (NASA) evaluates and disseminates its computer programs created under government contract. NASA's intent is to allow private industry to have access to as much government-originated work as possible. These programs are called COSMIC, named after the Computer Software Management and Information Center at the University of Georgia. They are referenced in "A Directory of Computer Programs Available from COSMIC." Both reference and programs may be obtained from COSMIC, Order Desk, Barrow Hall, University of Georgia, Athens, Ga. 30601.

The computer programs (source decks) are disseminated in tape or card form, depending on the requester's preference. The fee is $75 for programs furnished on a nonreturnable tape or for those in card form, not exceeding 2,000 cards. The fee for documentation is between $1.50 (for a document size not more than 25 sheets) to $15 for a document of over 300 sheets. Documentation and program must be ordered as separate items.

For example, COSMIC's Pipe Stress Program (MFS-1904) is used to calculate the stresses in pipe made up of bends and straights. It is limited to no more than 4 supports and no more than 29 pipe sections. The documentation for this routine gives a full description of the method of solution, including equations and mathematical definitions. Another example is program MFS-1677, which is applicable for calculations of transport properties of nonreacting mixtures of low-pressure gases. The properties generated are specific heat, viscosity, conductivity, and Prandtl number. MFS-1675 deals with orifice sizing for fluid systems, MFS-1910 deals with three-dimensional heat transfer, etc.

• *The BMD Programs* (Biomedical Computer Programs), an excellent source for statistical programs, is put out by the Health Sciences Computing Facility, Dept. of Biomathematics, School of Medicine, University of California, Los Angeles, Calif. 90024. This work is sponsored by a grant of the Div. of Research Resource Facilities of the National Institutes of Health. It includes, for instance, program BMDOID (entitled Simple Data Description) which calculates means, standard deviations, standard errors of means, maximum and minimum values, ranges, and sample sizes. Other BMD programs include those on multivariate analysis, regression analysis and variance analysis. The available programs are referenced in "BMD, Biomedical Computer Programs" edited by W. J. Dixon, which may be ordered from the UCLA Student Store, 308 Westwood Blvd., Los Angeles, Calif., at $5.50 per copy.

Using Manufacturers' Organizations

Several of the large computer manufacturers offer membership in program interchange organizations. For example, SHARE is an organization to promote the free interchange of information concerning the use of certain IBM computer systems to reduce redundant effort in the preparation of computer programs for general use. Program SDA 3055 of the Share Program Library lists a program written by the Langley Research Center of Hampton, Va., which computes the atmospheric density, pressure, temperature, and sound velocity for altitudes from 3 miles below sea level to 400 miles above.

The SWAP Pool Exchange is a user's organization of Control Data Systems; its *Program Catalog* is published by Control Data Corp., 3145 Porter Drive, Palo Alto, Calif., 94304. In addition, every time-sharing system offers a large catalog of programs for its users.

A large list of computer users' organizations is given in the Computer Yearbook and Directory published by American Data Processing, Inc., 19802 Mack Ave., Detroit, Mich. 48236. This volume also lists companies offering data-processing services, and associations that have data-processing interests. ■

Meet the Authors

◀ **Manny Meyers** is Chief, Editorial & Publications Section, Ammunition Engineering Directorate, Bldg. 355, Picatinny Arsenal, Dover, N.J. 07801. His work there involves helping engineers produce readable manuscripts. He has a B.A. in journalism, an M.A. (and is presently a candidate for a Ph.D.) in American Civilization. All his degrees are from New York University.

Sidney Kravitz, now in the data processing systems office of the Scientific and Engineering Applications Div. of Picatinny Arsenal, obtained his B. Mech. Eng. from Cooper Union in 1948. Shortly after obtaining his Master Aeronautical Eng. from New York University in 1948, he worked for four years on ENIAC, the first digital computer. A licensed professional engineer, he has also taught mathematics and mechanical engineering at Newark College of Engineering, and has worked for Vector Design Co. and Eastern Engineering Co. ▶

Trade Secrets, Industrial Espionage, And the Clam-Up Concept

After setting the stage with examples of legitimate, illegitimate and borderline ways of garnering industrial intelligence, the author introduces "The Proper Clam-Up"—an unusual formula that can help you decide just how much protection is enough.

HERBERT POPPER, *Chemical Engineering*

To illustrate the type of problem we will be discussing, I would like to cite a case involving a rather distraught research manager. His particular division—let's call it the Stardust Div.—hadn't come up with any significant advances in years, whereas a steady stream of product and process improvements seemed to pour out of the research lab of his division's chief competitor—the Turned-On Chemical Co.

In desperation, he turned to a firm of technical consultants for advice. He told them, "If I could only find out how the Turned-On people get their stardust to such uniform size, this might save my job. Our stardust is really purer than theirs, but we've been losing customers because theirs is more uniform."

When the technical consultants heard this, they pulled down the blinds, locked the door, searched the room for hidden microphones, turned on the radio very loud just in case they missed any, and then whispered to the research manager "Go see Joe Smith of Smith Research Services. Tell him we sent you. He has ways of getting such information."

"Are they ethical ways?" gulped the research manager. "My company would fire me on the spot if they ever found out I condoned anything unethical. Their good name means everything to them. They would boil me in oil."

"Don't worry," said the consultants. "The methods he uses are *his* business. His bill will look quite innocuous—just as if it came from a technical consultant."

Well, the research manager wasn't quite convinced, but after many sleepless nights he decided to bury his doubts and see Smith. After some more sleepless nights, Smith got him the information he needed. The next day, the research manager got a large but discreet-looking bill in the mail from Smith. With a trembling hand, the research manager gave it an account number that corresponded to "technical consulting services," and sent the bill through for payment.

All that night, he tossed and turned in his bed, despite aspirin, tranquilizers and bourbon, worrying whether anyone would question the bill, whether the truth would come out, whether he would have to leave town and start a new life, and so on. He was a wreck when he staggered in to work an hour late. And his state of mind did not improve when he found a note on his desk to call the company controller about the "Smith" bill. After fidgeting around the rest of the morning wondering whether the company would at least let him finish out the week, and whether he would need good references to get into the Peace Corps, he mustered up his courage and called the controller.

"Ah yes, the Smith bill," the controller said, somewhat ominously. "Your girl gave it the *Technical Consulting* account number—that's 960-120. I guess she must have been thinking of the wrong Smith—for Smith the industrial spy, we use 960-121."

The Bug in the Union's Hotel Room

As you may have gathered by now, I sort of made up that case history. Before I am suspected of anti-management bias, let me turn to another case history —this is a true one it is part of an actual news story filed by a McGraw-Hill correspondent in Canada in 1966:

"An internal investigation has been launched by the International Brotherhood of Pulp, Sulphite and Paper Mill Workers, at Vancouver, B. C., into the bugging of a hotel room occupied by officials of a rival union. In a prepared statement, the union said it is 'conducting its own investigation into the reported happening concerning the tapping of the other union's hotel room. Prompt and immediate remedial action will be taken if anyone connected with our organization has violated any statute or laws.'"

Apparently, someone in the international union had paid a private detective $250 to get information about whether alleged harassment of members of his union by members of the rival Pulp and Paper Workers of Canada was condoned by its officers. It seems the detective used hidden microphones to get the answer.

Having restored some degree of labor/management balance to the topic of espionage, let me go back to the research manager of the Stardust Div. and take a more serious look at his problem.

"We're managing to overcome our lack of skilled labour."

He knows his competitor is ahead of him in stardust technology. His own independent research doesn't seem able to narrow the gap. He needs some new lead—some new avenue to embark on. What methods can he use to get such a lead?

The methods that present themselves fall into three areas: white, grey and black.

The white-area methods are not only perfectly legal but perfectly ethical. Thus, if our research manager wants to get a new lead on stardust technology he can search the technical literature, hire a reputable technical consultant or research laboratory, analyze his competitor's products, go over his competitor's brochures and annual reports, and even interview the competitor's customers to find out why they prefer his product. The research manager can conduct a patent search—maybe he can get help from a recently expired patent, or make enough of an improvement or modification on someone else's existing patent to file for a patent of his own.

Let's move on to the grey areas. These are generally legal, but involve some questions of ethics. One grey method our research manager might use, for instance, is to send a friend of his from another company to attend a technical meeting to see what can be picked up by mingling with the conferees, asking the speakers leading questions, and so on. Since this person will bear the name-tag of some noncompeting company, he might be able to get some leads that the research manager would be unable to get if he himself asked the questions. The same would be true if our research manager sent this person on a plant tour of his competitor's facilities if such a tour were available.

Actually I'm not sure that such strategems as the above are terribly important as grey areas. I think the most important grey area is the one that comes up if our hypothetical research manager decides to hire an R&D engineer away from his competitor, hoping he will bring new know-how with him.[1] A variation of this is merely to interview the candidate without any real intent to hire, hoping to pick up some tidbit of information during the interview. For instance, in an employee bulletin, the 3M company states:

"It is common for competitors of 3M to use a job interview as a means of attempting to find out whether 3M research was investigating the possibility of a certain manufacturing process."

There are many other grey areas—like taking a picture of a competitor's plant from a helicopter or private plane but let's move on to the black areas. What *really* nasty, illegal ways can our hypothetical research manager use to find out how how Turned-On Chemical Co. makes its stardust?

He can pretend to be interested in licensing Turned-On's process. If Turned-On is careless, he can pick up some of its trade secrets before he makes any licensing commitment.[2]

(Here is a variation on this: If he knows that Turned-On has filed a patent application, he can also file one, with a number of very broad claims that cover what he believes Turned-On is patenting. The Patent Office may then declare an interference, which will give him access to Turned-On's application file. Having gained the information, he concedes priority, withdraws his application, and goes to work on ways of getting around or improving on Turned-On's pending patent.)

He can hire a professional investigator—someone like Smith, the industrial spy—who will get a specific piece of information even if it means trespassing on the competitor's property, passing himself off as a telephone repairman, and so on.

Our research manager can plant an agent on his competitor's payroll—let's say, a maintenance mechanic whom he pretends to fire, who then goes to work for the competitor, but who maintains friendly contacts with the old employer, thus drawing two salaries, plus bonus.

Then there is eavesdropping—electronic and otherwise. If our hypothetical Stardust man is ruthless enough, he can bribe one of his competitor's employees to reveal secrets, deliver copies of drawings, and so on. And if our man is even *more* ruthless, he can use some form of blackmail to achieve the same ends.

© Punch

"I know it's meant to be revolutionary, but I'm beginning to have the horrible feeling that British Motors **wanted** us to steal these plans."

The Other Side of the Coin

Before going further, let's see what Stardust's competitor, the Turned-On Chemical Co., can do to protect itself from the activities of our nefarious research manager. Let's say Turned-On wanted to go all out in protecting itself, regardless of what it cost.

First of all, there are the obvious physical methods —plant guards, fences, alarm systems, TV-monitored exits and entrances, company incinerators for waste paper, and so on.

Then there are the obvious procedural methods: an employee educational campaign, emphasis on locking desks and file cabinets, clearing desks before interviewing visitors, insisting on badges and personal escorts for plant visitors, limiting access to highly important documents, having rules about the type of work that can and cannot be taken home, using code names for as much technical data as possible, splitting up a big secret into many little secrets so that only one or two key people can piece the whole story together, establishing clearance procedures for talks and articles with guidance on the sort of questions the speaker should or should not try to answer after his talk, and so on.

Then there are the methods aimed at reducing loss of trade secrets via so-called mobile employees. These include carefully worded employment agreements that embody a clause against working for a direct competitor for a reasonable period of time, careful classification of documents so that the employee knows what's secret and what's not, exit interviews reminding an employee that he has had access to proprietary information, possibly a letter to his new employer with such a reminder, a checkout system to make sure all company notebooks have been returned, and so on.

To avoid disclosing information in patents—information that might help its competitors—Turned-On can try to keep the new information secret and not get a patent. This means that the person responsible for the innovation may never be free to discuss it in public, which may not help his morale one bit. However, remember we are discussing a theoretical situation here—one based on the proposition that Turned-On wants to go all the way in security, regardless of costs and side-effects.

To guard itself against "planted" employees or potentially dishonest employees, Turned-On can use psychological testing and thorough investigation of job applicants. It can even use polygraphs—lie detectors—if permitted in the state that Company X operates in.

To guard itself against hidden microphones and transmitters, it can arrange for searches to be made using electronic detection gear. Also it can use jamming devices, such as a recently introduced voice scrambler that renders radio and telephone conversations unintelligible by dividing the voice spectrum into five parts, and interchanging. The parts are sent either inverted or erect to protect conversations against snoopers.

If, despite all these precautions, it looks like a trade secret is about to get away or has gotten away, Turned-On can seek relief via a variety of civil and criminal laws. Questions involving mobile employees rely on civil law—in other words, on civil injunctions or possibly suits for damages. If outright theft is involved, Turned-On may be able to resort to the National Stolen Property Act. (American Cyanamid successfully used this act in 1965 against some ex-employees of Lederle Laboratories who had sold an outsider certain sample cultures along with copies of documents. A lawyer with a flair for puns coined the classic description of this theft of cultures: "The Great Strain Robbery.")

Finally, there are eight or nine states that have trade-secret-protection laws. If Turned-On is in one of these states, and if some tangible document or copy of a document has been taken from it, and if the company can prove that the culprit *knew* he was stealing trade secrets, it may be able to recover them.

How to Wind Up Safe, Yet Sorry

How far should a company go in its security program? Here I would like to paraphrase a story in an earlier article,[3] and to add some new twists:

Once upon a time, the executive vice president of a dynamic, prosperous firm—let's say, our old friend, the Turned-On Chemical Co.—attended a rather disturbing seminar on the "growing problem of industrial espionage and loss of trade secrets." Convinced that Turned-On's defenses against this sort of loss were inadequate, he proceeded to institute a new protection system. "We must be losing millions from all the know-how that has been leaking out," he told the executive committee. "But our new security setup will stop all that—I'm proud to say it's the most rigorous in the industry, and represents yet another fine example of Turned-On's leadership. Sooner or later, you'll see the benefits show up in our profit statements."

Unfortunately, it wasn't sooner. In fact, the first year after the system was instituted proved to be rather hectic. A couple of key engineers left rather than sign revised secrecy agreements that would have drastically restricted their future employment prospects. A chemist made a mistake deciphering a new supersecret code, and blew up half the lab. A colleague of his almost blew up the other half, when he made an uneducated guess rather than go through the red tape of asking for some locked-up reports to which he no longer had routine access.

The next years saw improvement in some areas, but new problems in others. A prized Ph.D. quit in a huff when he was mildly reprimanded for falling asleep during three successive Industrial Security lectures. Another valuable but temperamental researcher left when his name was omitted from the new, streamlined distribution list for certain types of confidential reports; despite all assurances, he felt the omission represented a downgrading of his status.

There was also a rather embarrassing plant accident involving a reactor that was handling a new product. The equipment supplier insisted that the accident would never have happened if he had been told what his equipment was going to be used for, but secrecy requirements had made this impossible. Reporters who wanted to interview the plant manager about this accident were given Turned-On's now-usual "no comment" treatment; they therefore had to base their story on the colorful diatribe of a slightly injured union steward, and of an excited passerby who saw "a blinding flash, and then a huge mushroom cloud rising from the plant and billowing out just like an atomic bomb." This story, and the local furor it precipitated, did nothing to endear Turned-On Chemicals to either the city fathers or the Zoning Commission.

On other fronts, a feeling of technical isolation began to set in. Constant "button your lip" warnings had dissuaded most engineers from discussing their work with anyone—even with colleagues within the company. The animated cross-fertilization of technical ideas that used to take place during lunch hours or coffee breaks was replaced by cautious small talk.

Turnover among the technical staff increased. And getting replacements involved costly recruiting campaigns because applicants did not exactly flock to the company's doors. Apparently, a rumor had started in the job market that Turned-On engineers were no longer making a name for themselves, that the company was bogged down in red tape, and that you had to bare your soul to a lie detector before they would even think of hiring you no matter how badly they needed you.

Turned-On also developed other troubles, but the "most rigorous security system in the industry" kept some of these troubles from leaking out. Some years later, therefore, there was considerable surprise in the trade when Turned-On, in one of its extremely rare and typically tight-lipped public announcements, revealed it was filing voluntary bankruptcy papers.

Presenting a New Ratio

The reason I made the above story end the way it did is not to ridicule industrial security measures. In fact, one purpose of this article is to underline the importance of security procedures—not only to prevent outright theft of secrets but also to avoid costly disclosures that may come about from carelessness or ignorance.

However, it is very important that industrial security measures be optimized rather than maximized—and I believe that the following optimization ratio can be helpful in this regard:

$$\frac{(T - H_e)\,(P_{ro})\,(P_{er})}{C + L + A + M + U + P}$$

T = Tangible worth of the secret (i.e., the approximate net present worth).

H_e = Hep factor, representing the portion of the secret that the competition is already hep to.

P_{ro} = Probability that part of the secret will be revealed without new restrictions.

P_{er} = Percentage of the secret vulnerable to such exposure.

C = Cost of protection in terms of hardware or direct labor.

L = Loss in efficiency through more-cumbersome communications (e.g., time lost by engineers who have to shuttle back and forth between their work areas and the departmental safe).

A = Arrested professional development; this can cost the company as well as the individual.

M = Morale factor—i.e., the cost of any lowering of morale that could result from new security measures.

U = Useful publicity that goes down the drain.

P = Popper's factor, which covers the possible cost of the reprisal effect.

Perhaps the last term requires further comment. If one company becomes unduly tight-lipped, its competitors may feel impelled to follow suit, so that eventually everybody either works in the dark or has to resort to expensive ways of garnering routine competitive information. In either case, this bears a price tag that has to be considered.

Please understand that the only reason I referred to the price tag as Popper's factor is that I wanted the last term in the denominator to begin with a *P*, so that the ratio becomes easier to remember;* it now spells out "The Proper Clam-Up." The degree of clam-up or protection that is proper is determined by the degree to which the ratio is more or less than unity.

The "Proper Clam-Up" in Action

Let us see how this would work out for an example. Suppose one of your subordinates wants to give a talk at the next National Meeting of the AIChE; he wants to discuss some general aspects of new techniques that he used during the development of a new process. Of course, if you wanted to *maximize* security, and the new process were confidential, you would immediately say "Not a chance—you can't give the talk." But if you wanted to *optimize* security rather than maximize it, you would go further. You would think of "The Proper Clam-Up" ratio, and in the numerator you would try to consider the:

- Tangible worth of the new process the engineer was working on.
- Part of the process technology that competitors are already hep to.
- Probability that the talk or discussion period will reveal secret information. If the engineer is going to limit his talk to broad techniques, maybe no secrets will be revealed.
- Percentage of the secret that is vulnerable. If your subordinate is a junior engineer who worked on only a small aspect of the process, perhaps he can only give away a fraction of the total secret no matter how many slips of the tongue he might make.

In the denominator, the cost of protection (C), and the loss in efficiency (L), would both be zero in this case. The "A" term, however, could be substantial. If the engineer gives the talk, and if it is published, his professional development is furthered, his communication skills may improve, he will gain in poise, he might derive a new insight into the topic, and so on. All this would definitely have some value to the company, as well as to him.

The morale factor (M) could also be important, especially if your engineer has a classmate who works for a university research laboratory and who is encouraged to make a name for himself by giving papers at the drop of a mortarboard.

Furthermore, the "U" term would have to be considered, because useful publicity could be lost. Conversely, giving the paper would help keep the company's name in front of actual and potential customers, as well as potential employees. The paper may help establish or maintain the company's image as a leader in technology.

Finally, the "P" term would also apply in this case. Engineers from competing companies who hear your subordinate's talk or read a published version of it may very well decide that they too have something worthwhile to contribute on this topic, and your own company would eventually benefit from this professional exchange. If your man's talk is squelched, this exchange might never come about.

After going through this mental manipulation, you may decide that the "T" term in the numerator is so formidable that the ratio is much bigger than unity and that you must make a "no-go" decision on the talk. But before you do, remember that the numerator is amenable to modification. For instance, by going over your man's outline, suggesting changes, and guiding him on the question-and-answer period, maybe you can reduce the P_{ro} factor further without negating the value of the presentation. (Of course, if you go too far and try to get the numerator close to zero, the paper may end up so bland that the would-be speaker may not get it accepted for delivery or publication.)

Other "Clam-Up" Applications

I believe that the "Proper Clam-Up" concept can be helpful not only with clearance problems but also in deciding how far to go in revising technical employment agreements, whether certain documents should be locked up in a safe rather than being readily available to researchers, and for similar security-optimization decisions. Although the terms in the ratio may not always lend themselves to precise computation, one can usually assign at least some relative order-of-magnitude values, when evaluating the need for a given security measure.

Incidentally, if someone figures out a computer program for the "Proper Clam-Up" concept, his company may eventually have to decide whether to let him publish his program, or whether it must be kept under wraps. The answer to this question could then be supplied by the program itself. It is rather interesting to think of a program deciding its own destiny, isn't it? ■

References

1. For fuller discussion, see: Arnold, T. H., Jr., "Are You Locked Into Your Job By What You Know?," *Chem. Eng.*, May 9, 1966, p. 141 *(see p. 193 of this volume)*.
2. Wade, W., "Industrial Espionage and Mis-Use of Trade Secrets," Advance House, 1965.
3. Popper, H., "How Safe Are Your Company's Secrets?," *Chem. Eng.*, May 23, 1966, p. 157.

* Readers who can think of a more appropriate term that begins with *P* are invited to send in their suggestions.

3

Appraisal, Development and Rewarding of Professional Employees

Closely related to the communications skills discussed in the previous section are the human-relations skills needed to deal effectively with one's subordinates and bosses. The emphasis here is on judging, paying, and promoting people according to sensible, objective criteria, and also on trying to get oneself judged that way.

Related articles, appearing in a later section, are: "How to Squelch Intraplant Politicking" (p. 325); and "What Do Foremen Need From Their Bosses?" (p. 329).

Performance Appraisal And the Engineer—I

Few things affect your career as directly as the way your performance is appraised. And few things are shrouded in as much controversy. In this first part of a straight talking, two-article series, appraisal and counseling interviews receive hard—and constructive—scrutiny.

HERBERT POPPER, Chemical Engineering

Henry and Harvey are two young engineers of contrasting personalities. (Who says all engineers are alike, anyway?)

Henry is always being rather critical of himself, although he has come up with some really worthwhile technical innovations. In fact, he is such a good engineer that his employer, the Tightlip Chemical Co., makes him a group leader and gives him an important part of a process development project. Although he is not given much guidance, he makes an excellent start and is soon well ahead of the schedule he has set up for himself and his group.

Then a technical problem that really couldn't have been foreseen crops up at a time when Henry is short-staffed because of sick leave and vacations. When Henry tells his boss that he has slipped a couple of weeks behind schedule, his boss gets edgy and points out that Henry's project will prove a serious bottleneck if the work falls behind any further. So, the boss now asks for frequent reports on progress and problems.

Actually, no major new problems do come up, and Henry brings the project to a successful conclusion, only a few weeks late. His boss is relieved that giving such an important assignment to a new group leader has worked out well. Recognizing that Henry has been working long hours under pressure, he now gives him an easier project.

Unfortunately, this hits Henry during a period of postcreative letdown, and reinforces some doubts that he has had about his performance. Believing that the new project is a downgrading of responsibility, and remembering that he had to issue more progress reports than was customary during the latter phase of the previous project, he reasons that management is displeased with him, and that he would be better off starting somewhere else with a clean slate. When he tells his boss that he has accepted an offer from another company, it is too late to convince Henry that the nearby grass is really greener than he realized.

Let's now turn to Harvey. He is a rather unproductive engineer, but has convinced himself and others that he is a brilliant one, and Harvey can be very persuasive—that's his forte. So, he too is made a group leader. Although the project assigned to him is much simpler than that given to Henry, he is soon two months behind schedule. This does not diminish Harvey's self-confidence one iota, and he tries to convince his boss that in less inspired hands, the project would be even further behind schedule.

His boss isn't quite convinced. "How can someone be more than two months behind schedule after only one month?" he ruefully asks himself. However, being reluctant to change group leaders in midstream, he rationalizes that the project isn't a crucial one anyway, so he lets Harvey stay on, and even finds something nice to say about some minor administrative aspect of Harvey's work, so as to give him encouragement.

Well Harvey does wind up the project eventually, but it is about six months late, and his contribution to the work is of questionable quality. Nevertheless, Harvey feels very good. After all, hasn't he kept his cool while surmounting technical difficulties that would have buried a lesser engineer? And hasn't his boss patted him on the back midway during the project? True, the boss has been rather curt in later interviews, but surely only because he envies Harvey's more-dynamic personality and wants to keep him from becoming too flamboyant (or perhaps even becoming an eventual rival).

Harvey's illusions are not immediately dispelled. So, when he hears of an opening in the Special Promotions Dept. that would be ideal for him, he doesn't go after it because he reasons that his engineering virtuosity would be wasted there, and that he has a greater future in technical management.

Overdrawn, You Say?

Certainly Henry's case is not exaggerated. "Before we started our new performance and counseling system, highly rated engineers would often leave the company for no other reason than because they thought they were failing," relates a personnel manager.

As for Harvey, perhaps his case was slightly overdrawn, but there are numerous instances in which engineers have been prevented from taking remedial action in regard to their careers because management hadn't told them how they were really doing until it was too late. They were supposed to have picked up certain day-to-day signals, but many people only pick up those signals that reinforce their own image of their performance.

And that's why you need some system of performance appraisal—one that goes beyond day-to-day comments on a particular project, and that gives an engineer enough information to decide how he's been doing over a stretch of time, and what goals he should set for himself.

This is all very well, you say, but telling people how they are doing and how they can improve isn't all that easy in real life. Isn't it true that some so-called performance-appraisal and counseling interviews do more harm than good? And that many others are polite exercises in evasion and double-talk?

Yes, in researching this project, several such cases were brought to light. But this does not constitute a sweeping "beyond hope" indictment of interviews. In fact, when we sent questionnaires to 1,000 CE subscribers asking them whether their own experience indicated that interviews were a good way of telling engineers how they are doing and how they can improve, the great majority of those who responded answered affirmatively, as discussed in the box on the next page.

Many comments were made on how these interviews could be more meaningful (only a sampling is shown in the box). In this first part of the performance appraisal series, we will enlarge on these comments, bring in the views of other engineers and managers, and of two practical-minded experts: Joseph H. McPherson and Edwin M. Glasscock.*

A Less Than Satisfactory Interview

Before proceeding to the suggestions on how the engineer and his boss can get the most good out of interviews, let us create an interview that does not meet its potential—one that is a free-form dramatization of comments from readers and others. Our cast is made up of an engineer (E), who is maintenance supervisor in a medium sized chemical plant, and the plant manager (M); their unspoken thoughts are printed in italics, their conversation in regular type.

M: (*I don't have to watch my words so carefully with this one. Ed has been with the company 20 years, and he'll stick around. Not like that whiz-kid who quit in a huff when all I said was that his great work on the operations research project didn't excuse his not coming in at the same time as everyone else.*) Ed, you know the ropes. This is the time of the year when I am required to discuss and record "employee performance on the job," to "identify those areas requiring immediate improvement," to "plan for future progress of the employee," and to "provide the employee an opportunity to voice his opinions on any aspect of his job."

E: (*Ah well, the same old malarky again. I've got to listen to him before he'll listen to me, so I won't say much at the start.*) Okay, Marty, give it to me straight. Fire away.

M: (*Good attitude. This should go fast.*) Ed, your performance on the job has been good. The plant equipment seems to be in fine shape, and we haven't had too much downtime except for the reactor failure last week. You stayed within your

* Dr. McPherson is Manager, Personnel Research and Development, at Dow Chemical Co.; Dr. Glasscock is a psychological consultant to management, with offices in Montclair, N. J.

budget most of the year, but were 10% over for the last two months, as you know.

E: (*Darn it, why do these interviews always come up at such a bad time?*) Well, we did have to start breaking in those new mechanics—it's lucky we could find any in this labor market. Spare parts and maintenance supplies also went up a lot in the last couple of months.

M: Yes, I know. Everything has been zooming up except the prices we get for our plant's products . . .

E: (*And except my salary . . .*)

M: . . . So we have to work all the harder to offset increased costs with increased efficiency. This brings me to the part of the interview where I'm supposed to identify areas calling for immediate improvement, and I would say the cost area is it. I'm not going to tell you how to cut costs—that's your baby—I just wanted to say that the pressure from headquarters is really going to be on during the next year. A word to the wise, as they say.

E: (*I think he's used that same cliche during the last six interviews I've had with him.*) You know, I made some cost-saving suggestions in that "Predictive Maintenance" memo I wrote six months ago A good predictive maintenance program could have prevented that reactor failure you mentioned before, and it could save us a lot of money.

M: (*Why did he have to go and bring that up!*). Ed, your memo has not been forgotten. Trouble is, it would cost a pile of dough to set such a program up, and with money as tight as it is, I don't know if we could justify it.

E: And then I wrote another memo about giving maintenance a bigger voice in equipment purchasing

Experience of CE Readers With Performance Appraisal Interviews

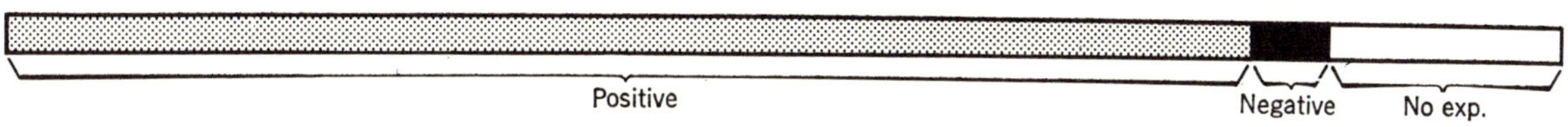

A week after every issue is published, our magazine sends out a readership survey to a group of 500 subscribers (selected on a rotating basis). The surveys for the May 19 and June 2 issues included the following question.

"In regard to a recent You & Your Job article on performance-counseling interviews,[1] does your experience indicate that face-to-face interviews are a good way of letting engineers know how they are doing and how they can improve?"

Readers were asked to check off one of three boxes. The 173 responses received led to the following tally:

- ☐ "Yes"—80%.
- ☐ "No"—5%.
- ☐ "Never involved in a performance interview"—15%.

Among the shorter comments sent in by readers were:

"Yes, if it's an honest interview—but these are few and far between."

"In a small company, frequent interviews are a must; they are rewarding to both employee and employer."

"Only effective if enough time is taken to prepare for interview."

"The supervisor and the engineer must be mature and honest about shortcomings, or such interviews can be worthless."

"Interviews are a necessity, particularly for inexperienced people so you don't lose them through neglect."

"Have used interviews successfully for years."

"Yes—only because interviews are better than nothing."

"Must be done three or four times a year. Very few managers or supervisors really tell their people the real facts about their performance."

"Results depend heavily on the interviewer's attitude and ability. Of prime importance are the supervisor's **continuing** observations. Last-minute observations only can be harmful. Interview should be scheduled and not hit-or-miss."

Among the longer comments received was the following letter from the chief engineer at a medium-sized CPI facility:

"Counseling interviews, in my experience, have been unpleasant affairs. If the employee has been doing a good job, the interview is very brief and he feels let down. If an employee is not doing a good job, he should not have to wait until a formal counseling interview to find out. The real problem is to have better and more frequent communication between supervisor and employee. Many years ago, at an annual appraisal, my boss told me that I had spent too much time with detail on a project that was completed six months before. My immediate thought was 'So **now** you tell me!'

"On the bright side, from the employee's viewpoint, is that his boss is forced to think about him at least once a year. Prior to our policy of annual reviews, the years could easily slide by without any significant progress being made regarding salary, responsibility, training or performance.

"In his advice to the employee, Erik Laurangel (author of the "Counseling Interviews . . . or Brainwashing?" article) suggests preparation for the appraisal interview. Why not go one step further? Why shouldn't the employee initiate an appraisal of his performance, his job-assignment progress, his salary progress, his training progress, his promotion potential and the future of the company itself? What an effect it would have on the boss if an employee made an appointment for next Wednesday morning at 10 a.m. to discuss 'job related matters.' Why not apply Laurangel's previously espoused 'resumes for employers' philosophy to annual appraisal interviews?"

decisions, so we can prevent trouble before it starts.

M: I hadn't forgotten about that one either. Trouble is, that's a sort of touchy one. If we go too far and get a lot of equipment that's easy to maintain but costs a fortune, we'll be accused of "gold plating." However, we certainly welcome your ideas on features we should look for in new equipment—or on any other aspect, for that matter.

E: (*Trouble is, six months after you express an idea, you still don't know what became of it.*) I'll try to come up with some suggestions.

M: (*He had me worried for a while, but I guess I handled him pretty well.*) Good. I know the suggestions will be good ones. Now, let's go on to the part where we discuss your future progress. Where do you see yourself ten years from now?

E: (*I'll tell him what I told him the last three years.*) I guess I see myself as manager of a plant somewhere in the company. I've made that statement before, but I never did ask you what my chances were, specifically.

M: (*That's a rough one. There are a lot of guys in operations competing with him. I don't have the foggiest idea if and when I'll get promoted out of this plant, and who knows how many other plants are going to be built? I better not say things that will sound like a commitment.*) Well, I certainly regard you very highly. You have an excellent background in one aspect of plant management. Of course, I don't have to tell you that there's more to plant management than keeping the equipment onstream. There's keeping the union happy, keeping the community happy, keeping the marketing and quality-control boys in headquarters happy, not to mention those guys in finance and accounting. Do you really want to take on all those headaches?

E: (*As far as headaches go, the maintenance department isn't exactly a bed of roses either.*) I think my maintenance background has prepared me for a lot of those headaches. I have to deal with the union, I have to check quality, and I don't have to tell you that I have to work with budgets.

M: Yes, but you don't have much background in production.

E: Well, how about giving me an assignment in production for a year or two?

M: (*I'll have to get out of that one—I need him in maintenance.*) Ed, that's not a bad idea—I wish I could give you such an assignment right now. But, frankly, you're the best man we have for the maintenance department, and we need you there. Without a good maintenance supervisor, this plant could go to pot in a week, and I can't take a chance on putting in a new guy—at least, right now. Besides, what if you decide you don't like it in production? Your responsibilities would be less—I'd have to start you off as a unit supervisor, which normally carries a lower salary classification than the one you have now. I tell you what, why don't you take these courses in industrial management that some of the production boys have taken—that way, you'll broaden your operations background without having to give up your present assignment.

E: I already took those courses. You even complimented me when I got my certificate.

M: Sorry. Stupid of me to forget. We'll have to find some way of putting these courses to work—I'm sure that some of them will be useful in maintenance, as well as production. Anyway, seeing that it's almost lunchtime, let's get to the last part—go ahead and sound off on any part of your job. Don't pull any punches.

E: (*Is he kidding? If I tell him what I really think, he just might put down that I'm not "managerial" or "enthusiastic" enough to be a candidate for plant manager, so I better not go too far. But I'm not going to let him off the hook entirely.*) I do have a few ideas beyond the ones I discussed in those memos—for instance, on long-range planning in regard to the equipment repair problem.

M: Fine. Tell you what, why don't you outline them in a memo, just to give me something concrete to look at?

E: (*And to give you something you can bury in your "Hold" box six months. Ah well, at least it will show I'm trying. Now for the point I've really wanted to raise all along. I hope I don't botch it.*) Seeing that you said I should feel free to discuss any aspect of my job, would you mind if I asked a question about salaries?

M: Ed, you know that this is not supposed to be a salary-review interview. (*Maybe that sounded too autocratic.*) But go ahead anyway.

E: Thanks. Well, as you yourself have pointed out, being responsible for maintenance of all equipment is a pretty important job. I like the job, even though it means long hours. But I guess I'd be even happier if I was making a somewhat higher salary. That way I could worry less about our household budget and concentrate on the maintenance budget.

M. Do you really feel you are doing all that badly salary wise? I believe you said you were happy with the $1,000 raise you got nine months ago.

E: (*Yes, but I didn't realize that $1,000 was a 6% raise in a year when the cost of living went up by about 6% and when my family's expenses went up by twice that much because the older kid started college. But I can't put it on that basis.*) I'm still not complaining. But you know, the difference between doing a good job and a mediocre job in maintenance can mean hundreds of thousands of dollars a year, and I guess I'd be less impatient about becoming a plant manager if I could do a little better right here in maintenance.

M: (*He may deserve a $2,000 raise, but I doubt whether salary administration would go along.*) Ed, let's be realistic. I've already mentioned that maintenance budgets are going to be under pressure next year. You'll get a decent raise when you become due, but I can't promise you more than the last—not unless you can come up with something pretty spectacular by then. You know, a program that is written up somewhere and copied by other plants in the company—that sort of thing.

E: Well, how about the "Predictive Maintenance" program I proposed in that memo?

M: (*Is he back on that jag again? He's beginning to sound like a broken phonograph record.*) Predictive Maintenance is a fine concept, but I doubt whether we could get the appropriation to start it off, what with money being as tight as it is.

E: (*That's what he said before. He's beginning to sound like a broken phonograph record. Also, he's beginning to sound grouchy. That's my cue to wind things up.*)

Appraising the Appraisal

We leave it to the reader to decide whether the interview, having gone full circle, is now going to come to a graceful or graceless conclusion. It really doesn't matter very much, because even a polite, graceful conclusion obviously can't remedy the shortcomings of what has gone before.

Just how badly did the plant manager handle the interview? I have tried to avoid having him do *everything* wrong; he deserves credit for trying to:

1. Maintain an informal, reasonably democratic atmosphere, rather than pulling his rank.
2. Give Ed a fairly well-balanced picture of his performance.
3. Avoid criticism of personality, or of trying to make Ed over in his own image. (Note that the plant manager was wise enough to avoid statements such as "you should be more aggressive" or "you should try to get along better with people"; such general statements lead to resentment and defensiveness more often than to constructive change.)

On the negative side, let's skip the points that are minor or self-evident, and concentrate on those best lending themselves to constructive criticism.

First of all, it seems that plant manager Marty has used the same old format on Ed year after year—starting with the pat on the back, then pointing out some obvious area for improvement, and so on. For a senior employee like Ed, it's no wonder the whole procedure has become empty. It seems wise to consider that men need different approaches at different times.[2]

For instance, in a performance-appraisal and counseling interview for someone who has held a particular assignment less than a year, the aim should be to develop a better understanding of the specific job responsibilities, the indicators by which the degree of success is being measured, and the resources or tools the man can use to meet his goals.

At the other end of the spectrum, the interview with someone like Ed can center on new ways in which the company could benefit from his years of experience—e.g., via enlargement of responsibilities—and on other ways in which the company could help him maintain professional vigor. In this connection, it is a mistake to assume that the only day-to-day "vitalizers" Marty can offer people like Ed are pats on the back. A sampling of general vitalizers is shown on the opposite page; the interview can bring out the ones that are most appropriate to Ed and his situation.

Of course, in talking with someone like Ed, the plant manager must be prepared to discuss career goals in somewhat more specific terms than were used in the interview. Marty must have had a pretty good idea of Ed's goal before the interview started, and he should have prepared himself to give a more meaningful answer, not just about plant management, but also about possibilities of promotion to a position such as corporate maintenance consultant. (More on this aspect in Part II, when we discuss how interviews fit into the general performance-evaluation picture.)

Problem-Solving vs. Sitting in Judgment

The foregoing is but a preamble to what I consider to be the greatest failing in the Marty-Ed interview: Their conversation did not generate any new insights or stimulate any new plans of action, other than the very general plan of trying to reduce costs. The relationship between Marty and Ed was the very sterile one of judge vs. appellant (or, occasionally, salesman vs. skeptical customer).

Ed was not expected to really participate in the first half of the interview; his function was to listen to the judgment, perhaps bring up some minor comments, and then absorb a pep talk on performance improvement. And this goes against some major—and healthy—trends in performance-appraisal interviews, the principal ones being:

1. Greater emphasis on self-appraisal or, at least, on fuller participation in the appraisal process. This means giving an employee enough information so he himself can determine whether or not he is doing a good job. He naturally must have a review with his superior, but this review is chiefly a matter of putting things in proper perspective.[3] The major effort in terms of performance analysis is put in by the interviewee, and such an analysis is much more likely to stimulate constructive, nondefensive action than an appraisal in which the employee did not participate.
2. Greater emphasis on setting specific objectives, and analyzing how well they have been met.

The idea of specific, quantified objectives is hardly new, but the last five years have seen much wider acceptance of this concept. One factor in this acceptance was a study by General Electric Co., published in 1965, which indicated that the traditional type of annual interview was of questionable value in getting people to improve performance. The praise at such interviews tended to have no effect, perhaps because it was regarded as the sandwich that surrounded the raw meat of criticism. And criticism itself brought on defensive reactions that were essentially denials of responsibility for a poor performance. General Electric found that superior results were achieved when goal-setting, not criticism, was used to improve performance—when the manager and the subordinate *together* set specific goals to be achieved, rather than merely discussed needed improvement.

Further, the study suggested that such sessions should be held as needs occur—every few months, if necessary—and that this produces much less of a "threat" feeling than the formal, annual appraisal and salary review discussions. (Some companies using this "as needed" approach have found that there is some value in having an annual discussion as a supplement and sort of summary.)

Goal-Setting in Action

"Performance-evaluation techniques based on praise and criticism have little meaning in an employment market under pressure from a shortage of applicants. Honest criticism must stand up to the possibility of losing an employee, particularly one who does not yet have a stake in pension plans, etc. Few supervisors are courageous enough to criticize the imperfections in a generally satisfactory young engineer when the department is already short of qualified technical personnel."

The above opinion was voiced by Steve Simpson—who, as Vice-President of Fanning Personnel Agency in New York City, has a good feel for the realities of the technical manpower market. Mr. Simpson thus confirms that goal-oriented interviews based on participation and problem solving are much more in tune with the times than interviews based on critical judgments.

Aside from the turnover question, individuals who have been able to influence their superior in planning the activities involved in their responsibilities are more likely to be committed to the accomplishment of these objectives. Furthermore, the men most intimately familiar with the job to be done can make a significant contribution to the planning process.[4]

This approach is illustrated in a new performance and development counseling program being put into effect for technical and professional employees in one

20 Vitalizers to Restore an Engineer's Vigor

All of us get into a performance rut sometimes. It is unrealistic to expect that an annual interview —all by itself—will act as a "vitalizer" for a whole year. However, an interview can offer clues as to what kind of other vitalizers are most appropriate.

If you are a technical manager, and want to help one of your subordinates maintain vigor in his job but can't offer him the hope of an immediate promotion, consider some of these day-to-day vitalizers, and perhaps add some of your own:

1. Help him add to his skills via a new or enlarged assignment.
2. Send him to a conference, and ask him to report back on matters of departmental interest.
3. Send him on a special trip (e.g., to suppliers, contractors or other plants), and ask him to report back.
4. Ask him to review special books and articles.
5. Encourage him to write an article or present a paper.
6. Assign him to a plant or departmental committee.
7. Relieve him of some routine work to make way for something new.
8. Ask him to help train someone else.
9. See to it that officials above you recognize his ability.
10. Show in the presence of others that you understand and support what he is doing.
11. Ask him for his advice on how operations could be impoved, or on other problems (assuming you are prepared to take some sort of action after you have solicited his views).
12. Get him together with others who could benefit from the mutual acquaintance.
13. See that the group he belongs to is recognized.
14. Help him see the results of his labors, or let him participate in the implementation of his ideas.
15. Give publicity to one of his achievements—e.g., in the company newspaper.
16. Help him see the relationship between his work and the work of others.
17. Give his group a special task that connotes confidence.
18. Show concern for his general welfare, and interest in him as a person.
19. Call his attention to something you have heard or read that he is interested in.
20. Increase his participation in setting up goals for his department, and in helping to measure how well the department has achieved these goals.

(Note: These 20 vitalizers stem from a list of 39 suggested by Dow's J. H. McPherson,[2] and are intended only as a sampling to stimulate your thinking.)

large petroleum company. Several years ago, its management recognized growing employee interest in having a better understanding of the basis for performance appraisals, as well as more voice in work and development planning. The new program was developed in response to these interests, and its unusual feature is the way in which the individual employee's work and development objectives are planned so as to support the organization's goals and objectives.

Here is how the program works: The employee fills out a form listing what he has accomplished in the period under review (typically six to twelve months) his tentative work objectives for the coming period, and his career-development interests and plans. After his supervisor has studied the information on the form, a performance and development discussion is held between supervisor and employee. An appraisal of the employee's performance, which is made following the discussion, is reviewed at several higher levels. This results-based appraisal becomes the basis for salary treatment and future advancement consideration.

Efforts to introduce the program began last year. These include a one-day training seminar for supervisors, which the company considers essential to the program's success. Well over half of its supervisors have attended the seminar and others are scheduled for it.

A Typical Set of Goals

Turning to a major chemical division of another company, a switch from trait-centered to goal-centered appraisal interviews seems to have worked out very well. The new approach is not only more popular with the interviewees, but also with the interviewers—they used to recoil from the task of telling an engineer (and noting on his personnel record) what they thought of his dependability, initiative and so on.

I asked the head of maintenance for this division what sort of goals he would set for someone having charge of maintenance for a particular plant or unit (i.e., someone like our friend Ed). He felt that the following set of goals, arrived at with the participation of the interviewee, would be typical:

1. Maintain or improve maintenance costs (relative to capital investment or some other criterion) despite projected increases in cost of labor and materials.
2. Complete the year with no disabling injuries, and with a 10% decrease in injuries requiring medical attention.
3. Add three more plant units to the preventive maintenance program.
4. Develop an objective and explicit method for evaluating the results of preventive maintenance.
5. Develop and place in operation a technical program of corrective maintenance, aimed at locating and solving recurring problems with equipment.
6. Broaden the nondestructive testing program to include more equipment.
7. Improve the tool room operation by (a) optimizing the number of tools via better usage information, (b) devising a more efficient checkout system, (c) using craft toolboxes to cut down on trips to the toolroom, and (d) developing a tool-condition inspection program to improve safety.
8. Improve manpower scheduling effectiveness by increasing to 70% the number of craftsmen who work on planned and scheduled jobs compared with the total craftsmen.
9. Develop a method of handling a fluctuating engineering workload using consultants; place this in operation and evaluate its success.

Benefits and Challenges

Summarizing some comments sent in to us by R. L. Howard of Du Pont's Beaumont (Tex.) Works, performance interviews are a good way to clear the air; they are vital for goals-directioning, and to promote understanding between supervisor and engineer. They permit the

- *Interviewee* to pause and firm up job goals, to air any problems or gripes, to review his own strengths and weaknesses in performance, and to make job suggestions.
- *Interviewer* to make suggestions in areas of the job or performance that may not be aired at group meetings, and to become aware of the engineer's attitudes and aspirations.

Reaping these advantages of goal-directed, participative interviews is by no means automatic. The main challenges are to keep the interview on course, to come up with meaningful goals and yardsticks, and to promote some degree of consistency with the department's managerial and salary administration philosophies. We will examine these aspects in Part II, scheduled for the next issue. ■

References

1. Laurangel, E. C., "Counseling Interviews . . . or Brainwashing?," *Chem. Eng.*, Apr. 7, 1969, p. 156.
2. McPherson, J. H., "How Am I Doing,?" a booklet published in 1968 by Pendell Co., P.O. Box 288, Midland, Mich.
3. Meyer, H. H., et al., "Split Roles in Performance Appraisals," *Harvard Bus. Rev.*, Jan.-Feb. 1965.
4. Glasscock, E. M., private communication.

Meet the Author

Herbert Popper joined CHEMICAL ENGINEERING's editorial staff in 1961, following five years as a chemical supervisor and plant engineer with the Clorox Co. He has been interested in performance appraisal for some time, having touched on that topic in the "Career Crossroads" conference-in-print that he developed in 1967. Popper is a director of the N. Y. section of AIChE, and is on the local professional-development and program committees of ASME.

Performance Appraisal And the Engineer—II

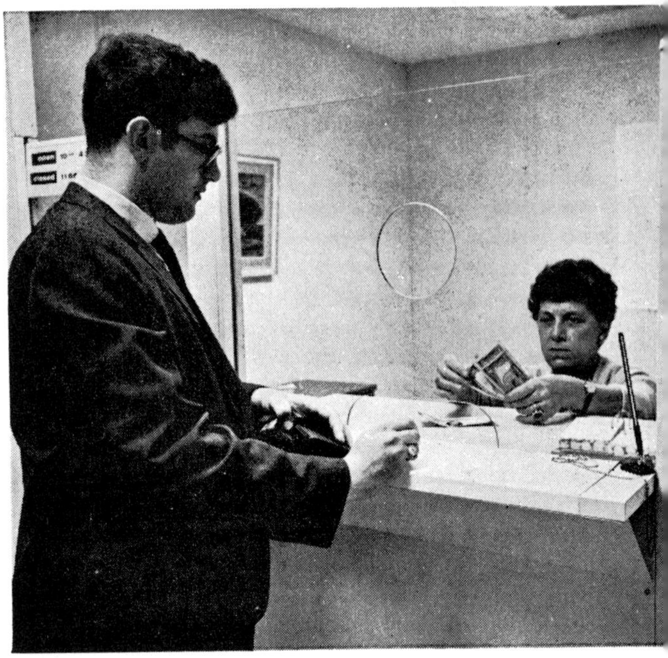

We explore several key relationships—including those between salary and performance, and between what is promised and what can be delivered in counseling.

HERBERT POPPER, Chemical Engineering

In the previous article, we examined the justification for any sort of performance-appraisal interview, illustrated some pitfalls, and then discussed the trend toward goal-centered performance appraisals in which the interviewee is an active participant.

We concluded by pointing out that the benefits from this approach—which can be substantial—depend on keeping the interview on course, coming up with meaningful goals and yardsticks, and maintaining some degree of consistency with the department's managerial and salary-administration philosophies. Let us now come to grips with these problems, one at a time.

Keeping the Interview on Course

In his influential book "The Appraisal Interview,"[1] Maier discusses three methods the interviewer might use:

1. *Tell and Sell.* The initial objective here is to communicate management's evaluation of the employee to him. It is assumed that the evaluation is accurate and fair; the supervisor's job is to let the employee—i.e., the engineer—know how he is doing, gain his acceptance of the evaluation, and get him to follow a plan outlined for his improvement.

2. *Tell and Listen.* This method provides more two-way communication than the previous one. The supervisor covers the strong and weak spots of the engineer's job performance during the first part of the interview, postponing any points of disagreement for the second part, which is devoted to thoroughly exploring the engineer's reaction about the interview, thus releasing any defensive feelings. In short, the supervisor is acting as a judge during the first part, and as a listener and counselor during the second.

3. *Listen and Problem-Solve.* Here, the emphasis is not on communicating a preformed evaluation, but on stimulating the growth and development of the engineer by having him examine his job and his problems. The interviewer is a helper rather than a judge; he should not try to fit solutions to problems until they have been thoroughly explored. That way, he may learn as much from the interviewee as the latter will learn from him.

The first method—tell and sell—may be successful if there is such a wide gap between the capability or experience-level of the supervisor and the engineer that a father-son type of relationship is acceptable. The second method requires that the engineer feel sufficiently uninhibited in the presence of his supervisor to eventually express, rather than suppress, his defensive feelings (and that the supervisor will approach the engineer's responses with a reasonably open mind, rather than with defensive feelings of his own).

As was illustrated in the previous article, these two approaches have serious pitfalls, even when used for the narrow purpose of communicating an evaluation and allowing a man to strengthen his weaknesses.

The third approach, the one that emphasizes problem-solving and that relies on an adult-to-adult rather than father-to-son relationship, is the one that is best suited to the goal-centered type of interview.

An interview of this kind can start off with a mutual examination of the engineer's job responsibilities in the light of developments since the past interview. (This is particularly important for relatively new engineers, who may not have a very clear picture of their total responsibilities, or for any others whose responsibilities have been in a state of flux.) The interview can then proceed to a mutual examination of the goals established at the previous interview—how well they were met, how realistic they were in the first place, and what problems were encountered in meeting and trying to meet them. This can then lead into a mutual exploration of how unsolved problems can be tackled, and what can be done to improve the engineer's productivity.

This productivity improvement has two facets—what the engineer can do himself (e.g., by trying new approaches, acquiring new skills), and what the supervisor can do to help (e.g., by providing better communications, refining responsibilities, giving the engineer better support). In turn, this can lead to the setting of new goals, and then to the career counseling part of the interview.

If the interview involves two reasonably articulate people operating in an atmosphere of mutual trust, there is no reason why either one should regard it as an ordeal. Each one realizes that the other is not a trained psychologist, and does not expect him to act with the smoothness of one. But each person should feel that the other one is really listening to him, and is at least evaluating his ideas even if he does not immediately agree with them.

To sum up this general approach, let me quote from a technical manager who has experimented with several approaches, and found this one to be best: "With the emphasis on results and performance, the individual is left with considerable freedom as to how he gets the job done. This avoids the tendency to mold each individual into a stereotype of the particular company or department. A good supervisor recognizes that a variety of personalities, backgrounds and approaches to getting a job done can be a real departmental asset. Also, a good supervisor will avoid advocating a course of action that would lead an engineer to spend an inordinate amount of time remedying his minor weaknesses when he should instead be spending it exploiting his major strengths."

Goal Achievement vs. Salaries

Every system of performance appraisal ever devised has probably run into some problems in the salary-administration area, and goal-directed appraisal is not immune from such problems. However, the problems are less formidable, and the solutions appear more logical than for other systems, particularly in a company whose top management is goals oriented.

Obviously, there must be a strong tie-in between goal achievement and salaries: A supervisor cannot judge a man on the basis of how well he has achieved a mutually agreed-upon set of goals on the one hand, and then turn in an evaluation based upon other factors. But then again, if the tie-in is so strong that it excludes the factors not covered by any specific goals, a warped sense of priorities may ensue wherein the engineer will try to meet his goals at all costs, even if it means neglecting some basic parts of his job, or postponing important new projects merely because they are not included in his formalized goals.

Furthermore, as H. Houser points out on the next page, the engineer who feels that his salary depends on how well he meets his goals and nothing else will try to make sure that his goals are as easy as possible.

To avoid either of these extremes, it is well to:

- Make sure the goals cover key facets of the job, rather than fringe areas or frills.
- Make goal-achievement an important factor, but not the only factor, in determining the size of a raise.
- Review the validity of the goals when necessary, to reflect new priorities—being flexible enough so as not to restrain the individual's creativity. This may mean holding several short goal-setting interviews during the course of a year. (E.g., in some types of work, goals may be discussed at the beginning, critical stages and conclusion of each project, with the annual review serving a summing-up and looking-ahead function.)
- Take into account the difficulty of the goals, as well as all circumstances within and beyond the person's control.

As an example of how such requirements can be met, Monsanto's Organics Div. bases salary administration on three factors:[2]

1. The engineer's responsibility bracket, which is a function of the job he is filling.
2. His experience level, and potential for advancement.
3. His performance rating, based on the goals and coaching systems. An engineer's rating will reflect:

Difficulty of goals	weighted 25%
Amount of goals met	25%
Quality of results	25%
Timing of completion	10%
Supervisory assistance required	10%
Extra activities	5%

Talking in general terms now, and not about any particular company, there is one implied requirement for the complete success of a goals-oriented salary administration system: Management must be willing to give an outstanding raise to a man who has done an outstanding job in meeting difficult goals. If the difference between his raise and that of a merely routine performer is going to be of a token nature, the whole procedure becomes a "token" policy. And since denying a generally capable but routine performer any sort of raise is hard to condone in an economy

How Young Engineers Feel About Goals Programs

H. T. HOUSER
Training and Development Manager, Monsanto Co.

More and more companies are becoming convinced that an effective goals program is essential to participative management; their training and development activities are based on this conviction.

In general, the young engineers I encounter also feel that goals programs are helpful, both as a guide to future action and to tell them how they are progressing. However, there are occasional problems in defining and measuring job-related goals. For instance, if the young engineer is assigned to a design or other technical staff organization, his goals cannot be as neatly defined as if he were assigned to a production unit.

A partial list of goals for a design engineer might be:

1. Complete a heat-exchanger design manual by Dec. 31.
2. Complete the process design work on Project X by Oct. 1.
3. With appropriate production people, complete dry-run and startup procedures at Plant Y by July 1.
4. Prepare economic evaluations of Project Z by Sept. 1.

These goals all have a deadline and are measurable, but completion by the date set might be more difficult. For example, market conditions might dictate completion of Project X six months earlier, meaning extra effort must be concentrated on it at the expense of other goals. Or, Project X might be cancelled entirely, making completion of Goal 2 above impossible. Also, completion of the heat-exchanger design manual might mean different things to the engineer and his supervisor; one may consider it complete and the other may see needs for further revisions.

Because of such problems, the young engineers' approach to their goals program is determined, in large part, by their relationship with their supervisors. If the supervisor indicates that an engineer's year-end rating will depend entirely on whether or not the goals have been met, the engineer will obviously list no goals that he is not reasonably certain he can achieve. As Dean James L. Hayes of Duquesne University puts it, "No man will willingly help fashion the club with which he is to be beaten to death." Even well-meaning supervisors sometimes fail to consider all the circumstances involved when evaluating the performance of engineers assigned to them; the engineer's performance rating may suffer as a result. After a few such unfortunate experiences, the engineer is naturally reluctant to be specific in putting future goals into writing.

Finally, I have heard young engineers complain that goals programs often make no provisions for career goals and career counseling. The engineer recognizes that job-related goals are important to him and to the company, but he is equally, if not more, concerned with his career goals. Supervisors who recognize this, who demonstrate the company's interest in the engineer's future, and who include career counseling in their appraisals, are rewarded with a much more enthusiastic attitude from their young engineers, and lower turnover.

With such counseling, the engineer can write his personal goals. A list might include:

1. Begin work on an advanced degree and take two courses per year.
2. Make application for registration and take the necessary exams.
3. Join the appropriate professional society (if not already a member) and seek an active role in the local chapter.

Personal goals such as these are relatively easy to define and measure, and the supervisor should encourage the engineer to attain them.

---

where not getting a raise means a substantial cut in one's standard of living, this means that the company must have an enlightened view on the total amount available for raises.

As Steve Simpson, vice-president of Fanning Personnel Agency puts it:

"Salary evaluations must be released from a rigid, once-a-year stereotype that is easily recognizable for what it is by the engineer. The company that pays the extraordinary producer only marginally more than the routine practitioner runs the risk of losing him: In today's inflationary environment and tight job market, he is under great pressure to "get a piece of the action"—either via nonroutine increases in his salary level, or via profit-sharing, stock plans, special bonuses or other incentives. He wants more than just professional pride to distinguish him from some of the mechanical trades that pay $7-8/hr., plus overtime."

Many companies, of course, have been moving in this "big raise for the big producer" direction. In 1966, for instance, Standard Oil Co. of California scrapped its old pay policies for 9,000 technical and administrative personnel because those policies were insufficiently sensitive to performance factors. The company had long had a merit salary plan on paper, but executives were so preoccupied with the spiral in starting wages for new employees that they tended to give standard-size pay hikes to everyone else. As in many other firms, the main concern was to keep the salaries of older workers ahead of those of the newcomers, rather than to establish a meaningful performance/salary relationship. As one executive put it, "There was a gray orderliness about it all, but there was also a tendency to lose the guy you needed the most."[3]

Under the new plan, which was adopted following a four-year experiment with 1,000 employees, supervisors must recommend the size of each subordinate's raise, ranging from nothing to 20%. The 20% jump is twice the maximum allowed under the old policy. To ensure that the pay in one department does not shoot ahead of that in another merely because the department heads have different standards, top executives regularly review the overall salary growth.

Frank W. Piersol, in charge of salary adminstration, tells *CE* that the merit pay system has been working satisfactorily during the last couple of years. Rapidly rising economic factors have made it necessary to accelerate individual salary progress relative to the past: The range of merit increases is still zero to 20% of salary, but the intervals between increases has been shortened—i.e., the best employees receive the largest increases at shorter intervals. Although Mr. Piersol points out that factors other than pay are also important, he does feel that the pay plan is reducing turnover among outstanding employees because management now has much greater flexibility in rewarding them with fast salary progress.

Career Counseling

The case for career counseling, and for the discussion of personal goals during performance-appraisal interviews has been well made by H. Houser, on p. 140.

During such discussions, a standard question—particularly for younger engineers—has been "What would you like to be doing in ten years?" As McPherson points out,[4] persons are set to be asked that question, and are set to answer it so they won't appear foolish and lacking in direction. Often the questioner assumes that the interviewee is secure and informed in his answer, and will use it as a basis for action without taking into account the real ambivalence or lack of information that may dwell in the person. It might be better to say "Tell me some of your views about the various directions you are considering" or "Most of us are periodically vague about the directions we should take; if you feel that way, maybe our discussion might help you reduce the vagueness."

A Career-Development and Counseling Program: Aims, Procedures, Benefits

J. W. BOWERS
Manager, Personnel Administration
Chemicals and Plastics Operations Div.
Union Carbide Corp.

Although performance appraisal and career development are interrelated, this does not mean that both topics have to be covered during the same engineer/manager interview. Under the comprehensive career-development program for professional employees, which Union Carbide's chemicals and plastics divisions formalized three years ago, we do not review career aspirations in depth during a performance appraisal, but use a separate process.

To start, there is an information-gathering interview between the engineer and his manager. A series of questions that cover all aspects of career developments serve as a guide; they are known in advance by both the interviewer and interviewee, and include such topics as: why the individual is in his current job, the type of assignments on which he has excelled, how he feels about his current assignment (including salary treatment), areas of mutual confidence, clarity of goals, personal development needs, positions of interest in the future, general performance level, etc.

The questions are designed to help the engineer assess his performance, his goals and what he will need to do to accelerate his growth. There is no need for the engineer and his manager to agree on the answers. Where there is disagreement, each man's answer goes into the interview report; the engineer knows that his views have been understood and he knows exactly what his manager's views are.

Next, to obtain the benefits of diversified experience and breadth of knowledge, the manager presents this information to a review group that includes his own superior, one or two others at least one level of management above the man being reviewed, and also a professional personnel man.

In all cases, the manager is responsible for giving feedback from the group review to the engineer, including a clear understanding of the group's feelings and commitments. Together, the engineer and his manager explore more thoroughly his specific training and development needs and work out followup plans.

The career development program is designed as a long-range program for career planning purposes, and it requires periodic updating. The updating may be done at any time at the initiative of either the man or his manager, but it should be done at least every three years.

As a matter of philosophy, neither formal programs nor managers can "develop" individuals per se, but both can contribute to their development not only by encouraging them to learn new skills and to participate in training courses but also by involving them in new experiences that expose them to increasingly difficult problem-solving and decision-making situations. Managers must provide a climate conducive to high performance and growth, but the real effort must come from the individual.

According to A. B. Burns, Vice-President of Engineering, this career-development process works because it is fully supported by top management; the information developed is available to all managers when making decisions about people and jobs. As a result:

• Movements across functions and divisions have become much more frequent because professional people are becoming known on a wider basis.

• When job opportunities arise, candidates from all functions are reviewed.

• Attention is given to people who need and want broader experience.

• The understanding between an engineer and his supervisor has been substantially enhanced.

Such a discussion will be most successful when both participants have taken the time to prepare themselves. The engineer should think about the various issues he should bring up, to avoid having afterthoughts along the lines of "What I really should have asked about is . . ." The supervisor should anticipate some of these questions, and make sure he has enough background on them so that he does not have to say "I'll have to let you know later" to everything. He may have to have preliminary discussion with his own boss or with the personnel department to get the latest information on the actual and projected needs for various types of technical talent within the company.

As Kipp pointed out recently,[5] engineers are rightly becoming much less inhibited in their career questions, and the interviewer should be prepared to give a meaningful answer to someone of clear ability who asks about a career-building approach that may eventually qualify him for top management. If all the interviewer can offer is a platitude along the lines of "Do good work and you will eventually be recognized," the career-counseling part of the interview may do more harm than good.

If the company or division has some sort of formalized career-development plan, some of the "maybes" can be taken out of counseling. For instance, Monsanto's Organics Div. has a formal policy to find new or expanded responsibilities for supervisory people at least every five years if they so desire. Each salaried man is encouraged to "examine his adequacy with relationship to the needs of his job and to his future job needs. He examines himself in areas of education, experience, etc., and lays out a running three-year plan for personal development—with appropriate priorities. The company supports his plan with necessary time off and money for education, and with job-related development activities for experience and exposure."[2]

Many other good examples could be cited, but let's move on to a "bad" example—one that shows what happens when the climate is entirely wrong for meaningful career counseling and performance appraisal.

A Sour View on Interviews

Here is a condensation of one of the contributions we obtained in researching this article. The author, who is in R&D work, gave us his name and company, but asked not to be identified.

I am a department head, and have conducted several performance-appraisal interviews. I myself have been reviewed in other jobs, but only twice in six years of employment here.

In my own case, when I tried to make post-interview improvements where suggested, and then found that neither my salary nor my position improved, I came to the conclusion that the reviews were not serving any useful purpose.

When I review someone here, I am required to "discuss and record employee performance on the job," to "identify those areas requiring immediate improvement," to "plan for future progress of the employee," and to "provide the employee an opportunity to voice his opinions on any aspect of his job."

We do not have a salary-review interview, and this poses a serious handicap. The employee is not unreasonable in feeling that performance interviews are disguised salary reviews. He is told that they aren't, but it takes either a brave or a foolhardy man to criticize his boss or company at his only annual review.

I do not like to give a man a single annual review of his performance, but I suppose that, if not for a hard and fast date for an annual review, some men would never receive any review at all. I think I can keep a man fairly well informed on a continuous basis, so it is no terrible strain to put something down in writing once a year.

However, what satisfies me might not satisfy another supervisor. I may tell my subordinate to continue doing exactly what he is doing, but he may be in for a shock if he moves into another department.

This brings up a vital point: promising more than one is authorized to promise. I have to be very careful to explain to an employee that the review is strictly my own appraisal; I am not able to say, "The company wants . . ." In some cases, I have company policy to rely on, but in the performance features that are most clearly related to the employee's rating, I feel that it is really what I want that counts. Nevertheless, he must understand that there are certain limits on the improvements I am capable of making for his situation.

What happens if I tell him, "You're doing a good job. I depend upon you to do things right and on time, and you always come through," and then he gets no salary raise? What will he think? It has happened to me, and I know what I thought; after the raise did not appear, I would ask what had gone wrong, and then receive some lame excuse about extending reviews to a greater period than a year. The explanation would have satisfied me if told to me in advance, but after the fact, it was worthless.

Planning for the future progress of an employee is also in this category of unwarranted promises. There is nothing I can plan for him without first clearing it with higher levels of management, and even then, management tends to reverse its decisions. Our company follows an erratic course, and neither my subordinate nor I can feel any confidence that a long-range program will be carried out.

I can specifically encourage a subordinate to take certain classes or correspondence courses, for which he will be partially reimbursed. But I can't assure him that his paycheck will reflect his efforts at self-improvement.

As for the employee expressing his opinions . . . well, he has to sign that review form, and it goes into the permanent file. Who wants to see anything critical or disloyal get in writing? Does anyone think that I would put criticism in writing, if I thought the man was doing a good job, with certain exceptions?

The way it works out, about one year out of every three I try to conjure up some rather mild statement regarding a possible improvement. If I have anything stronger to say, I tell a man verbally, and nobody else ever knows about it.

System for Total Integration of Goal-Setting, Performance Evaluation and Feedback

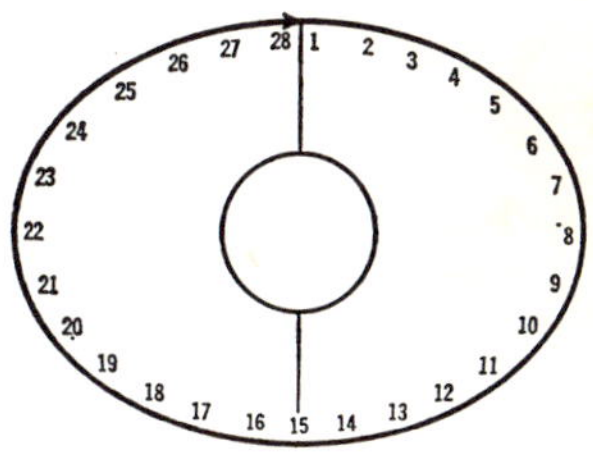

1. Set corporate goals, such as sales and earnings (in interaction with divisional levels).
2. Set divisional goals (in interaction with corporate level and with units).
3. Set unit goals (in interaction with division and person).
4. Set individual responsibilities, goals and indicators (in interaction with unit and persons).
5. Assess corporate progress toward goals.
6. (Feedback through system.)
7. Take corrective corporate action.
8. Assess progress toward division goals.
9. (Feedback through system.)
10. Take corrective divisional action.
11. Assess progress toward unit goals.
12. (Feedback through system).
13. Take corrective unit action.
14. Assess individual progress toward goals.
15. Supply feedback to individual via job performance review.
16. Take individual corrective action.
17. Evaluate corporation's achievement of its goals for the year.
18. Supply corporate feedback—"state of the corporation."
19. Evaluate the division's achievement of its goals for the year.
20. Supply division feedback—"state of the division."
21. Evaluate unit's goal achievement for the year.
22. Supply feedback to unit—"our group's progress."
23. Evaluate individual's goal achievement for the year.
24. Supply feedback to the individual.
25. Supply administrative feedback to the corporation about performance of divisions, units and individuals, as basis for reward-setting.
26. At corporate level, pass out rewards to stockholders and divisions, with appropriate comments.
27. At divisional level, pass out rewards to units, with appropriate comments.
28. At unit level, pass out rewards to individuals with discussion of relationship to goal-achievement level and economic conditions.

Note: Some of these activities may occur simultaneously. The process is a continuous one and, after Step 28, reverts back to Step 1, marking the beginning of a new round of goal setting, appraisal, feedback, correction, and ultimate reward.[4]

Managerial Philosophy

I regard the critical contribution not as an indictment of this or that kind of interview, but rather as an indictment of the atmosphere in which interviews are sometimes being held. Specifically, a company that follows an extremely erratic course in regard to wages and promotions—one that lacks effective communication between supervisors and higher management, and that also seems to lack any sort of long-range business philosophy—cannot expect the simple act of holding annual performance interviews to lead to high morale and motivation.

There are many "management styles"—authoritarian, benevolently autocratic, consultive, participative, and management by objectives. Each of these styles is probably useful in certain situations, and each undoubtedly affects the methods used in the job-appraisal interview. If the style is one of control, there is apt to be a concentration of "How did you do?" in the interview. If the style involves participation, objectives and goals, there is apt to be a concentration on "Where have we been and where are we going?"[3] These latter styles have been growing the fastest in recent years—hence the trend toward participative, goal-directed interviews.

There are other management styles that are usually not mentioned in the literature. As McPherson points out,[4] there is "management by crony," a political method whereby whom you know is more important than the achievement of goals. There is "management by cue," a system that operates on a constant search for the cues about how top management feels about an issue as a guide to behavior. Then there is "management by vendetta," which has as its goal the reduction of one's enemies and the validation of one's past judgments. There is "off-balance management," designed to keep employees guessing so they will cater to the throne in search for information to use in the reduction of their anxiety.

So the context of the total organization must be considered as the background for the behaviors that occur in the performance-appraisal process. The stated business objectives help a person fit his work in with organizational goals. The value system of the organization and the management style used to implement these values gives the emotional context for the performance-appraisal system.

Consequently, those who wish to improve their appraisal systems or those who wonder why their current systems are inadequate may find it wise to look at the organizational context and managerial style.

The managerial style of the company for whom the anti-interview contributor of the previous section works seems to be a mixture of "authoritarian" and "off-balance" management, with a dash of "management by erratica." (I don't think the last style needs to be defined.)

Could goal-directed interviews be a success in such a setting? Not very easily, although it is possible to conceive of a "tail wagging the dog" situation in which a goal-directed appraisal-interview and reward system that managed to establish itself in one department despite all obstacles might influence the company to move closer to the "management by objectives" style.

Space does not permit a detailed discussion of management by objectives. Perhaps the ideal system of this type is one that integrates the goal-setting, per-

To Tell or Not To Tell

If you are a technical manager and you use numerical performance ratings for your engineers (e.g., 1 = outstanding, 2 = excellent, 3 = good, etc.), should you tell each man his actual rating? Here are some opinions pro and con.

NO

1. All men will be resentful unless they're high on the ladder.
2. The top performers all want to be 1's. None of them want to be 2's.
3. Knowing he has a low rating will only depress the person and will not increase his motivation.
4. If you tell him he is a 3, he'll ask you why, and if it's because you used a forced distribution you can't really explain why he was given a 3 and another man a 2. You get into trouble because the numbers are discrete, yet the differences between the people are not.
5. Men will talk over their ratings with each other. The 4's will be disgruntled, the 2's will scapegoat the 1's, and the 3's might "settle in."
6. You might lose some good 2's just because they are not 1's.
7. There is a place for some good 3's but they don't want to be known as 3's.
8. This rating is an administrative one that management uses for decision-making purposes and should not be revealed to the employees.
9. The rating system hasn't been perfected yet and explaining it would be premature.
10. The employees are bright enough to pick up the cues about their rating, you don't need to tell them.
11. It's good to keep them searching---it's part of management's control system.

YES

1. Many men kid themselves and maintain the illusion that they are doing well. They can do this for years and kindly management can reinforce their illusions. Often a rude awakening occurs too late for them to make a transfer to other occupations.
2. Top performers that the company very much wants to keep do not get the feedback they need to see the rosy future pictured for them.
3. A communicated low rating is not devastating if it is accompanied by a mutual specific plan for improvement.
4. The disagreements between a man and his boss about how he is rated can lead to a resolution that is beneficial to both of them.
5. Management may need its "protected rating," but whatever the rating, it should be consistent with the information given the man in order to reduce the hypocrisy in the system.
6. If the system is developing, why not say so and explain some of the unsolved problems that are being worked on.
7. Some employees are cue sensitive, others are not. The ones who are most in danger may be the least sensitive.
8. When an item as important as a rating is maintained on a secretive basis, human nature dictates that doubts will arise and integrity will be questioned.

formance appraisal and feedback systems from the corporate level all the way down to the individual level, as shown in the table on p. 143.

While companies vary in how closely they approach this total system, the concept of goal integration between all levels of the company (including the individual level) via joint goal-setting sessions is becoming widely accepted.[6]

With this in mind, let us give "salary" the last word.

Salary Administration Revisited

As we have seen, the goals-directed approach assumes that the engineer and his supervisor can jointly establish a set of responsibilities, goals and goal-attainment indicators appropriate to the job, and to the department's objectives.

There are certain skills and approaches to problems that are instrumental to the achievement of goals—e.g., communication skills, specific technical knowledge, ability to work smoothly with line and staff people, etc. These instrumental factors, as well as the goals themselves, are of course discussed in goal-directed appraisal interviews: in fact, a major part of the interview can be spent on various facets of these instrumental factors. But in making salary recommendations, the emphasis should be on goal attainment, not on the instrumental factors. If an engineer has skills and talents that he has not used in meeting his goals, it is up to him and his supervisor to set up new goals that will enable him and the company to capitalize on these talents.

As for how the spoils should be divided, we have already discussed one system for translating goal-achievement into salary ratings. Various other systems are in use or are possible. McPherson[4] discusses a relatively simple system: A group has achieved 90% of its goals and objectives, hence the group manager gets a larger slice of money than if the group had been less proficient. There are 12 men in the group; 8 of them have achieved 90% of their goals, as measured by the agreed-on indicators. They each get the same, above-average, increase. Of the others, 2 achieved 50% of their goals, and get only an average percentage rise. Unusual circumstances make it unwise to evaluate the eleventh man at the usual time. The twelfth man has achieved only 10% of his goals. He is transferred or released.

In some systems, the supervisor has to supply a numerical administrative rating to express a man's performance. Furthermore, managers are sometimes asked to prepare a forced distribution of their overall ratings—for instance:

Percent of Staff	Rating	
10	1	Outstanding
30	2	Excellent
40	3	Good
18	4	Adequate
2	5	Inadequate

Such ratings may tie in with internal guidelines where a "1" can be given a raise of up to $x\%$ of his salary at a minimum interval of a months; a "2" can be given up to $y\%$ of his salary at a minimum interval of b months, and so on.

Since the rating affects pay, promotional outlook, etc., one can argue that the engineer should be told his rating. However, not everyone agrees, and the issue causes heated arguments. The table on p. 144 looks at some of the pros and cons.

Some managers resolve the dilemma by forthrightly giving the employee the actual number of his rating, and taking it from there in the discussion. Others try to communicate the number in an oblique fashion. In the goal-directed system, the solution to the dilemma might consist of the supervisor and engineer agreeing that the engineer has met, say, 90% of his goals, the supervisor then saying that he will therefore turn in a high evaluation, and letting it go at that.

This brings us to salary-review discussions. Should a salary discussion be part of the goals-directed appraisal interview? Most people say no, because salary talk tends to inhibit or overshadow the other discussions. Once the engineer knows how his boss views his goal achievements in regard to difficulty, timing and special problems—and the engineer can usually pick this up from the appraisal interview even if he is not told specifically—he should be able to get a fair idea of how he will do in terms of salary.

At the time he gets his raise, the relationship between the amount and his performance should be made apparent if it is not already so, the social justice of the raise may need to be related, any corporate or economic factors that affect the amount given must be explained, and the engineer must be given a chance to express his reaction to the raise, to which the manager must listen carefully.[4]

There is a school of thought that says no matter how big the raise, the recipient should be reserved in his acceptance, to imply that it was no more and perhaps less than he deserved, and that the company should regard it as a floor rather than a ceiling for future raises. There is another school that says such a policy may label the recipient as a mercenary, politicking ingrate if the raise is an unusually big one, and if his boss went to some trouble to get it for him.

Having tried to give or cite positive advice on most matters related to performance appraisal, I think I will leave this particular matter to the reader's best judgment, in the hope that decisions on how to react to bigger-than-expected raises will plague him many, many times. ■

References

1. Maier, N. R. F., "The Appraisal Interview: Objectives, Methods and Skills," Wiley, 1958.
2. Schneider, H. L., "Motivation in Plant Engineering," paper presented at the 20th Plant Engineering and Maintenance Conference, Cleveland, Mar. 24, 1969.
3. Lawrence, J. F., "Some Firms Abandon Uniform Pay Increases, Reward Big Producers," *Wall St. J.*, Apr. 22, 1968.
4. McPherson, J. H., "How Am I Doing?", a booklet published in 1968 by Pendell Co., P.O. Box 288, Midland, Mich. 48640. (Excerpts reprinted with author's permission.)
5. Kipp, E. M., "Custom-Building Your Management Career," *Chem. Eng.*, Apr. 21, 1969, p. 116.
6. Glasscock, E. M. (management consultant), private communication.

How Job Descriptions Can Help You

Job descriptions can benefit the engineer as well as his supervisor. With their help, the engineer can analyze and evaluate his work, understand his relationship with others, and determine his lines of promotion.

CONRAD BERENSON, *The City University of New York*

Job descriptions have been a useful management and self-development tool for nearly 30 years, but they are still neither fully understood nor very widely used in the chemical process industries. For instance, few engineers are aware of the personal benefits to be obtained from a set of job descriptions. These personal, as well as managerial applications, will be described below.

The engineer may be fortunate enough to be given a set of job descriptions by his firm. If he can't obtain them, he will still find it very rewarding to write the description for his own job and for the jobs of people with whom he works closely (his supervisor, for example).

Unfortunately, the term "job descriptions" is used differently by various people. We will define job descriptions as: written reports that outline the duties, responsibilities and conditions attendant to the work assignment—and that specify the work performed, the skill and training period required, and the personal qualifications required by the holder of the job.

Bear in mind that the job description is a description of the job and not of the individual who happens to hold the job or who will hold the job. When a description is written so that it emphasizes the personnel requirements in order to facilitate employee selection and placement, it is frequently called a job specification.

Personal Benefits of Job Descriptions

The engineer or scientist in the chemical process industries can gain important personal benefits from job descriptions by using them in:

1. Analyzing personal job duties. As an example, let us consider Fig. 1, which is the description for a group leader in a thermosetting plastics laboratory. Assume for the moment that our group leader, Kandel, has just obtained his position. Among the questions he is likely to have are many that can be answered by his job description, e.g.: How much time should I spend on providing technical service to the firm's customers? What importance does my organization put upon attendance at technical meetings? Reference to Fig. 1 will show that the latter question is answered by Job Duty 5; "Attends frequent scientific meetings on thermosetting polymers at universities . . . occupies 10% of Group Leader's time." Job Duty 6 answers the former question.

The job description will enable Kandel to develop a better and faster understanding of his job, and that of his subordinates. He will be able to refresh his perspective anytime doubts arise as to the scope of his job, or the relative importance of his various responsibilities.

This periodic refresher can be very important to some people. Take Brecon, a plant superintendent, and let's say he rose to this responsible position by doing a top-notch job of managing plant maintenance. He still enjoys spending time on maintenance matters; he feels more secure making decisions in this field than in any other. If he is not careful, he may find himself short-changing his many other duties (and possibly cramping the style of the people with direct maintenance responsibilities). Periodic study of his job description, Fig. 2, will help him correct his perspective, and allocate his time and energies in accordance with the real needs of the job.

2. Understanding the relationships between the jobs of those with whom the engineer must work. For instance, if the product development engineer reports only to the director of research, just what is this engineer's responsibility to the marketing manager when the latter asks him to increase the average molecular weight of a polymer? Examination of his own job description plus those of the other individuals mentioned above should provide the chemical engineer with the basis for the proper decisions in this area.

3. Determining lines of promotion within the engineering organization, as well as within the firm as a whole. Many job descriptions have (and all should have) a statement as to the next step and alternate next steps on the promotional ladder. Thus, referring first to the description of his own job and then to the descriptions of jobs one step above will provide the chemical engineer with a basis for determining the resources that he will have to bring to the next job,

i.e., resources of education, skill, experience, physical ability, and so on.

He can thus formulate a realistic program to obtain the necessary background. Or, if upon reading the description of the job he wants, the chemical engineer finds that he already has the necessary background, or feels that certain of the specifications are unreasonable and not needed for the job, he can try to sway management to his point of view.

The important fact to remember is that approaching the right job should not be treated in a haphazard fashion by the applicant—long-range planning is necessary. And, in order that this long-range planning be properly directed, the characteristics of the job goal must be defined, i.e., the job description must be known and understood.

4. Evaluating your job performance by comparing what you do on the job with what the job description says you should do. It is always extremely difficult to evaluate one's own performance. The job is made a great deal simpler by having a description to use as an objective basis for this evaluation.

Referring once again to Fig. 1, if Kandel, the group leader, is not providing enough technical consultation to customers or enough reports to the director of plastics research, he will ultimately hear about it. But, it is best that he have a yardstick for budgeting his time for these duties, so that any feedback he gets about his job performance is positive, not negative. On the other hand, if Kandel finds that too much of his time is spent on customer consultation or in report writing, he has to decide among several alternatives: Am I attempting to do more than is necessary? Am I too slow in my work? Is the job description no longer suitable and therefore must it be revised?

JOB TITLE: Group Leader-Chemical Research
JOB CODE: A-456 DEPT.: Thermosetting Plastics
DATE: Nov. 13, 1962 LOCATION: N. Y. Laboratory
DESCRIPTION PREPARED BY: M. Kandel

JOB SUMMARY
Provides first-line technical and administrative supervision for chemists and chemical engineers engaged in innovating and developing applications for thermosetting polymers.

JOB DUTIES
1. Assigns areas of investigation to chemists and to chemical engineers.
2. Selects important variables to be studied in the different research problems (occupies 10% of the Group Leader's time).
3. Consults with the workers in his group to provide technical guidance (occupies 20% of the Group Leader's time).
4. Reads technical journals in the field of thermosetting polymers (approximately 12 monthly) to remain in the forefront of the state of the science (occupies about 10% of the Group Leader's time).
5. Attends frequent scientific meetings on thermosetting polymers at universities, industry seminars and trade shows (takes 10% of Group Leader's time).
6. Provides technical consultation to customers with severe thermosetting resin applications problems (occupies 10% of Group Leader's time).
7. Writes weekly and monthly reports to the Director of Plastics Research regarding research and development progress (takes 20% of Group Leader's time).
8. Performs routine administrative duties, i.e., signs time sheets for his group, keeps attendance records, rates his workers, purchases supplies for his group (occupies 20% of Group Leader's time).

JOB SPECIFICATIONS
EDUCATION: Ph.D. in Polymer Engineering or Polymer Chemistry; courses should have included macromolecular theory, differential and integral calculus, differential and partial differential equations (mathematics beyond Fourier analysis is not required), 3 years of physics, 3 years of organic chemistry, 3 years of physical chemistry. Should be conversant with computer applications and mathematical statistics.
EXPERIENCE: 10 years of total relevant work experience beyond the Master's degree date; 5 years of education (student or teacher) may be included in the above. Experience should be in high polymer theory and applications.
PHYSICAL DEMANDS: Good health to meet the demands of frequent travel to and contact with personnel outside the company at seminars, conventions, consulting assignments, etc. Must have vigor and stamina to perform the manifold job duties.
SUPERVISION: will supervise 6-10 chemists and chemical engineers; reports to Director of Plastics Research. Must be able to communicate orally and in writing--both vertically and horizontally in the company organization.
WORKING CONDITIONS: Good. 80% sitting at a desk in a comfortable, quiet, air-conditioned office.
RESPONSIBILITIES: to be able to keep the company in its outstanding position in thermosetting polymers research through his skill in leadership and research.

Technical-job description and prerequisites—Fig. 1

Each of these questions will, in turn, lead the chemical engineer to other questions about the nature of his work and how well he is performing it. In the example just cited, if Kandel had found that 25% of his time was taken up in satisfying the demands of the firm's customers for data about its thermosetting resins, then he should consider such choices as:

- Delegating part of this duty to another group.
- Recommending that his own group be given funds to hire an experienced engineer to handle the consultation work.
- Suggesting that some of his other duties be withdrawn to provide the time for this consultation.
- Initiating a study to see whether or not it is still in the company's interest to provide this customer service, etc.

5. Understanding the basis for the firm's salary structure. The job descriptions enable the engineer to study the requirements of the various jobs. We must bear in mind that it is not the job title that provides the basis for the job's wage scale; rather it is the specific duties, responsibilities, and working conditions by which management justifies its existing wages.

When the engineer seeks to examine why he gets paid what he does, he can gain valuable insight from comparing the descriptions of various jobs—not the titles. Careful analysis of the job requirements and the wages paid, especially correlation of the differences between the various requirements and their financial rewards, will show the engineer just what he must emphasize in his work or background in order to present a valid case for a wage increase.

One trouble spot in this analysis is that usually the engineer cannot determine the firm's wage scale through official channels. But, conversations with other employees, judgment, and help-wanted advertisements for similar jobs in newspapers and professional magazines will provide the sought-after data with reasonable accuracy.

How Management Can Use Job Descriptions

The uses discussed so far are the ones most important to the nonsupervisory chemical engineer. There are other uses of job descriptions that are of primary interest to the engineer's supervisors. However, even these need to be recognized and understood by the chemical engineer. For one thing, they provide his supervisors with a tool with which to rate him. Not knowing how one's work is to be evaluated is to work with a severe handicap.

Supervisors and managers find job descriptions useful in:

1. Assisting in hiring and placing engineers in the proper job. The management of the engineering organization uses the job description as a yardstick for placement purposes. Thus, when an applicant for a group leader's job is interviewed, his background is matched against the job requirements outlined in the description. If the job calls for ten years' experience, a Ph.D., managerial background, etc., then those applicants who do not have these qualifications can be rejected. If the applicants' qualifications match the requirements of some other job in the firm, they can be considered for that.

2. Forecasting the training needs of the engineering department. A vital part of management's job is to determine the type of personnel that will be needed in the years ahead. In making such an estimate, management starts out with a forecast of what future areas of activity the enterprise will be involved in. Following this, preliminary job descriptions are written for those functions that must be performed in these new activities. Comparison of these preliminary job descriptions with the descriptions for present jobs is the next stage. If the comparison shows that skills will be needed in the future that are not used in the enterprise at the present time, training is going to be needed; the study will help management decide what the objective of the training ought to be, how much training will have to be done, etc.

By publicizing the skills that will be needed in the future, management can get employees started on educational programs of their own. An example of this can be seen among those discerning chemical engineers who, in the early 1950's, trained themselves in advanced concepts of the statistical design of experiments. As their firms' need for this knowledge increased, so did the opportunities and pay of those who were ready with the proper statistical skills.

3. Improving the flow of work. By using the job descriptions, it is possible to trace the flow of work from one individual to another, from one organizational

PLANT SUPERINTENDENT

I. BASIC FUNCTION

Responsible for planning and directing the efficient and profitable operation of the manufacturing activities of the division. Provide required product at the lowest possible cost consistent with schedule requirements and quality specifications. Reports and is accountable to the Regional Director.

II. RESPONSIBILITIES

1. Analyze projected short-range sales volume and forecast production schedules required to maintain the sales.

2. Analyze projected long-range sales volume and forecast production requirements to enable the forecasting of required facilities and equipment.

3. Plan, direct, and control the installation schedule of such equipment and facilities and recommend capital expenditures necessary to maintain the forecasted program.

6. Organize, direct, and control the proper use of facilities and personnel to meet forecast production goals, and to achieve the maximum quality and efficiency at minimum operating cost.

7. Accountable for the selection, training, development, motivation, promotion, assignment, and other conditions of employment for all supervisory and hourly employees under his jurisdiction.

11. Coordinates with the Trade and Industrial Quality Control Managers to insure the production of standard quality material with a maximum turnover of manufacturing equipment.

12. Directs the preparation of the Inventory Control Report, Monthly Operating Report, Annual Operating R port, Inventory Control Meeting Reports, Capital Expenditure Estimates, and the Manufacturing Short and Long-Range Planning Report, and coordinates these reports with the Regional Director noting exceptions and with recommendations to correct deficiencies if necessary.

16. Assist the Regional Director in the negotiation of collective bargaining contracts with the advice and assistance of the Corporate Personnel Relations Staff Representative and coordinate all labor problems, employee relations programs, personnel practices, and changes in conditions of employment with the Regional Director and Staff Representative.

17. Serve as a member of the Division Planning Committee, Cost Control Committee, Finished Stock Inventory Control Committee, and other such committees that would pertain to this position.

III. AUTHORITY

1. Capital expenditures in excess of $100 are subje to the approval of the Regional Director.

2. Salary adjustments for supervisory personnel and their staffs and additions and deletions to the staffs are subject to the review and approval of the Regional Director

IV. STANDARDS OF EFFECTIVE PERFORMANCE

1. Accomplishing manufacturing goals and objectives as jointly established by the Coatings & Resins Superintend ent and the Regional Director.

2. The degree to which the Plant Superintendent provides required products at minimum cost, of proper quality and time.

Managerial job description—Fig. 2

unit to another, and from inception to completion. It can then be analyzed with a view toward greater efficiency.

4. Revising the engineering department's organization structure. Should there be more or fewer junior engineers, senior engineers, engineering aides, etc., on the staff? Should the work be organized according to the unit operations performed or according to the nature of the product, e.g., polymers, insecticides or fertilizers?

For example, examination of the job descriptions might show that it is impossible for a chemical engineer to be fully aware of all of the latest developments related to his work in fields as dissimilar as plastics and organic intermediates. The technology in each of these fields, as they are approached by the corporation, may simply be too extensive. Yet the organization chart might reveal that plastics and organic intermediates utilize the same pilot-plant engineers because much of the same equipment can be used in both fields. Logic, as well as competitive pressure to keep abreast of the latest developments in each field, would then dictate an organizational separation into two pilot-plant units—one for polymers, the other for organic intermediates.

5. Reassigning and fixing functions and responsibilities in the engineering organization. This is the sort of activity in which such decisions are made as: Who sizes the heat exchangers—the chemical or the mechanical engineering group? Does the laboratory analyst working in the pilot plant report to the engineering group head or to the leader of the analytical group? To whom does the maintenance man in the pilot plant report—the project engineer or the maintenance foreman?

6. Providing data as to proper channels of communication. To whom does the chief engineering executive report? What is the relationship between the research director and the purchasing director? When a bench chemist finishes a semiannual report, to whom shall he send copies? All of this is the sort of information that is extremely helpful to the chemist or engineer in his day-to-day duties, and it is important for management to provide such data. The last illustration (report copies) is particularly important. Great care must be taken to make sure that every individual who must know the report's contents should get a copy; conversely, only those with a need to know should get copies. Careful study of the job descriptions will indicate the proper people to whom reports should be sent.

Obtaining Job Descriptions

A company can use either its own personnel or management consultants in preparing a set of job descriptions.

A comprehensive source of published job descriptions specific to the chemical process industries is now available.* This, as well as other published descriptions, can be used as a starting point for the development of job descriptions that fill the specific needs of the company, department and individual engineers.

Once prepared, descriptions must be kept up-to-date. The environment is which engineers operate is a dynamic one. Changes in personnel, corporate objectives, product lines, methods and materials of manufacture, budgets, control techniques, technology, competitors, suppliers, etc., all create the necessity for periodic review and revision. (For these and other reasons, job descriptions should not be terribly detailed, and they should not be interpreted too rigidly.)

Engineers should have easy access to the file of job descriptions. Unfortunately some CPI firms treat their job descriptions as top secret. It need hardly be said that this is a waste of good opportunity to improve the engineering department's performance.

There is seldom a well-founded excuse for not supplying each employee in the department with a complete compendium of job descriptions for the engineering department as well as for the personnel of other departments with whom he may come into contact. Unless this is done, the employee knows neither the dimensions of his job nor how he will be evaluated, nor who in the firm can aid him in his work. Not having knowledge of his opportunities for advancement, he is reduced, in general, to working in the dark. The engineer must make every effort to see that the management of his firm provides him with these job descriptions.

If there is no way of obtaining a set, the engineer should consider writing a description for at least his own job—and reviewing it periodically, making revisions as necessary. This will help him allocate his time, set goals for himself, and give him some of the other advantages mentioned earlier. ■

* Berenson, C., Ruhnke, H. O., "Job Descriptions in the Chemical Process Industries," Corporate Publications, New York, 1964. Vol. I: Production and Research Personnel; Vol. II: Managerial and Administrative Personnel; Vol. III (to be published shortly): Marketing Personnel. All three volumes include a compilation of job descriptions obtained from a broad cross-section of CPI firms.

Meet the Author

Conrad Berenson is Associate Professor of Marketing at the Baruch School of Business and Public Administration of The City College of The City University of New York, and Executive Director of The Research Foundation of The City University of New York. He received a B. Ch.E. from The City College and a Ph.D. in marketing and management from New York University. In addition to his teaching and research administration background, Dr. Berenson has had extensive industrial experience as an engineer.

Whether you are considering candidates for promotion or evaluating your own chances, you should know more about . . .

The Mechanisms Of Promotion

CONRAD BERENSON, *The City University of New York*

Did you miss out on that last promotion? Or are you perhaps agonizing over which of your subordinates to recommend for a vacant Area Supervisor job?

In either case—whether you are on the giving or receiving end—a better knowledge of the mechanisms of promotion will improve your future performance.

For the supervisor, there are many rewards to a well thought out promotion system. They include highly motivated subordinates, low turnover and easier recruiting. The penalties for a poor policy are the opposites. Furthermore, the supervisor will himself be rated by his superiors on how well he provides opportunities for promotion for those who work for him.

As for the nonsupervisory chemical engineer, the importance of understanding the factors that influence his promotion opportunities is also obvious. Unless he appreciates these factors fully, his chances for advancement will be lessened.

Some promotion factors are based directly on an evaluation of the individuals being considered; others depend upon the firm's needs and are primarily external to the individual. The first of the individual-centered factors that we will discuss is seniority.

Seniority: An Imperfect Yardstick

Seniority is a fundamental criterion for promotion in almost every society, and there is no doubt that it counts heavily with the engineering manager, too. There are pros and cons to seniority as a criterion, however, and there is little doubt that the disadvantages outweigh the advantages. Let's look at both:

The primary advantage to seniority as a tool for making promotions is that it lends itself to exact measurement. That is, if we know that an engineer has been in our plant for three years, or ten or eleven, we then have something firm on which to base a decision. Of course, even this apparently precise measurement can be obfuscated—for example, what is the seniority of an engineer who comes into the plant after three years at a plant in another division of the same company? Does he get credit for these three years? If so, is it full credit, half credit, or some other amount?

Take another example: suppose an engineer has been working for you for ten years, five years in production and five years in pilot-plant work. How will you weigh his seniority? Does he have five years in each field, or does he have ten years seniority, if a better production job opens up?

A second advantage to seniority as a yardstick is that it is accepted as such by many employees, particularly the technicians and other blue-collar workers. Its appeal to them is that it is subject to measurement and permits them always to know where they stand.

The major drawback to basing promotions on seniority is that it encourages the wrong people to stay with the company. The first-class engineer does not have to rely on such a crutch to get ahead and he has enough faith in his capabilities to go where they are rewarded.

Another problem can be illustrated by the engineer who has had ten years experience in the production of organic compounds. Has he had ten years of pro-

Examine your psychological fit for promotion—Table I

Will you have to act differently from your normal pattern?
Do you want to work closely with others?
Will you be able to discipline your personnel?
Will you have to cut ties with friends and associates?
Are these ties important to you?
Will there be new ties available as replacements?
Can you transmit your drive through others?
Do you prefer to do jobs yourself?
Will the job bore you or make you restless?
Can you work independently?
Do you depend upon your superiors or subordinates for guidance?
Have prior promotions made you anxious?
Would you be happier as a follower than as a leader?
Do you want the promotion so that you can *do* something?
Do you want the promotion so that you can *be* someone?

gressive experience or has he done the same year ten times over? If the latter is the case, how can we compare him to an individual who has had only five years, but in diversified and progressive work?

Finally, we need only recognize that there is no correlation between ability and years of experience. Merely putting in time and lasting a long while cannot magically put people with lesser ability at the head of the class.

Merit: Difficult to Measure

Another frequently used criterion is merit—unfortunately, it too has its drawbacks. Measuring merit has not yet been reduced to a simple and readily acceptable form, and the use of this criterion is subject to favoritism, either consciously or unconsciously.

Of course, a merit system that is properly applied goes a long way toward attracting and keeping engineering personnel. What is needed is a formal system of merit rating that will be fair and effective. If you are the supervisor, you must do a good job of observing the plant's personnel. You must inform them precisely how the rating system works and what is expected of them. (A good job description is invaluable here.) You must rate them at intervals frequent enough that the information can be fed back to improve their performance. You must help them not only to achieve the aims of the plant but also their own objectives. And bear in mind that poor performance might be a reflection of poor supervision and not an inherent lack in the individual concerned.

If you are a nonsupervisory engineer, you should examine any new job situation carefully to evaluate the merit system in use.

Ability: A Measure of Future Potential

A merit evaluation is based primarily on past and present performance. Ability is a promotion yardstick that can be used to predict a candidate's future potential. It is even more difficult to measure ability accurately than merit.

As a result of this difficulty, the individual's education, both before and after graduation, is often used as a guide. The premise is that what he has done in the past to prepare himself is an indication of what his ability to perform will be in the future. Unfortunately, this assumption fails to consider the effect of motivation on past performance. Someone who has coasted may excel if given a higher motivation.

Another way to spot ability is to look for those members of the rank and file who exert the informal authority, that is, the natural leaders. These people can have more influence over their colleagues than their official group leader or superintendent and therefore are logical candidates for advancement, assuming they have the other necessary abilities. We will discuss these lines of informal authority in a future article.

We mentioned earlier that there are also promotion criteria that are external to the individual. One of these is the age distribution within the organization.

A Wide Age Distribution Is Desirable

It would be unwise for a supervisor to hire in such a way that all those who report to him are within the same age group. An age distribution like that stifles opportunity among those who are lower in the department's hierarchy because they may have to wait a long time for anyone above them to retire or leave.

A variety of ages in the various positions beneath the manager provides a range of opportunities for people of different experience and ability levels. It also means that the department is getting constant infusions of young people with new ideas who have received their training at different times and from different schools of thought.

The possibility of introducing a fresh face from outside the organization ought to be considered every time a promotion is in the offing. For one thing, this approach avoids the dangers of inbreeding, and, furthermore, it is frequently cheaper to bring in a qualified outsider than to train an existing employee.

This practice also allows you to attract good new people to the company without obliging them to start at the bottom and work their way up.

Each promotion creates the necessity, and the opportunity, for additional promotions. Thus it is important to have a string of good subordinates so that, if the polymer research manager is to be promoted, one of his group leaders will be capable of assuming the manager's job, the assistant group leader will advance to group leader, and someone within the group will take the assistant's spot. The disadvantages to this chain system are that many people must be trained and a great deal of internal paper work is involved.

The rewards of the system outweigh these drawbacks, however. The primary attraction is that it provides many opportunities for those in the path of the chain, not just for one individual. The job that must ultimately be filled from the outside is the lowest link and can easily be filled by inexperienced personnel.

There is another group of factors that should be taken into account when formulating a promotion

Are engineers satisfied with promotions?—Table II

	Engineers, %	Supervisory Engineers, %
Very satisfied	8	16
Moderately satisfied	29	37
Neutral	22	28
Moderately dissatisfied	25	11
Very dissatisfied	15	6
Not responding	1	2

Source: *Machine Design*, March 26, 1964, p. 152

policy. These are the human factors—the effects on the people involved.

Effects On the Chosen One

Not everyone who is promoted will perform well in the new job. Even if he does, the emotional toll on him may not be worth it to him. It is important to remember, therefore, that many people actually fear promotions—they are so afraid of failure in a new job that they will be happier without the added load.

On the other hand, many people avidly desire advancement as a means for earning more money, realizing their potential, and enhancing their self-esteem. To judge whether or not a particular promotion would be to your advantage, try answering Table I.

If a promotion has been a mistake, often the only way to remedy it is to demote the individual or discharge him. If you demote him, you are likely to have a disgruntled engineer spreading discontent all around him. It pays to avoid such mistakes.

Effects on Those He Leaves Behind

Every time one man is promoted, there will be a number of others who are not. These unsuccessful candidates must be given a good explanation of why the chosen one got the job and they did not. If you are doing the promoting, remember that you must continue to live with those left behind and that their feelings must be taken into account. One time-tested approach is to let the unlikely candidates know that they are not in the running even before the opening arises. This avoids dismaying a great many people with high expectations.

If you have missed out on a promotion that you felt you should have received, you will do yourself a great disservice if you don't make a serious effort to find out just why you were not selected. Find out the criteria used to make the selection, find out how you measure up to them, determine your chances for the future, and decide what you will have to do to get that better job. You may have to leave the company.

Effects on the Outside World

The possible effects of internal promotion policies (and other policies, for that matter) on those outside the confines of the company itself is often neglected.

One of the first things potential employees will want to know about your firm is "How easy, or tough, are promotions at Marvelous Chemical Co.?" If they learn that the promotion system is biased, that it neglects merit, or that promotions are few and far between, they will probably pass you by. A reputation for poor internal policies can also damage chances with customers and suppliers.

Effects on Those Who Do the Promoting

Managers usually try to avoid two things in making promotions: promoting unqualified men and passing over qualified ones. However, this practice falls down in the attempt to define "qualified." Each company has an image of the ideally promotable engineer, an image that perpetuates the present "company type."

One executive, quoted in *Harvard Business Review*, explained it this way: "I would like to recommend people for promotion on merit alone, but I am afraid that my own judgment will be called into question if I recommend anyone who deviates too markedly from the kind of person I see getting ahead in my company."

In short, those making promotions observe an "unwritten law", consciously or unconsciously, that perpetuates the company type in a vicious circle. If you are one of these decision-makers, perhaps understanding this failing may help you break out of it.

If you are a nonsupervisory chemical engineer and feel you do not fit the image in your firm, chances are that promotions will not come your way, regardless of your merit or ability. Your best bet is to look for a company or industry segment whose image of the promotable engineer you can satisfy.

What do engineers actually think of their chances for promotion? Are they satisfied with their opportunities? What about those engineers who are also functioning as supervisors? Results of a survey of 1,000 engineers and managers are shown in Table II.

This article has explored the factors that must be taken into account when considering candidates for promotion on the one hand, or when evaluating your own prospects for advancement on the other. The consequences of doing this job badly can be severe in either case, while the rewards of doing it wisely are correspondingly great. ■

Meet the Author

Conrad Berenson is Associate Professor of Marketing at the Baruch School of Business and Public Administration of the City College of the City University of New York, and Executive Director of the University's Research Foundation. A frequent contributor to this magazine, Dr. Berenson has also had experience in research administration and in industry as a chemical engineer.

Reward Systems for Technical Men

Part 1

This realistic approach to establishing professional salary scales depends upon combining meaningful data on what your competitors are paying, plus an objective ranking of your own employees' contributions.

P. W. MALONEY, *Armour and Co.*

Managers, including even technically trained ones, seem to store up their irrationality for the days when they're considering compensation matters. Why is it that people who demonstrate dedicated objectivity in their mainstream activities behave so strangely when trying to decide how to pay their people!

I mean attitudes like these:

• "Actually, you know, we shouldn't really worry whether we're paying our highest-rated people well enough. They are the guys who know they're going to advance fast, so money isn't that important."

• "I got out of this one quite easily. I just told Jackhammer that I wanted to give him an increase, but that the Personnel Department wouldn't let me."

• "All right, fellows, it's taken most of the morning, but I guess we all agree that Psmythe is the better man. We'll raise him to $975 and only bring Glockenspiel up to $970."

It's time to bring a bit of realism into rewarding the technical workforce. Can we strip away emotion, hypotheses, and self-taught expertise in personnel psychology? Let's try.

The Bare Bones of Compensation

Although the total reward system includes many non-pay ways to recognize performance positively or negatively, let's concentrate on money for now, starting with a simple portrayal of a compensation system.

Fig. 1 might be called the "Steerhorn Model of Compensation Administration." It includes all real elements of any pay program for professional personnel. The two inputs are shown at the top—a representation of competitive pay practices, and an attempt to discriminate among one's employees on the basis of their relative contribution. These data feed into whatever compensation system the organization has devised, and the salary budget is the output. Let's examine each element in the Steerhorn.

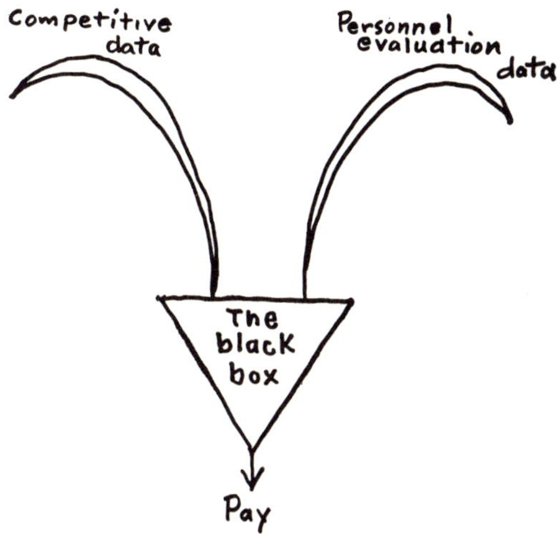

The Steerhorn: the basic elements of a compensation system for professionals—Fig. 1

Competitive Surveying: Know Thy Enemy

We're going to back into the how-to's on collecting competitive data by looking at a very important how-not-to: The approach taken by the El Grosso Analyzers. These are the compensation administrators who proudly point to their participation in an annual salary survey of 8,000 firms. With this large sample, you can bet their salaries are competitively derived. They point to the fairness of a survey that has such broad representation and statistical significance.

Of course the statistics are impressive, but so would be data on the number of icicles that fell from the roofs in Novosibirsk in 1937. These large sample surveys include everyone from Colossus, Ltd., down to the Ma and Pa store down the street. And size isn't the

This is the first of two articles on technical compensation, and first in a series on technical administration in general. The next installment will tell how to arrive at pay levels after all the data are available. The subsequent articles will discuss promotion and recruiting.

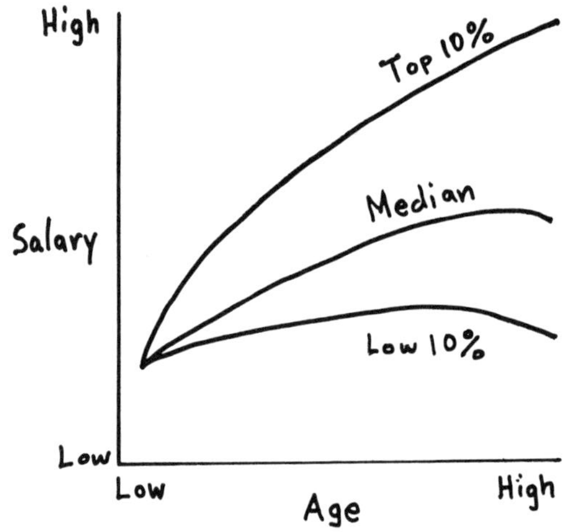

This is the absolute minimum amount of data required for salary survey reports—Fig. 2

only variable; firms exactly like yours may be included, but so are many companies who employ quite different types of people.

What you need are data on your competition in the labor market. Who is trying to hire and retain the kinds of people you want? How much are they paying them?

You have two readily available checks on who your competitors are. When fellows refuse your offers, whose offers do they accept? There may be a pattern in the kinds of companies they join. The second check is at the other end of the line. When professionals leave your employ, where do they go? Answers to these two questions should give you clues to use as one basis for planning salary comparisons.

For many types of jobs, it is adequate for the survey analyzers to prepare competitive medians and ranges. But for technical jobs, these data can be useless, or worse, as will be explained later. For now, the minimum data you should need is shown in Fig. 2.

Note that the abscissa is age. Some analysts prefer "years of experience" or "years since first degree," but I have found that, there being no recent disruptions from military service, age correlates closely enough. Its advantage lies in allowing more comparisons; just about everyone has age available in some accessible record, but the other two values can be harder to ferret out.

If you are overhauling your salary survey techniques and the suggestions offered here seem to have some merit, there are some loose ends that should also be controlled:

- The survey data should be reported separately by degree level.
- There should be a discipline breakdown (ideally, "chemist" would be broken down because it has become almost as inclusive a title as "engineer").
- Supervisory and nonsupervisory data should be separate, and the definition of "supervisory" must be carefully spelled out. The first level of management tends to be a fuzzy thing in many technical organizations; the surveyor must be sure that only true-blue managers are included.

The best composed set of survey directions will not insure uniform compliance by all the firms in the survey; there are too many variables. There's nothing that beats a conversation to be sure you're on the same wave length, as regards the kinds of people being included in the data. "Engineer" to you might mean "draftsman" to the fellow across the street; and his systems expert might be your B. A. mathematician.

The Ultimate: Exchanging Individual Data

The way to do the best job of survey analysis is to exchange raw data. It is inappropriate (and unnecessary) to identify individuals by name, but each salary, with the age, supervisory status, degree, etc., that go with it, can be traded. Because most companies have these data on EDP equipment, the job can be done by an exchange and/or pooling of cards or tape. (When we dive into that black box at the base of the Steerhorn, we'll figure out what to do with the stuff once you have it.)

I hear you muttering that all this is fine for some cartel with lots of loose cash lying around, but too luxurious for your company. Please don't abandon me until you do one little mental exercise:

1. Estimate your materials purchasing budget.
2. Add up the salaries of those employees who make sure you're not getting taken in the materials market place—the pricing people, the clerks who support the buyers, etc.
3. Divide (2) by (1).
4. Now, do the same for your salary budget and its watchdogs.

Food for thought, I'll wager. If the ratio you calculated in (4) is as low as I think it is, relative to (3), think about it for a moment. In a technical operation, the salary cost is too large a segment of the operating budget for inadequate administration.

The pressures to attract and keep good technical talent have blinded many employers to another competitive force that must influence compensation planning: product competition. R&D and engineering costs cannot be very much higher than your competitors' before they put your final product price out of line. This provides an excuse to digress into a fascinating related hypothesis:

Productivity vs. Experience

If you plot technical salaries against age, the odds are high that they will follow the familiar "maturity curve" regression line. (More later, on whether this is good or bad. For now, we'll just work with the empirical evidence.) Wherever I've seen the salary vs. age plot for a technical unit of any size, the median shows a steeply rising curve up to age 40 or so, with deceleration to about age 50, and a leveling out from there to the terminal age. But, and remember that we're talking about the medians, this may not be congruent with the productivity curve. It's freely admitted that technical employees just out of school don't earn their inflated starting salaries for the first year. To my knowledge, no one has figured out where productivity does catch up with pay, but it must be something later than a year and a day after graduation.

So there is a disparity between pay and output in the early years, to the employer's disadvantage. But what intrigues me is the potential disparity at the other end.

As a benchmark, let's look at M (management) types, and how their productivity changes in their senior years. Unless senility makes inroads, the manager's performance remains near level, or even improves toward the end of his career. His manipulation of men and materials benefits from each year of experience. Granted, it may take special effort from higher levels to keep him motivated as he gets older, but this can be and usually is accomplished.

The M type has something else going for him too—the skills and knowledge he has acquired are largely transferable; he can be moved into different areas without much loss in his contribution. Indeed in many cases the move provides him with a new stimulus that regenerates.

Now the average bench chemist or design engineer is very likely to decline in value in his later years. No one has yet found a feasible way to keep the mass of the technical workforce from becoming technically obsolete.

Over the years, our average T (technical) man certainly builds up a body of knowledge far greater than is possessed by the green college graduate. So long as this knowledge can be profitably used by the employer, his productivity curve should be a healthy one. But such technological stability must be infrequent, judging by the rate at which technology changes.

Coupled with this failing is the relative degree of transferability of T types and M types. The T type is primarily managing things that are within himself. Unlike the M, who can often buy or rent a new expertise when he is moved from work area A to area B, the T cannot just plug new units of knowledge into himself. And learning doesn't get easier with age.

There may be yet a third force that stacks the deck against the aging T type. Successful leadership experience with a unit of three men is genuine preparation for the M type to manage twelve; and what he learns there helps equip him to lead twenty-five, etc. (Experience suggests that most managers who are arrested at middle levels represent selection errors in earlier years). But can one follow this reasoning with the T type? Does conceptual experience with routine technical problems go very far in preparing one for conceptual success at a more creative level of engineering? Perhaps not; it may be a matter of either having or not having the higher conceptual ability. Thus, only specific, applicable experiences prevents an early leveling off in the contribution rate. And if events make that experience inapplicable . . .

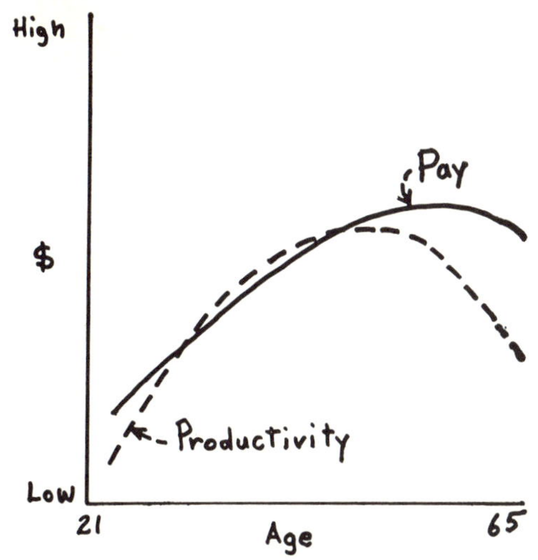

Productivity does not keep pace with salary in an average technical man's career—Fig. 3

This leads to the hypothetical picture in Fig. 3, which in turn brings us back to the competitive aspects of compensation administration. The technical organization with an older population is probably getting fewer units of technological output per salary dollar than are its more youthful competitors.

There is no pat solution to this problem, but knowledge of it might influence decision-making; if nothing else, it might lead to restraint in coping with the bleeding hearts who always want to throw Old Charley a bone. ("After all, even if he is overpaid, it's been three years since he had a merit increase. We don't want him to think we don't appreciate the fact that he's trying.")

Performance Evaluation is for Realists

Meanwhile, back at the Steerhorn, there awaits the second input to our compensation system: rating data on employees. And few are the areas of personnel administration where the smoke is so thick and the temperature so low.

A few background comments before we get specific: One basic consideration is the need to separate compensation evaluation from the ratings needed for good personnel development planning. The latter can become awfully complex and might have a hard time staying at the top of management's priority list. But the ratings for compensation purposes can (and many claim, should) be handled separately, because compensation decisions are of more immediacy than development decisions.

Developing Standards

Another bit of background: A purchasing agent can't operate unless he knows the specs of the material he is to buy; neither can a manager pay for performance in accord with company standards when those standards don't exist. To say, "Pay the better people more" is about as helpful as "Buy the pump that works best."

Developing performance standards is not a task for staff people; at best, they can serve a catalytic role. People are nowhere so influential to the organization's success or failure as they are in a professional group. Performance standards must therefore have the same degree of attention from the people who make the final accounting—line management at all levels—as they would give to the physical standards in a process unit.

If this seems to be much to do about very little, consider some of the issues that must be resolved when performance standards are set:

- How is potential considered in the system? What part of a man's value is his seeming capability to fill key jobs tomorrow?
- How do you balance the dollar value of a technical man's project results with the technical skill employed to get the answer? This one has lots of pitfalls. Because of his expertise, a highly able T man can be assigned for long periods to a toughie that must be solved even though the immediate dollar return is low. Conversely, a less able man can be put on work that has low technical demands but involves heaps of dollars.
- Which of the following get more credit and how big are the differences: fundamental conceptual work; specialization in a narrow applied field; general capability in several technical fields?
- Is there differentiation, and if so in which direction and how far, for technical vs. managerial contribution?
- Has your organization faced up to the "negative findings" issue? Are there clearly defined ground rules for situations where the "forget the whole thing" answer is rewarded? (It takes very little of the wrong kind of pressure from on high to induce middle management to show displeasure with negative findings.)
- How much weight do communications skills have in your environment? Relative to his technical capability, how important is the man's strength in getting his message over to others?
- And how about political skills?

These questions, and others like them, must be considered in deciding what kind of behavior the organization will reward or penalize. Decisions won't be easy, and top management must be involved in making them. But would you rather have people guessing at the answers?

Tight Reins on Procedural Approach

One of the common inhibitors to developing good rating data is failure of top management to police the activity. There is a convenient screen behind which they can hide: "How can I hold him accountable for what his engineers do if I don't let him rate their performance?" which is sound as far as it goes. Unfortunately, it is sometimes used as a bar to any involvement at all.

Look at a few of the difficulties the organization can expect if the rating procedure isn't strongly coordinated:

- Inconsistency. Some managers will favor the Old Charleys, and others will be enamored of the young hotshots; in either case, the bias can color the real performance data.
- Other biases. There's no predicting what behavior will get to the heart of a given manager, with positive or negative effects on the employee's rating. Anything from necktie style to his wife's liberal tendencies can sway the poor fellow's performance rating.
- General tendencies. Some managers will rate all their men high, and others all of them low.
- Limitations of experience. A manager can only rate in terms of the people he has known. In a new situation, if people perform above or below his experience scale, he can stumble in evaluating them.

No rating system, no matter how cagily developed, can outwit your shrewder managers. But centralized guidance can at least combat unintentional difficulties.

What goes into sound central guidance? First, a viable operating manual. Performance standards, the rating procedures to be undertaken each year, and the way the data are to be fed into the Black Box must all be documented and distributed. As the rating techniques acquire sophistication, they become unavoidably more complicated, and new people entering managerial ranks must therefore be trained. The manual can go a long way, but more personal help from the central staff should also be provided.

The Rating Technique: A Specific Proposal

As hordes of personnel administrators have proved, it is always possible to build a rating system that is a little more complicated than its predecessor. Unfortunately these well-meaning folks are designing micrometers to measure cordwood. Given the current status of the behavioral sciences, a highly refined, accurate measure of performance just isn't possible. Remember what we are seeking from the compensation ratings: just some representation of performance that will tell us how this man should be paid relative to his colleagues.

There are at least two fundamental reasons for the emphasis being on relativity: first, a prime goal in salary administration is to achieve discrimination for its own sake; who should get the bigger and shinier medals? Second, the pot of salary dollars is not infinite; who should get a bigger share?

It will also emerge that only a technique based on relativity will enable the type of compensation system that we will espouse later. Without ranking, the entire compensation system for a technical group has one very shaky leg.

If it is a relative picture of performance we want, the way to get it is absurdly obvious: rank the employees from best to worst (worst to best is an allowed alternative). Your basis for the ranking is the sum of all the performance standards discussed earlier.

By now, the anti-ranking types are heading for their typewriters to dash off messages to the editor about the difficulties ranking presents. To forestall these hecklers, let's get the major problems out in the open:

Mixed Background—Can chemists, electronics types and project leaders be ranked in one list? Certainly. Apples and oranges can be compared, if your standard is weight. The standard here is value to the organization. In arriving at this value, we use the standards described earlier, the ones developed by line management. Most of these can cut across differences in discipline, type of assignment, etc.

Combining Units—Who is going to blend the discrete sections, departments, or what have you? Where these sections are part of a "working family," it's a matter of getting the head men together. If there is no day-to-day interaction among the groups, the problem becomes more difficult.

There are at least three routes of attack; first, if the family groups are large enough, one may be able to assume that they are equivalent in value, and that quality has the same statistical distribution within them. Then merging is a simple statistical process.

Alternately, a search for common denominators—managers who have transferred from one unit to another, or whose work cuts across units—will yield more than you might suspect.

The third alternative is having the top-level managers make over-all judgments on the relative merit of the departments. For instance, they might decide that Units A and C are equal in worth, so there should be a merging of the type discussed first. But if they see a big difference between Units A and B, they would decree that the top man in the rank list for Unit A is equivalent to the 75th percentile man in Unit B, and so on down the two rank lists. (This allocation by the top group is sometimes regarded as being a trifle arbitrary.)

Sure, none of these answers is wholly satisfactory. But don't forget the inadequacies in the alternatives.

You'll Be Glad You Ranked

The profits from a good ranking system justify the effort and frustrations that go into it. The most immediate gain is the reality (or honesty) it brings to your rating data. No longer must you puzzle over the differences among ten people who are all labeled "Excellent." The best one is first, the least excellent man is tenth, and that's it. At the other extreme, the least valuable people are in the spotlight; there can no longer be an amorphous mass of "satisfactory" employees, a situation that defies you to identify the cases where it might be time for rigorous action.

The people doing your manpower planning will cherish the ranking data because now they have tangible measures of the quality of the replacement resources.

The very problems to be surmounted when ranking is installed will yield their own byproduct benefits. When unit heads try to merge their populations, it will become apparent that performance is being measured against different sets of standards, some necessary and others just reflecting differences in the viewpoints held by the top men. In some cases, it will be found that performance is being measured today in light of a peculiar aspect of the work that disappeared years ago.

Whatever the reasons for the differences, and whether or not common standards are later adopted, it is healthy to get the situation out in the open. If nothing else, investigation may explain such strange things as why people who are highly rated in Division A fail to succeed when they are moved elsewhere.

People enroute to general management will benefit from movement around the organization. If the degree of this movement is inadequate, the problem will be spotlighted during a thorough ranking operation; when there are not enough managers with multidivisional experience, the merging procedure becomes a bit frustrating. The responsible reaction to this frustration is concern for the apparent lack of mobility in the organization. Steadfast insistence on merged rankings can, therefore, yield two more benefits: periodic assessment of inter-unit managerial movement, and stimulation to increase same.

But the preceding describes the benefits of ranking above and beyond its assistance to realistic compensation administration. The virtue of ranking data for pay purposes is its quantitative nature. Sure, the data do not have the validity of temperature and pressure readings (and that's why any compensation system has to allow for managerial judgment after all the analytical work is done). But the technique does yield data that are amenable to a certain systematic approach. This approach is what we shall cover in the next installment. ■

Meet the Author

P. W. Maloney is manager of manpower planning for Armour and Co., Chicago. Prior to joining Armour in 1964, he held several industrial relations positions with Esso Research and Engineering Co., the final one of which was manager of employee relations. He holds a B.A. and M.A. from the University of Minnesota in industrial relations and personnel psychology.

The American Management Assn. has published two works by Maloney: "Management's Talent Search," a book on recruiting professionals, and "The Potential Employee," a manual and an LP record on interviewing techniques. Maloney is a member of the American Psychological Assn., the American Soc. for Personnel Administration, and the Soc. for Applied Anthropology.

Here's how to set up an equitable pay scale for the scientists and engineers in your company once you have collected the necessary competitive and rating data.

Reward Systems for Technical Men

Part 2

P. W. MALONEY, *Armour and Co.*

In the first installment of this two-part article (on page 153 of this volume), we discussed how the elements of a pay program can be likened to a Steerhorn. Two prongs of data feed down into the Black Box in which the salary program is developed. Having disposed of how to collect the data in the first article, we'll now take a look at what goes on in the Black Box.

Two types of data, the competitive picture and performance evaluations, come together in the Black Box, and are manipulated by whoever administers compensation. The objective of this manipulation is clear-cut: to yield pay that, within competitive limits, makes the optimum contribution to attracting, retaining and motivating the kinds of personnel best suited for the jobs to be done.

There are many ways to fulfill this objective, and nothing that follows is submitted as an ideal solution. But one possible technique will be presented just to give some kind of benchmark.

First we must accept the rigors that are encountered when any system is developed.

Before getting down to specifics, we must conduct a defogging operation and face a few unkind realities. Although our data lack the degree of calibration we might prefer, their inadequacies don't mean that we should use them in a hit-or-miss fashion. Our job is to systematize the way we use the Steerhorn output so as to get the best answers relative to the stated objective. This will be done in the proposed system example we'll get to eventually.

To evolve a healthy reward system, we must recognize and cope with those variables that could contaminate it. The first of these is the delusion that all men are created equal. Any manager will admit that some individuals perform better than others. But there are other levels of difference: for instance, the credentials the man brings to the job. A Ph.D. comes at a very high price, and if we want to be in the Ph.D. marketplace, we have to admit it. Holding the total economy constant, three years on the job are worth about $100/month for the average B.S. graduate. Those same three years are worth twice or three times that for the man who used them to get his doctorate. We're now seeing this at the M.S. level too; these people are moving further away from B.S. and closer to Ph.D. starting salaries. Only the individual employer can decide whether or not the additional training has increased the man's value proportionate to the price increase.

Many recruiters say that they are working the Ph.D. market because most of the good men (and increasingly chemical engineers) are today going on for their doctorates. This means that a good B.S. or M.S. man is hard to find. The rising M.S. salary levels, however, may reverse this trend some day.

Ph.D.'s Present a Longer Range Problem

People planning salary administration must face this question: When do Ph.D. salaries merge into the pay for the general technical (T) population? Some people answer "Never." That is, there should be continuing separate, higher compensation guides for the Ph.D.'s. It's not hard to find supporting data for this; survey reports that show the data separately, usually indicate Ph.D. salaries consistently higher than B.S. or M.S. at all experience levels.

But after 20 years, any gain from the degree really should wash out. The survey data to the contrary must reflect the fact that 20 years ago the better men went on for the doctorate.

All right, so at 20 years we'll let the performance ranking, rather than degree level, separate people's pay. But we can't do this during the second year, and you can't even hold the Ph.D.'s pay constant during the early years to let the lower degree men catch up. At what rate, then, do you taper in? What is the point of merging, in age or years of experience? And try to think of a way to make your system plausible to the B.S. man who is a big contributor but is short of the merge point.

There are other ways that an employee's saleability is increased besides being a good performer or having a Ph.D. He can be a patent attorney or have some other skill with a lopsided supply-demand ratio. If you are hiring such people, just add up the compensation problems your Ph.D.'s give you and square them. Here you don't even have the excuse of additional training time to fall back on, as you do with the doctoral lads.

Your Laboratory Is Not a Machine Shop

Another delusion is the idea that in a technical group there is any kind of orderly job progression or hierarchy.

For contrast, consider Mother Murphy's Meatball Co. Here you have a neat hierarchy of jobs. The lowest-level plant people report to a foreman, several of the latter report to a general foreman, and the half dozen or so of these report to the superintendent. Each level has describably bigger responsibilities and calls for additional skills and experience. The result is a symmetrical, pyramidal, organization chart. The assortment of technical groups we have in industry today evolved late in the game after the Mother Murphy hierarchy was thoroughly entrenched in textbook and mind. It was natural for many to stamp R&D, and engineering, with the same die. So we have another pyramidal establishment of groups, sections, departments, etc. If we could give the problem of organizing technical groups to someone whose brain were washed clean of this precedent, he might evolve something entirely different. Why? Because the nature of the work is different.

Fig. 1 is a hypothetical comparison of several jobs, showing the degree to which the positions allow growth, that is, increased productivity, through the application of experience and native capability. On the far left, we have the production · positions at Mother Murphy's. The fellows rolling the meatballs just can't improve beyond a certain point because there are only so many ways to make a meatball. But eventually the better men are rewarded by promotion to foreman. This move results in a discrete jump in the depth, complexity, etc., of the man's job; he is now doing something different, and at a higher level of contribution. But here, too, he can only progress so far before moving up to general foreman, where again he at once takes on bigger responsibilities. This job allows more room to grow than the foreman position, but even here is a limit to what degree a man can increase his contribution.

The same sort of growth potential exists for the craftsmen in the plant, as also shown in Fig. 1. They can make an increased contribution as they get better, but only up to a point. In time, the electrician may make a discrete jump from his craftsman's level to a foreman's position.

Now consider the technical men who are doing fully professional work. The Fig. 1 portrayal of their jobs intends to show that the growth potential is open-ended. The theoretical limit on their contribution is quite high and in many situations may really not exist at all. Their principal output is a stream of ideas and creative problem solutions. There is never an oversupply of either, nor a ceiling on how creative a given item can be.

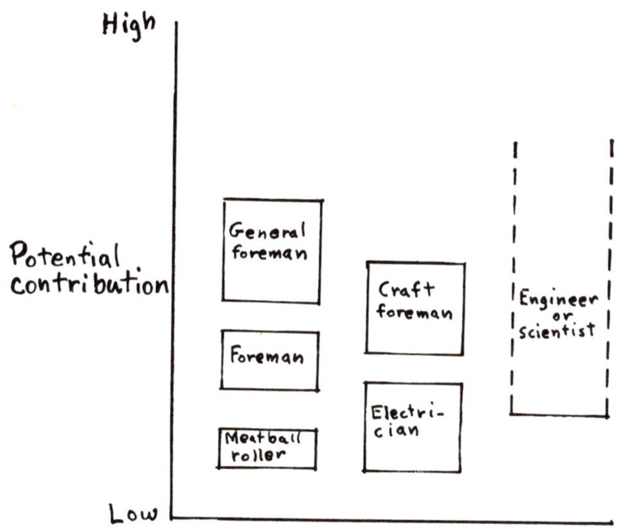

Different types of jobs exhibit different types of growth potential—Fig. 1

Note another difference between the T's and the production/craft comparisons: there are no discrete moves to higher level jobs; the T man's growth is incremental. Each day, if he is still growing, he becomes a slightly better contributor. He cannot be made a bigger contributor by management edict.

The exception, of course, is the move from bench work to supervision of a T group. But our present concern is not with the M (management) types; we're addressing ourselves to compensation of the career T people.

What bearing does all this have on compensation? First, it shows that among the individuals in a T unit, there can be more variance in individual contribution than among all foremen or among all meatball rollers. It's not inconceivable that one or more of the T people will be of greater value—and should therefore be paid more—than the man who is supervising the technical group.

A second aspect of the T job configuration has major implications for compensation administration. This is the absence of discrete job levels, the gradual increase in a man's value. There is no logic in conventional stair-step, promotional salary progression as the man moves to higher-level positions. To be realistic, the system must be structured in terms of the man spending his career in a single job, albeit one that can continually grow bigger.

The organization can, of course, use "technical ladder" job titles to denote a T man reaching a higher value. But it should be recognized that this is an artificial situation exactly opposite to title changing for the M type, who is promoted from section head, for example, to department head, and then starts doing higher-level work. The T man works up to the higher contributory level first, and then gets the honorary title. We shall return to this single-job phenomenon shortly.

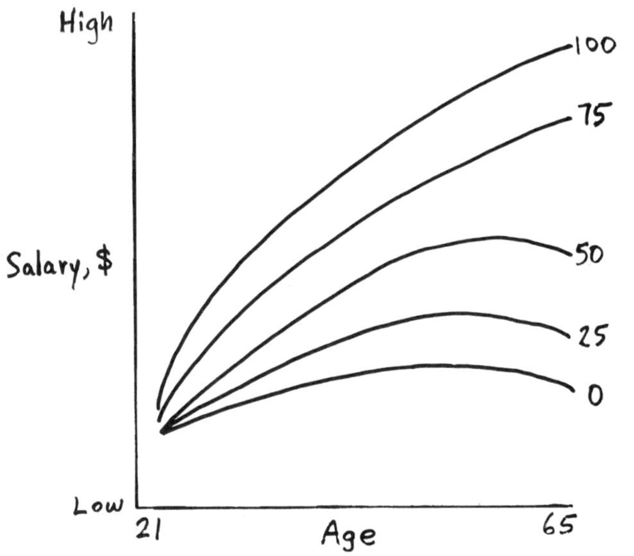

These curves show the characteristic configuration of pay in a technical group—Fig. 2

Finally, the relationships displayed in Fig. 1 call for new thinking on the compensation for supervisors of T's. At Mother Murphy's, all foremen can be paid about the same; the fellow supervising ten Meatball Rollers, Breaded, has the same responsibility and makes the same contribution as the foremen over ten Meatball Rollers, Plain. Not so in the technical organization. The leader whose people have progressed far up the growth scale has a task that is different in many ways from the job facing the supervisor of ten less fully grown T's. So automatically putting all first-level T supervisors in one classification, all second-level bosses in another, etc., might constitute improper treatment.

Maturity Curves Are Here To Stay

This leads us to the last big defogging operation: a bit of straight thinking on so-called maturity curves as a basis for compensating the T population.

In the preceding section, I suggested that in reality there is a single, career-long job classification for the T types. However, published salary-survey data indicate that T compensation goes up to $25,000, or more. How can one cope with this kind of spread? How do you move T's of varying value through this broad salary range?

Some people delude themselves by "objectively" measuring performance and paying accordingly. They know what a $10,000 T is like, how good you have to be to merit $15,000, etc. But our knowledge of human behavior is years away from this kind of measurement, and the managers who practice it are going entirely by intuition. I can't say they're wrong, but I can say that they can't transfer their feel to other people. So when you change managers, you also have a new intuition for a measuring device.

Many managers try to achieve a more quantitative approach by turning to job evaluation. Though this is useful elsewhere, it doesn't belong in the professional-technical world of work. How can you slap parameters on a job that is almost entirely a function of the man in it? Doing job evaluation properly for the T's would establish a kind of perpetual WPA for the personnel department. Every man would have to be evaluated, and at frequent intervals.

The job evaluation approach is more than just invalid when used on the T's; it is potentially dangerous because if the rates established are noncompetitive, a good T man who is not yet caught in the trap of the company's benefit plans can just pack up and leave. And well-evaluated jobs filled with inadequate people won't please anybody.

A Way Out of the Dilemma

Salary survey after salary survey shows a characteristic relationship between salary and age. There is a sharp increase up into the early 40's, deceleration into the 50's, and then flattening out to age 65. (And for less-than-average performance, a drop from the 50's to career-end.) These relationships are shown in Fig. 2, (a snapshot in time, not a portrayal of individual salary histories).

My recommendation is to adopt guidelines like these as the foundation for your pay system. It's not that these guides are intrinsically right (although their proper use will assure that you are paying competitively), it's just that there is no satisfactory way to put dollar values directly on various levels of T performance.

Most opposition to this guideline approach centers around two arguments. First, that it's wrong to pay people for getting older. Second, that it induces automatic salary administration, with no real managerial thought. Let's examine both of these positions for a moment.

If you object to paying people for getting older, chart your present T salaries, dollars against age. If you have more than a handful of people, and if your pay practices are anything at all like everyone else's, your points will form the characteristic neat pattern shown in Fig. 2. There are any number of ways you may have arrived there; the end result is the same. The beauty of maturity guidelines is the systemization they bring and the way they predict where you will be x years hence.

Now to the second objection—that guideline salary administration becomes automated. As a matter of fact, I will recommend a degree of automation later that might make you wince. But there are still two control points where human judgment is brought to bear. (In my opinion, human judgment shouldn't be allowed anywhere else in the system; it's too dangerous.)

The primary control point is in the performance rating prong of the Steerhorn. Here is where automation will probably never play more than a supportive role. Only the human being can look at two other human beings and evaluate one as being more worthy than the other. The second control is at the other end of the line. Once the system pumps out proposed salary treatment, the manager must apply his intuition and knowledge of special situations to it. If not abused, "managerial statesmanship" is a legitimate basis for discarding the product of the system.

But in between these two control points, judgment is neither needed nor desirable once the system is established. As we shall see, all elements in the system can be expressed quantitatively.

The Proposed System

First let us lay out the approach in cookbook fashion and then examine it in terms of some obvious criteria for good compensation administration. Over-all, you pour two cups of competitive data and a gallon of rating data into the Black Box, stir well, and pour out your salary program for the next year or so. Unfortunately we shall have to cover the specific steps in highly abbreviated form.

We've already discussed the acquisition and interpretation of competitive salary data. Now your job is to decide how competitive you want to be. Do you want to compete with the best paying of your chosen competitors? The lowest paying? An average? The choice is an important one. Suggestion: evolve a few alternatives, and rough out what it would cost to do each. This is not a complicated statistical problem.

Now your competitive data are narrowed down to those you wish to match. If you've done the survey job right, you have individual competitive salaries, which makes the rest of the job easy. If you only have age vs. dollar plots, you'll have to make some rough guesses. And if you only have over-all average salaries, forget the whole thing.

You should now be in a position to go into the Black Box and mesh the competitive data with your internal performance evaluation data. (The methodology must be developed by a statistician.) You have data from which you can evolve two further sets of data, expressed in identical terms:

- For each age, you can develop what the competition pays the top man, the 99th percentile man, the 98th percentile man, etc.
- For each age, you can identify your top man, your 99th percentile man, etc.,

This sounds as if it would only make sense with a T population in the thousands. Not so; smaller groups just call for the use of deciles or quartiles, rather than percentiles.

The Black Box treatment is a mathematical breeze; what does it take to bring the salary paid your 23rd percentile man, age 26, to the competitive salary at the 23rd percentile, age 26?

Aside from the time it takes to get good ratings, the line management burden in instituting this procedure is light. Given the right input, the system cranks out tentative salary treatment, which replaces a lot of calculation, speculation and perturbation by the organization's managers.

You'll want to put in several controlling ground rules; for example:

- No increase will be more than $x\%$ of present salary, or after less than y months since the last increase.
- If an employee's current percentile position is different from his position last year by more than x places, there will be restraint on the resultant treatment.
- No increase will be more than $\pm x\%$ (and/or y dollars) different from an employee's previous one.
- A new employee will always get an initial increase after x months, regardless of his percentile position, unless he is on probation.

Computer Application

Note that your input data and your control factors can all be quantified. Therefore you have an obvious computer application. With a little work, the machine can give you your annual salary program—amount and timing of all increases—subject, of course, to the application of managerial statesmanship.

This neither takes care of your early year Ph.D. problem nor the compensation for other people with high-demand skills. By one means or another, special guidelines have to be added for these situations. There will be other problems peculiar to your industry or company that will also call for special ground rules.

The preceding has been of course a gross oversimplication. Unless one is dealing with a specific situation, it's difficult to narrow down very far from such a generalized description. In essence, however, the entire operation is simply a matter of getting competitive and rating data expressed in the same quantitative terms so they can be brought together, and then bringing them together.

Proposal Viewed in Light of Criteria

A compensation system must satisfy several demands. Let's look at the approach set forth above, in terms of the more important criteria.

We earlier concluded that there are no measurement tools to show that one chemist has an absolute value of $10,000 and another $17,500. But our compensation system can be expected to allocate money fairly on a relative basis. The better man should get more, and this is one of the prime features of the proposed system; the starting point is the relative ranking of all professionals.

Any maturity guide system is vulnerable here in one respect: there are ceilings on what the young hotshot can earn; he could be making less than older, apparently less valuable people. In considering this criticism, be sure first that the problem isn't the other way around. It may be that the oldtimers are overpaid, either through poor rating, or sudden disintegration of performance. Their being overpaid doesn't justify raising other people up to their level.

But if the case is legitimate, try this rationale: if the compensation guides were appropriately geared against competition in the first place, there is no need to pay the man more. He is getting as much as people of his value do in any organization like yours.

The system must flag individual cases where compensation administration has gone awry. In this respect too, the proposed program meets the test. Ratings and competitive data are expressed quantitatively and on the same basis (percentile) so that any discrepancy between the two stands out. Also, a history of percentile ratings can be reported when each year's salary budget is prepared. Big rating shifts of individuals or groups will stand out quite conspicuously.

The system must be understandable to the users. For technically trained people, there should be no difficulty in learning the concepts and procedure involved in this quantitative approach.

The system must be palatable to the people affected by it. Specifically, both its administrators and the employees paid under it must feel that the program is equitable. Intellectually, the system should be acceptable; it has at least as many built-in equity guarantees as any other system. At the emotional level there might be negative reactions to the seeming elements of predestination.

It must survive changes in management. This is the real pitfall in more freewheeling compensation programs; because the T's are managed by different people, their pay can get erratic treatment.

Under the proposed approach, if there are changes under new management, they are quickly known and can be analyzed before any real damage is done. The only ways to change a man's pay pattern are to move him in the ranking list, or to pay contrary to what the system indicates he should receive.

The system should be easy to use. That is, an inordinate amount of management time should not be consumed in its operation. The ranking portion of the approach proposed here is not an easy venture, but achieving a sound ranking of professional employees is fundamental to managing them intelligently. The usefulness of these data in compensation and a host of other purposes justifies all the effort involved.

This approach will also call for a bigger time investment by the compensation staff people, but if the program gives the benefits it is designed to supply, it is cheap at the price. In a technical group of any size, the salary budget is a hefty part of the total operating cost; it deserves able management.

Middle Management's Role

So far, my comments have been addressed to men with managerial responsibilities. This has to be, because compensation philosophies and systems are not set by people in subordinate positions. But what should concern the manager who is closest to the firing line?

His most important compensation role is in the rating process. No one else has a first-hand knowledge of performance on a day-to-day basis, and a lot depends on the reliability with which this performance is observed and transmitted up the line.

Concluding Considerations

Perhaps it is appropriate to conclude with a few questions and observations about the staff people administering your program.

- Who is primarily responsible for running your compensation program? It is important enough to demand the best talent you can apply, and on a full-time basis.
- Is there analytical strength in your compensation group? The art has taken on some sophistication and there's room for a lot more. You won't get the benefits of this unless people with statistical capability are doing the work.
- Is there sound experience in T compensation in the staff group? In this article, we've noted several areas where tools that work well on the rest of the labor force won't suffice for T compensation administration. Beware of the man who has had decades of experience with nonprofessional groups.
- Do you have an active salary committee, with representatives from all major segments of the organization? It's the best insurance you can buy against inconsistency among the units.
- Finally, do your people have access to, training for, and an orientation towards the use of electronic data processing. They must.

So much for the world of T compensation! It's an exciting place, and in large part, still unexplored. ■

Ladders for Technical Status Seekers

Part 1

One technique used to keep morale high in large technical organizations is a ladder of arbitrary job titles that parallel those in the usual management hierarchy. Here's a discussion of the pitfalls and benefits of such a scheme.

P. W. MALONEY, *Armour & Co.*

The last decade has seen rapid growth in the use of dual ladders of advancement in technical organizations. This method provides a hierarchy of technical titles for the successful professional man who is not following an administrative career.

The technique has been acclaimed as an almost miraculous answer to several of the human-relations problems that plague technical organizations—problems that become increasingly troublesome as the organization grows larger. But now, the pendulum seems to be returning; users of the technique have encountered too many bad side-effects. In some quarters, there is even talk of abandoning the entire concept. This would be unfortunate because the original idea has some value, and the most serious problems that have come about are only the result of inadequate planning in installing and using the technique.

First we shall review the principal factors that led to the evolution of the concept, then we'll examine the problems that have developed and finally, in the next installment, I shall propose solutions to the more important of these problems.

After World War II, technical organizations mushroomed both in number and size, and the industrial society was faced with a new phenomenon: the presence of highly organized, complex groups of professional personnel. For the most part, these groups were engaged in technical work that could not be carried out by individuals. Rather, the jobs to be done called for integrated team efforts—and team efforts call for leadership, control and administration.

Previously, when organizations had relatively small numbers of engineers and scientists, these people could enjoy many of the status symbols and operating methods that are common to other professions. But with their endeavors organized into conventionally large work entities, things could not remain the same.

One of the more significant changes was the emergence of leadership levels. In the traditional professional group—the college faculty, the law firm, the medical clinic—levels of authority are difficult to identify. Any staff member can tell himself that, in the public eye at least, he has virtual equality with his colleagues. But the modern conglomeration of R&D teams must be organized, the teams must have leaders, and the leaders must have leaders.

In other words, organizational hierarchies have been created that are similar to those in other parts of the business enterprise. Now there are professional groups where we have foremen, general foremen and general superintendents, although the actual titles might be more flattering.

Those who have been given the mantle of leadership are happy with this change because they are no different from other people in their desire for power over others. The losers are those whose status has been reduced in comparison with their peers.

Being human, those who have succeeded administratively are not reluctant to add to their prerogatives and status symbols. Because they run the organization, it is not difficult for them to do so.

(Some large technical organizations have tried to perpetuate the truly professional atmosphere by pre-

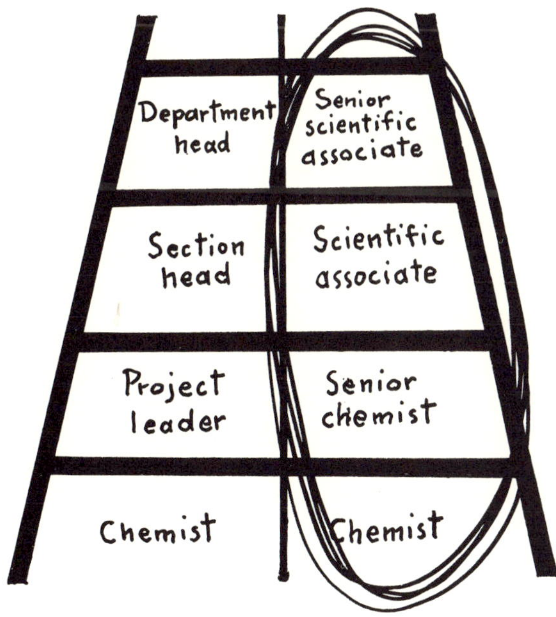

"A hierarchy of technical titles for the successful technical man."

This is the third in a series of articles on technical administration. The first two dealt with reward systems for technical men.

venting public recognition of staff members as managers, directors, etc. This does avoid some of the problems described below, but not all of them; if the organization is doing team research and development, there is little alternative to having leaders of the teams and leaders of the leaders, just as in any other organization. Some people are obviously better in the leadership role and, therefore, tend to advance in leadership positions. The hierarchy is still there; it is just somewhat more obscure. And this effort to hide a managerial situation that actually exists can have its own bad effects, just as serious as those in more structured organizations.)

With Structuring Came Problems

The administrators who run these structured technical organizations face many problems—problems that grow with the size and complexity of their organization. Some of these problems are:

1. Many professionals aspire to the management hierarchy. There isn't room for all of them, someone has to stay on the bench, and many of the aspirants are not qualified for management anyway. Nevertheless, they do aspire. How are they to be kept motivated in one role when they seek another?
2. Technical managers sometimes make mistakes when they promote other professionals to the managerial ladder. The victims of their errors are still "professional brethren." How to return them to the ranks without loss of face?
3. It is not unlikely that the professional ties among all scientists and engineers create some feeling of guilt for the rising administrators. What can be done to give their non-administrative colleagues more status? (This has been called the "bridge party problem." The department head's wife responds with pride when asked what her husband does. But her partner, whose husband was a classmate of the other, cannot feel that "chemist" has anywhere near the status value in our society as "head of department XYZ.")
4. Administrators are sincerely distraught at their inability to find a means of publicly rewarding the contributions of their professional people. Even though these contributions might get scant attention from the technical communities, they still represent truly professional innovation and creativity, and the man deserves something he can wear on his sleeve. A salary increase doesn't do the whole job.
5. Technical (T) people who become managers can't be expected to be proficient in the behavioral sciences. They often overrate the value of a title as a motivating device. Plagued with the need to increase T productivity, they can mistakenly feel that giving out titles will magically solve their leadership problems.

The Technical Ladder Is Built

It was these five problems and others like them that built the dual ladder of progression. In this system, families of job titles are established, usually in direct parallel with the existing administrative ladders.

Note one highly critical element: equivalence between the two ladders is usually either implied or explicitly stated. The public is informed that the technical man in a given position is as valuable as his administrative peer and has the same status and prerogatives; the only difference between them is the T man's lack of administrative responsibility.

With such a system, it is believed that a man can maintain a purely professional career and still pride himself on his publicly equal value to his M (managerial) counterpart.

It hasn't worked too well in practice. Let's see why it hasn't:

Reaction Is Sometimes Surprising

The builders of technical ladders are sometimes dismayed to find that the device is not gleefully received by the T population. These disappointed managers were probably deluded by the ivory-tower writers who solemnly state that the technical man wants no part of management, has a passionate affection for technical work, and is just looking for a way to stay in the professional field without losing status.

In the industrial R&D organization, this generally isn't true. Most of the technical people who enter industry do so for the same reasons that motivate their fellow students in business administration or finance, namely, they want to rise to positions of prominence in a significant economic enterprise. They wisely perceive that a technical degree (preferably an advanced one) is a topnotch union card for the career they seek. They are quite willing to begin in R&D or in engineering because this presents the easiest transition from academic life, but few of them plan to spend the rest of their careers in the cloisters.

Small wonder then that the technical ladder creates something less than enthusiasm among their ranks, and small wonder that the appointee to the T ladder might regard it as the kiss of death. The world is advised that he has been measured and found wanting as far as managerial advancement is concerned.

Selection Presents a Real Challenge

In considering the ways in which people will be selected for technical ladder positions, management should realize that this is not a matter to be taken lightly. Appointees who are undeserving in the eyes of the professional work force create negative feelings toward the ladder. Also, it is difficult or impossible to remove a technical title from a man. It is easier to shelve an M type who doesn't make out; some kind of hazy title can always be supplied or he can even be dropped back into the professional workforce. But depriving a professional of his technical title has an air of disbarment about it that is odious to most technical administrators. Therefore, when a man is assigned a given position on the ladder, he should be put there with the expectation that he is going to stay (unless he goes higher) for the rest of his career.

The first appointments to a new ladder don't present

much of a problem because the outstanding professionals are easy to identify and are generally recognized by everyone. If there is any difficulty at all, it is in deciding to which levels they should be appointed. But now the fun begins. Now the problems appear and with geometrically increasing speed.

1. Selecting T's for technical ladder positions is complicated by the difficulty in measuring their true T value. It is this measurement that determines whether or not a person makes the ladder and, if so, how high he goes on it.

Comparing M jobs and T jobs illustrates the problem created by this need for measurement. On the M side, a man moves to the project leader level if a job is open, and when it is determined that he has the necessary qualifications. The "necessary qualifications" in this instance are that he can, in the judgment of his superiors, effectively lead X men and control Y dollars. There are performance data at hand to facilitate the making of this go/no-go judgment.

Now, let's consider the parallel technical ladder position. What determines at what point a man has reached the "senior scientist" level of value? What do you count, what do you measure, what do you evaluate? What kinds of goodness make him equivalent to the project leader? There just isn't any quantitative, objective answer to this question, let alone to the same question regarding higher-level T positions. We just aren't this good at appraisal.

Review if you will the reasoning put forth in an earlier article (*Chem. Eng.*, Aug. 2, 1965, p. 117) to the effect that there is really only a single T position in the organization. Any slices made in this single job are wholly arbitrary and can't be supported by the type of logic that makes a section head different from a research director.

2. The problems in accommodating different types of professional personnel on the ladder demonstrate further the measurement difficulty. Are all types of technical contribution to be recognized? If so, are some to be given more weight than others? Basic researchers will resent technologists who achieve higher positions on the ladder; both will resent the engineers, and all will resent any nontechnical professionals who are given recognition for their contributions in administration.

3. The very process of selection presents challenging problems. If a lot of discretion is left in the hands of separate technical units, the lack of measurement tools leads to nothing short of chaos; there can be incredible inconsistencies in the ways in which people are recognized by appointment to various positions.

If, on the other hand, reliance is placed on some central committee whose membership doesn't have personal knowledge of the work being done by the nominees, the measurement problem rears its head in another way. Because the recommending sponsors usually cannot submit nice, clean numbers to justify their choices, they resort to literary endeavors. The amount of effort, time, and sheer paper than can be put forth is astounding.

4. When the first T's are placed on the ladder, the management is usually conservative. It is desirable to avoid using all the motivational ammunition in one broadside, and the organization should try to differentiate between the clearly outstanding professionals and those who are excellent but not quite as good. But now we must decide how many more people should make the grade. If too many make various levels of the ladder, obviously the device will lose its impact and it will no longer meet the goals for which it was designed.

An alternative is to set an arbitrary number of personnel who can make each rung of the ladder. But this contradicts the spirit of the endeavor. The ladder was established as a system of career advancement for people who are succeeding in their profession. Theoretically, all members of the staff should be able to advance; the limitation that exists on the administrative ladder where you only need so many supervisors and so many supervisors of supervisors is not needed here.

Once again, the problem is one of measurement. There are no absolute standards for the value of T contribution.

Administering the Ladder Also Brings Problems

Management's problems don't end when it has figured out effective (or the least dangerous) ways to choose appointees to the technical ladder. There are at least three further groups of problems:

- Those resulting from the aging of the technical population.
- Those from the aging of the organization itself.
- Those from a critical fallacy in the assumptions that are fundamental to most technical ladders.

Let's look first at the problems that come about as the T people get older:

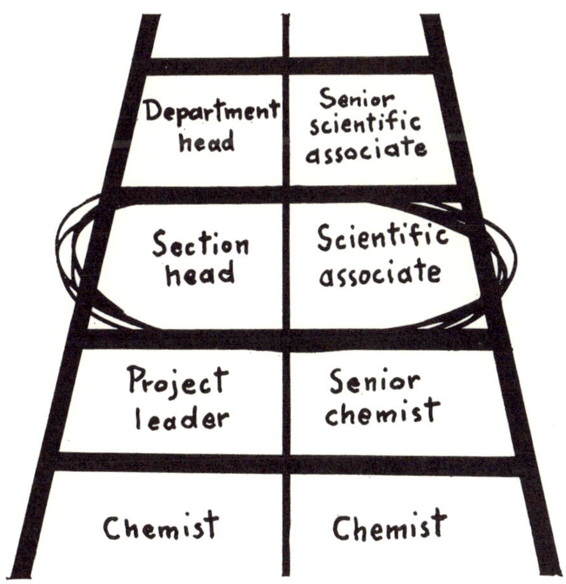

"What can be done about the bridge-party problem?"

1. In an earlier article in this series *(Chem. Eng.,* Aug. 2, p. 117), I presented the hypothesis that in later years one's technical ability tends to fall off at a much more rapid rate than one's managerial ability. As a matter of fact, by the time an individual gets close to retirement, his rate of technical contribution is falling off at a very fast rate unless he is truly exceptional. Those who just barely made the lower rungs of the technical ladder earlier in their careers consequently present an annoying problem. If selection was done well originally, they had value that was recognized as being worth the granted title. Remember that we are talking about the marginal appointee, the fellow who just barely made it. Now he is on the decline, and eventually passes far below that marginal level he once attained. The younger T people on the staff don't have a historical viewpoint. They question the significance of a technical ladder that gives recognition to people they perceive as being inadequate.

2. Every technical organization of any size has its one-shot contributors. These are the T people who came up with one astounding finding, one invention, or one engineering feat, and then don't accomplish anything of comparable worth during the rest of their careers. In some of these cases, the single accomplishment was not really a function of the man's personal capability; in others, he began to rest on his oars after his accomplishment. In either case, if there is recognition of the accomplishment through technical ladder advancement, as the years go by there is an increasingly glaring inequity. Someone is at a higher T level who apparently is of little value to the organization.

3. Given a technical ladder, the organization finds cases that seemingly justify its use for other than purely technical recognition, i.e., where an appointment provides a way out of a particularly sticky personnel problem.

It is fine to state in the program charter that this will not be done; it is fine to have heads of the organization pontificate repeatedly that it shall not be done. It will still happen in moments of crisis. Then the years roll by, the exigencies of the situation are forgotten, and once again people ask what significance the ladder really bears.

4. In its early years with the technical ladder, the organization can get carried away with it. One result is a tendency to move the most able T people upwards too quickly. The next generation of managers inherits a group of T people who have hit their titular ceiling when they still have 20 years of their career ahead of them. That's a long time to sit with no advancement if a man is steadily increasing in value.

Problems of Organizational Maturity

As the individual ages, so does the company. In most segments of the economy, there are not yet any really "old" technical organizations. The large technical unit is in itself an outgrowth of World War II; those that existed prior to that time have mushroomed so much in the last 15 to 20 years that the new input has kept them relatively "young." However, the day of reckoning is coming. Eventually there has to be some stabilization in the growth of technical organizations, and with it will come certain pains of middle age. In many units, there are already preliminary twinges, such as the following technical ladder problems:

1. As indicated earlier, arbitrary limitation of the numbers of people who can reach a given rung of the technical ladder is contrary to the spirit of the program. But in our society, a higher title spells some degree of "chiefdom," no matter how it is guised, and people who have a chance to assume the cloak of power will do so. So, as the T organization grows older and a higher proportion of its members have technical titles at varying levels, it can find itself with more chiefs than indians.

This situation is not compatible with the needs of most industrial research where there must be organized teams rather than individual thinkers.

2. The all-chiefs situation is also incompatible with the fact that in any industrial organization there will always be a reasonable portion of nonglamorous, somewhat grubby research and development projects. Somebody has to do these; and almost everyone must handle his share. In the older organization, where a high proportion of the people regard themselves above the level of journeyman chemist or engineer, the mundane work can go begging.

3. Perhaps most serious is the "reporting to" problem. If there is supposed to be equivalence between the two ladders, a T man at a given level presumably must report to an M at a higher level. In older technical organizations, this will leave no one reporting to the lower level managers, and far too many people reporting to intermediate and higher level ones. The latter cannot handle the administration for so many people and they cannot lead them effectively. There is also further damage to the necessary team organization.

Some organizations meet this problem, they think, by keeping work leadership at the appropriate level while giving the T individual his "personnel" supervision from a level on the M ladder that is above him. This means that the T man's equivalence will be preserved relative to his receipt of salary increases and performance-review interviews. In time, this gets to be ludicrous too; the administrator who is talking about performance and pay can be so far away from the man's work that he cannot interact sensibly. Perhaps more important, the T man can easily see the artificiality of the whole situation.

4. Those climbing a technical ladder expect public recognition of the differences between the various levels on it. This recognition is often stimulated by discriminate use of office space, carpets, parking spaces and similar status symbols. Indeed, some companies go so far as to give the T man prerogatives that are equivalent to those granted an M at the same level. At first, this looks like a really good idea, and there is no question about the public impression that the T man is equivalent to his M peer.

In the mature technical organization, however, continued granting of these prerogatives becomes impossible. The proportion of T ladder appointees gradually

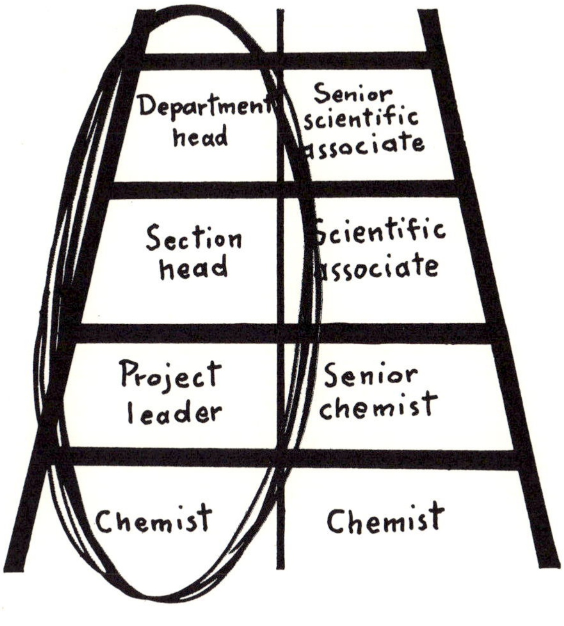

"Why does a man get chosen for the managerial ladder?"

swells to the point where no company could afford the cost. The retreat to a more rational approach is painful; it's always hard to take things away once they've been given.

Faulty Assumption Leads to Problems

The final group of administrative problems are those that result from a faulty assumption about the nature of the device. Namely, that there is equivalence between the two ladders, that a person at level A on the administrative ladder is truly equivalent to an individual at level A on the technical ladder.

Why does a man get chosen for the administrative ladder? First of all, because he is one of the most highly capable technical men in the group from which he is drawn. He is put in the management job so that he can multiply his own creative capabilities through the minds and hands of others. This he cannot do unless he is at least their equal in his technical competence. True, as he moves further up the managerial ladder, he is bound to lose some of his association with the world of technology; the increasing burdens of administrative chores will force him to give lower priority to his technical updating. But keep in mind that when he first made the grade he was a top-notch T man.

He also has whatever it takes to lead men, he can organize his own efforts and the efforts of others, he can be effective in relationships with nontechnical managers, he can design, sell and execute programs.

How, then, are M's & T's equal? Certainly a few scientists and engineers make outstanding contributions worth fortunes to their employers. But the technical ladder is built to accommodate a much larger population than these occasional extreme contributors. It is used for recognizing the "Steady Eddy," the highly dependable technical man who can be given a technical assignment with assurance that he will do the job in a professional manner. It is difficult in my mind to equate this man with a person having equal or better technical competence, and a flair for management besides.

It is also difficult for employers to equate the two, when you get down to the brass tacks of business management. If you hold age constant and compare the average salary of managers and T people, you will find in most cases a substantial difference in favor of the managers. The average pay of the T people at a given level will probably be the same as the M people, but this is illusory because T people advance less quickly and therefore are older than the M people at the same level. Salary, as we have seen, has a strong age component in its makeup.

These professionals are certainly not unintelligent and they recognize this pay differential; therein lies one of the effects of the fallacy about the nature of the dual ladder. The T's look with tongue-in-cheek upon statements of "equivalence" when they see the M people making more than they do.

For the same reason (lack of true equivalence), the M people move much more quickly up their ladder than do the T people. This again leads the T man to take a cynical view of the technical ladder. Actually, the technical ladder points up his "second-class" status, because it is now obvious that the M man is moving to higher levels more quickly than the T man. In the same vein, the M man is usually susceptible to being transferred (with a promotion) into non-R&D portions of the organization. The pure T man is of little use to other operational areas and therefore tends to be quite immobile. Once again, the difference between the two groups is spotlighted; the technical ladder seems to lead nowhere.

The lack of real equivalence in ability creates yet another problem. As noted earlier, there are strong forces that operate in our society when you give a man higher-level titles. He tends to believe, regardless of what he is told, that he is in a higher position of authority in the organization. Therefore, higher-level T ladder people tend to seek and use authority. They are rebuffed by the M's who feel, usually correctly, that the T's don't have the types of abilities that would let them use this power effectively. Therefore, the T's are denied the power and can become embittered. In many organizations, the titled T's are even restrained from influencing the nomination of new members to the "club"; the M's don't trust their managerial discretion.

In the next installment, I'll suggest a new approach to T ladder administration. ■

Ladders for Technical Status Seekers

Part 2

Title giving could be administered more objectively and be a more meaningful form of public recognition if it were dissociated from the management ladder of advancement and associated, instead, with a valid program of compensation.

P. W. MALONEY, *Armour & Co.*

How dual ladders of progression evolved as a way to publicly recognize, by the use of titles, the more-skilled nonmanagerial technical (T) men was covered in the previous article (*Chem. Eng.*, Sept. 13, p. 205). The problems associated with using these ladders were also explored. Now we will look at a new proposal for achieving the ends pursued with existing versions of the technical ladder. The obvious goal is a ladder free of the defects of its predecessors.

Let's pinpoint what led to the development of technical ladders. The growth of these ladders really stemmed from a single need: management had to find some way to signal, publicly and meaningfully, the attainment of higher levels of technical value by members of the nonadministrative T population. How can we accomplish this goal and avoid the difficulties that were described in the preceding article?

First, T work must be dissociated from the managerial ladder. The misfortunes that have accompanied

In Part 1 of this two-part article on advancement ladders (page 163 of this volume), Mr. Maloney dwelt upon the drawbacks of having the path of promotion for technical men modelled after the promotion ladder for management.

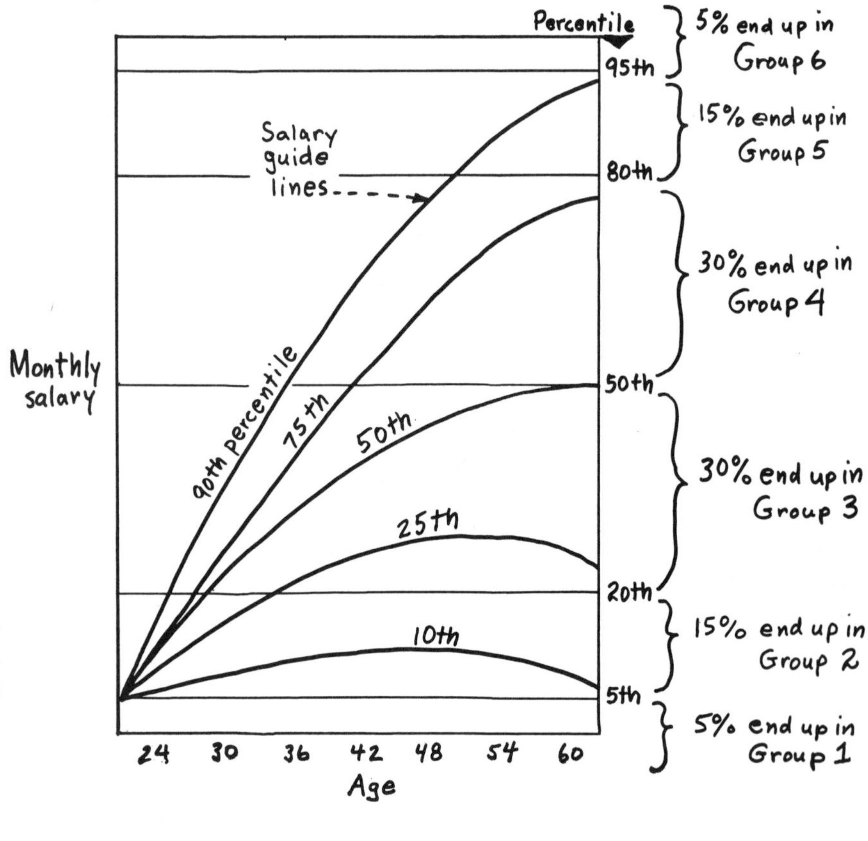

A technical man's eligibility for a title is related by this graph to his company's salary program. A T man's salary and age identify the salary guide line along which he is moving. His age and guide line place him in a percentile group, making him eligible for that group's title.

attempts to create two parallel structures make it necessary to acknowledge that there cannot be two equivalent hierarchies. Furthermore, there is no need for such equivalence. The professional career can attain dignity, and gain for itself public recognition of value rankings without being tied to the ladder for M people.

The technical ladder must not be a hierachy of positions. The articles that dealt with compensation (*Chem. Eng.*, Aug. 2, p. 117 and Aug. 16, p. 121) showed that in the nonmanagerial T population, there is only one position. It isn't possible to slice the work being done by professional T's into meaningful job levels. It isn't possible to create a hierarchy by writing job descriptions. The T ladder is not analogous to the M ladder, where section heads are clearly different from division heads, etc. As was shown, however, management does try to impose position levels on the T work force, with results ranging from the ineffective to the injurious.

There is no need to burden ourselves with the impossible. We can have a recognition system in the absence of discrete position levels. Because it is possible (the technique will be reviewed shortly), let's stop talking "ladders" and speak of "value recognition programs." We do want to discriminately gage value to the organization. But we don't want to try to create position levels in an activity that just isn't amenable to their use.

Standards for Value Recognition

Our "value recognition program" must meet these criteria:

1. The program must discriminate among the various levels of value in two ways. First, the most valuable people must obviously reach the highest levels of the recognition system, and the less valuable contributors must be held to appropriately lower levels. Second, those who are to move furthest up the recognition hierarchy should receive their earlier recognitions at a quicker rate than the people who seem slated to end their careers at lower levels. This provides for the desired public recognition of value differences early in the T career.

2. The recognitions must be spaced sensibly throughout a T man's career. There must not be a spurt of recognitions followed by twenty years of silence (unless, of course, an individual's performance varies in this fashion).

3. The system must have a minimum of inherent administrative headaches. It must not take mountains of paper and endless meetings to arrive at decisions on the movement of personnel up the value recognition structure.

4. There must be sensible, consistent criteria behind the decision to move a T man to higher levels of recognition. The criteria must be stable. They must be such that, in a given case, a new administrator would have come up with the same decision as did his predecessors. The criteria must clearly reflect technical value to the organization, regardless of the type of work being done by its scientists and engineers.

5. Above all, the total program must be meaningful and palatable to the population it affects most—the nonadministrative T people. They must feel that there is logic and justice in the decisions affecting their recognition. They must not, as is often the case now, be bewildered by many of the appointments that are made.

Two Fundamentals of the Proposed Approach

The proposed technique rests on two foundations. The first is the belief that value is normally distributive. There is no shortage of research to prove that the value of any sizable group of people will fall into the familiar bell-shaped curve when measured on the same scale. We shall assume, then, that when the T's hired this year wind up their careers, their relative values to the organization will fit the pattern. A small proportion will have progressed very little from their status of today, or will have even lost ground. A few will have achieved technical excellence. The bulk will fall between the extremes, with the model group at the median level of value. Therefore—and this is a key point—recognition for T accomplishment should be distributed normally, with a only few getting the highest recognition.

Our second fundamental relates to the problem of establishing criteria: What measuring devices should be used to decide which T's are more valuable and which are less valuable? The easiest way out is to use a measure that already exists, and one that is probably proper: the amount of money the organization is willing to pay for the T man's services. Although it has its inadequacies, the dollar figure does reflect some degree of hard-nosed evaluation of contribution. As long as we already have this—and in light of the previously described fallacies in other techniques for measuring T value—let's consider salary to be a useful criterion.

There are those who would argue against this approach. They validly point out the inadequacies of their pay systems, which make these systems poor criteria. To these objectors, we must say: Get your compensation programs in order. There's more reason for doing this than just to facilitate the installing of a sensible "value recognition program."

An intriguing facet to the use of pay as a criterion for recognition is that it points up the impossibility of equating M and T careers. The M man moves to a new titular level and then become eligible for higher incomes. One hundred and eighty degrees from this lies the path of T advancement; the T men are given money first and then become eligible for a title!

Tie-In to Compensation Administration

To understand the value recognition program, it is necessary to refer to the proposed compensation system for technical organizations in the August articles. The program espoused a salary treatment that would follow a series of career salary guidelines. The administrative guide would be a plot of dollars against age. An example of this is shown in the graph. The salary of the individual consistently rated as better than ninety percent of the T's of his age would tend to follow the ninetieth percentile line. The salary of the technical man rated in the middle of his age group would follow the fiftieth percentile line, etc.

Two comments are necessary about the configuration of the graph. First, the number of percentile lines can be more or less than the five shown. For a large organization, there could be 100 lines; the large number of T men in each age group allow this degree of discrimination. On the other hand, in the small organization, where there is only a handful in a given age group, only a few lines are necessary. Usually, an individual T man's percentile position must be interpolated.

Second, there is the problem of changes as the total wage structure moves upward. In this situation, the curves and the horizontal lines are not changed. Only the dollar figures on the ordinate need be altered.

Theoretically, a group of new hires coming into the organization at the same time, and at the same age, would end their careers with their salaries distributed as shown by the end-points of the percentile lines at the right side of the graph.

Now examine the horizontal lines in the graph. Where they end at the right side of the graph, they divide the people nearing retirement into six groups. For example, the top horizontal line terminates at about the ninety-fifth percentile point at the right edge of the graph. Therefore, the salaries of five percent of the people at the end of careers would be above this horizontal line and in group number six. Similarly, fifteen percent would be in group number five; these are the people whose final salaries are between the eightieth and ninety-fifth percentiles. And there are, of course, thirty per cent in groups four and five, fifteen percent in group two, and five percent in group one. Naturally, at any earlier time, as this particular age group was moving toward age sixty-five, the horizontal lines would not divide it in this proportionate manner. It only does so when they reach retirement age.

This end-of-career distribution is the basis of the value recognition program. Under this system, when a given age group of T people reach retiring age, the people in group number six—the top five percent—would have the highest technical title; the fifteen percent of the group in group number five would have the next highest technical title, etc. Thus, we are on our first foundation—that, at career end, the available titles will be allocated among the T's in roughly a statistically normal fashion.

Following from this, the horizontal lines have been extended to the left to provide a technique for deciding when a man gets a given title. When a person's salary crosses one of the horizontal lines, he is qualified for the title that is assigned by the organization to that zone. So we see now the place for our second foundation—the use of salary as a criterion for recognition.

Let's follow the career of a hypothetical T man who is consistently rated as being better than 90% of the people in his age group.

This exceptional man's salary would follow the ninetieth percentile guide line. Therefore, it would cross the first horizontal line very shortly after he joined the organization—perhaps in a year or two. He would then be given the number two title. In a few years, following the graph, his salary would take him into the zone of the group three people. At this time, he would be given the number three title. In the same way, at roughly age thirty-six, he would receive title number four, and at age fifty title number five, the last title he would receive. (A man would have to be rated above the ninety-fifth percentile befor his salary would ever qualify him for the highest title in the organization.)

The people in the bottom five percent would never qualify for even title number two. Since their salary line wouldn't cross the first horizontal line, they would end their careers as they started, as chemists or engineers. A man in the fiftieth percentile would progress steadily and would probably make title number four shortly before he retired.

System Meets the Criteria

Earlier we presented several criteria that a value recognition program should satisfy. Let's see how the proposed technique shapes up against these criteria.

1. Any system should enable periodic recognition throughout the career of the topnotch T man. As we indicated by our example with the ninetieth percentile individual, we have this under the value recognition program, as portrayed in the graph.

There must be a second type of discrimination, i.e., the better performaners must move faster through the available levels of recognition than do the poorer performers. We're in good shape here, since the man whose salary increases at a faster rate acquires higher titles more quickly.

2. The system has to have safeguards so that advancements do not come in bunches, to insure that whatever titles a man is going to get during his career will be spread out in a sensible fashion. This is the case here. For example, the people making only one or two advancements will do so relatively late in their careers.

3. The system must have a minimum of administrative headaches. In the last article, we pointed out the many administrative difficulties of a technical ladder system that is dependent upon subjective measurement. In our value recognition program, the theoretical basis for administration is clear-cut. When a man makes a given salary, he gets a given title.

We should note that there will be administrative complexities in the early career years. Starting salaries create temporary inequities between Ph.D.'s and B.S./M.S. people; similarly, certain skilled categories tend to receive artificially high salaries in the earliest years. Administrative guides can be built to accommodate these situations until the technical population gets into the early to middle thirties and its salaries stabilize.

When the system is first installed, there will be overpaid people whom the management doesn't wish to see given a title; so be it. This does point up an incidential benefit that is derived from the proposed system. The manager who tends to "buy" his way out of personnel problems will be restrained. If he gives enough money to the man, he also has to give him the title. His actions thus come to public attention.

4. Our fourth criterion is perhaps the most critical: namely, that criteria for advancement be valid, consistent measures of value. The proposed system enables us to ignore: whether a given type of work deserves recognition and another type doesn't; the handling of one-shot performers; and whether Man A, who is being considered for a given title, is really as good as Man B who already has it. These problems have already been resolved by the appraisal effort that is necessary to administer the compensation system. Matters like these had to be considered when it was being decided how pay should be allocated to the technical work force. We are simply making another use of the data that have already been acquired.

5. Finally, we noted that any system must be palatable and understandable to the T population it serves. No system is ever going to be wholly satisfactory to the individual who doesn't rise to the top under it. However, the value recognition program, because of its simplicity, is less likely to confuse and confound than the technical ladders that are built upon hazy assumptions. To the extent that an organization's pay structure is defensible, its decisions based on value-recognition-program appointments are also defensible. If the compensation system is not defensible, it probably wouldn't make much difference what kind of system is used to give recognition.

Additional Administrative Guides

1. The degree to which the workings of a system will be made public knowledge would, of course, be up to the management to decide in terms of its over-all policies on disclosing compensation and related matters. Opening up the entire picture to the T population is a titillating prospect; then, the fact that a man has a given title makes it more or less public information that he is worth a given amount of money to the organization. What type of public recognition can be more to the point?

2. If this type of program is to succeed, it must be made clear that these titles do not represent position changes. The individual is not in a higher rank; rather, he has a new medal. It shows technical value to the organization and is independent of position and type of work. It does not mean that the T man will move into a new and larger office, that he has a new parking place, or that he has acquired any degree of power over other people in the company. This must be reemphasized continually if the troubles pointed out in the previous article are to be avoided.

3. The number of percentile groups (and, therefore, titles) chosen and shown in the graph is six. Any number can be used. Six seems a happy compromise between having so many levels that the program is diluted and having so few that there are not enough goals for the good technical man.

4. No attention is given here to the title naming; each organization will have its own reasons for using or not using a particular appellation. It is suggested, though, that the titles have broad enough meaning to encompass the entire population. In this context, "research associate" is awkward when it comes to honoring an engineer; if "associate" alone were used, this difficulty is avoided. Some organizations, of course, have equivalent "research" and "engineering" associate titles. This will work well until the day an agronomist merits equivalent recognition.

5. These medals should be something the man will keep forever. If he had a particular T title when he moves into an M job, he retains it; thus, if he chooses to use it, he is publicizing that he has a certain level of T expertise in addition to his M capability.

This aspect brings out another fringe benefit of the proposed approach. The young T man who is headed for an M career, but for whom there are no immediately available positions, can be rewarded with interim recognitions at ascending T levels while he waits for his M appointment. Since the value recognition program is now dissociated from the M ladder, there won't be problems in trying to slot him into an M job at a level equivalent to his T title.

6. It is appropriate to reiterate the vital importance of having a valid compensation program if this approach is employed. This is an asset rather than a liability. If it takes this kind of pressure to generate a cleaning up on the pay fronts, a useful purpose will have been served.

Concluding Considerations

Many faults can be found with this system. In practice, it obviously cannot work out as neatly as theory indicates it should. However, its adoption is recommended for at least three reasons. First, the need for some type of differential recognition of technical value is necessary in technical organizations. Second, the systems that have been tried have so many inadequacies that something new must be developed. In theory, at least, the system proposed here appears to meet many of the objections raised. Finally, the simplicity of the system has intrinsic appeal. Far too much managerial time in the technical organizations of most companies is consumed in the administration of personnel matters. ■

This Business of Recruiting

Developing a comprehensive program to meet an organization's present and future manpower requirements is not an afternoon chore; it calls for time-consuming analysis and systematic planning by an experienced professional.

P. W. MALONEY, *Armour & Co.*

"What? Mother Murphy's is hiring 25 chemical engineers this year? O.K., fellows, let's up our quota to 30."

"You heard the Old Man; he wants to double sales in the next five years. Let's see, then. We'll have to increase manpower about 15% each year. So our quota for new hires this year will be . . ."

"Whadya mean, do I know what my long range manpower needs are? You think I'm not on top of something this important? Let's see, I need a process engineer for the B unit, next month my secretary goes on maternity leave, and oh, yeah, we might want to put on one more draftsman . . ."

"Say, Jack, I just got my invitation to Joe Zilch's retirement party. I think we ought to start planning how we're going to replace him . . ."

Developing a comprehensive picture of an organization's present and future manpower needs is no afternoon chore; it calls for time-consuming, painstaking analysis. Our thesis is this: recruiting can and should be put on a businesslike basis; it should be run about as efficiently as other operations in an organization. So let's stop to look at some reasons for going to the trouble of systematic manpower planning.

Instead of indulging in platitudes about the virtues of good manpower planning, let's ask for answers to five questions:

1. In your organization, which of today's more serious and expensive problems could have been avoided if better manpower planning had been done at an earlier date?

2. What truly significant changes have been made in your organization's manpower planning process since the days when the inadequate planning was done that lead to the problems referred to in the first question?

3. Is your organization like others in that the contribution of each employee is having an increasingly greater effect on productivity? In other words, is more and more of the lower level work being done by non-human means, thereby making each unit of the residual human effort more contributory to success or failure?

4. Similarly, is your organization like others in that achieving a competitive edge by means of its available physical resources is becoming progressively harder? (The point is this: Improved industrial communications and skillful merchandising now permit everyone quick access to innovations in hardware. For an organization to be successful, its leaders must place increasing reliance on its human resources for advantages over competitors.)

5. For your organization, like others, is it becoming increasingly more difficult to go into the marketplace and get "instant people" when new manpower needs develop?

These questions usually yield answers that suggest a need for a growing up in personnel administration. It's time for company officials to stop writing speeches and composing slogans about their passion for those wonderful people, their employees. It's time that the managing of this resource be put on a business-like footing. It's time for the folksy back-slapper to give way to the analyst and the achiever.

Manpower planning is not terribly complex, nor does it require skills not already possessed by persons in most organizations. But effective planning doesn't come about without a conscious effort on the part of some able person or persons in an organization. If they have only a vague feeling that manpower planning is necessary and think it will come about normally as a part of the managerial function, the leaders of an organization are indulging a dangerous self-delusion.

Now let's look at the components of effective manpower planning. Let's see what must be done by the leaders of an organization before they can realistically feel confident that they have done their best to identify and solve their most pressing short and long range manpower problems.

Two Basic Elements in Manpower Planning

At any instant there is a set of "job demands" in any organization; i.e., there is a total quantity of work

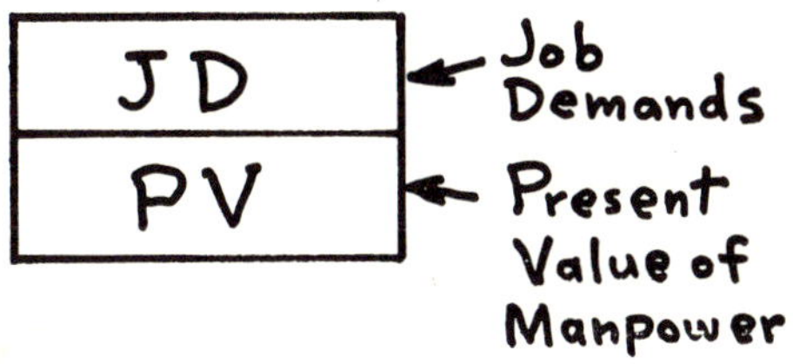

Job demands and resources ideally balanced in an instantaneous work and manpower situation—Fig. 1

that at a certain moment must be done in an organization if it is to be successful. At this moment there is available to an organization the services of a specific number of employees. The sum of the capabilities of these employees for performing the work of this moment represents the resource value of an organization (Fig. 1).

As time elapses, the "job demands" and the "present value" of an organization are both affected by innumerable forces. These forces come from both inside and outside of an organization. For lack of a better term, let's call the sum of these internal and external forces the "environment." By selecting a few of the more obvious forces and expanding the model, we have a situation such as is shown in Fig. 2.

Changes to the Static Picture

As time goes on, the officials of an organization usually make decisions that change *JD* (job demand). However, *PV* (present value—the total value of today's human resources of an organization) will be affected during this elapse of time by factors not wholly within the control of these officials.

Employees have a potential value as well as a present value. The degree to which this potential will be realized is dependent upon the different forces in the "environment." The opportunity is always there for making *PV* into something different.

The movement of personnel in and out of an organization will also affect *PV*. If manpower acquisitions for an organization are quantitatively different from its losses, *PV* will be changed.

The sum of these forces in portrayed in Fig. 3.

Since change is never-ending, a model such is shown in Fig. 4 is necessary to illustrate these forces over a period of time.

A Tripartite Function

Now that we have all the important elements for the manpower planner's chemistry, let's see what he can do with them.

Manpower planning might be separated into three project areas:

- Enabling work: projects having little or no intrinsic value, but which are necessary to provide a base for research or action.
- Research: learning more about the elements in the model and their relationships.
- Action: evolving changes within the elements and/or their relationships.

The "research" and "action" projects can be faulty or even impossible to do unless some foundation work is first done. At least two types of enabling work must be done at an early stage.

Without first centralizing the data on employees, the most elementary research will be difficult to accomplish. (In an organization of any size, personnel records must be mechanized if the data on employees are to be usable.)

Secondly, the manpower planner must stay in close touch with the operational planners, because the operational planners control influential areas of *E* (environment). If these *E* forces are to be integrated into the planning process, they must be understood by the manpower planner. If manpower planning is in the hands of staff people, there must be ways of keeping them in contact with the operational planners.

Inviting Avenues in Research

The second broad category of possible work is in "research." Let's examine our model (Fig. 4).

Certain aspects of today's manpower situation (viz., the balance between JD_1 and PV_1, or JD_1/PV_1) are worth investigating. In general, is today's PV_1 in balance with the current JD_1? How does the balance vary from function to function, level to level, and between professional and managerial positions? Is PV_1 distributed in the best way to handle JD_1.

Environment (E)

Factors affecting job demands

External	Internal
Changes in the economy	Acquisitions, divestitures
Changes in consumer demand	Changes in product lines
Legislation, decrees, agency interpretations	Cost reduction programs

Factors affecting present value of manpower

External	Internal
Manpower supply and demand	Development programs
Expectations in the society	Management philosophy
Legislation	Compensation programs

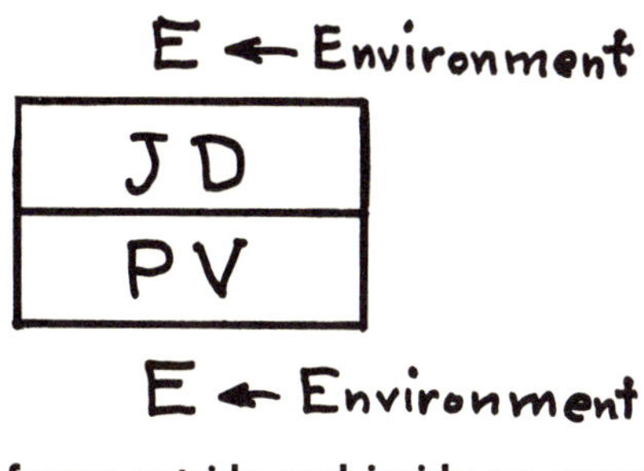

Many forces outside and inside an organization affect the work and manpower situation—Fig 2

What is the outlook for JD_2/PV_2? How good a balance will there be x years from now, extrapolating from current conditions and trends?

How might PoV_1 influence PV_2, i.e., what growth material is there?

What are the pertinent E_1 forces; how strong are they; and how will they influence JD_1 and PV_1?

What will future ins and outs be, and what will be the net effect on JD_2/PV_2?

Such a study program will give the planner a better knowledge of the manpower situation, both present and projected.

"Action" Alternatives Available

The following are examples of programs in the third category, direct action, which might result from the research:

In general, targets can be set for the JD_2/PV_2 balance to get a better frame of reference for the decision making that will affect their future relationship.

There could be reshuffling of present resources (PV_1) to improve the balance between JD_1 and PV_1.

Portions of E_1 are manipulable; for instance, organizational leaders can control their actions that affect the realization of PoV_1. Also, since JD is set by company officials, there is the possibility of better cognizance of PV_1 and PV_2.

Personnel ins and outs can be manipulated to make the JD_2/PV_2 better than JD_1/PV_1.

Back to Recruiting

We have seen that recruiting is not an isolated function. It is one factor in a complex system of related events that influence a manpower situation. Having the type of manpower planning described, where do we go from here?

With this type of planning, both immediate and longer range targets for the acquisition of personnel can now be established. These targets should be both qualitative and quantitative. A suggested framework for target setting is given in the table.

This eight-way classification of manpower is suggested as a prerequisite for intelligent planning of recruiting. When it comes to filling specific jobs, of course, a detailed job description is also necessary.

But we're now talking about the planner who looks a few years ahead. The competent employment administrator will realize that it takes different techniques and proficiencies to acquire each of the eight types of employees. He also knows that he won't make his company a prime competitor in any one of these labor markets overnight; penetrating some markets takes months, others take years. The experienced employment man will demand lead time for those labor markets in which he will be most active.

Major Recruiting Pitfalls

Let's now move to a few specifics on recruiting. Some of these comments apply to all types of employment. Others are pointed more towards the recruiting of professional technical people.

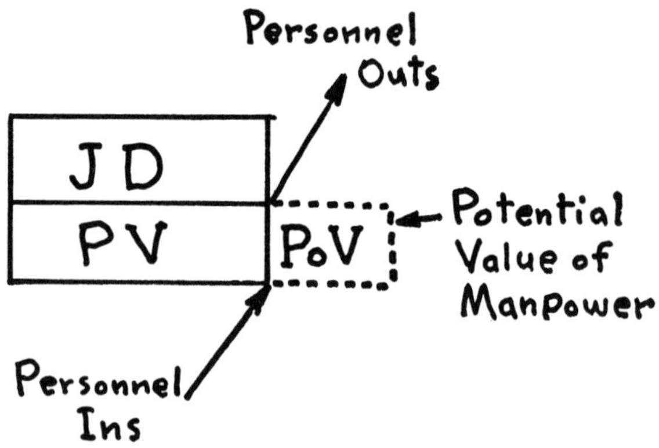

Manpower losses may be offset by hiring new employees or by developing replacements—Fig. 3

There is one step that can be taken in any organization that will take it a long way toward solving its recruiting problems. This is the obvious measure of putting into employment work only those people who are qualified to handle the job in the way it should be done.

Too often recruiting is looked upon as something just about anyone can do. The chore is therefore given to some technical retread who flunked out of other parts of the organization.

Just as dangerous is the practice of blindly assigning the job to the personnel staff without a careful analysis of their capability for doing the job.

The job candidates that employment people don't bring to the attention of management can't possibly be hired. In the last few decades, the highly skilled practitioner in the art of recruiting has come into being. Much has been learned about the right ways and wrong ways of trying to get the right kind of people for an organization; and this knowledge is held only by those who have extensive experience in recruiting.

Running a recruiting program of any size is one of the more demanding administrative jobs in the company. It calls for persuasive skills, ability in establishing and administrating highly complex systems, and an aggressiveness in pursuing a target, often in the face of managerial apathy or resistance.

The recruiting administrator must be a highly skilled manager, having intelligence, insight and experience. If this task is given to the most qualified man and if he is given the support he needs, many organizational problems will disappear. Without a competent recruiting manager, an organization cannot possibly be successfully competitive in a labor market where the supply and demand situation is not in its favor.

Management Commitment Essential

A topnotch recruiting administrator can with time build some involvement by line managers with his activity. He will have the skill to persuade these people that the welfare of their organization depends upon them giving some of their time to the recruiting function.

This job will be easier if top management helps. As with any activity, recruiting will not be taken seriously by line managers unless success in it is rewarded and failure penalized. Responsibility for successful recruiting must be written into the job descriptions of the line managers. Their success in hiring able people must be audited along with their success in the main activities of their jobs.

One situation where formal managerial responsibility for recruiting is particularly necessary is when manpower planning has made it clear that a certain quota of people of high potential, whether managerial or technical, must be hired now for positions that will be open in the future.

Why does the hiring of employees of high potential call for much closer attention to the activities of employment managers? Because hiring such employees will be resisted.

Don't forget that a lower level manager's success is usually gauged by his performance during a particular month or year; he knows this, and behaves accordingly. Let's say there is a job vacancy that needs filling badly. If he can find someone who can handle the immediate responsibilities, why should he leave the position open to go looking for someone with more potential? And why should he hurt this year's profits by hiring an expensive employee when he can get the job filled with someone who comes cheaper? Furthermore, why should he trouble himself with one of those hot-shots who do nothing but ask questions and suggest changes that cost money?

Since there are these forces that combat an intent to hire high caliber people, managerial reward and punishment must be brought to bear.

While we are considering the role of line people in recruiting, it is appropriate to caution against allowing a decentralizing of the employment function. There are many managers who are not willing to give up any part of the employment procedure to a staff unit; they will try to do the whole job alone if they are not curbed.

Certainly the final decision must rest in the hands of line personnel. When the search is for a specialist, the line man might have good ideas about sources that should be used. But people who are skilled and experienced in recruiting can almost always do a more efficient job of finding and screening candidates. This is an art, and a complex one, that they have mastered.

Planning Recruiting Strategy

It is incredible how often management will walk into a recruiting program of major dimensions and play it off the cuff. The results usually are no better than in other endeavors that are pursued in this fashion. There are always plenty of competitors who do the job right.

A critical element in a recruiting program is establishing selection criteria. For a particular manpower plan, how important is potential value as opposed to present value in the hiring program? What relative weights should be given to the candidates' capabilities for basic technical work, for development, or for business acumen in a functional analysis.

Sound strategy also depends on clear ground rules. In making a decision, what weight should be given to experience, academic achievement, test data, and reference checks? And, given these broad criteria, there must be a more precise definition of standards when it comes to hiring a person. Who is to establish these standards? Is this task left wholly to line people, or should other persons of the organization be involved? If the man hired today by unit *A* will someday be the property of units *B*, *C* or *D*, should the latter units, either directly or represented by a staff unit, be involved in selecting the criteria used by unit *A*?

Should evidence of accomplishments be standardized? Should everyone be required to ask applicants to supply transcripts, to fill in the standard application blank, and to supply information from which a reference check can be made?

The starting salary policy of an organization constitutes an important element of strategy. For the people who are leaving college to enter the labor market, salary policy is a minor problem (if the right

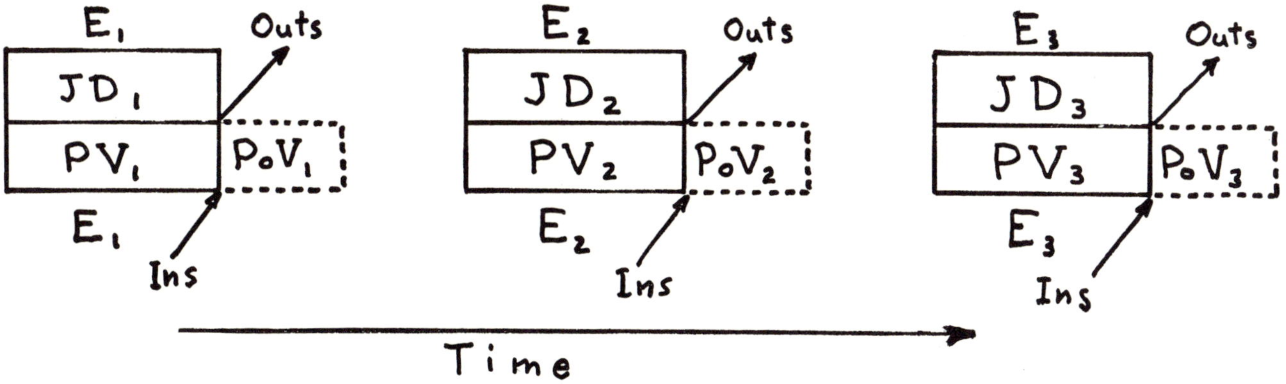

Changes in the job demands and the manpower resources of an organization are never-ending—Fig. 4

Categories of manpower requirements for short and long range target setting

Type & Example	Labor Market Geography	Recruiting Sources
1. Hourly (laborers)	Local	Walk-ins, local agencies
2. Salaried non-professional (clerical)	Local	Local schools, walk-ins, agencies, newspaper advertising
3. Salaried semi-professional (technicians)	Regional	Two-year institutes, newspaper advertising, agencies, walk-ins
4. Professional, inexperienced (engineers, chemists)	National with regional emphasis	Campus recruiting
5. Professional, some experience (1–5 years)	National with regional emphasis	Journal advertising, alumni placement, write-ins, select agencies
6. Professional, highly experienced in specialty	National	Personal contact, journal advertising, select agencies
7. Managerial (under $20,000)	National	Personal contact, journal and newspaper advertising, select agencies
8. Executive (over $20,000)	National	Personal contact, search agencies, sometimes newspaper advertising

kinds of competitive data have been accumulated); there isn't much difference in the starting salaries offered to people of different quality. Starting salary decisions become difficult when experienced people are being hired. The salary range between high and low quality performers becomes appreciable five or more years after the first degree.

With a salary system similar to the one described in earlier articles, the chore of deciding upon an appropriate offer can be easier. Under this system, salary is determined by age and quality ranking. For instance, there is a target salary for a person of a particular age who is ranked at the 10th percentile, and another target salary for a person of the same age who is at the 50th percentile. A salary offer is, therefore, made by relating the quality of the candidate to the people already in the organization who are at his approximate age.

For example, if the candidate is a 35-year-old engineer whose value seems to be about equivalent to that of the average performing engineer of 35 already on the payroll, his starting salary should be near the 50th percentile salary for engineers about age 35. And if a candidate seems to be better than nine-tenths of the engineers of his age group, he could be offered the 90 percentile salary for that age level.

This is, of course, only a base point. The need for a particular candidate's talents might be so great that it may be necessary to ignore the salary structure and to pay more than the evaluation of him indicates he should receive. Contrarily, it might be felt that, as a matter of policy, any new employee should be hired at a conservatively derived salary to avoid the possibility of his being overpaid if the evaluation of him turns out to have been too high.

A lot of wasted time and embarrassments can be avoided by having a clearly defined salary policy. Nothing is so silly as going through the whole employment process with a man, and having him scorn an offer because it is below what he had set as a minimum at the outset of negotiations.

Letting the managers of line units decide the amounts of starting salaries can lead to serious problems. Employment staff people must be involved in setting salaries and they must have the right to appeal differences with line managers. Without the active participation of the staff people, time and effort will be wasted because salary offers will be too low, or people will be hired at salary levels that will distort the existing salary structure.

A thorough review of the starting salary policy should lead to a consideration of other problems. For instance, should bargaining with the man someone wants to hire be allowed? And under what specific circumstances will it be all right to hire a man at a salary so high that it disturbs the existing salary structure?

Communicate Realistic Expectations

There was a time when recruiting programs took on an aura of sales campaigns. Hopefully, most employers are now sophisticated enough to perceive a difference between overwhelming consumers and selling engineers on coming to work for them.

Once the man has bought his Stanley Steamer, he is stuck with it. But an engineer who accepts a position sold by false advertisements has little difficulty in leaving and finding equal or better employment elsewhere. There isn't much to be gained from acquiring people on the basis of information that is not wholly true. Although the top managers may fully realize this, the individual line manager may not. Also, in his need to get somebody on his payroll, he may get carried away with his recruiting techniques. Avoiding false expectations must be a part of the over-all strategy.

In considering what expectations should be communicated to candidates, an analysis of individual growth opportunities should not be overlooked. Regardless of his level—from the starting engineer to the highest level executive—the professional man is keenly interested in how much a job will enhance his quality as an employee.

Of course, an analysis of a company's growth potential might yield results that are rather disturbing. People may be hired into positions with little room for personal development. (This suggests a need for organizational changes if good people are to be attracted.) Be assured that your competitors are showing the candidates how their positions will make them more-qualified men.

Part of pre-program strategic planning should be a structuring of the ways to evaluate the program. What kinds of people accept our offers and what kind refuse them? What are their reasons for their doing either?

How do we compare with our competition in getting candidates to accept offers? How many candidates are lost because of errors in timing or communication by the employment unit? How much money was wasted because candidates were brought in to discuss one job, and upon their arrival, it was discovered that something else was really being offered? How many and which managers hurt the employment effort because of their poor planning or indifferent attention? The means for answering such questions should be worked out before a program is actually used.

Notes on Tactics

Even with a good recruiting strategy, tactical errors are still possible.

1. A key to good recruiting is identifying and penetrating the right manpower sources. Here is where having skilled recruiting personnel makes a difference. If they know how to find the type of people that will meet the organization's needs, they will both reduce recruiting costs and upgrade the caliber of personnel brought in for consideration.

Too often managers despair of finding exactly what they are looking for and accept a compromise candidate because of the pressure to fill the job. The labor market is mobile and has millions of people in it. In this market there is always someone who would be very pleased to take the position that is open and who could perform well in it. The job is to find him. Good recruiting personnel will find him.

2. Recruiting has several highly personal aspects to it; if the candidate perceives that he is being treated as a unit on the assembly line, he will not feel warmly towards the organization. But, without some system, a program of larger size will fall apart. Good candidates will be lost; messages will not go out soon enough and candidates will take other jobs; and people will be hired who would have been avoided had their history been more thoroughly evaluated. Avoiding such problems lies in having a straightforward system.

3. Effective line involvement will not come about unless responsibility for recruiting is a part of a manager's job. The priority of this responsibility must be restated periodically throughout the life of the program before line managers will give the day-to-day attention to this function that it demands.

4. Today's candidate is becoming quite sophisticated in the art of job hunting. He will not accept on faith the suitability of a particular position for him. He will seek the best data about a job, its growth opportunities, and the psychological environment in which he will be employed.

Because of these desires, he feels strongly that during his visit he should spend time with the people who might be his peers. When he talks to the people who might be his equals, he gets a view that he accepts more readily than that which is given by top executives.

Talking to his equals also makes him confident that nothing is being hid from him. (He especially needs this reassurance if he has had an unhappy experience with another company.)

Not providing interviews with peers—half of the man's time could profitably be spent this way during a visit—is one of the most common and more serious mistakes made by companies.

5. Finally, the strategic and tactical elements we have reviewed must be observed in all levels of employment. Some companies run top-notch programs through the first six levels shown in the table and then forget all the rules when hiring managers and executives.

Certainly a different approach is necessary with these candidates; their stature and record of accomplishment entitle them to considerations not given to inexperienced graduates. But this doesn't mean that recruiting should be conducted out of some executive's vest pocket. There is still a need for criteria, for ground rules and for a system. And if the employment unit is staffed as it should be, there is still room for this expertise in the recruiting process. ■

Why Do People Leave Your Company?

Who are the engineers leaving your company—random individuals, mostly misfits encouraged to resign, or those whose loss is regretted? If you do not know, you may be missing clues about the professional environment in your company.

P. W. MALONEY, *Armour & Co.*

The company that runs a sharp employment program will harvest its crop of new-hires every year. But, as its management proudly watches the new acquisitions come on board, it had best remember a dismal fact: within five years, half or more of these people will be working for someone else.

The termination of professional people is a thriving business. Unfortunately, it is not a well tended one. Both voluntary and encouraged losses just happen; seldom is there any programming.

Are you approaching the problem of personnel terminations with the same structured thinking you are probably applying to recruitment and other personnel programs?

Do you really have a good fix on the types of people you are losing? Is it a random slice? Or is it only the misfits? Or are you performing a service for the industry by furnishing postgraduate training, losing (only) your best people to other employers?

Do you even know how many people you are losing? Have you ever calculated what proportion of your recruitment program is necessary just to replace losses? Have you ever estimated how many recruitment dollars can be directly charged to personnel attrition?

You might be missing valuable clues to improving your selecting of personnel. What are the differences between those people who survive in your environment and those who don't? What types don't seem to come up to your standards? What types decide that you don't really have what they are seeking?

Sound answers to these questions can save a lot of the money and motion wasted on the recruitment front. (And when you find out what type of professionals tend to decide you're not their kettle of fish, you might get your eyes opened about your organizational environment.)

You might not be aware of serious managerial inadequacies that might be the cause of your attrition. Do you know, for sure, that the losses you encouraged were really among the people who failed because of their own deficiencies? Do you know how many of the people you were sorry to lose could have been saved with better managerial performance?

You probably don't know how consistent your departments and divisions are in deciding that a professional is not shaping up. Are people being let go from one part of the company, when in other places professionals of equivalent caliber are being retained? And even if you convince yourself that this is appropriate because job demands differ, wouldn't transfer rather than termination be a less costly solution?

Do you know how many people have been kept on the payroll who shouldn't be? For an indicator, do you know how many professionals with five or more years of company service were encouraged to leave last year?

Perhaps the changing nature of your technology made some of them obsolete. But do you know how many were not identified earlier as being inadequate, just because no one tried to identify them? And how many professionals have so much service that you are loathe to let them go, despite their incompatibility with your needs? Perhaps they could have been measured and found lacking ten years ago?

Do you know how much the loss of confidential information that leaves with terminating professionals is costing you? Are you confident that you have taken every reasonable measure to minimize this loss?

Indeed, there are some worthwhile reasons for wondering whether your termination process is adequately programmed. Now let's see what such programming would involve.

Programming Personnel Terminations

There are five ways that management can structure its role in the termination process. None of them calls for excessive expenditures of time or money. When compared with the waste that may have been revealed by the questions already asked, the investment is infinitesimal.

1. Keep better records. Retaining some data about the people who terminate permits a degree of analysis.

This is the sixth and last article of a series on technical administration.

2. Conduct termination interviews. This, it is true, will take some time but the return may be well rewarding.

3. Do some research. Termination records and interviews must be systematically analyzed periodically.

4. Program your involuntary losses. A company should coordinate what its branches are doing about separating inadequate professionals.

5. Plan to prevent the loss of confidential data. There should be clearly defined, rigorously enforced rules that protect the company's proprietary information.

Record Keeping Calls for Precise Classification

An impending termination should trigger the collecting of data about the employee who is leaving. This should be done by the personnel department. For research purposes that will be described later, these data should be kept for several years.

First get the data that will classify the employees who leave. Obviously, it is easy to classify terminations caused by layoffs, deaths and retirements (although the last should be categorized as voluntary or encouraged). Of all the people who leave, those who resign should be of most concern. They should be separated into four groups:

1. "Not encouraged, sorry to lose."
2. "Not encouraged, not sorry to lose." Some discretion is necessary here. It could be said that some in this group were headed toward encouraged resignations, that they are the ones who can be easily replaced with people of more ability. For better or worse, this provides an escape for the manager who is leery of indicating too often that he is losing people he should be retaining.
3. "Encouraged." If anyone is actually discharged, he should be included in this category.
4. "Encouraged, sorry to lose." Although a rarity, it can happen. It describes the capable person who is asked to leave because his particular skill can no longer be used. He should be categorized this way, rather than as an encouraged loss, so that the data on those who are judged unsatisfactory are kept apart.

The file on those who terminate should include background information. The appropriate manager should explain in writing why a particular resignation was encouraged. He should pay particular attention to any of the man's faults that could or should have been identified when he was being hired.

When the loss is not an encouraged one, regardless of whether it is a "sorry" or a "not sorry" loss, the supervisor should record the reasons the man gives for leaving. For whatever value it might have, the supervisor should also indicate what might have been done to prevent the loss, if a transfer should have been considered, if other assignments might have kept him from resigning, and if the man clearly understood how the company regarded him.

The man who is resigning can even be asked to write some of the data. Have him fill out a questionnaire that elicits from him what he liked or disliked about his job and the company. This is often a part of a programmed termination, which is covered next.

Interviews: An Information Gold Mine

A termination interview should be a structured approach to learning from a departing employee the facts, attitudes and opinions that are relative to his leaving. It should be handled by a staff member who has no direct organizational relationship to the person who is leaving.

Is it worthwhile? Will enough be learned, beyond the information that can be supplied by line supervisers, to justify someone spending an hour or more with the departing employee?

Refer back to our reasons for getting business-like about terminations. You will note that properly satisfying our need for knowledge depends on getting valid information about why people leave. Institutionalizing the interview is fully justified because it brings a third party into the process of getting this information. This third party is not present to employ Gestapo tactics on line managers. There are valid reasons for having a staff person get these data.

First of all, the events leading up to the termination, regardless of who initiated it, are likely to have built barriers that will impede effective communication between the professional and his supervisor. Secondly, one particular staff man can be trained to do the interviewing job more effectively (this will be discussed further); it isn't possible to train all line managers to do a professional job of interviewing. Similarly, by handling more and more of the interviews, the staff interviewer becomes more adept at it. Finally, the staff representative is probably closer to the "big picture" than the manager of any department. He can interpret the terminator's comments against a larger frame of reference.

Let's bring out a few cardinal points about termination interviews.

First of all, the interviewer should have been trained. A session with a person who is resigning will only be valuable if truly basic issues are uncovered and analyzed. Actually, an experienced counseling or clinical psychologist should be brought into the game for this play but, at the very least, the interviewer should have had a degree of training. As little as ten hours of such training with an expert will put the interviewer steps ahead.

If the organization has systematized its data collection on the people who resign, there is usually a pack of information that the interviewer should assimilate before he walks into an interview. When the case is unusual or puzzling, he should talk privately with the supervisor of the person who is resigning before he starts the interview.

The interview must be structured. A little later, we'll touch on the reports that should come out of termination interviews. A report will lack validity if the interviewer hasn't followed a specific outline. A company should decide what information it wants most, and make sure this is elicited in every interview.

The results of the interview must be written up well. Doing an objective job of this is not easy, and training for the task should be a part of the interviewer's education. Any psychologist who knows how to train interviewers is aware of what should and shouldn't be in an interview report.

Research Needs Careful Guidance

The company that does what has been outlined is well ahead of the game. But it must be sure that effort won't go to waste. Data that fill the personnel department's file drawers won't influence many decisions.

The first "research" job hardly deserves the name: it's simply a matter of reporting the statistics on all the different types of terminations that occur. When a company first goes into programmed termination planning, it should dig back into its records for a year or two and label earlier terminations. These data will not only be interesting but will also supply a base against which new trends may be compared.

There is one immediate and obvious application for these data: projecting the company's recruiting needs to replace anticipated losses. The recruitment program based on some foreknowledge of targets is bound to be better than those of the competitors.

Statistical reports usually arouse management's curiosity about why there are so many losses of each type. The intensity of management's questioning varies exponentially with the closeness of the time for reviewing the recruiting budget. In such situations, staff units had better be ready with analyses of the termination interviews. When reasonably objective evidence is absent, the void will be filled with defensive arguments about there being plenty of manpower around. However, interview reports can be dangerously misleading when done improperly.

Encouraged Resignations Merit Programming

At the beginning of the article, we touched on some of the problems that can arise if the classifying of encouraged terminations is not guided. These problems stem from the same shortcoming that generates so many difficulties in the administration of technical personnel: our inability to measure performance against a set of absolute standards. In earlier articles, we saw why this weakness should make us wary of having department heads alone select persons for top-level pay or recognition. For the same reason, we should also be wary of uncoordinated designating of flunk-outs.

We wiggled out of the earlier problem by leaning on interdepartmental rankings. With this technique, most or all of the professional personnel are put into a single best-to-worst ranking. We can use these same data to improve our classifying of encouraged resignations.

We particularly want to get a fix on the shorter service performers, so let's look at the ranking of all the professionals under the age of 35. Of course, in any ranking of performance the younger people will tend to be at the bottom, because they haven't had the time to acquire the experience that will make them as valuable as older employees. However, those in this younger group can be compared with each other.

The departments having those people who are in the bottom 10% of this ranking should be asked to explain why these people should not be terminated. If this kind of appraisal is done annually, the company can be more certain that it is fully aware of the situation concerning those who are inadequate performers. Steps can be taken to improve their performance, if this is possible, or the terminations can be expedited.

A case can be made for forcing the same type of scrutiny on people who are over 55. Again, it is widely accepted that every company has to encourage some of its people to retire at an early age. Their performance goes down hill so rapidly, or their technological skills are so out of date, that their retention is a burden to the company. Therefore, a company can also look at the ranking of its older people, and particularly scrutinize those at the bottom of the list.

Security of Confidential Information

When a person leaves the company, specific steps must be taken to prevent proprietary information from being acquired by the company he is joining. First of all, the organization should not assume that the engineer or chemist is fully aware of his obligation to prevent the transfer of information to his new employer. This should have been made clear to him before he left.

Secondly, a person should be asked to sign a statement indicating that he has returned all confidential materials, including handbooks and laboratory books, to his supervisor. This statement should be obtained by the supervisor, so that he is personally involved.

The company should routinely write a letter to the man's new employer, advising in what general areas their new employee possesses information that is of proprietary value. This puts the new employer on notice, and removes any basis for later claims that the confidential information was picked up inadvertently. If the employee has signed an agreement with the company he is leaving that obligates him even further on security, a copy of this contract should accompany the letter to his new employer.

There is a lot of room for better programming of termination procedures. When you figure that it costs from $2,000 to $5,000 to replace each professional who leaves, it seems that the dollars alone supply enough incentive for better planning. As in other areas of personnel administration, it's time that the handling of terminations be put on a businesslike basis. ■

4

The Technical Management Aspects of Business Law

What do a construction contract, an employment agreement, a union organization campaign, an equipment warranty, a patent, a supervisor's reference, and a product-liability dispute have in common? They all involve aspects of business law with which technical managers and most engineers should be familiar.

PART 1

What Engineers Should Know About CONTRACTS

In this first of two articles, the author discusses the essential elements of all contracts. In the sequel, to be published in an early issue, he will deal with specific types of contracts —e.g., those for equipment purchase, real-estate transfer, engineering services, and construction.

J. J. STEINKE, *Lummus Co.*

The contract is probably the most widely used instrument for the conduct of business in our modern society. Not only do contractual agreements figure prominently in our daily personal lives, but the engineer is constantly performing his services under contract, or utilizing contracts on behalf of his company.

Yet it is generally only the law or business student who is given the opportunity to study contracts in the undergraduate or professional school. Is it any wonder then that there are many misunderstandings in the normal pursuit of life and business?

All purchases constitute some form of contract, whether conducted in a dime store, in a real estate agency, or in a chemical plant. Many contracts are oral agreements, while others must be written to be valid. Common examples of contracts are transportation, parking and entertainment tickets, time purchase agreements, charge accounts, service agreements, house repair contracts, rental or lease agreements, purchase orders, real estate transfers, employment agreements, engineering-service or construction contracts.

One purpose served by written agreements is to improve the common understanding between the parties to a transaction. Many people are reluctant to ask for a written agreement when conducting a purchase, feeling that it may detract from the personal relationship they enjoy with the seller. But a written agreement can often avoid misunderstandings, and can thereby actually improve personal relationships.

While a well-executed agreement is not essential to the conduct of business (perhaps the majority of all business is carried on through poorly executed or unenforceable agreements), it does contribute to greater satisfaction to both parties in any transaction, and gives them the full protection of the law. It is therefore important that all engineers learn the fundamentals of contractual procedures for the practice of their profession.

While we will discuss some of the legal as well as economic phases of contracts, there is of course no intent to deliver authoritative opinions with respect to the practice of law; for such, the engineer is referred to law literature and legal counsel. Rather, this series is intended to improve the engineers' understanding of a necessary discipline, and to show what both the engineering and legal professions can contribute to successful contractual agreements.

Before going further, I would like to stress the importance of good faith. We will see that the signed contract is a mere scrap of paper if not entered into with intent to conclude a genuine workable agreement.

Definition and Requirements of a Contract

Canfield and Bowman[1] state "A contract is an agreement enforceable at law between two or more competent parties for a legal consideration or in writing under seal, to do or not to do certain things." While a number of other definitions have been given, all contain several elements that are requisite to a proper contract:

1. *Offer and Acceptance*—There must be an offer and an acceptance, either implied or by words, to form the basis for an agreement. The offer may be a formal proposal or bid, or an offer to purchase merchandise on display. The acceptance may be written, or may be implied from the action of the recipient. Since reality of consent is requisite to an agreement, the offer and acceptance must be communicated directly between the principals.

2. *Agreement*—There must be a "meeting of the minds" on the terms of the contract. This also implies direct communication between the principles. If there is evidence that the parties have not reached essential agreement on the terms, the contract may be a nullity (i.e., without legal force or effect). This does not mean that all details must be finitely fixed; for example, if a purchaser makes modifications to a proposal and the vendor accepts them at extra charge, the agreement is still valid even though the exact

amount of the charge is not spelled out at the time.

3. *Valid Consideration*—There must be a worthy consideration, usually in the form of money payment, in return for the promise to do certain things. The amount of the consideration need not be adequate, but only of real value in the eyes of the law. A consideration shall also be legal, possible, and present or future. Exceptions to this are agreements under seal, which require no consideration if properly executed.

4. *Competent Parties*—The principals to a contract must be legally competent. The parties shall have reached their majority (be adults), shall be mentally competent and not under the influence of drugs or liquor.

Agreements may be made with a person under legal age, which is usually 21 years, but they are voidable at his option until he reaches his majority. The law does not enforce contracts against a minor during his tender years, but the minor may enforce the agreement with the other party.

5. *Legal Form and Content*—The last requirement is that the contract be legal in both form and content. The engineer who is not schooled in contract law is urged to obtain advice on this aspect before concluding agreements. A detailed discussion of legal form and content is beyond the scope of this section, except to state a few general principles:

Contracts under seal owe their validity primarily to their form and, as we have said, are valid even though there be no consideration. Thus they are used for gratuitous promises and for conveyance of land. A sealed instrument must be in writing, must be sealed and signed, and takes effect as of the date of delivery. A property deed is this type of instrument.

Other contracts are governed by sections 4 and 17 of the "Statute of Frauds." This English statute has been embraced by most states in the U. S.; applications include: (1) contracts relating to the sale or transfer of real property; (2) any agreement that is not to be performed within the space of one year from its drawing.

A tavern or nightclub is not a particularly good place to conclude a contract. If you do choose to conduct business over a cocktail, you had best delay the contract signing until the next morning, since the agreement may otherwise be declared void.

All such contracts must be in writing to be enforceable. A valid contract may not contain any statements of a fraudulent nature, must not contain anything in the nature of a tort (a violation of human rights), shall not be made under duress or undue influence, nor contain provisions that are contrary to common law or statute law, and shall not be contrary to public policy.

Examples of contracts that are void or voidable (subject to being ruled lacking in legal force) due to legal content would be agreements pertaining to gambling, commission of crime, restraint of trade, violations of public trust or welfare, and agreements that do not conform to licensing laws.

This last area is of particular interest to engineers. It is mandatory that the registration laws for professional engineers be complied with by the contracting parties, or the contract may be found illegal in content. When the contract is stated to be governed by laws of a state other than the one in which the engineer's design is to be utilized, the implication is that the engineer need be registered in both states, since he is practicing in both localities. A ruling should be obtained in each instance. Also, certain states restrict the practice of engineering by corporations and regulate such practice by a separate registration.

These legal requirements are most likely to become an issue in the event of a structural failing or a disaster during or subsequent to construction. An engineer or engineer-contractor would have difficulty enforcing payment for services if he were found to be practicing illegally, whether the failing were due to negligence or otherwise.

Where violation of the law occurred without willful intent, the agreement may be amended so as to conform to the law, and thus become revalidated. Until such conformance, however, the offending party cannot enforce the agreement. Thus the fees for professional services rendered may not be collectible if the services were performed without legal sanction.

Termination of Contract

A contract may be terminated in five different ways:

1. *By Performance*—When both parties have completed all of their obligations, the contract has been discharged. If one party has substantially performed his obligations, and thereby enriched the other party, he is entitled to remuneration less a deduction for the deficiencies of performance.

2. *By Agreement*—A contract may be terminated by agreement, which may take the form of complete nullification, or replacement with a new agreement. Many contracts contain provisions for cancellation under certain terms and conditions.

3. *By Legal Process*—This would include actions brought as a result of deficiencies in legal form or content. Bankruptcy and dissolution of partnerships

will also lead to contract termination by legal process.

4. *By Breach*—This occurs when one of the parties does not fulfill his obligations. When breach occurs, the injured party may institute suit for damages, and the contract may be terminated by action of the court.

5. *By Impossibility of Performance*—Circumstances arising subsequent to the signing of the contract may force termination. Thus if a building for which alterations were contracted burned down prior to the start of work, the contract is terminated. Dissolution of partnerships and bankruptcy are also causes for termination in this category.

Of course, a majority of contracts are terminated by specific performance. This is not to say that conclusion always results in complete satisfaction to both parties, since it is practically impossible to cover all facets of a complex transaction in words. This is where the engineer fills a most important function, and where good faith must be relied upon.

Most differences can be resolved by the engineers representing both parties. It is essential that they maintain an objective attitude, and do not become emotionally involved in a dispute. If a dispute must go to litigation, then the engineers have failed to some degree in the preparation and execution of the contract.

Settling of Disputes

In the event of unsatisfactory performance under contract, several courses of action are possible, namely: (1) conciliation, (2) arbitration, (3) suit in court for damages or specific performance. The engineer is instrumental in bringing each of these actions to bear.

Conciliation is most often used. The multiplicity and complexity of agreements necessary to the normal conduct of business requires the constant review, interpretation and resolution of contract terms and conditions. It is good business to settle differences by conciliation, resorting to legal action only as a last resort. Thus it is the responsibility—the duty—of the engineers representing the two parties to an agreement to make it work.

Many contracts contain a provision for settlement by arbitration. Arbitration can also be used in the absence of such a provision, if both parties agree to it. One party is not enough, since the other cannot be deprived of protection under the law by this procedure. If it can be demonstrated that the findings were arrived at erroneously or have overlooked lawful content, relief may still be sought in the courts.

Either party may institute suit in court for injuries sustained in the performance of the contract. The courts may award damages or insist on specific performance, depending on the circumstances. In court cases, the written evidence is most important, although "parole" or oral evidence is also admissible.

Since technical terminology is construed in the sense of normal usage, it is important that the technical exhibits to a contract be prepared by the most experienced engineers. For instance, let's say Brown purchased a plant site from Smith with the condition that the fronting street would be "hard surfaced" within a year, at Smith's expense. The street was subsequently surfaced with gravel. Brown had assumed that the terminology meant a macadam surface. Since the term "hard surfaced" is nonspecific and has many meanings, Brown cannot hold Smith to a macadam surface.

It should also be kept in mind that the courts will only enforce the specific terms of a contract if they do not violate legal precedence. Thus the mere fact that a condition is stated in a contract does not necessarily make it enforceable.

These everyday examples illustrate the five essentials of a proper contract:

1. Offer and Acceptance

Brown purchased a house through a real estate agent. The agent advised that he would arrange for a mortgage loan. At the title closing, a mortgage broker, Smith, presented his bill to Brown. Brown refused to pay. There was clearly no communicated offer and acceptance between Brown and Smith, therefore, Brown was not obliged to pay for a service he did not accept and was not aware of.

2. Agreement on the Terms

White agreed to paint Black's house for a given sum of money, which Black accepted. Subsequently it developed that White intended to paint a one-story house, while Black had in mind his two-story dwelling. This lack of agreement would make the contract a nullity.

3. Valid Consideration

Brown saved the life of Smith by the application of first aid following an accident. Out of gratitude Smith signed a promisory note of $500 to Brown. Payment of the note can not be enforced since the consideration, namely saving a life, is not present or future.

4. Competent Parties

White and Black concluded an agreement in a bar room. Black breached the agreement, claiming that he was under the influence of liquor when it was made. The agreement is not enforceable if the claim is supported by evidence, since the law does not consider an intoxicated person competent.

5. Legal Form and Content

White contracted with Black to install plumbing in Black's new house. White advised that he was not licensed but would have a friend obtain the approval for him. White was restrained from doing the work by the plumbing inspector. Black could not sue for breach of contract since he knowingly concluded an agreement contrary to law.

Financial Aspects

The consideration in a contract is usually in the form of money. The amount to be paid and the manner of payment may generally be stated in very simple terms. However, this part of the agreement is very often misunderstood and misapplied.

The consideration is made in return for a specific promise as contained in the offer. Therefore *no chanye whatsoever* may be made in conditions relating to the promise without affecting the consideration. When a change is made in the requirements for performance, the contract must be amended by mutual agreement with respect to the price. Without such agreement, the contract remains in force as originally stated.

From the above, it is evident that in the strict sense there can be no such thing as a "fixed price" on a complex construction project. Changes, if only minor, are usually inevitable. The customary approach is to agree beforehand to a method for handling modifications in the work, either by cost estimate and reimbursement on either a cost-plus-fee basis, or by a quoted unit-price arrangement. If there is no previously agreed-to procedure, the promisee is largely at the mercy of the promisor.

One of the most common faults of contract procedure is lack of agreement on the price when concluding the agreement. Thus if several bids are considered, and an order is given to the low bidder, the order frequently contains requirements different from the original specification. The order does not constitute an acceptance, and a contract is not in force until the promisor accepts the order. If he accepts with the notation that the new requirements will affect the price, the contract is in force, but there has been no specific agreement on the price. It is always in the interest of both parties to agree on the price before concluding the revised agreement.

Contract Form

All written contracts, whether for the sale of real and personal property or for the supply of goods and services, include provision for defining these basic elements:

- Parties to the agreement.
- Purpose.
- Things to be performed.
- Consideration for performance.
- Conditions for payment.
- Acceptance.

In the simpler agreements, these elements are readily identifiable as short concise statements, whereas, in the more-complex contracts, many pages may be used to define any one of them. The order in which the subjects are covered is not important.

The contract usually begins with the names and addresses of the parties to the agreement. Thereafter follows a short statement outlining the purpose that the agreement serves.

Usually, a definition of things to be performed comes next. In the case of a land transfer, a metes and bounds description of the land to be sold is included. A simple purchase agreement includes a description of the articles covered in the sale. (Specifications, since they define quality and quantity, relate to this part of the contract.) The corresponding section in an insurance contract would define the amount of insurance, and the conditions under which it would be paid.

The consideration for performance is the amount of money, the goods, the forebearance or the promise that is given in return for the things performed. The type of payment—e.g., lump sum, cost plus fee, unit price or term payment—is covered here.

The conditions for the performance, and the conditions for payment in return, include the manner and means by which these acts are to be accomplished. Such conditions would include timing, liabilities, changes, confidence, assignability, cancellation, accounting procedures, method of payment, force majeure (Acts of God), liquidated damages (payment for injuries sustained) and any other factor that modifies or relates to the method of performance and payment. If no conditions are stated, the operating procedures are left to be worked out by the participants.

The acceptance of the performance may be considered a part of conditions for performance, but is usually stated separately to designate the type and extent of approvals required. A land transfer usually requires a clear title to the land and delivery of the deed in a specific form to the purchaser. Acceptance occurs when the title is closed. In a simple purchase agreement, provisions are usually stated for inspection and acceptance of goods. Construction contracts invariably contain language covering the acceptance of the work; usually care and custody of the property is assumed by the purchaser upon acceptance, although title does not necessarily pass at that time.

We will discuss some of these elements in greater detail when we review specific types of contracts—i.e., transfers of real or personal property, contracts for engineering services, and construction contracts. ■

References

1. Canfield, D. T., Bowman, J. H., "Business, Legal, and Ethical Phases of Engineering," McGraw-Hill, New York, 1954.
2. Abbett, R. W., "Engineering Contracts and Specifications," Wiley, New York, 1963.
3. McCullough, C. B., "The Engineer at Law," Iowa State College Press (under cooperative agreement with the Oregon State Highway Dept.).
4. Dunham, C. W., Young, R. D., "Contracts, Specifications and Law for Engineers," McGraw-Hill, New York, 1958.

Meet the Author

John J. Steinke, who is a project manager for Lummus Co., in Newark, N. J., has considerable experience in contract engineering. He has managed the engineering and construction for a variety of chemical and petroleum plants, including several in Canada, Europe, the Near East and South Africa. His previous experience was with M. W. Kellogg Co. and American Brake Shoe Co.

A graduate of Brooklyn Polytechnic Institute, from where he has B.M.E. and M.E. degrees, he has also attended Columbia University. He is a member of AIChE and the American Assn. of Cost Engineers, and is a licensed engineer in N. Y., N. J. and Del.

PART 2

What Engineers Should Know About CONTRACTS

This engineer's-eye view of three important types of contracts —for purchase and sale of goods, engineering services, and construction—focuses on the economic as well as legal aspects of contractual procedures and decisions.

J. J. STEINKE,
Lummus Co.

We are going to examine the three types of contractual agreements most often encountered by the engineer in the chemical process industries. We will not dwell at length on such phases as legal precedent, but instead will emphasize some of the economic factors that in a given situation may favor one type of agreement over another, or one procedure over a less efficient or less suitable alternative.

When property of any kind is sold, the seller can convey no better title, no higher degree of ownership, than he himself possesses.

Of course with personal property, such as materials or equipment, the degree of ownership is usually not in question. A solvent business firm may be assumed to be the undisputed owner of the products it sells, and the purchaser places his reliance for clear title on the reputation of the company.

However, with real property (which includes land and things permanently annexed to it, such as buildings, trees, wells, pavements and fences), the degree of ownership cannot be assumed and there are other factors that must be defined.

For instance, let us say that *A* wishes to sell a parcel of land to *B* for a sum of money—apparently a very simple transaction. If *B* gives this some thought, he will want to ascertain that *A* is in full possession of the land, to what degree *A* may be limited in his use of the property, the definition of the land boundaries, and what the land includes in the way of buildings or facilities.

For his part, *A* will want assurance that he will receive payment in accordance with given conditions, and that he will not be required to expend these monies in connection with the transfer. Translating all these terms into language that truly represents the intent of the parties, in a form understandable at law, constitutes the true essence of a real-property contract.

To return to contracts involving personal property —and these are our main concern here—the essential elements include the description of the merchandise, warranties expressed or implied, terms of payment, terms of delivery, and the time and place at which title is to transfer.

Making the Purchase

In the writer's experience, the greatest stumbling block in concluding sales agreements usually centers around definition of the article in question, particularly when complicated equipment is involved. Usually the purchaser issues an inquiry complete with specifications, whereupon quotations are received, a bid analysis is made, and a purchase order is placed with the successful vendor. If the purchase order agrees completely with the quotation and contains an acceptance of the bid, a valid agreement is in force. However if the purchase order deviates in any major respect from the quotation, it becomes a counter offer, which must be accepted by the vendor to be valid.

Very often, the vendor does not bother to formally accept the revised terms of the purchase order, or to indicate that the revisions will affect his quotation, but will proceed with manufacture and delivery before indicating a change in price. In such an event, the entire transaction would be consummated without a binding agreement, and a dispute might ensue over the change in price. While business may be conducted on this basis, and very often is, the result may lead to misunderstanding and dissatisfaction.

To avoid these problems, either the purchase order must contain the buyer's full acceptance of the quotation, including general terms; or the seller, in accepting a purchase order that contains revisions, must state his exceptions and changes in price. It is then up to the buyer to accept these.

Another common misconception is that a vendor's price as contained in a bid is firm for the stated time limit. In many cases, a contractor predicates his bid price on a vendor's quotation, hoping that he may purchase at the quoted price. While the vendor's repu-

The views expressed in this and the preceding article are solely those of the author and are not intended to represent those of the Lummus Co. These articles present guidelines for engineers and the lay views expressed do not represent legal opinions, interpretations of the law or definitive procedures in commercial transactions.

Glossary of Terms

Barter	A trade or an exchange
Consignment	Handing over for keeping or shipping.
Fungible Goods	Bulk material that can be measured or weighed.
Liquidated Damages	Money settlement for loss incurred.
Mechanic's Lien	A detention of property by a workman for the nonpayment of services rendered.
Sales Act	An act passed in England in 1893 that supplements the Statute of Frauds and lays down positive rules for sales transactions.
Statute of Frauds	An act passed in England in 1676 that covers contracts for the sale of goods and is aimed at preventing fraud and perjury by requiring material (written) evidence of the contract.

tation may hinge upon his willingness to sell at the bid price, he is not legally obligated to do so until the bid has been accepted. If he decides to revise his bid before the contractor accepts, the contractor has no recourse except to seek another vendor.

The contractor may obviate this difficulty by obtaining an option to buy for a specified period of time, or by precommitting the purchase. An option to purchase must be in the form of an offer by the vendor and a written acceptance by the purchaser. Precommittment is a conditional acceptance based upon the purchaser concluding a contract with a third party.

Implied and Expressed Warranties

Usually a warranty is implicit in the sale of manufactured articles. The purchaser may assume that the article is new, unused and free of manufacturing defects unless otherwise stated. Thus he does have legal recourse if the article is unacceptable on these counts.

In addition, reputable venders usually express a warranty, providing for replacement if the article is found defective within a period of 12 or 18 months. An expressed warranty will be limited in extent to that stated. Thus it is important to review the terms of expressed warranties before concluding a purchase.

Other Terms of Sale

The price, and terms of payment and delivery should be designated in a quotation. The price quoted is one of the terms of the vendor's offer and may or may not coincide with the terms of the purchase inquiry. The purchaser must review the vendor's offer in all respects, to see that it does conform with the inquiry, thereby insuring a firm price.

If there are any questions on the quotation relative to the inquiry, these should be resolved with the vendor before concluding the purchase. All too often, the vendee claims that he intended one thing in the inquiry, and the vendor claims that this was not made clear, and hence not included in the quotation. Under such circumstances, the weight of evidence lies with the vendor unless the vendee had made clear his assumptions at the time the quotation was received. It is advisable to record all agreements in writing, since all recorded events form evidence to prove the contract.

The terms of payment may provide for progress payments or for part payment equivalent to partial delivery. In the case of "entire contracts," delivery must be consummated, and payments made in full. In the case of fungible goods, the transaction may be terminated at some intermediate point of delivery by notice, and payment for that part received, if the intent is so stated in the agreement. Thus if the purchaser contracts to buy concrete at $15 per cubic yard, he will purchase as much as he needs at this price. If he contracts for 1,000 cu. yards of concrete at $15 per cubic yard, the contract requires full delivery and full payment.

Passing of Title

The time at which title passes is a very significant point. The general principle is that title passes according to the intent of the parties. In the absence of expressed intent, title passes when the sale has been executed if the article is ready for delivery. If something remains to be done to the article, title does not pass until it is done.

As a practical matter, title usually passes to the purchaser when a manufactured article has been turned over to a common carrier on an fob. order. If the order includes delivery, e.g., fob. point of destination, title passes upon delivery. Therefore possession of the goods is not synonymous with ownership.

The significance of passing of title lies in liability attached to loss prior to receipt. To cite an example, if *A* purchases furniture at a store and concludes an agreement, title passes as soon as the furniture is designated for delivery. If a loss occurs prior to delivery, ownership of the furniture resides in *A* and he must take action against the seller or carrier for the loss. However, if the purchase agreement calls for delivery, that is, fob. his residence, title does not pass until it is delivered and accepted.

Usually, manufactured commodities are quoted fob. point of manufacture. Therefore, if any loss occurs in transit, the buyer must take action against the carrier to recover his loss.

Rights of Buyers and Sellers

Rights of the purchaser include the right to reject defective goods as defined by the warranty. Defective goods must be replaced by the seller. The purchaser is also entitled to receive exact measure. In an entire or indivisible contract, payment may be deferred until the last shipment is made. Overages may be returned to the seller or accepted at the stipulated unit price.

Rights of the seller include prompt receipt of payment for the goods as called for in the contract. In the event of default, the seller has the following remedies available to him:

- If possession and title rest with the purchaser, the seller may sue for the contract price.
- If possession rests with the purchaser and title

How to obtain competitive bids

The price quoted by a vendor may be made firm and therefore meaningful in a competitive market if the definition of the product and the terms of sale are agreed upon. If the parties have not reached a meeting of the minds on definition, the whole procedure of obtaining competitive bids is pointless.

Consider the following case. The purchaser issues an incomplete inquiry (hoping to save schedule time) in order to select a low bidder. He later modifies the inquiry and asks the successful bidder to make a price adjustment. The bidder, knowing that the offer is no longer competitive may add some fat to the adjustment.

The correct procedure for concluding a purchase includes:

1. Issuance of a complete inquiry.
2. Receipt of firm and complete quotations.
3. Conference between buyer and seller to resolve differences, to be confirmed by memorandum.
4. Issuance of a purchase order incorporating all terms agreed upon.
5. Acceptance of the order by the vendor.

with the seller, the seller may sue for damage due to the buyer's refusal to take title.

• If both possession and title resides with the seller, he may either (1) sue for the contract price, (2) sell the goods and sue for the difference between the selling price and the contract price, or (3) hold the goods and sue for the difference between the contract price and the market value at the time delivery is refused.

There are other remedies available to the seller, such as stoppage in transit, which need not be discussed here.

Legal Basis for Property Transfers

Personal property includes not only fabricated or processed goods but also such items as mined ores and minerals, and cultivated crops. Most localities have drafted statutes patterned after the Sales Act and the Statute of Frauds that govern the transfer of personal property and of real property.

Transactions involving land or other real property are governed by laws covering property rights. Since these transactions constitute a rather specialized area as far as most engineers are concerned, the reader is referred to the literature for a detailed discussion, and we will move on to contracts involving engineering services and construction.

Contracts for Engineering Services

Engineering service contracts are those requiring the specialized skill of a professionally trained consultant for their performance. The products of these contracts include expert advice or testimony, drawings, specifications, economic evaluations, recommendations and reports.

These contracts differ from others in that competitive bidding is not employed, financial liabilities are limited, and the reputation of the practitioner is of great significance.

The engineering consultant is usually retained by the owner to protect his interests in a project. As is the case with a lawyer or physician, the profit derived from the consultant's employment is difficult to equate, but the potential loss in his absence may be considerable.

Besides the preparation of drawings and specifications, the consultant's duties may involve invitations to bid, bid analyses and the letting of contracts. He will also arrange for property transfers, searches and surveys; he will supply lines and grades, establish design criteria, supervise construction, and serve as the arbiter of disputes between the owner and second parties.

The consultant must be knowledgeable in costs to accurately predict project expenditures. He must be able to predict time schedules and plan the project's execution within a certain period.

Consultant's Fees

It is obvious from the foregoing that the responsibilities attached to engineering consulting are great. A small saving in a consultant's fee may prove to be very costly if he cannot properly discharge all of the responsibilities that he has undertaken in accepting the job.

Since consultants may be retained for almost any purpose, the amount of fee should be a function of the degree of responsibility assumed. Naturally, a drafting shop is not entitled to the same fee as a consultant handling an entire project. But, in either case, the fee is predicated on the estimated expenditure of man-hours. Specialists retained on a daily basis are paid a per diem rate that may average $100 per day, plus expenses, more or less. Other consultants are retained on a hourly rate, which may range from $5 per hour of expended time for simple drafting services, to $7.50 per hour plus a percentage (book cost plus 100% being an old formula) for complete consulting engineering services.

Consulting fees are also stated as a percentage of total project cost or as a fixed fee. Recommended percentage fees are published by most engineering societies, and will range from 6% for large uncomplicated projects to 12% or more for small projects requiring process designs.

Guarantees may be included in an engineering contract; but unless stipulated, the consultant's fees are not modified by failures in the finished product. The courts have repeatedly held that an architect or engineer who has conscientiously applied himself in accordance with proved scientific principles cannot be deprived of payment because of the inadequacy of the final design. (The principle is somewhat analogous to the one involved in paying a surgeon for an operation that was "successful," but in which the patient died.)

Contract Form

The form of engineering contracts is similar to others in its elements. The terms of performance and results expected must be stated. The consideration and terms of payment are included. Time limi-

Are lump sum contracts always advisable?

It is a common error to assume that the lump sum contract offers the lowest cost under all conditions. Unless the project is very well defined, this may not be so. Furthermore, although lump sum contracts contain certain guarantees of plant performance, these guarantees cover only the barest elements of sufficiency, and satisfaction is sometimes difficult to obtain under the circumstances.

As an example, consider the case where a plant is guaranteed against faulty material and workmanship for one year, and meets the constructor's performance guarantee covering three days of operation. Should this plant require a yearly maintenance expenditure of 10% of investment as opposed to 4% for a higher quality unit, 6% saving in contract price would just offset the difference in one year's maintenance cost, to say nothing of downtime expenses, lost production and increased utility costs.

tations and other conditions of performance are defined. The contract will be interpreted according to the law of the state wherein the contract is made, unless otherwise indicated. Of course, the licensing law of the state where the work is to be carried out must be complied with.

It is particularly important for the consultant to protect himself against default in payment, since he is not able to file a mechanic's lien. Further, the title to all specifications, drawings, etc., will rest with the party stated in the agreement.

Total Contracts

In the case of complicated processing plants, the engineering is often offered with the construction. These contracts are termed total contracts. There has been much discussion in engineering circles as to the propriety of such arrangements. Without debating the point here, suffice it to say that many engineers owe their professional development to "package contractors," and many of the technical advances in the petroleum and chemical fields that we enjoy today would have been impossible without the continuing efforts and research of such engineering contractors.

The combining of the engineering and contracting functions can lead to abuses if not held in check, but in a competitive market many advantages may also accrue therefrom. Large sums of money are expended in research and in the pooling of engineering talent in order that a "guaranteed" process may be offered. Also, large sums are spent in estimating guaranteed lump-sum proposals. These two offerings would be impossible without the package contractor. Such contracts are usually accompanied by guaranteed performance, lump-sum price, and may include a guaranteed time schedule. In these instances, the owner's interests are guarded by his own engineering staff and his project manager.

Construction Contracts

Construction contracts exemplify the most highly developed type of agreement.

The relationship of the contractor to the owner is a fundamental one, taking either the form of agency or independent contractor. In the latter case, the contractor is entirely responsible for the outcome, and in order for him to carry out his responsibility, he must be permitted to act independently of the owner and engineer. Most contracts contain a statement to this effect.

While the stated intention of the parties to an agreement will usually govern, if other conditions are present that preclude the fulfillment of these intentions, such intentions may be voided. For instance, if in the preparation of specifications and designs, the owner and engineer dictate construction procedures to the contractor to such a degree that he can no longer meet the performance requirements, he will in effect become their agent, and they become responsible to a degree for the outcome of the project.

The degree of responsibility placed upon the contractor must be confined to his sphere of influence. It may be incorrectly assumed that inclusion of a statement placing responsibility on the contractor for engineering, as an example, would relieve others of their obligations. Actually failure of an engineering design could under no circumstances be held against the contractor when he performs a construction service only.

Frequently a phrase is inserted making the contractor responsible for checking omissions in design. It is doubtful that even these terms would hold up under court test.

Another area of uncertainty frequently referred to the contractor involves soil conditions or obstructions under the surface of the ground. It would be unreasonable to require anyone to provide for subsurface eventualities in a congested area. However, such language is sometimes inserted in an inquiry or draft contract. In order to avoid confusion and possible litigation, the prudent contractor will take exception to any unreasonable demands about subsurface work. Owners are ill advised to prejudice a bid on this basis, since such conditions probably would not be binding if unreasonable in content. The American Society of Civil Engineers, The American Institute of Architects and the U. S. Government (Standard Form of Construction Contracts) have each prepared suggested language to be used in a contract to provide for unexpected subsurface conditions.

Some contracts contain bonus-penalty clauses, properly referred to as liquidated-damage clauses. The courts have repeatedly held that liquidated damages for tardy performance are not valid unless a corresponding bonus is given for early completion. Also the extent of liquidated damage must be commensurate with the actual loss suffered for tardy performance. The principle followed here is that the payment made cannot be a penalty on the contractor, but rather a levy to compensate the owner for his loss.

The above-stated conditions, and many others of a similar nature, have taken a more or less set pattern as a result of precedent through the years. It is therefore prudent to examine standard contract forms for guidance in the preparation of contracts. While these forms cannot be used in their entirety for any given purpose, typical clauses may be borrowed to cover a particular situation.

Features of various types of construction contracts

Type of Contract	Basic Example	Best-Suited Situation	Advantages, Disadvantages, Comments
Lump Sum	Build plant, as per specification, for a fixed price.	General specifications: complete. Process design: complete. Scope: equipment, piping and instruments are well defined. Layout: tentatively established. Performance: requirements indicated; operating cost to be specified. Guarantees: complete mechanical and performance guarantees required.	May result in lowest price and in most-efficient execution for well-defined projects. Any additions to work are priced on a cost-plus basis.
Guaranteed Maximum Sum	Build plant for a cost reimbursable price not to exceed a max. sum, any savings resulting from further design studies may be split between owner and contractor.	Same as for lump sum contract, except that scope and layout need not be so well defined.	Both the owner's and the contractor's engineers have equal incentive to reduce costs.
Cost plus Fixed or Percentage Fee	Build plant for cost of materials and labor, plus a fixed or percentage fee.	General specifications: complete. Process design: preliminary. Scope: not well defined. Layout: preliminary or nonexistent. Performance: tentative requirements and economics indicated. Guarantees: mechanical only.	Contractor has less incentive to improve efficiency; owner has complete control over quality, but runs the risk of over-specifying the project and exceeding the budget. Conversion to a lump sum or guaranteed maximum sum contract is possible.
Unit Price	Lay pipeline at \$X/ft. (Used in contracts involving large quantities of material or labor that may be priced on a unit basis.)	General specifications: complete. Design: complete. Scope: well defined but subject to variation. Performance: requirements indicated. Guarantees: complete guarantees required.	Allows variation in quantity at a fixed price per unit. (Caution: Guard against unbalanced bids. If the quantities are incorrectly estimated in the inquiry, a contractor may overprice underestimated units and underprice overestimated units. When the units are installed, the total cost will be higher than on a balanced bid.)

Form of Construction Contracts

The general form of construction contract does not deviate substantially from a pattern except insofar as content is concerned. The following elements will be recognized in most contracts.

First, a general statement of the agreement is given with a listing of the names and addresses of the principals. These clauses are followed by a statement regarding the purpose, and a definition of terms is also included in the preamble.

The body of the contract outlines the terms of performance and the terms of payment. These will include (not necessarily in this order): Services performed by each party, approvals, price, terms of payment, cost-accounting procedures, changes in the work, insurance, liability for losses, acceptance of the work, patent infringement, guarantees or warranties, secrecy agreements, force majeure, assignability, right of suspension, arbitration, law governing the agreement, notices and acceptance of the contract.

The formal contract will refer to technical exhibits that will include specifications, drawings, lists defining specific services, and cost schedules. These are the areas where engineers usually contribute the most in preparing a contract.

The language covering much of the body has been standardized. For instance, protection against patent infringement suits may be afforded both parties by a mutual clause designed to place the responsibility in the event of suit.

The usual mechanical warranty covers replacement of defective material or workmanship within one year of installation. Process or operating guarantees (performance guarantees) generally require a trial run of several days at the option of the owner. It is important to specify the remedies in the event of inadequate performance.

Secrecy or nondisclosure agreements will protect the owner and contractor-engineer with respect to proprietary know-how for a period of time.

Assignability limitations will prevent either party from transferring his rights or responsibilities to a third party. (The right to subcontract should be carefully defined.)

A force majeure clause relieves the contractor, and the owner, of responsibility for failure to perform if prevented from doing so by conditions outside their control, such as war, acts of God and the like.

The right of suspension clause defines the terms for suspension of the contract by the owner.

The arbitration clause designates the form that arbitration will take in the event of disagreement. However, the findings of an arbitration board will not necessarily discharge the contract, since either party may still resort to suit in law or equity if rights have been violated.

The remaining elements of the contracts are tailor-made to the work at hand.

Types of Reimbursement

Construction contracts are usually classified according to the consideration or payment agreed upon. The lump sum, cost plus fixed fee, cost plus percentage fee, guaranteed maximum sum, and the unit price contract each refers to a particular type of reimbursement, according to the requirements of the project.

As shown in the table on this page, each type of contract has its incentives and disadvantages; the latter can sometimes be counteracted by control measures. Combinations are often employed. The correct use of each type of contract may result in the complete satisfaction of both parties; whereas a selection that is unsuited to the requirements of the project can only result in mutual dissatisfaction.

Specifications, and Their Effect on Cost

Specifications should adequately describe the project and be sufficiently detailed to form a reasonable basis for the comparison and evaluation of bids. At the same time, the owner should be objective about exceptions taken (i.e., alternatives suggested by the contractor) and about sources of supply.

If an owner sets up very rigid specifications for equipment and materials, limiting the source of supply to a specific vendor or vendors, he must recognize that he pays a price for this decision. He should be convinced that this course of action will result in savings (such as in maintenance or in the reduction of warehouse spares) commensurate with the costs incurred.

The way specifications are written can have a tremendous influence on the upset cost of a project. Contractors are usually more sensitive to this effect than are owners. Engineers who are not knowledgeable in cost estimating should not be entrusted to the job of specification writing.

Truly competitive bids cannot be obtained with varying specifications. On occasion, contractors are asked to bid according to their own specifications, the intent being that the owner will compare the offers and reduce them to a common denominator. To do so is virtually impossible for a complicated project.

First of all, bid specifications are so general that many features having a great effect upon cost cannot be compared. For instance, one of the most costly classes of materials and labor is that of piping. The specification for piping materials is often deleted in a contract bid.

Second, the bid specifications presented by contractors are usually quite voluminous and not readily subject to comparison. When comparison is possible, the correct denominator to use is in question. For example, should two contractors be held to the same size of tower if they both guarantee the plant with respect to operability and operating cost?

Thus, if truly competitive bids are desired, it is best to issue adequate specifications with the inquiry and to request the bidders to state any exceptions taken, so that the alternatives can be evaluated on a case-to-case basis.

Legal Basis for Construction Contracts

The last and perhaps most-important phase of construction contracts involves the laws by which they are governed. The form of the law varies from state to state and from country to country. Most of the U. S. and the British Commonwealth follow the form of law derived from English common law. The state of Louisiana, most countries in Europe and their possessions, are governed by law derived from Roman civil law. The question arises as to which law governs a contract drawn between parties residing in different states or countries, and which court has jurisdiction in the event of malperformance on an executory contract.

The general rule is that unless otherwise stated, the law governing contract is *lex loci contractus* and the enforcement is *lex fori.* This means that the governing law is that existing where the contract is made (i.e., accepted), and the enforcement is by the jurisdiction where carried out.

A contract that is binding in one place is binding in every place. Usually the law governing is stated in the contract. Many elements of law, however, serve to limit the rights and remedies available to the parties. This is further modified by international agreements between countries.

Within the boundaries of the U. S., the state may not be sued except with its permission. There are other modifying principles that affect government contracts, such as the ability of the Federal Government to renegotiate contracts.

These complexities need not be detailed further, since they are the province of lawyers rather than engineers. Any business firm is well advised to employ legal counsel familiar with the law in all areas covered by its operations.

Formal Contract Is Lawyer's Responsibility

In conclusion, it is evident that the preparation of contracts encountered in engineering practice requires a great deal of background experience and training. The writing of specifications and definitions should be entrusted to experienced engineers. The preparation of the formal contract is the responsibility of the lawyers. Since each of us may be required to assist in the preparation of contracts, a better understanding of the principles better enables us to contribute.

Those who are intimately concerned with the preparation of engineering contracts are urged to continue their review of the literature, since the guiding principles of contracts are constantly changing to meet new situations as the times demand. ■

References

1. Tucker, J. I., "Contracts in Engineering," McGraw-Hill, 1947.
2. Vaughn, R. C., "Legal Aspects of Engineering," Prentice-Hall, 1962.
3. Kerr, T. S., "Business Law," Wiley, 1939.
4. Dykstra, G. O., Dykstra, L. G., "The Business Law of Real Estate," MacMillan, 1956.

Are You Locked Into Your Job By What You Know?

"Technical men wish as individuals to be able to live in the great American tradition of personal mobility. They want to be able to change jobs without sacrificing personal and financial status. How can they do this without advertently or inadvertently bestowing on a new employer the benefit of valuable research and technology developed by a former employer? The question is sociological as well as financial and legal, and poses great difficulty to the courts in solving trade-secret problems."

CHAIRMAN AT A RECENT SEMINAR FOR PRACTICING LAWYERS

THOMAS H. ARNOLD, JR., *Esso Chemical Co.*

Regardless of what his employer may be—corporation, research institute, university, governmental agency, consulting firm—the engineer who works for a salary is legally classified as a "servant" of the "master" who employs him.

The blunt terms "master" and "servant" may at first seem demeaning and shockingly unprofessional. But don't stay shocked too long. Ponder some definitions given by Black's Law Dictionary (4th ed.), and see if you recognize yourself:

> **Master and Servant: "The relation of master and servant exists where one person, for pay or other valuable consideration, enters into the service of another and devotes to him his personal labor for an agreed period."**
>
> **Master: "A principal who employs another to perform service in his affairs and who controls or has right to control the physical conduct of the other in performance of the service."**
>
> **Servant: "One employed to perform service in the master's affairs, whose physical conduct in performance of the service is controlled, or is subject to the right to control, by the master."**
>
> **"One who is employed to render personal service to another otherwise than in the pursuit of an independent calling, and who, in such service, remains entirely under control and direction of employer."**
>
> **"A person of whatever rank or position in employ and subject to direction or control of another, in any department of labor or business."**

In today's heady technological climate, temptations abound for engineer-servants to see if they can better their lot by changing their employer-masters. When making such a change, what kind of knowledge may properly be appropriated and transported from one job to another, and what knowledge must be left behind for the use of the former master and his remaining servants?

Just where does the appropriation shade into a *misappropriation* of knowledge—when can a former master legally restrain the ex-servant, and perhaps his new master, from using the knowledge?

How onerous does the misappropriation have to be in order to collect tribute-money for damages?

Then, in what cases does the misappropriation so offend society's prescribed norms of right-and-wrong that the ex-servant can be dealt with as a criminal, and have meted out to him sanctions such as jail sentences or fines?

I cannot provide cut-and-dried answers to these questions, nor can anyone else. If the law were so clear and certain, there would be no further cases tried in the courts, no more appeals taken to higher courts, and no more interest by Congress and state legislatures in passing trade-secret statutes. In point of fact, however, more and more cases are being tried and appealed, and an increasing number of federal and state statutes in this field are being enacted or considered.

Interest in the problems of the mobile employee has been mounting explosively. Witness the increasing number of articles appearing in legal, technical, management and even popular publications—as well as symposia at professional meetings. Since CHEMICAL ENGINEERING's landmark article, "Does Your Employee Own Your Knowledge?" (July 28, 1958), publications from *Fortune* to *Reader's Digest* have featured stories on the subject (not to mention a television "drama" last year in which Perry Mason cleared the reputation of a mobile engineer).

Why, then, write yet another article? Because most engineers are not nearly as knowledgeable in this fast-moving field as they should be, considering how much there is at stake. I hope that what follows will fill in some of the gaps, and will put the engineer in a better position to participate in policy formulation.

In the final analysis, every employed engineer, and every employer of engineers, is being affected by the course the evolving law is taking in this area. It would be to their advantage—and to society's—if both groups were to become more effectively outspoken on this subject than they have been so far.

Will the Real Misappropriator Please Rise?

By considering some actual cases, and thinking them through for himself, the engineer can gain a better insight into the trade-secret protection problem. Not long ago, two rather instructive cases involving mobile employees were decided in Connecticut. I will summarize the facts of each case, and you try to decide the outcome, and the reasons for it.

In Case No. 1, the complaining party—a corporation that we will call P (for plaintiff)—began an action against defendant D and seven other individuals, all of whom were P's former employees. Corporation P sought an injunction and damages against the defendants for making wrongful use of P's trade secrets.

P Corp.'s business is primarily in the field of electronics, and the defendants had been employed in P's semiconductor division in scientific and engineering capacities, with job titles such as technical director, R&D group head, transistor group head, product engineering group head, and product engineer.

Defendant D, before actually working for P Corp., recommended that the firm go into the semiconductor business, which it subsequently did, with D in charge of the new division. Three years later, however, a reorganization of the division resulted in another employee being placed in charge of all administrative and manufacturing functions, with D on the same level having responsibility for the entire technical staff, including the defendants.

P maintained strict security precautions to safeguard its processes for the manufacture of transistors and diodes. All defendants signed employment agreements, which read in part as follows:

"I will not, either during or subsequent to the term of my employment, directly or indirectly divulge to unauthorized persons any information designated as top secret, secret, or confidential by the Company or by any Department of the United States Government, or designated by the Company as limited, restricted, private, or as a trade secret or not known to the general public as recognized standard practice, whether acquired or developed by me in the course of my employment or obtained from other employees; and that I will not, either during or subsequent to the term of my employment, directly or indirectly publish or otherwise divulge any such information without written authorization from the Company to do so.

"I will not during the term of my employment except with the consent of my immediate superior, take out of the Company's plants, or disclose to unauthorized persons, any such Company property relating or peculiar to the Company's products, research or development, and that if at the termination of my employment I am in possession of any such Company property, I will return the same at the time I surrender my identification card."

Security provisions were made and enforced concerning details relating to materials, equipment, procedures, blueprints, drawings, and reports in the research, development and production of semiconductor devices that had or were likely to have any value to a competitor. These measures were supplemented by specific instructions to all employees, the restricted disclosure of P's activities to outsiders, the barring of any competitor's representatives from production areas, the requirement that all visitors sign a log book and be escorted at all times while visiting P's facilities.

In January, 1959, P appointed a new person as sole head of its semiconductor division. Although his technical knowledge and skills were apparently inferior to D's, D now had to report to him. Irked at this appointment, D decided to leave P and form a competing company of his own. From February until May, 1959, D discussed his plans with other key employees, and persuaded the other defendants to eventually leave the company along with him. During this time, D and the other defendants held many meetings among themselves and with financial backers, and completed arrangements necessary to organize the new company.

While still in P Corp.'s employ, the defendants delayed informing the firm of a development made on a certain device, apparently in order to utilize the development later themselves.

On May 25, 1959, D and the other defendants resigned from P. After leaving, they attempted to hire other employees away from P, and did not seek employees anywhere else. One ex-employee took with him P's "black book" containing the list of vendors for P's semiconductor division.

By December, 1959, 32 of P's employees had left, and 28 had gone to work for the defendants' new company. Only one-fourth the original number of key people were left at P's semiconductor division.

The court had to decide the question of "whether the defendants, former employees, misappropriated trade secrets of the plaintiff employer, which they learned during and in consequence of the confidential relationship of employment."

The defendants denied that they had misappropriated trade secrets, and claimed that many of the steps in the process under dispute were known in principle —and that some of the equipment, materials and subprocesses used were known in the semiconductor as well as other fields.

In fact, the defendants were able to show that each element of each sub-process had been mentioned in treatises and scientific writings before. Consequently, they argued, the process was "a teaching of prior art and . . . something obvious to a good mechanic skilled in the trade."

Furthermore, the defendants stated that their process represented a considerable modification over that used by P, and hence could really not be considered "the same." They claimed that no secret processing information had been physically removed from P, and that "an employee, after leaving his employer is at liberty to use for his own benefit any information he remembers," unless it is not disclosed by prior art or is not readily discernible by others skilled in the field.

How would *you* find—for plaintiff or defendants—based on the facts stated here?

Case 2: The Wheels of Fortune

In this second Connecticut case, let us again use P and D terminology. Company P claimed that D and two other defendants, all of whom were P's former employees, misappropriated P's "trade secrets, consisting of the exact nature of supplies purchased from various companies, its formulas, processes and techniques for manufacturing, and information relating to the identity and needs of its customers," resulting "in an unfair competitive advantage" to a company established by defendants.

Company P, through one of its divisions, manufactures abrasive cutting wheels. Its action in the Connecticut state court sought: (1) an injunction restraining defendants from using or revealing to others P's trade secrets, or from manufacturing or selling abrasive cutting wheels using such trade secrets, (2) an accounting of all profits realized by defendants from the sale of such wheels, and (3) money damages.

Defendant D had worked for P for twenty-four years, and when he resigned in October, 1960, he was P's plant manager and earned $35,000/yr. Another defendant worked thirteen years for P, and when he resigned was P's plant foreman. The third individual defendant had been a salesman for P for eight years before he resigned.

In the spring of 1959, defendant D had agreed with an investor to go into the abrasive wheel business, and later that year, P's plant foreman joined in the agreement. During the fall of 1959, a new factory building was built, and equipment was installed for the new abrasive business. Before D severed his connection with company P, he gave checks to the investor for the purchase of supplies and materials for the new firm.

The first meeting of this new firm was held in November, 1960; D and his former plant foreman were two of its three incorporators and officers. The first wheels were ready for sale in March, 1961, at which time P's salesman resigned to join the firm.

Company P had been in the abrasive wheel business since 1919, and its sales reached about $3.5 million annually when D's company began competing with it. Until that time, P had been the only company in the U. S. *specializing* in the manufacture of abrasive cut-off wheels; it had 400 to 500 different formulas for cutting wheels, tailored to individual customer needs.

Ingredients, such as abrasives, bonding materials, and liquids used as fillers, are commonly known and used in the industry. P sought to restrain the defendants' manufacture and sale of wheels made similar to its products, and use of its production methods, techniques and formulas.

Finally, though the defendants "were skilled technicians in a specialized industry," they had not been working for P under any contracts of employment.

Based on these facts, how would you have decided this case, using the same legal principles as the court in Case 1 was required to use?

The *Plaintiff* Who Was Left With No Defense

The trade-secret misappropriation problem is not confined to technical employees, as noted in Sports Illustrated, Jan. 24, 1966. George Halas, owner of the Chicago Bears professional football team, brought court action against ex-Bear defensive coach George Allen who left Chicago to join the rival Los Angeles Rams.

Halas, as the story is told, was "not solely concerned with losing the knowledge, skill and team spirit Allen has acquired through his association with the Bears. More specifically, Halas asserts that Allen has in his possession the Bears' Defense Manual No. 2, the Bears' Defense Textbook Number 2, and numerous motion picture films of plays."

Allen, so Sports Illustrated concludes, could "jump his contract and become coach of any Peewee League team he wants, but those classroom aids are not going to leave the state of Illinois without a court fight!"

Covenants and Competition

Restrictive covenants in employment contracts—i.e., where the employee agrees not to work for a competitor of his employer until a certain length of time has passed, or within a certain geographical area if the employer's business is a regional one—generally have been held legal, if their scope is "not unreasonable." The nature of the information or trade secrets must, first of all, be of the protectible type. The employer must make all efforts to keep them confidential and secret himself, and out of the public domain. The activity from which an ex-employee is restrained from working in must be only narrow enough to reasonably protect the former employer's trade secrets from misappropriation. A chemical engineer could not be restrained from working in "the petrochemical industry" for example.

The time during which an ex-employee must have his job freedom restricted must be no longer than reasonably necessary to allow the employer to reap the benefits that might accrue from the use of his trade secrets; and the better rule of law would have as a maximum time-limit "so long as the secrets do not become known to others in the field." Finally, since industry today is widely competitive, any geographical restraint is not as enforceable as earlier when industry was narrowly localized.

In summary, the courts tend to scrutinize all such covenants closely, and the underlying criteria involve the "reasonableness" of any restraints that the covenant attempts to impose after an employee is no longer the "servant" of a particular "master."

The decisions in these two cases reached opposite results. The court awarded damages against the former employees in Case 1, while the court in Case 2 ruled for the defendant ex-employees. Both opinions have been published in the *United States Patent Quarterly*, which collects court opinions in cases involving "industrial intellectual property"—patents, trademarks and trade secrets. (I would suggest that your time might be profitably spent browsing through some volumes in your company, county, municipal or public law library.) Let's compare the bases of each court's rulings, and observe the courts' language and logic.

• Case 1: *Sperry Rand Corp.* v. *Rothlein et al.*, 143 U.S.P.Q. 172, (U. S. Dist. Ct., Conn., 1964).

The court first held that Sperry Rand's process for manufacturing certain transistors fell within the category of a "protectible trade secret," and that the company had "expended a great deal in time, effort and money to develop a process for manufacturing an unusually high-quality transistor."

The judge rejected the argument that the defendants could not have been expected to "wipe clean the slate of their memory," and held that Sperry's process constituted a trade secret that could be legally protected.

As for the "prior art" being "something obvious to a good mechanic," this defense argument was disposed of by quoting an earlier, controlling case:

> **". . . there is positive danger in resort to . . . (this argument) to justify the marauding instincts of trust violators, since thereby the carefully built law of trade secrets may be destroyed as a practical matter and such business assets be left open to hijacking from all sides."** ***Franke* v. *Wiltschek*, 99 U.S.P.Q. 431 (1953).**

The employment contract and its express terms were emphasized by the court as evidence of the confidential relationship between Sperry and defendants, and the fiduciary duty owed to Sperry. The judge gave a particularly good summary of the dilemma facing the courts in trade-secrets cases involving mobile technical employees. Consider the opposing "social and economic interests" cited in his words:

> **"The defendants protest that they should not be penalized or barred from using their skills and knowledge. This issue must be resolved in a balancing of social and economic interests. It is desirable, on the one hand, to encourage competition and facilitate the exercise by an individual of all his skill and knowledge. It is also desirable that established businesses be given reasonable protection against unfair trade practices. It is obvious that encouragement of technological advancement is vital to the maintenance and growth of our economic order in general and our industrial productivity, in particular. If protection of trade secrets is withheld, however, research and development will be impaired, for no employer would be willing to spend large sums on research and development of new ideas, methods and processes, if they can be freely taken and used. But if the protection afforded is too broad, intimidation will result, the dissemination of ideas will be impaired, competition will be fettered and the public will suffer."**

The judge then noted that there were two important policies to be resolved:

1. The policy that encourages employees to seek better jobs from other employers or to go into business for themselves.

2. The policy that is designed to protect employers against improper disclosures of information that their employees have received in confidence.

It is significant that in reaching his decision here, the judge alluded to principles of fair play, ethics, and morality. He quoted with approval another court's opinion that the second policy stated above "may perhaps be receiving increased recognition in the light of the marked changes in the attitude of the law towards the need for commercial morality." Later, in discussing the specific acts of the defendants before and after leaving Sperry Rand, he observed that such behavior and its consequences had "a bearing on the degree to which the defendants violated business ethics and morality."

• Case 2: *American Chain & Cable Co.* v. *Avery et al.*, 143 U.S.P.Q. 126, (Conn. Superior Ct., 1964).

The defendant Avery had been American Chain's abrasive-wheel plant manager, and became president of his own firm after he left.

Basically, the court in this case held that: (1) plaintiff's wheels were not materially different than those produced by its competitors, nor did its formulas produce wheels with superior advances or differences; (2) although defendants were "considered trusted employees" of American Chain; had access to its manufacturing formulas, processes, and techniques; were acquainted with suppliers of ingredients used by the plaintiff; and knew the names of its customers and the types and quantities of wheels each purchased, they were nevertheless not employed under a contract nor had they promised plaintiff to keep its information secret and to use it only as authorized; and (3) the evidence did not show that "the plaintiff has been unlawfully harmed or that the defendants have unlawfully benefited."

In stating its view of "the law" governing the case, the court noted that an "employee has a right to grow with his experience, and he can carry away for general use his skill and everything that he has learned at his place of emlpoyment, except trade secrets." But, the court then concluded that the evidence showed that the process details and suppliers and customers information were not trade secrets, even though American Chain considered and claimed them to be such.

Observe the similarity of the court's language here to that in the Sperry Rand case:

> **"It does not seem proper that the jurisdiction of a court should be invoked to prevent competition and employees from utilizing their skills, aptitudes and subjective general knowledge after they have terminated their employment, unless their former employer has made known to them while in employment that the things, subsequently claimed to be trade secrets, should in fact be regarded as trade secrets. . . ."**
>
> **". . . The totality of the relief demanded by the plaintiff would prevent them from pursuing their occupations and would entirely stifle competition with the plaintiff. Competition must not be stifled unless unfair. In our industrial and economic system, allegiance to an employer is not forever . . . Of course, the longer the employee has served his principal and the more intense the loyalty has been, the greater the irritation to the principal against whom the former employee must complete. This irritation or any resulting harm therefrom, in the absence of an appropriation of property based upon breach of confidence, does not, however, establish a cause of action."**

It is not easy to reconcile the rationale and outcome of these two cases, which have very similar fact-patterns. Moreover, each court used the same body of trade-secrets law (i.e., Connecticut's) to formulate its decision. Two factors that may have caused the Sperry Rand case to be decided as it was are: (1) the defendants had signed employment contracts that explicitly dealt with trade secrets or confidential information, and (2) the end-products developed at Sperry were unique and had qualities superior to any others on the market at that time, and the entire industry was in its infancy.

Other Recent Cases

As we've just seen, one cannot easily predict with any certainty the point at which a master's trade secrets begin and a servant's intellectual talents and assets end. Every case may have certain facets to it that push the final decision onto one side of the line or the other. We'll briefly take some other recent trade-secrets cases, and listen to the courts' language as they "make law" in this field.

• *Van Products Co.* v. *General Welding & Fabricating Co. et al.*, 147 U.S.P.Q. 221 (Pa. Supreme Court, 1965).

The trial court had granted a permanent injunction against Van's ex-employee (Rapp), who later worked for a competitor. The court enjoined Rapp and his new employer from making deliquescent desiccant air dryers, on the ground that Van's trade secrets had been misappropriated by Rapp.

On appeal, however, the lower court was reversed, and the higher court held that there were no protectible trade secrets in the dryers to begin with. In reaching its decision, the court noted:

> **". . . One rather salient point runs steadfastly throughout decisions in this area in most jurisdictions, and that is that the employee, upon terminating his employment relationship with his employer, is entitled to take with him 'the experience, knowledge, memory, and skill, which he had gained while there employed . . . A man's aptitude, his skill, his dexterity, his manual and mental ability, and such other subjective knowledge as he obtains while in the course of his employment, are not the property of his employer and the right to use and expand these powers remains his property unless curtailed through some restrictive covenants entered into with the employer' . . . There is no restrictive covenant involved in the present case."**
>
> **". . . The question still remains to be answered whether the information is such here that we must act as a judicial eraser to blot out Rapp's knowledge and skill."**
>
> **". . . It would seem, then, that this generalized knowledge in the field of deliquescent dessicant air dryers was a labor asset which Rapp was privileged to carry with him to his new employer . . . They are not of such character as to be classified as trade secrets."**
>
> **". . . We are convinced that Van has failed to show that this information was any more than the recorded**

results of Rapp's acquisition of general knowledge and skill in this field . . . We do not consider the result herein reached as harsh or out of tune with accepted commercial morality as it may, at first blush, appear. . . ."

• *Welex Jet Services, Inc.* v. *Owen*, 325 Southwestern Reporter 2d 856 (Texas Court of Civil Appeal, 1959).

This appellate court held that "a former employee is not under obligation to forego the exercise of his inventive powers, even if they were incited by knowledge gained in the performance of his contractual duties. The efficiency and skills which he developed through that experience belong to him, not to his former employer."

• *Holiday Hill Stone Products, Inc.* v. *Peek*, 146 U.S.P.Q. 505 (Tex. Ct. of Civ. Appeals, 1965).

Peek, Holiday Hill's plant manager and salesman, had restrictive covenants in his employment contract that forbade him to: (1) engage in the manufacture of cast stone or cast brick for three years within 700 miles of San Antonio, after his employment with Holiday Hill ended; (2) ever sell anywhere in the U. S. to Holiday Hill's customers; and (3) ever disclose his employer's trade secrets or formulas, and names of customers, suppliers, employees or agents.

The trial court enforced the terms of this contract when Peek went to work for a competitor after leaving Holiday Hill.

However, on appeal, this court noted that Peek had a large family to support, and that he had worked mainly in the cast stone manufacturing business most of his adult life. Then, it modified the injunction and held that the contract terms were unreasonable. Result: Peek was restrained from being engaged in the manufacture of cast stone for only one year; from selling cast stone to Holiday Hill's customers for three years; and permanently from disclosing his ex-employer's trade secrets and other business information.

• *Schulenburg* v. *Signatrol, Inc.*, 147 U.S.P.Q. 167 (Ill. Supreme Court, 1965).

Former employees set up a competing company for the manufacture of electric signal flashers, and the trial court enjoined them permanently from use of blueprints and drawings of plaintiff, which the court held were protectible trade secrets, whether they were "stolen, copied, traced, or memorized by defendants." Even though this appellate court confirmed the trial court's ruling, it reduced the period of the injunction to "that period of time required by defendants to duplicate plaintiff's product by lawful means" of design and development.

The court also made this interesting observation of the facts as found by the trial court:

"This record is replete with evidence leading to but one conclusion: that these defendants have surreptitiously taken the particularized information contained in plaintiff's blueprints and have employed it to establish a competing business. Even today's 'morals of the market place' are more demanding than this, and should not be relaxed by this court."

• *Seismograph Service Corp.* v. *Offshore Raydist, Inc.*, 135 *Federal Supplement* 342 (U. S. District Court, Louisiana, 1955).

Again, we have a court that astutely and succinctly posed the dilemma of trade-secrets policy balancing:

"No single test can be applied in all cases where improper acquisition of business information is charged. The inventiveness of the human mind staggers the imagination. It is simply the difference between right and wrong, honesty and dishonesty, which is the touchstone in an issue of this kind."

Practical Needs of the Engineer

We could continue our trip through the law reports involving civil actions between masters and their ex-servants who have gone to greener pastures. Some courts have darkened the grass therein as quickly as the proverbial prairie fire, while others have left the greenery intact.

Certainly all well-informed chemical engineers have some knowledge of the most talked-about cases: *Du Pont* (and Hirsch, its ex-titanium oxide specialist); *Goodrich* (and its former space-suit engineer, Wohlgemuth); *Carter Products* (and Fine, its one-time aerosol shaving-cream researcher); *Kodak* (and Warren, its former employee who knew much about secret film-emulsion formulas); *Monsanto* (and Miller, once its engineer and phosphorus expert); and *American Cyanamid* (and Fox, one of its drug researchers). And now, you know about *Sperry Rand* and Rothlein et al., and *American Chain* and Avery et al.

Not too far in the future, you'll be hearing about *Armour* and Csendes (and nine other ex-Armour fertilizer specialists); *Grace-DuBois* and Hargodine; *Abbott* and Braaten and Mueller; *Occidental* and Berry, et al.; *Nalco* and Hall.

The list will obviously never end. And each of you must surely ask, "How can I make sure *I* stay off future tabulations of this kind?"

I'll not attempt to catalog the answers to this important query, for there may be as many answers as there are questions! Aside from reviewing your Boy Scout Oath, recalling the Eighth and Tenth Commandments, or just staying put in your present job, you might:

• Acquaint yourself with Sections 1, 2, and 9 of AIChE's Code of Ethics.

• Read and get to understand all provisions in any employment-connected agreements you sign or have already signed.

• Read some of the material cited in the references.

• Discuss with relevant management representatives your present and potential employers' trade-secret policies as applied to your job.

• Discuss any specific questions with your present and potential employers' lawyers.

• Hire or consult your own lawyer.

You must remember that trade secrets present one of the least predictable areas of jurisprudence. The "law of trade secrets" is primarily court-made, rather than legislature-enacted. Ultimately, the problem must be resolved by judges, with the help of lawyers from both sides of the case. Even judges who have considerable experience in trade-secrets cases have difficulty keeping up with and evaluating all the current developments of this fast-moving, highly technical field.

For this reason, I believe there is much we can do as engineers to help educate the members of our legal profession to the practical needs of our own engineering profession. The problem of trade secrets and today's more-mobile technical employee demands the best efforts of *both* professions.

It is currently fashionable to feature sessions dealing with this topic at engineering meetings—conventions, luncheons, company seminars, etc. Both technical and legal publications are almost inundated with manuscripts for articles on the subject. But, I notice the typical pattern is for *lawyers* to talk to, write for, and take the initiative in formulating policies affecting, *technical employees.* I do *not* observe *engineers* speaking to bar association luncheons or at lawyers' conventions or seminars, or writing in legal journals.

The customary practice today calls for a group of engineers to listen to lawyers expound on our subject topic, and then to ask confused, groping questions that usually lend themselves to several alternative, equally valid answers anyway. It seems to me that there should be corresponding tableaux in which the roles are reversed! For engineers certainly have much to say that would be valuable to lawyers (and judges and legislators)—on *all* sides of the trade-secrets controversy. Engineers have a vital stake in the formulations of social, economic, and legal policies that affect them as well as the whole of society.

It's interesting that there are practically no reported cases involving actions by law firms or law departments against an ex-employee (or partner) who goes into practice for himself or joins another law firm or company. Surely, each law firm and corporate legal department has trade secrets or other confidential, proprietary information that it does not want revealed or used by competitors. (And I'm referring not only to specific client or case "secrets" but also to techniques, strategies, legal-business methods, etc., all of which go to make some firms more successful or more profitable than others.)

Perhaps it's the threat of disbarment, or the loss of *all* rights to practice the profession, which keeps such cases out of the courts. Such self-policing, of course, would be much more difficult for the technical professions where licensing is not universally required. Also, it may be that there are not many lawyers who can carry away "secrets" whose potential value can reach the sizable levels of those found in modern, complex, technical-based industries.

What's Coming Next?

Many commentators have suggested that all the "competitive torts" (i.e., civil wrongs or injuries arising from economically competitive acts or conduct in

Ultimate Restraints for Ultimate Wrongs

In foreign countries, and increasingly in the U. S., "the state" has been alerted to the potential dangers of trade-secrets misappropriation. Consequently, more and more statutes imposing criminal sanctions on wrongdoers are appearing on the law books (i.e., jail sentences and fines).

The Georgia legislature is the latest to pass such a statute. Some other states that have passed or considered similar statutes includes New York, New Jersey, Pennsylvania, Wisconsin and California.

The recent Lederle Laboratories ex-employees case was tried in Federal court under the National Stolen Property Act (Section 2314, Title 18, United States Code), as discussed in the Apr. 25 issue of CHEMICAL ENGINEERING. This statute does not mention trade secrets explicity, but covers "any goods, wares, merchandise . . . stolen, converted, or taken by fraud."

Because criminal sanctions are stringent ones, statutes usually require that specific property (tangible and defined) be taken, and that the taker **know** he is stealing trade secrets and he **intends** to take them illegally. The New Jersey law, a typical one, reads in part:

"Any person who, with intent to deprive or withhold from the owner thereof the control of a trade secret, or with an intent to appropriate a trade secret to his own use or to the use of another, (a) steals or embezzles an article representing a trade secret, or (b) without authority makes or causes to be made a copy of an article representing a trade secret, is guilty of a misdemeanor . . . [or] a high misdemeanor."

"(a) The word 'article' means any object, material, device or substance or copy thereof, including any writing, record, recording, drawing, sample, specimen, prototype, model, photograph, microorganism, blueprint or map. . . ."

"(c) The term 'trade secret' means the whole or any portion or phase of any scientific or technical information, design, process, procedure, formula or improvement which is secret and of value; and a trade secret shall be presumed to be secret when the owner thereof takes measures to prevent it from becoming available to persons other than those selected by the owner to have access thereto for limited purposes. . . ."

the field of business) be grouped into an overall tort classification of unfair competition. Others have disagreed. They argue that the criteria of "unfairness" connotes moral or ethical bases of decision, rather than the more objective basis of the economic impact, or effect, of alternate decisions in a case.

In Europe, the "unfair competition" concept has received widespread acceptance. For example, in 1909 Germany enacted a "Law Against Unfair Competition." Section 1 outlaws generally any competitive conduct found to be *contra bonos mores* (contrary to honest practices):

> **"A person who, for purposes of competition, acts in his commercial relations *contra bonos mores* can be held liable for the damages he causes and be compelled to restrain from further interference."**

Subsequent sections list and prohibit particular acts of unfair competition, including unauthorized disclosure and misuse of manufacturing or trade secrets.

From the general tone of recent court opinions, it's my feeling that the ultimate compromise will take into account moral, ethical, *and* economic values, as applied to individual employees, to corporate employers, and to the important third-party: society itself. The wrongful taking of trade secrets has been recognized in our Anglo-American legal system since at least 1820. In the subsequent nearly 150 years, no one has come up with a simple criteria to apply in every controversy that may arise.

Perhaps a fresh approach may be suggested by an intensive campaign to cross-fertilize the legal and technical professions. The street runs at least two ways between them, but one lane is currently more travelled than the other. There is plenty of room on this other lane—and technical men should not be reluctant to venture onto it.

Specifically, I think engineers need to acquire a wider background on trade-secret protection than can be obtained merely by reading their company's security manual or regulations. With this wider background, they can be opinion molders within their own company—e.g., when talking to legal, management, employee-relations and R&D people. They can influence their societies. They can offer to give talks at lawyers' meetings, and to present their views to judges and legislators. They can even offer to write articles about the technical man's trade-secret views in legal publications, such as the American Bar Assn. Journal.

Selected Reading

Here is a suggested reading list that engineers should find helpful in gaining the sort of background referred to in the last section. Most good law schools will have these books in their libraries. Periodicals may be ordered from the school or company publishing it. State and Federal Court cases are usually reported in the *United States Patent Quarterly*, published as weekly pamphlets, and as bound volumes periodically.

Some of the cases cited in this article, or the references below, should make interesting topics for discussion by engineering groups, and perhaps serve as the basis of dialogues between groups of lawyers and engineers.

One word of advice: *Don't* be reluctant or hesitant to read law books or articles, or legal opinions. You will understand more of what you read there than most lawyers understand about the technical details of your trade secrets! ■

1. Callmann, R., "The Law of Unfair Competition and Trade-Marks," 2d ed., Chicago, 1950, and 1965 supplement, p. 62, sec. 51.4, esp.
2. Ellis, R., "Trade Secrets," Baker, Voorhis & Co., New York, 1953.
3. Turner, A. E., "The Law of Trade Secrets," Maxwell, Ltd., London, 1962. Distributed in U. S. by Fred B. Rothman & Co., South Hackensack, N. J. This is an excellent book on "the law" of this subject.
4. Kalinowski, J. O. von, Key Employees and Trade Secrets, *Virginia Law Review* 47:583, May, 1961, Univ. of Virginia Law School.
5. McTiernan, C. E., Employees and Trade Secrets, *J. of the Patent Office Society* 41:820, Dec., 1959, Washington, D. C.
6. Coleman, L. A., Cole, C. B., The Effect of Shifting Employment on Trade Secrets, *The Business Lawyer* 14:319, Jan. 1959, Amer. Bar Assn., Chicago.
7. Stedman, J. C., Trade Secrets, *Ohio State Law J.* 23:4, No. 1, 1962, Columbus, Ohio.
8. Klein, H. D., The Technical Trade Secret Quadrangle: A Survey, *Northwestern U. Law Review* 55:437, Sept.-Oct. 1960, Chicago.
9. Kopko, A. J., Protection of Trade Secrets in the Employer-Employee Relationship, *Notre Dame Lawyer* 39:200, Feb. 1964, South Bend, Ind.
10. Leydig, C. F., Protecting Trade Secrets When Employees Move, *The Business Lawyer* 31:325, Jan. 1966, Amer. Bar Assn., Chicago.
11. Liebhafsky, D. S., Industrial Secrets and the Skilled Employee, *New York U. Law Review*, 38:324, Apr. 1963, New York.
12. Note, Developments in the Law—Competitive Torts, *Harvard Law Review* 77:888, esp. 932-959, Mar. 1964, Cambridge, Mass.
13. Blake, H. M., Employee Agreements Not to Compete, *Harvard Law Review* 73:625, Feb. 1960.
14. Note, Injunctions to Protect Trade Secrets—The Goodrich and Du Pont Cases, *Virginia Law Review* 51:917, June, 1965.
15. O'Meara, J. R., "Employee Patent and Secrecy Agreements," National Industrial Conference Board, 1965, New York.
16. Munster, J. H., Smith, J. C., The Care and Feeding of Intellectual Property, *Science*, 148:739, May, 1965.

Meet the Author

Thomas H. Arnold, Jr., is well qualified to discuss the interplay between the legal and chemical engineering professions—he has degrees and experience in both fields. As a chemical engineering graduate of Louisiana State University, he has worked for FMC Corp., Humble Oil & Refining Co. and Ethyl Corp.; he has also spent four years on the editorial staff of CHEMICAL ENGINEERING. Arnold went on to graduate from Yale Law School in 1964. After a year with a West Coast law firm, he joined Esso Chemical Co. (45 Rockefeller Plaza, N. Y., N.Y.).

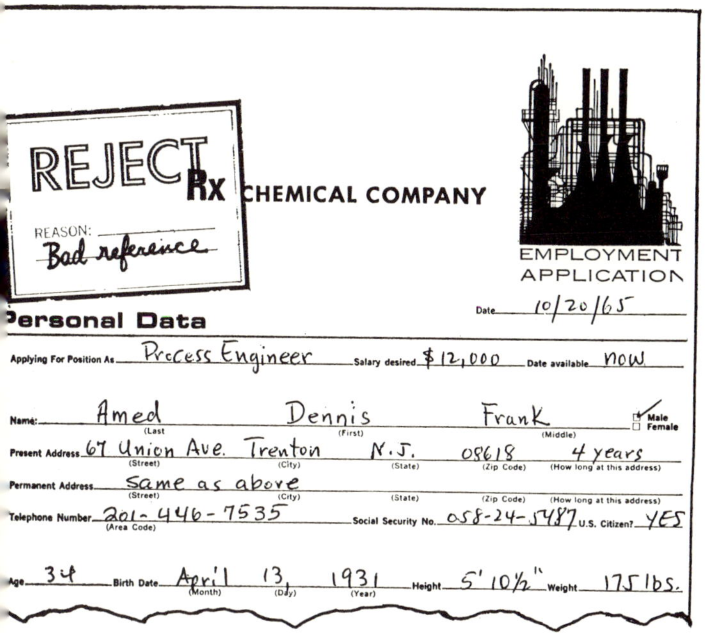
REJECT
Rx CHEMICAL COMPANY
REASON: Bad reference
EMPLOYMENT APPLICATION
Date 10/20/65
Personal Data
Applying For Position As Process Engineer Salary desired $12,000 Date available now
Name: Amed (Last) Dennis (First) Frank (Middle) ☑ Male ☐ Female
Present Address 67 Union Ave. (Street) Trenton (City) N.J. (State) 08618 (Zip Code) 4 years (How long at this address)
Permanent Address Same as above (Street) (City) (State) (Zip Code) (How long at this address)
Telephone Number 201-446-7535 (Area Code) Social Security No. 058-24-5487 U.S. Citizen? YES
Age 34 Birth Date April (Month) 13, (Day) 1931 (Year) Height 5' 10½" Weight 175 lbs.

Defamation by References

A bad reference from a former employer can play havoc with an engineer's career. A lawsuit is one possible way to get relief, but there are also other approaches.

DONALD G. DAUS, *U. S. Patent Office**

If an engineer has difficulty finding employment because of an undeservedly bad reference from a company he worked for, what would happen if he brought suit against that company?

Here are two contrasting examples of what *could* happen:

In a recent case,[1] Engineer Jones left the *A* Engineering Co. voluntarily after four years to work for *B* Co. Jones claimed that *A* then communicated to *B* that he was undesirable. Jones was fired. A month later, he accepted a position at *C* Engineering Co. but was rejected following another communication from *A*—similarly with *D* Chemical Co., after another bad reference.

Allegedly, Company *A*'s acts were based on an affidavit from an employee that Jones had company documents in his possession. It developed that these documents, voluntarily returned, were standard engineering aids; *A* Co. admitted these were neither secret nor confidential.

Jones sued for libel and slander. He received $15,000 in an out-of-court settlement.

The second example[2] has a different ending:

Engineer Smith had been employed in Argentina, Asia and the U. S. by *X* Co. and its foreign subsidiaries for some 17 years. He resigned in 1950 because he believed his prospects with the company were unsatisfactory. He was given a service letter by *X* Co., listing his various assignments and positions in some detail. The letter, which contained references to his personality and his professional ability in rather glowing terms, concluded with hopes for his success in the future.

Smith's complaint stated that there was an "express understanding" between him and *X* Co. that (1) Smith could use this letter for the purpose of obtaining employment, that (2) the company would provide prospective employers with the same facts as were recited in this letter, and (3) would answer any other pertinent inquiries in accordance with these facts.

Smith stated that all the persons to whom he reported during his employment were alive and that he had sought employment in 1956 from four different firms, the agents of which read the above letter and sought further information from *X* Co.

The firm's response: "This is in reply to your letter of March 12 in which you request information concerning Mr. Smith. The official to whom he reported passed away several years ago, and I am unable to give you first-hand information concerning Mr. Smith. His personnel record indicates that he had approximately 17 years of service with the company. The records further show that he submitted his resignation on Aug. 9, 1950, which was accepted.

"Insofar as I can determine from his records, we would not be prepared to consider him for re-engagement." The company subsequently refused to explain this further. Smith then brought suit.

The court refused to order damages, holding that "silence is not libel," and that a former employee cannot force an employer to give him a favorable recommendation, even if the records show that such a recommendation is deserved.

A Look at the Defamation Laws

From an analytical viewpoint, there is a great deal of the law of defamation that makes little sense. The explanation is part history and partly the result of a headlong clash between the freedom-of-expression concept on one hand, and sympathy for the victim of defamotion on the other.

Slander, or oral defamation, was in medieval times a "spiritual" matter, of concern to the ecclesiastical courts that punished slanderers by meting out penance. Eventually, slander became part of English common law, the basis for American law.

Libel, or written defamation, was originally made a crime in England in order to suppress seditious publications. This legal concept developed with the spread

*This article should not be interpreted as giving legal advice, nor as necessarily reflecting the official viewpoint of the U. S. Patent Office.

of printing—damages being a substitute for duels.

The major distinction between slander and libel is that libel is considered the more serious, and that certain written words can be the basis of suit without a showing of actual harm. This includes words affecting practice of a business, trade or profession. The other distinctions between libel and slander are becoming blurred, partly because of the advent of television and radio. Libel and slander will be discussed together under the term of defamation.

When the Engineer Goes to Court

In order to have a sound basis for a defamation suit: information must have been "published"* to a third person concerning the injured party, and understood by the third person to refer to this party in a defamatory sense; furthermore, it must have been communicated in the absence of a privilege or other accepted defense (such as "truth," where applicable). Let's look at these requirements more closely.

Possible defamation of the engineer can refer either to his character (in which case proof of damage is required), or to his fitness for his profession. Defamation of character would include charges of insubordination, forgery, stealing or "Communism." Terms such as these have been held actionable; other terms, such as "labor agitator," have not.

Defamation involving an engineer's fitness for his profession does not require proof of actual damages. Terms that have been held actionable include gross incapacity or inefficiency, neglect of duty, and disloyalty. However, terms such as carelessness, mediocrity, "not eligible for rehiring" or "desirable to terminate" have not been a sound basis for a lawsuit.

(The earliest cases of defamation of professionals go back to 1619, and not surprisingly involved lawyers. They could get damages if called ignorant of their profession, but not if called undistinguished.)

The words in question may require an explanation —"inducement"—to show how they are defamatory. The sense in which the innocent-sounding words are understood is termed the "innuendo." For example, being called "scholarly" can be defamatory to a plant or field engineer. In libel cases, the engineer must show how the innuendo has actually hurt him.

The statements must be false, according to the common law. However, some states have modified this requirement so that damaging statements that are quite true are still held actionable if, for instance, they are made from malicious or otherwise improper motives.

The defense that the defamatory statement is true is termed "justification." The party claiming truth must prove it, exactly as stated, and this can be difficult. Since the keystone of a professional is his judgment, his errors are not always easy to prove to the satisfaction of a jury. Furthermore, asserting in the preliminary legal papers that the defamation is true is again "publishing," and is sometimes taken by the court as proof of malice.

*This is a legal term for either written or oral communication.

Failure to prove alleged dishonesty can be quite expensive. In one case, an employee accused of theft, sued for libel. During the trial, the police arrested a third party for this crime, and the defendant company paid $10,000 in damages for the libel.

Not all libel or slander cases are settled by jury verdict. During the trial, the judge may dismiss the complaint if he feels that, as a matter of law, a reasonable person would not regard the words as defamatory, or that the words are capable of "innocent interpretation." This prevents the jury from even considering how the words were actually understood.

In one case,[3] a jury's award was set aside on appeal under the innocent-interpretation rule. A patent agent had voluntarily left his law firm. The firm sent all its customers a telegram saying that the company had found it desirable to terminate him. The jury considered this to be libel, and ordered a $20,000 award. But on appeal, the judge held that the words in question could mean a voluntary separation, and reversed the award.

Proof of Communication

Proving that some communication took place between a former employer and a third party is usually not difficult in itself. The engineer may work for a month or so and be suddenly laid off for no apparent reason. If this happens two or three times, there is little doubt that something is being published—either verbally or in writing.

To be actionable, this publication can be to the prospective employer or client, a bonding company, a state engineering registration board, a government security-investigation agency, employment agencies, or even credit bureaus and detective agencies.

For slander, the spoken words must be known exactly to bring a successful suit. If a sham phone call is arranged to get evidence, that call is not a good basis for a suit, since the court probably would consider that the particular slander was consented to.

When written reports have been used, a copy is usually needed. If the engineer is certain that he is being libeled, his lawyer can use the Federal Court procedure of "Discovery" to obtain a copy. This is expensive and should be considered only if all the other elements of a defamation suit can be proved.

The trend to using the telephone[4] makes proof difficult. The prospective employer is not likely to disclose the exact wording of the phone call voluntarily (i.e., without Discovery procedures), since a reputation for violating confidences would make his future reference-checking more difficult.

When Does "Privilege" Exist?

One of the most important aspects of defamation is the concept of "privilege." This privilege, which is an immunity from lawsuit that is given for the furtherance or protection of a public or private interest, can be either absolute or qualified.

Absolute immunity from responsibility, without regard to the publisher's motive or reasonableness of his acts, includes such things as judicial proceedings, legislative proceedings (e.g., the famous McCarthy hearings), governmental executive acts, and words consented to by the injured party.

Qualified privilege is conditioned upon good motives and reasonable behavior. It includes:

- Publications made to protect or advance a legitimate interest of the "publisher."
- Publications made to protect or advance an interest of the recipient or other third person, where the ordinary reasonable person would be expected to speak.
- Communications between those having a common interest, for protecting or advancing that interest. This includes having a common employer or employee (e.g., two executives in Company *X* exchanging opinions about an engineer's performance or character) —or the relation of employment agency, surety company, credit agency, physician or detective, to an engineer's potential employer.
- Publications made to proper persons in the public interest, such as reporting a robbery.
- Reports, "fair" comment or criticism on matters of public interest or concern, e.g., newspaper or magazine articles.

The doctrine of qualified privilege rests on the principle that the public interest and advantage of publication in each particular class of cases outweighs the occasional private and personal damage caused. It is considered more advantageous to the community that individuals should occasionally be damaged than that men under the privileged circumstances should not be at liberty to speak and publish what they believe to be true, although it may be defamatory.

The immunity of qualified privilege is probably the biggest obstacle towards recovery by the defamed engineer. It is not insurmountable, but in 90 cases surveyed by the writer the potential employee won only 17 times (after appeal), and 9 of these wins involved situations where there has been no privilege. In such situations, the defamed party had better than an even chance; whereas in cases where a privilege was involved, he had one chance in five to win in the trial court, and an even chance to lose that on appeal.

Under the defense of "qualified privilege," two main questions are raised, which are entirely distinct:

1. Does the occasion of the speaking or writing give rise to immunity?
2. Has the publisher lost this immunity by abuse?

In regard to the first question, a prospective employer must have intended to hire (or at least seriously consider for hire) in order to establish a common-interest privilege between himself and an engineer's former employer.

If the potential employer was specifically not authorized to contact references, and did anyway, is privilege present? The courts have not yet ruled on these points; the dividing line is hazy and may vary from state to state. As a practical matter, the courts would likely consider that any company following up on an application is within the privilege. Followups on names submitted by employment agencies or "search firms," without specific applications, are probably outside the privilege. With no precedents, it is not safe to bet on the results, and hence the preceding comments are merely an opinion.

Once a qualified privilege is found, the case may be dismissed unless the defamed engineer can show abuse of the privilege, misuse, "malice," or conditions creating a presumption of malice.

Indications of malice on the part of former employers include nonbelief of the statement's truth, deliberate publication to nonprivileged parties, and lack of reasonable grounds for the statements made. Malice has been proven by unreasonable refusal to explain reasons, extra voluntary acts explainable only by malice, and by publishing more than necessary to accomplish the purpose for which the privilege is given. Gross negligence in checking facts is not malice.

Malice won't defeat an *absolute* privilege. This is important to government employees and former employees, since they can be defamed with no chance for recourse. Voluntary statements to a governmental agency are not absolutely privileged, but required statements as to security-clearance investigations usually are. School teachers reportedly have never won suits on references by their principals. State University professors are probably also included in the absolutely privileged area.

Consent is seldom a basis for absolute privilege in the type of employment situations we have been discussing. It has been held that a request for a recommendation is not an invitation to publish defamation.

Corporate vs. Individual Liability

To this point, no distinction has been made between liability of the corporation and of the employee who actually says or writes the defamatory words. Generally, if one is liable, so also is the other.

A corporation is liable for defamation expressed by its employee while acting in the scope of his employment, even though the statement was neither authorized before nor approved afterwards—and even if the statement serves the malicious personal interests of the employee. A minority of jurisdictions upholds this liability only where defamation is authorized.

When the corporation is liable, it is jointly so with the employee. Usually the corporation alone is sued, since few employees could pay the $10,000 to $20,000 that may be awarded.

There have been occasional instances where a line manager was held liable as an individual while his corporation was exonerated, but these cases are rare, and usually limited to instances where a personal grudge is shown to exist between a manager and the former subordinate.

Legal Action as a Remedy

All in all, the chances of a defamed engineer recovering damages aren't too good, particularly if he has to prove abuse of privilege. The awards are small, relative to engineers' salaries and possible legal costs.

Nevertheless, engineers should be aware of the remote possibilities of recovery, and watch for a possible trend to make it less remote. Since an unknown number of cases are settled out of court, the figures cited earlier (which show the former employee winning only 19% of the cases) should not be interpreted with undue pessimism. The trend toward holding an engineer to a strict duty on trade secrets may be giving rise to a related trend—one that holds the former employer to a duty of being as fair to the departed engineer as to the potential new employer.

Individual problems should, of course, be discussed with an attorney.

But legal action is obviously not a panacea for the defamed engineer, and won't solve the problem caused by employers whose intentional or unintentional silence or "faint praise" is damning in itself.* In view of this, are there any other ways in which the engineer can protect himself from the effects of a damaging reference?

Alternatives to Legal Action

The best way to avoid having to go to court over a damaging reference is, of course, not to get such a reference in the first place. Although most suggestions in this area merely fall into the realm of common sense or good judgment, a few brief reminders (based on advice of personnel consultants) may be in order:

If you leave your company because you have a better job lined up, give enough notice to avoid undue hardship and then don't loaf or strut during the notice period. Be sure to return any company literature or engineering aids in your possession. If you think your supervisor will face difficulty replacing you or explaining your resignation, you should be particularly careful not to rub salt into his wounds or act thoughtlessly during your final weeks—otherwise you will be creating a desire on his part to "get even, sometime."

When in doubt, it may be a good idea for you to ask your supervisor, some time before you leave, whether he will give you a good reference should you need one in the future. If you have acted responsibly in giving notice, and have phrased your request tactfully, the chances are he will say yes even if he resents your leaving or has criticized your work in the past. Having said yes, he will probably not go back on his word later.

(A request for a service letter of the "to whom it may concern" type may also persuade your supervisor to think positively, although such letters are of limited value in themselves, as brought out in a case cited earlier.[2])

Incidentally, when you are deciding whether to accept a new job, you may want to size up your prospective supervisor's attitude to people who have left his department. If he can't stop telling you how disloyal or incompetent your predecessor was, be on your guard: He may be the type of supervisor who takes resignation or request for a transfer as a personal affront.

But what if you have already got, or know that you are going to get, an undeservedly bad reference? Other than bringing suit, what can you do about it? One personnel man makes these recommendations:

- Try to pay a visit to the person involved. If you have something to apologize for—e.g., insufficient notice when you left—by all means apologize. After the man has let off steam, and after you have tactfully presented your side of the story, at least a partial reconciliation may be possible.
- If the above doesn't work, or if you don't want to try it, then give some indication to a prospective employer (at a late stage in the interviewing) that your old supervisor may not give you a brilliant reference. This will minimize the shock value of the negative comments when they arrive.
- If there are people at your old company who will give you a good reference although they were not your direct supervisors, also give their names to your prospective employer so that he can hear both sides of the story if he wants to.
- Don't lose perspective. When you are being interviewed, remember that the immediate objective is to impress the prospective employer with your qualifications and ability to do the job. If you don't establish this positive impression, the matter of references will prove academic. So don't let undue worry about what your old employer is going to say about you prevent you doing justice to the vital parts of the interview. ■

*See Ref. 5 for a discussion of how employers can communicate with each other without fear of law suits by using adjectives with "faint praise" connotations.

References

1. *Chem. Eng.*, Mar. 18, 1963, p. 81, 84.
2. 256 F.2d 824 (9 Cir. 1958).
3. 165 Ohio St. 549, 138 N.E.2d 391 (1958).
4. Magee, R. H., *Personnel J.*, **43**, pp. 551-555 (Nov. 1964).
5. Peres, S. H., Garcia, J. R., *Employee Relations Bulletin*, Report #837, Dec. 1962, pp. 6-7. (Pub. by Natl. Foremen's Inst.).

Meet the Author

Donald G. Daus is a patent examiner in the U.S. Patent Office, Washington, D.C. 20231. His industry experience includes work as a process engineer for Eli Lilly & Co., and A. E. Staley Mfg. Co. A registered professional engineer in Illinois, he has a B.S.Ch.E. degree from the University of Illinois, and an M.S.(Ch.E.) from Michigan State University. He is currently in the final year at George Washington University School of Law. Society memberships include AIChE, ACS and Alpha Chi Sigma.

The Foreman vs. the Union Organizer

When an aspiring union tries to organize a nonunion plant, or when it stages a membership raid on the established bargaining unit, all sorts of pitfalls can beset the uninformed or unwary foreman. Here's how to stay clear of trouble.

JOHN W. WHITTLESEY, *Union Carbide Corp.*

"Uneasy lies the head that wears a crown" is a maxim born in the days of kings, but it can be applied quite well to a modern supervisor faced with a union organizing campaign. Even if he has gone through a number of such campaigns, or if his men already belong to a union of some kind, he may fall prey to this uneasiness and be led into unwise courses of action.

Especially vulnerable is the foreman who has little or no experience with labor unions, and who suddenly realizes that, of all things, a union is organizing the hourly employees of his very own department. His uneasiness will usually be mixed with resentment, partly occasioned by uncertainty of what this unknown quantity will bring, and partly by the feeling that the situation is a direct challenge to the prestige and respect that he believes he enjoys.

Furthermore, most foremen tend to get confused about the Taft-Hartley Act. At best, this law is imperfectly understood by the vast bulk of supervision, and at worst, it is construed to mean that the federal government backs all union activity, right or wrong, and that any dissenters had better watch out—a view that most union organizers are too human to resist exploiting.

The chaotic state of mind that often results from these uncertainties may lead the supervisor into any one of a number of undesirable choices, from reprisal at one extreme to virtual abdication of his authority at the other.

Management's Responsibility

Of course, once aware that an organizing campaign has begun, top management should step in with the necessary guidance, if only as a matter of protecting its own self-interest. It should, as early as possible, make sure that all supervisors are fully aware of: (1) what their legal rights and limitations are in these matters; (2) precisely what policy the company is adopting towards the aspiring union and how supervisors are to carry it out; (3) where and to whom they can go in order to get any questions answered that may come up.

Such initial information should be given not only orally via supervisors' meetings but also placed in writing for ready reference. And, of course, supervision should be kept constantly informed throughout the campaign, so that the foreman need not rely on Dame Rumor or the grapevine to find out what's going on.

However, not all companies are this sophisticated in labor matters; some have only limited experience with organizing or "raiding" campaigns. Thus, the supervisor may, on any given occasion, be left guideless at such a critical juncture, damned if he does and goes too far, damned if he doesn't and falls short. In this dilemma, what should he do?

Five Basic Rules

First and foremost, the supervisor should realize that top management, whether it tells him so or not, expects him to continue exercising his full responsibilities for doing his job; for getting out the production efficiently and with a minimum of upheaval.

Second, he should realize that he can lawfully do so, under the Taft-Hartley Act or other pertinent labor laws, without any legal restraint whatever *except* that he cannot run his shift or department in such a way as to favor nonunion employees over union adherents, or members of one union over those of another.

Third, the mere fact that one of his group is a union member, or active for the union, does not in any way insulate the man from observance of customary rules of good conduct, or from doing his job to the best of his ability. A loafer sporting a union button is still a loafer; a fist in an employee's teeth is no less repulsive because it comes from a good union man. The signature of an employee on a union card is not a free pass to insubordination, nor does an organizing campaign warrant an automatic reduction in output to accommodate those who wish to discuss their problems on company time.

Fourth, the supervisor should realize that the organizing campaign is not directed at him personally, but stems from many other reasons. Thus, it is a fair assumption that the supervisor will retain all—or virtually all—of the relationship with his men that he had before the outset of the campaign, if he continues to carry on as before or even to improve the caliber of his supervision in those areas that need it.

Finally, the foreman should tread warily in certain areas that are beset with legal implications. Before elaborating on these, let's take a look at the basic law on what can and cannot, should and should not, be done with respect to aspiring unions.

The Taft-Hartley Act

Although the Taft-Hartley Act is a federal statute, almost any business that is not strictly local in character is subject to its terms. Essentially the Act was and is designed to encourage and protect union organization of employees, as well as to place certain limits on the sort of thing that unions can do in conducting their affairs in relation to employers.

The law was first passed as the Wagner Act in 1935, amended by the Taft-Hartley Act in 1947, and again by the Landrum-Griffin Act of 1959—which latter law also prescribed a so-called Bill of Rights for union members in their internal union affairs (which need not concern us here).

The core of the statute is section 7, which guarantees to employees the right to organize and bargain collectively with employers through their unions, and to engage in union activity; or, by the same token, to refrain therefrom. Accordingly, the Act provides for the orderly selection of union representatives through an election procedure that gives all affected employees the chance to vote on whether they want a union or not. These elections are conducted by the National Labor Relations Board, which is given the general function of administering the law, and which, if the union wins the election, certifies the winner as the "exclusive bargaining agent" with whom the employer must deal.

Certain acts (whether during an election campaign or not) by either employers or unions are prohibited as "unfair labor practices." Generally, the law forbids employers (1) to interfere with, restrain or coerce employees in respect to union membership or activity; (2) to discriminate against employees as to hire or tenure of employment in the same respect; (3) to foster company unions; (4) to refuse to bargain with a union chosen by the majority of employees.

As for unions, they cannot conduct secondary boycotts; cannot engage in jurisdictional strikes or some types of recognition picketing; cannot restrain or coerce employees, discriminate or force an employer to discriminate, refuse to bargain or force an employer to choose or reject any bargaining representative.

Applying Taft-Hartley

Plainly the foregoing is a highly general view of the law, but enough has been said to make its basic aims clear. More specifically, as it affects a supervisor in his conduct during a union organizing drive, it makes him an agent of the employer, so that whatever he does may be attributable to the employer, and may thus enmesh the company in the tangles of the law. Thus, the firing of an employee because he is a union organizer is an unlawful act; the National Labor Relations Board may order him reinstated with full back pay, even though the supervisor acted on his own and without consulting his superiors in taking such action.

Furthermore, conduct that constitutes an unfair labor practice is generally good grounds for a union to object to the results of an election and to claim the outcome was illegitimately influenced. The supervisor can therefore, through lack of care, place his company in the thoroughly discouraging position of having had a union lose an election, which the Board may then set aside and hold again. This time, though, the union has had an additional campaigning issue handed to it on a silver platter, which can be (and has been) exploited to swing the necessary number of votes into the union column.

An unwise firing may therefore not only clip the company's pocketbook, but may turn an election victory into an eventual defeat. Similarly, such acts as promotions, demotions, discipline, assignments of undesirable work, refusal of overtime opportunities, or even refusal to hire a new employee, if based on union connections or lack of them, are unfair labor practices that the Board can (and frequently does) remedy.

Suppose that the supervisor has mixed motives, say for demoting an employee. One, he is a union organizer; two, he was doing a job not quite up to snuff. This type of action is illegal also *so long as* consideration of the union aspect is a motive in the action taken.

On the other hand, the discharge or demotion of a union member, even an active one, during an organizing campaign, is perfectly legitimate if the reason for the action is unrelated to such membership. The basic problem here, of course, is the proof of the case; the Board is much more likely to be convinced of the foremen's impartiality if a demotion is based on demonstrable shortcomings, rather than on personal dislike. During an election campaign especially, any action taken to or for any employee should be squarely based on objective, business motives that stand up under scrutiny and not on whims or vague dislikes that will more than likely be regarded as mere pretexts to clobber a union adherent even if that was not the intent.

Legitimate and Illegitimate Union Activities

By no means must all activities on behalf of the union be tolerated by the supervisor. He may forbid the men to leave their jobs to discuss the union during working hours. He may prevent such congregating unless it is on the employee's own time—and enforce this by discipline if necessary. He may also prevent a union advocate who himself is on his lunch hour from talking to another employee who is at work. He may prevent the passing out of leaflets by employees in working areas. And, he may escort any non-employee organizer off the premises; unless this person has company permission to be there, he is in no better situation than any other trespasser.

But the foreman cannot legitimately prohibit a union adherent from doing the above prescribed things if he permits members of a rival union or group to do them. Also, he cannot order employees to stop wearing union buttons, or to stop talking about the union on company premises on their own time.

Discussing the Union

This brings us to what can or cannot be said about unions to the hourly employees. Plainly this is a subject where great care must be exercised, in view of the rather broad language of the law and the shifting interpretations placed on it by the Board.

The law only prohibits those statements that contain a promise of benefit or a threat of reprisal if an employee were to get out of or join a union, or vote for or against it. It is thus easy enough to refrain from telling John Doe that he'll get a raise in pay if he votes no-union, or that he'll get laid off or otherwise hurt if he does the opposite. But—and this has especially been true under recent Board decisions—what constitutes a promise or a threat has been broadly interpreted to cover almost any language from which an indirect threat or promise can be implied.

In one case, for example, the Board held that the showing to employees of a film that was a fictionalized account of some unhappy results of union success in another election in another company, was enough to warrant setting an election aside. In another case, the company's distribution of photographs tending to inflame racial prejudices had a similar result. In a third, an employer stated that even if the union won, he could not be forced to sign a union contract in the absence of agreement; this was held to be unlawful, even though the Board admitted that the employer was merely restating his legal rights.

All these cases represent a tightening up of Board policy toward speeches and statements in recent months, and a reversal of previous policies. Whether or not the cases were correctly decided is not in question here; the point is that virtually anything a supervisor says may come eventually under Board scrutiny where a union campaign is involved.

It is easy to control one's tongue when one knows in advance that questioning an employee as to his union membership or how he is going to vote is unlawful. It is not equally easy, however, in casual conversation with an employee to refrain from asking him in a spirit of pure friendliness how the union meeting went last night. But it is equally unlawful; and the distinction between this sort of question and merely listening to an employee volunteering such information without prompting (which *is* lawful) is sometimes hard to make.

Must the Foreman Remain Silent?

Now the above does not mean that the foreman must sound completely neutral about the union.

First of all, the supervisor can and should be prepared to give factual and straightforward answers to union statements, rumors or gossip. Union organizing campaigns are not always models of truthfulness, nor do they always elicit the *whole* story about any particular matter. The best weapons the supervisor has to combat erroneous assertions (whether deliberate or not) are the actual facts. For instance, if the union hands out a flier at the gates in which certain deficiencies are cited in the group insurance benefits, and if these deficiencies do not actually exist, there is nothing wrong with telling the men that the union doesn't know what it is talking about.

Secondly, the law does not forbid the supervisor from expressing a preference for continuing the plant non-union, or for dealing with employees directly, without third-party intervention, on matters affecting them. He can, likewise, remind employees that they currently enjoy such things as good wages and other benefits that the company *has* given them without union demands in the past. He cannot, however, go on from there and promise future goodies if only the union would go away.

The foreman, can of course, tell employees that they need not vote for the union even if they have signed cards; that they can join the union or not as they choose. And he can express the hope that their choice will be not to do so, as long as he does not imply that undesirable results will stem from their acting to the contrary.

In principle, the foreman is free to voice any misgivings that he may have about unions—as long as he stays within the framework we have discussed, and avoids making unsupported, potentially slanderous statements about individuals.

How Far to Go?

Since the supervisor does have these prerogatives, should he become an active anti-union campaigner? The answer to this will generally be "Only to the extent indicated by higher management."

The supervisor, especially when he meets up with a union for the first time is normally short on experience in such areas, and is almost sure, unless properly guided, to stumble and fall. Some of the problems he may encounter have already been alluded to. Many others may arise from attempts on his part to carry out an off-the-premises campaign.

For instance, he may decide to park his car across from the union meeting hall to see who attends the rally. This, of course, is unlawful surveillance. Or he may visit the homes of the hourly employees to try and sell not only them but their families on the advantages of a non-union plant. This, too, is forbidden. A few unlucky foremen have even attended union meetings, with unfortunate consequences. The risks connected with random off-plant activities of this sort are wholly disproportionate to the foreseeable results.

But what happens when management itself elects to institute a no-union campaign? This, of course, is not infrequently done, and successfully, because there are very few companies that welcome the advent of a union into the plant. Nor is there anything wrong or unlawful in top management seeking to persuade its plant employees that they do not need labor-union representation. Carl Hageman, a vice president of Union Carbide, undoubtedly spoke for a majority of management when he said, in a recent *Chemical Week* article: "All other things being equal, I am convinced that without unions there is more in it for the employee, the company and the customers. . . . We'll resist them to the best of our abilities. However, if one is certified, we will bargain with it in good faith."

Inevitably a campaign aimed at resisting unionization is going to entail supervisory participation. The foreman who has the regular and continuous day-to-day contact with the employees is plainly the person who is most likely to convey the management message effectively. But here again, it must be emphasized that going beyond what top management has prescribed is fraught with dangers. Management has an obligation in these circumstances to lay down carefully thought-out and detailed policies, procedures, do-and-don't lists and the like; and the foreman must resist what at times is an almost irresistible temptation to improve on what the company has asked for.

Thus, for example, it may be necessary to answer a certain untruthful piece of union propaganda. The foreman is told what to say when employees approach him and confront him with the union tale. But he may decide to go the company one better and call each of his people individually into the office and give each the answer (and perhaps a few more comments). This, however, has been proscribed by the National Labor Relations Board as constituting "undue influence" (possibly because it can be so effective), although the same statements made at the work place in response to a question, or even casually on the supervisor's initiative, are perfectly permissible.

The supervisor may be asked by higher management for his opinion of union strength in his department. This is normally a difficult question to answer accurately; most supervisors normally underrate union strength among their own people, overrate it among the men under other supervisors. Now there is nothing unlawful about asking a supervisor what he already knows, or about a supervisor relaying such information, assuming that he has acquired it lawfully. But if zeal for the company prompts the foreman to nail down the answer, chapter and verse, by polling his shift or department to get an exact head-count, then he's off base under Taft-Hartley—and will probably not get an accurate picture of current feeling in any case.

Again, suppose the election day has at last arrived. Campaigning has been stopped for 24 hours, as required; the polls are set up at prescribed places and the voting hours established. The only problem now is to see that everyone has a chance to vote, so as to ensure that, win, lose or draw, the result represents the feelings of as many eligible employees as possible. The supervisor has been told to make sure that all of his people get to the polls. So far, so good; no problem arises about arranging for each employee to have time to vote.

But the supervisor should not under any circumstances try to make sure that each man votes by personally escorting him to the polling booth, or worse yet, by waiting around until he casts his ballot and then escorting him (or a group) back onto the job. The only employer representatives allowed near the polls during voting hours are the selected observers who, by definition, must be nonsupervisory.

Championing the Union

Of course, the foreman should hold aloof from any personal support of the union. Entirely apart from the strong likelihood of management displeasure, it is unlawful for a foreman to give any material or financial aid to any union, whether authorized by his superior or not. Here again, he must remember that the law makes him an agent of his company. He cannot, of course, join up in the hourly employee's union. Nor is his activity protected by law as in the case of the hourly employees; thus he has little if any legal recourse if he does incur management displeasure.

When Unions Vie With Each Other

The temptation to lend support to a union frequently arises in a situation where two unions are competing for representation rights. Typically, one of them will be the incumbent; the other will be a "raiding" union. The supervisor may well have enjoyed excellent relationships with the incumbent's representatives and even may, at times, have sympathized with its efforts at material betterment of its members, especially if they happen to be employees whom he supervises. He may, likewise, have personal friendships among the hourly people and may even have gone to bat for them, individually or as a group, to get them more money, better conditions and the like. It is perfectly natural for him to side with the present union against the rival; and he may all too frequently take an active part in the campaign.

In principle, there is nothing wrong or unlawful about a supervisor expressing a preference for one union over the other, but this principle has highly limited application. An expression of preference may not only be contrary to his superior's thoughts on the matter, but it may lead to a more active involvement on behalf of the preferred union, and this can lead to trouble. The closer the race, the more likely the union is to seek the foreman's active help. The results can hardly fail to be illegal discrimination in handling rival union adherents, and may backfire in favor of the other union.

The consequences of taking sides may also be disastrous for future departmental harmony in the event the other side wins out: retaliation by way of grievances, turmoil, slowdowns in production and other familiar tactics will certainly not help the supervisor in effectively running his department in the future.

Even if the foreman's side wins, he may have become too deeply involved with his own people to remember clearly and at all times that he represents management, whose interests on any given question may be unlike those of the employees. In the last analysis, his company loyalties will have to control and he should avoid compromising them.

Neutrality as regards rival unions is thus the most logical and practical course to follow in the absence of any clear dictate from top management.

The Foreman vs. the Picket Line

Union organizing campaigns are tending over the years to involve less and less use of actual union muscle to bring the target plant into the fold. In the early days of the Wagner Act, violence, picketing, strikes and other unpleasant manifestations of union drives were more frequent, as was corresponding employer reaction to them. One favorable result of the existence of a federal labor statute has been the decided increase in orderly campaigns of persuasion as the appropriate way of winning a representation effort, and a diminution of the sometimes spectacular roughhouses of the late thirties.

This, of course, doesn't mean that such things don't happen any more—merely that they are less likely. Occasionally, a supervisor is faced with an organizing strike or picket-line, or even threats of violence to himself or his family. In such situations, he clearly should not attempt to go it alone. Police protection should be available to fend off violence away from plant premises, and keep down the more rugged bubblings of a picket-line.

Whether or not a supervisor is expected to continue on his job during a strike or picketing, or whether he will be asked to pitch in and work in the event his company decides to run the plant during the strike, is a matter for top management decision. It is enough to say that a verdict either way—including a decision to operate a plant in an economic or recognition strike with new-hire replacements—is by itself perfectly legal, as is a management request for supervision to conform with the decision.

The Campaign in Perspective

In summary, a supervisor confronted with a union organizing campaign should remember that the worst thing he can do is panic. The advent of a union would hardly represent the end of the world for him, and would not change his supervisory responsibilities in any essential fashion. Management policies will continue to be carried out; any change in the rules of the game occasioned by a union victory and the eventual establishment of contractual bargaining relationships will not alter basic principles, but only the details of administration.

It is hoped that this article has clarified some of the foreman's prerogatives in an organizing campaign, and shown how he can avoid transgressing federal labor law. When confronted with specific problems, he should seek additional guidance from his superiors.

Perhaps one should keep in mind the story of the luckless foreman who some years ago bet one of his hourly people $50 that the union would lose the election. The union lost; the foreman won his bet; but when he tried to collect it the night of the election, he handed the union a gilt-edged case to object to the results and to get the vote set aside. When the second election was held shortly thereafter, the $50 went down the drain. ■

Meet the Author

John W. Whittlesey is labor relations attorney for Union Carbide Corp., in New York City. His ten years at Carbide have involved him in many different phases of corporate labor relations, and were preceded by a number of related assignments, including ten years as labor attorney for the U. S. Chamber of Commerce.

A Hard Look at Equipment Warranties

It is comforting to know that a new piece of equipment has some sort of warranty to fall back on if things go wrong. But how much protection does such a warranty afford? In this first of two articles, the authors examine current practice, as revealed in their own survey of the warranty policies of 53 industrial firms.

SIDNEY A. BRESLER, *Consulting Engineer*, and MARTIN J. HERTZ, *House, Grossman, Vorhaus & Hemley*

The Tall-Tower Chemical Co.'s brand-new plant has just undergone a successful startup. But no sooner has Tall-Tower paid the last $1-million installment, and the contractor's men have packed up and left, then the boiler breaks down, and production stops for three days.

At this point, Tall-Tower probably thinks in the following stages:

1. What's wrong?
2. How long before it's fixed?
3. Who's going to pay for it?
4. What does my contract say?

The project manager takes out the contract and looks at it to see "what our rights are." He then discovers two things:

- The contract doesn't have the provisions he thought it had.
- He is not sure he understands what the provisions mean.

He looks through 5 pages of specifications and 4 pages of the typed contract that mention "warranties" in some places, "guaranties" in others; "representations" also crop up. Are these terms synonymous or interchangeable? Do they have the same meaning each time they are used? What, in fact, is their meaning as commonly used in a contract?

Some Key Definitions

A contract is a bundle of rights and obligations with regard to a particular subject that could be the purchase of a turbine or construction of a $100-million chemical complex. When a particular clause of the contract is not lived up to, the contract is said to be "breached," and the wronged party who has not gotten what he bargained for is entitled to lawful damages.

The contract may specify the measure of damages or the method of measuring damages (commonly referred to as *liquidated damages*). If not, the purchaser is entitled to be reimbursed for whatever the cost to make him whole again (called *direct damages*). This would include the cost of repair, and the profit he would have made on orders that he could not fill and that were cancelled. However, the operator would not be paid the profit he would have made on future orders that would have been placed with him had the customer not become disgruntled. Such "losses" are speculative and are called *consequential damages*.

In the preferred usage of the terms, a *representation* refers to a present state of facts. The contractor represents that the boiler is steel plate, ½-in. thick, the turbine is new and was manufactured by Northern Electric Corp., and the plant was built according to the plans and specifications annexed to the contract. Thus, the representation is either true or false at a particular moment in time, generally when final delivery is accepted.

The *warranty* normally will deal with a period of time and performance during such period; e.g., that:

- For 350 days after acceptance under given conditions, the plant will produce 1,000 tons of ammonia per day.
- The boiler will exhibit no defect in material or workmanship for one year; under proper conditions of use, the synthesis-compressor crankshaft will last two years.

It may be noted that representations and warranties are what we bargain and pay for. The *guaranty,* in this context, is what the contractor or vendor is obligated to do when he does not live up to the representations or warranties. In other words, a guaranty is a liquidated damage clause that refers to the broken representations or warranties.

Incidentally, liquidated damage clauses may also apply to any other provisions of the contract. A frequent source of confusion is a "guaranteed delivery or completion date." The use of the word "guaranteed" in this sense is unnecessary except for the purposes of emphasis. It takes the place of Tall-Tower writing in the words "We mean it!" What normally follows is something commonly referred to as a bonus incentive clause that is usually intended to reduce the purchase price if the plant is completed after the expected date. In extreme cases, the bonus incentive clause may be followed by or combined with a *cancellation clause.*

The last item requiring definition is the phrase *penalty clause*—a provision that seeks to punish one of the parties for breaking the contract by making him pay damages out of reasonable relation to the damages actually suffered or contemplated at the time of contract. Penalty clauses are generally illegal except in a government contract. But beware! Just because a clause is labeled "penalty clause" doesn't mean that it has no legal basis; a fairly standard liquidated-damage clause may be given the "penalty" label as a deterrent or for emphasis.

Warranties vs. Guaranties

A good deal of trouble can normally be avoided by precise use of legal words that are understood by the user. The most common problems arise where two different words are used interchangeably in a contract and have a different meaning sometimes and the same meaning other times.

We have already noted the unfortunate and sometimes intentional confusion in the use of the words penalty, damages, and liquidated damages.

The same type of confusion exists in the use of the words guaranty and warranty, and this has a historical basis (see "Black's Law Dictionary," 3rd ed., West Pub. Co., p. 860):

"*Guaranty* and *warranty* are derived from the same root, and are in fact etymologically the same word, the 'g' of the Norman French being interchangeable with the English 'w'. They are often used colloquially and in commercial transactions as having the same signification, as where a piece of machinery . . . is 'guarantied' for a term of years, 'warranted' being the more appropriate term in such case. . . ."

A guaranty classically referred, in legal usage, to a promise by some third person to "make good" if someone else did not adequately live up to his contract. Thus, while a dealer would make a *contract* with a purchaser, the manufacturer would deliver a *guaranty* to the purchaser. If anything went wrong, the pur-

Guaranties: How the Conditions Can Be Expressed

I—Time or Usage. The period of validity can be expressed as:

- A period of time (a year, a month) unrelated to usage.
- Extent of usage (e.g., one year for a certain catalyst, or for a gas turbine before major overhaul).
- A combination or alternate form of the time and usage.
- A relation to time or usage of another facility or event (while, e.g., the machine is used in the factory). This type of clause usually occurs in relation to temporary repairs.

II—Conditions of Usage. The conditions may require the item to be:

- Serviced by a specified servicer (e.g., inspected monthly by the seller's authorized representative).
- Used within a certain geographic area, or within specified limits of temperature, humidity or pressure (e.g., to be used only within temperatures of 40 F. to 120 F., or only within continental U.S. or in any other similar temperature zone, or only when the barometric pressure does not fall below 14.5 psia.).
- Used pursuant to stipulated conditions (e.g., not to be operated in excess of 2,000 rpm., or for more than 2 hr. at a time).

III—Place and Method of Execution. Normally, the seller insists that he will only return the money, or supply the parts, labor or materials at places he will specify. As its most obvious motive, this clause thrusts the burden and expense of field labor, freight charges and insurance upon the buyer. Beyond this, however, is the desire of the seller that he control the cost of repairs, and have some check on the honesty and reliability of the person doing the repairs. At the very least, the seller can limit the profit of an authorized repair man—or else eliminate the profit by requiring that his own franchised dealers make certain repairs at no profit, and supply parts at no profit.

IV—Complaint-Reporting Requirement. This type of clause is usually found in insurance policies. However, its inclusion sometimes serves to minimize consequential damage to a machine resulting from failure to present small malfunctions for repair. Suppose a machine does not self-lubricate properly. Continued operation may then cause a generator to burn out. If the user must notify the manufacturer within two days after he first has trouble, the manufacturer has a chance to service the machine before greater damage to the generator occurs. Otherwise, the user, particularly when he has an order from an important customer to deliver by a certain deadline, might be tempted to run the machine because "the guaranty will cover any damage to the production line caused by the machine's malfunction." Once the notice of defect is given, if the user operates the defective machine thereafter, he does so at his peril, and the guaranty may not apply.

chaser would look to the manufacturer for satisfaction and not to the dealer, although the actual contract of sale was with the dealer. If the purchaser obtained the sales agreement from a contractor rather than a dealer, the result was still the same.

In time, however, the purchaser decided that he wanted to rely on the engineering-construction contractor, the party he paid his money to, rather than upon an unknown manufacturer. Therefore, when he asked for a clause that would bind the contractor to give the purchaser satisfaction, he would say to the contractor "I want the guaranty from *you*." Thus the term "guaranty," although improperly applied, stuck.

Likewise, warranty was formerly an absolute undertaking. "If the turbine doesn't run for one year without breakdown, we will refund your money."

Modern commercial usage has obscured the classic legal meanings and distinctions. Although the preferred title today seems to be "warranty," the two terms are so often confused or used so interchangeably, sometimes even in the same paragraph, that it is impossible to know the meaning intended by the person who wrote the purchase-order form or sales-confirmation form, unless we read the word in the actual context used. However, as a general guideline, for modern commercial usage, a warranty tells us what quality or performance we have a right to rely upon, while a guaranty tells us what will be done by the seller, contractor or manufacturer in the event the goods are defective or do not meet certain standards of performance. The guaranty costs the seller or contractor or manufacturer money—or its equivalent in materials, parts or services.

What One Finds in Industrial Guaranties

The usual types of guaranties are as follows:

1. The return of the purchase price or part of the purchase price. The "satisfaction or money back" offer, when properly phrased and complied with, results in a return of the purchase price. For instance, the seller of catalysts usually offers to refund the purchase price if the catalyst does not last a certain period of time, as one of his options.

2. The replacement of a defective part. Normally a sale of an item of machinery will require that any defective part be replaced without charge for the part.

3. Free labor for repairs. Because it is difficult to determine whose fault it is that a machine has broken down, there is normally no charge for the labor involved in replacing a part, although the labor charge for disassembling and reassembling the machine the part came from is sometimes subject to negotiation.

In addition to the above three items, and sometimes as a concomitant, the seller may agree to service the unit for a period of time free of charge. Sometimes this service "extra" is of more benefit to the seller than to the buyer: the seller catches problems before they involve major repairs, and avoids the needless "bad will" that may arise if the customer suspects a defective part, whereas he is actually having trouble because he is misusing the equipment.

It is generally wise for the seller to state that his liability is limited to any of the above items except as expressly set forth in addition. To the extent that a guaranty goes beyond the conditions already outlined, it becomes a substitute for an insurance policy. Indeed, the damages to which the seller will expose himself should be covered by a reserve fund (amounting to being a self-insurer) or by an actual insurance policy. (The purchaser's reserve fund generally appears on the balance sheet as depreciation.)

Some additional clauses of guaranties may provide allowances for:

- Cost of merchandise affected by the defective machine or material (for example, parts improperly machined may ruin a portion of a device into which they are being assembled, or a defective catalyst may ruin a batch of chemical product).
- Cost of overhead while a machine is not running.
- Loss of profits while machine is not running.
- Reduction in price or a certain amount per day while a machine or factory is not in production or use.
- Extra expenses in connection with the faulty performance.

In any event, so far as the seller is concerned, none of the nonstandard guaranty clauses should be used without setting a specified limit of liability in dollars.

The guaranty will also determine the conditions under which the seller will be liable, or under which the guaranty will expire. The ways in which these expiration provisions are normally expressed are shown in the "box" opposite this page.

Survey of Current Industrial Practice

We can obtain a better perspective on the warranty procedures of the chemical process industries (CPI) by first examining what the practice is in American industry as a whole.

One may normally assume that the manufacturer or contractor will try to limit his liability and, therefore, have minimal financial exposure, while the user will seek the broadest possible terms and coverage. We have a further contrast in that the salesmen of the manufacturer or contractor are always putting pressure on their companies to grant wider, more liberal warranties in order to beat out the competitor. Understandably, there might be a wide divergence between the warranties sought in a purchase order and the terms granted in a confirmation, or "terms of sale," or "standard warranty."

To obtain a fair sampling of current practice in general industry, we queried national companies in a variety of fields, asking them to furnish lists of standard terms used in their equipment warranties. These firms had all sent representatives to attend a seminar on product service management held in New York City in February 1967, and we selected them on that basis.

The 33 companies replying are in the following fields: computers, construction equipment, processing and production-line equipment, business machines and typewriters, components for automated or moving equipment, induction-heating equipment, instrumentation, agricultural equipment, communications, containers and storage, appliances, machine tools, scientific and electrical apparatus. The companies sell to contractors, dealers, original-equipment manufacturers, and, in some cases, directly to consumers.

Although some of these companies include the chemical process industries among their customers, we did not send out any requests to operating companies in the CPI because such companies are treated at greater length in the CPI survey that we will discuss later.

The types of forms on which the warranty terms appear are described in the note accompanying Table I. In some cases, the descriptions are based on the title given to the form by the company; in others, they are based on our analysis.

In their letters of transmittal, some companies pointed out that the same terms appear in their sales orders, quotation requests and standard warranties. Other companies use one set of terms when they are selling or bidding, and another set when they are buying or requesting bids. Still other companies jumble the descriptive title at the head of the clause, so that the title might be "terms and conditions of quotations or sales."

Terms of the Warranty

The period during which the warranty is effective may be expressed in terms of months or in terms of use or operation. Frequently there is a dual measure within the same warranty—i. e., one year from date of shipment or 400 hours of actual use. This may be stated as "whichever occurs last" or "whichever occurs first."

Moreover, there are various events that signify the commencement of the period; these are shown below, along with the number of companies (out of the 33) that base the commencement on this event:

Event	*Distribution*
Shipment	9
Installation	4
Sale (or invoice)	3
Delivery (to eventual user)	8
Purchase	4

Of the companies surveyed, 28 firms based the term on a calendar period, while 11 companies used time of operation or use, as shown in Table I. It is to be noted that the total in both classes adds up to 39 although only 33 companies were surveyed; this is because some companies expressed their cutoff point as being either calendar period or running or performance time, whichever occurred first.

We conclude that the representative warranty period is based upon one calendar year from date of shipment or delivery. No pattern could be noted from analysis of the nature of the business, or from whether the sale was a small unit to a consumer or a large complex fabrication to a producer.

Of the companies surveyed, 27 routinely applied the warranty to defects of material and workmanship, one limited the warranty to specific parts, one referred to manufacturer's defects, one to nonexpendable parts, one only to latent defects in material and workmanship, and one didn't specify at all. Only one firm (a chemical company) required that the equipment ordered must be fit for the purpose specified and conform to specifications, but preferred to describe the equipment as first class and made in a skillful manner instead of using the words "free of defects in material and workmanship."

Of the three that discussed performance or capacity or specifications at all, not one required that the installation meet all three requirements, i. e.:

- Conform to specifications.
- Perform at 100% $\pm X\%$ of effectiveness or at 100% $\pm Y\%$ of capacity specified.
- Be fit for the purpose specified.

The latter provision was in one of the purchase-order forms. But in the terms of sale, it is more frequent to find an express disclaimer that the equipment is fit for any particular purpose. Interestingly enough, one warranty states that any claims made in the catalog are merely descriptive.

In this particular aspect of the warranty, the fine hand of the attorney for the vendor may be seen with greatest clarity, limiting the warranties to the classic, often litigated and very narrow phrase "defects of material and workmanship." This places the lid on possibly extravagant or exaggerated claims that the salesman may have made. Few companies are willing (and probably with good reason) to venture forth into assurances that will increase or extend their liability; for if they are not careful, such language might be interpreted as referring to consequential damages.

Remedial Action

As for remedial action promised, 30 companies agreed to, or required, repairs or replacement of the defective part, and 24 specifically or by implication stated the repair-or-replace choice was up to the seller. In 15 cases, the purchaser was required to pay shipping charges to the factory, but 2 companies agreed to pay these charges. The warranties of 3 companies were limited to replacement of parts only, 3 expressly stated that labor was at the seller's expense and 3 more stated that labor was at the purchaser's expense.

In other miscellaneous clauses, provision was made for labor, but not for mechanic's travel time; and an allowance was made toward purchase of a new replacement as an alternative to repair. One government order form required the seller to pay cost of packing, transportation, rigging, droppage and insurance. One purchase order had the ambiguous requirement that the seller make good any defect. Such a clause could only result in litigation if a serious dispute arose.

Interestingly enough, most of the linage of a warranty deals with "escape clauses"—i. e., circumstances under which warranties will not apply. This part is known as the "disclaimer" section.

But even before that section, warranties will be inapplicable except under *normal use and service*

Length of warranty periods encountered in a survey of general industry—Table I

Calendar Period	No. of Firms	Comments
3 mo.	2	Agricultural equipment; phrased alternately as 500 hr. of use
4 mo.	1	Sophisticated machine tools; phrased alternately as 1,000 hr. of use
6 mo.	2	Induction-heating equipment
12 mo.	21	Normal period in industry
18 mo.	2	
Total	28	
Time-of-Use Basis		
400 hr.	1	Agricultural equipment
500 hr.	2	Heat exchangers; agricultural equipment (original user only)
1,000 hr.	2	Electronic equipment (original user only); aviation equipment

(Continued) Calendar Period	No. of Firms	Comments
1,500 hr.	1	Agricultural equipment
1 yr.	2	
Lifetime	2	Tires (sliding-scale offset)
90 days during successful performance period	1	Electronic data processing systems, government contract
Total	11	

Note: The 39 items tabulated above were taken from 33 sets of standard terms. (In some cases, the length of the warranty period was presented on more than one basis in a set of terms.) The actual forms in which these terms appeared are distributed as follows:

Form	Number	Form	Number
Standard sales warranty	9	Instructions to, or via, dealer or distributor	3
Terms of sale	3	Acknowledgment	1
Conditions of sale	5	Authorized govt.-supplier warranty	2
Sales contract	1	Purchase orders	2
Quotations	5	Total	33
Catalog	2		

(15).* Equipment intended for a particular use will operate at varying degrees of efficiency if the conditions are altered, but unless it is designed to deal with changing factors within a particular range, the performance not only may be inefficient outside the design range, but permanent damage may be done to the equipment. Understandably, the manufacturer uses the term "normal use and service" to refer to the expected range. Because he cannot know what permanent damage has been done during the time the equipment operates outside of the expected range, he will disclaim responsibility or terminate the warranty the instant the equipment is operated outside the expected range or *is not used for the purpose designed* (3).

Since equipment must be properly cared for, any *repairs must be done by authorized representatives of the seller* (8). In this way, the seller knows that the facility has not been *altered* (9) or improperly repaired, and that he is not being called upon to repair the result of some other person's damage to the equipment. In line with this, most companies do not apply their warranty where equipment has been:

- Altered (changed, modified) or tampered with (9).
- Abused or misused (11).
- Subjected to negligence (8).
- Subjected to accident (7).

In short, manufacturers want to make sure that the equipment is in the same state as it was when delivered by them (normal wear and tear excepted) before they will repair or replace pursuant to their warranty. The equipment must have been *properly installed and operated according to instructions* (6). In some cases the warranties even specify particularly recurring causes of breakdown, and "fluctuating power supply" in the case of instrumentation.

Disclaimer Clauses

In many cases, an adjustment with reference to the equipment—whether repair, replacement, refund in whole or in part, etc.—may only cover a part of the damages suffered by the purchaser. If the machine is not fit for the purpose he intended generally, the seller assumes that the purchaser will investigate the applicability and not buy the machine if it is not suitable for the intended purpose. Therefore, the seller disclaims that he warrants the machine will be fit for a particular purpose or use.

Moreover, if the machine is defective, production may stop, raw material or goods in process may be ruined, and customers of the purchaser may be irrevocably lost. Therefore, the disclaimers point out that except for repair or replacement of the installation or its equivalent in money, the seller will not be liable for anything else, listing specifically *special* (5), *indirect* (3), or *consequental damages* (18), and that *no other warranty* (2) is made or *implied in law* (2) or *fact* (1), and that damages in any event are *limited to the purchase price or cost of replacement of the particular defective part* (8). (However, certain government agencies will not accept a waiver of so-called "consequential damages" from major equipment suppliers.)

*Note—Hereafter, a number shown in parenthesis indicates how many of the 33 companies use the term (or the italicized portions thereof) in their warranties.

Current Practice in the CPI

Having discussed the general legal principles that govern and apply under statutory and common law in the typical state, and then surveyed representative practices in U. S. industry as a whole, we are now ready to zero in on the practices in the CPI.

It could be said that the CPI and the process-equipment suppliers use generally accepted and standardized terms and conditions covering warranties and guaranties. The trouble is that several sets of such standards are used.

One set has been accepted by buyers, and is found in most purchase orders. A second set has been accepted by most equipment manufacturers, and becomes incorporated in their terms of sale.

It is quite interesting that many of the provisions and implications of the first set of standards is in direct contradiction to those specifically stated in the second set. There are very good reasons for these differences. However, this situation at best leads to an interesting exchange of correspondence between buyer and seller. At worst, processing of the order is delayed, and both prospective purchaser and seller suffer.

Hence the emergence of a third set of standards—a set that is deliberately vague so that it may be readily accepted in the knowledge that regardless of the terms of the signed contract, all claims for equipment malfunction will be settled in a negotiating session.

From the legal viewpoint, the terms of the purchase order govern, unless the buyer formally accedes to the seller's proposal. Even when the seller's quotation contains wording stating that his terms prevail, that previous agreements are cancelled, etc., the mere fact that he has manufactured or shipped the material described in a purchase order carries with it the implication that the provisions of the purchase order have been accepted.

Table II shows some typical provisions of purchase and sales contracts used by CPI firms. The table is based on a survey of purchase orders used by four operating companies and four engineering contractors, and of a dozen terms-of-sale forms used by equipment and catalyst manufacturers. Let us examine some of these provisions, noting areas of agreement and disagreement, and some of the reasons for the differences.

Quality and Performance

The guaranty with respect to performance is the most critical and most controversial item.

Manufacturers' sales terms frequently are limited to "warrants that materials and workmanship are the best obtainable" or are free of "defects" when shipped, with guaranties limited to replacing or repairing improperly fabricated parts, fob. shop. In some instances, the performance of the unit also will be warranted, although no statement is made as to what remedy the purchaser can claim if the equipment does not function as specified. In rarer instances, the manufacturing company may offer to take back all its equipment and refund the selling price.

The manufacturer may state that the performance specification is descriptive but not warranted. Under such conditions, if performance were substantially as anticipated, the purchaser would have no legal claim for small deviations. If the purchaser wanted a greater degree of certitude, his price might be increased to cover the greater risk undertaken by the seller, or he might be requested to pay for a performance test at the manufacturer's shop.

Purchasers expect a warranty that the equipment supplied is of "standard," "good" or "best" quality, produced in a workmanlike manner; and also that equipment will perform in accordance with the specifications written in the purchase order. Sometimes the purchase order has no statement regarding remedies to be applied if equipment does not comply with the terms of the order. The implication may be that the supplier will make good regardless of cost to him. At other times, it is stated that the vendor will be required to repair or replace inadequate equipment, and to assume all labor and freight charges.

Some purchase orders distinguish between breach of contract caused by failure of a material specification and failure of a performance specification.

Material that is defective in quality or workmanship must be replaced. Usually, although not always, it is relatively simple to determine whether or not such defects exist and where the liability rests. The cost of replacing a defective component may not be too great. (To avoid controversy, the purchase order should list materials to be used, to the extent that this is practical—e.g., for all parts of a heat exchange, but not for elements of a gas turbine. Furthermore, responsibility for selection of materials, either by purchaser or by supplier, should be unambiguous. Where possible, reference should be made to codes such as ASME or TEMA.)

Matters are more serious if equipment fails to perform as expected. Frequently, it is difficult to prove the poor performance. A replacement may be very expensive and, worse, impossible to obtain in a reasonable time.

Furthermore, the owner may prefer to run the plant at lower capacity than have it shut down completely while repairs are being made. Therefore, unless an off-the-shelf replacement for the equipment is readily available, both parties may prefer to reach a compromise settlement that results in the retention of the equipment rather than its replacement.

In general, both buyer and seller benefit if methods to be used in establishing performance are defined. It is usually sufficient to refer to established codes. Where there is no generally accepted industry standard—as, for example, with respect to noise level or to vibration—it is advisable to agree on a quantitative definition of acceptable standards as early as possible.

On long delivery items, when there is good faith on both sides, it is accepted practice to have a generally phrased order placed and accepted, with the understanding that the original agreement will be supplanted by a more detailed agreement establishing binding criteria at a later date. Although such a procedure does leave one or both parties in a position of maximum legal exposure, it may be a lesser evil than loading up a purchase order with very detailed specifications that are hastily conceived under pressure of time. The buyer must be aware of the implications and consequences of specifications that are written into the purchase order, particularly if there is a possibility of corrosion, erosion, or reaction between process fluid and equipment. If the specifications are so detailed as to remove all discretion from the seller, and the specifications are not correct, the buyer may find that he has equipment that complies with his specifications but does not perform in accordance with his requirements.

Warranty Period

In recent years, as plant size has increased and construction time lengthened, there has been increasing tendency to extend the time interval over which the warranty is held to apply. Although terms vary from company to company, frequently an interval centering around a one-year period is used in the CPI.

As might be anticipated, suppliers try to limit the time during which they may be held liable. They may allow 12 to 15 mo. after shipment or installation of equipment, and they may also limit the warranty period to not more than, say, 3 mo. or 1 yr. after the start of operations.

Purchasers usually ask that suppliers accept liability for 1 yr. after the start of operation, sometimes extending this to 18 mo. after shipment if the equipment is not placed in operation within 6 mo. after receipt. Under special circumstances, as when equipment is being exported, the purchaser may ask that the warranty period be extended for a longer interval than 18 mo.

We have already noted that the supplier would prefer to limit claims for material and workmanship to the condition of the equipment when it leaves his shop, because he has no control over what happens to it in the field. However even though the purchaser inspects the item carefully, he is naturally reluctant to agree to waive claims for defects that might be discovered during the course of operations.

Consequential Damage

Consequential or contingent liabilities may arise if malfunction of a piece of equipment prevents operation of a plant, and if the owner of the plant is forced to buy the product in the market, at a higher price than that at which he could have produced it, in order to satisfy commitments that he has already made.

Under common law in some states, and in many contracts with federal agencies, a seller of equipment may be held liable for such consequential damages. A

Frequency with which typical warranty clauses appear in a survey of: 12 "terms of sale" of CPI equipment and catalyst manufacturers; 4 purchase orders of CPI operating companies; 4 purchase orders of engineering contractors*—Table II

Clause of Warranty	Terms of Sale	Purchase Orders Operator	Purchase Orders Contractor
Quality and Performance			
1. Product is well made and of good material, or as specified	3	2	0
2. Product's quality is standard or better, or first-grade, or best for purpose	1	3	3
3. Product is free of defects in material or workmanship	9	4	2
4. Product is free of defects when shipped	1	1	1
None of the above mentioned in warranty	**3**	**0**	**1**
1. Equipment complies with performance specifications	2	3	4
2. Equipment is "Guaranteed" to comply with specifications, but extent of liability is not stated	5	0	0
3. Specifications are considered only descriptive	1	0	0
None of the above mentioned in warranty	**5**	**1**	**0**
Terms under which the seller will repair or replace defective parts:			
1. Fob. supplier's plant	11	0	1
2. At plant site, absorbing all costs	1	2	3
3. Only if notice is given immediately on appearance of defect	7	0	0
4. Only if equipment has been operated under normal or specified conditions	6	0	0
5. Supplier will not pay for the cost of repairs unless his prior consent is obtained	5	0	0
6. Supplier's maximum liability is no more than the price of the defective part	6	0	0
7. Supplier's maximum liability is the cost of the replacement part	0	1	0
None of the above mentioned in warranty of merchantability or fitness	**0**	**1**	**0**
1. Statement of merchantability or fitness for specified use	0	2	1
2. Statement that there is no warranty as to merchantability or fitness for use	2	0	0
Neither of the above mentioned in warranty	**10**	**2**	**3**
Warranty Period			
Extends 12 mo. after shipment	6	0	0
Extends 15 mo. after shipment	3	0	0
Extends 18 mo. after shipment	0	0	3
Not specified	**3**	**4**	**1**
Extends 3 mo. after start of operations	2	0	0
Extends 12 mo. after start of operations	4	0	0
Not specified	**6**	**2**	**0**
Consequential Damages			
Seller has no responsibility for consequential or indirect damages	12	0	2
Buyer retains all statutory warranties	0	3	0
Not mentioned in warranty	**0**	**1**	**2**
Ownership, Local Laws, Patent Infringement			
1. Statement that seller has title to merchandise	3	0	0
2. Warrant that equipment and material:			
Comply with all federal, state and local laws	0	3	2
Are made in conformance with Fair Labor Standards Act	1	1	0
Item 2 not mentioned in warranty	**11**	**1**	**2**
3. Buyer is held harmless against patent liability arising from use in accordance with offer of seller	3	3	3
4. Extent of seller's patent liability is to obtain license, replace equipment or refund purchase price	3	0	0
Items 3 and 4 not mentioned in warranty	**9**	**1**	**1**
Delivery			
1. Seller is not responsible if delivery is delayed: For any cause	1	0	0
For causes beyond his control	2	1	1
For causes beyond his reasonable control	7	0	0
2. If delivery is delayed, the buyer may: Cancel order	0	2	1
Cancel order and charge seller excess cost of alternate purchase	0	0	1
Items 1 and 2 not mentioned in warranty	**2**	**1**	**1**

* Note: Some of the frequency-of-appearance columns for the major sections of the table (i.e., for the major terms) add up to more than these numbers, because some purchase orders or terms of sale may incorporate more than one of the clauses listed in that section.

manufacturer who accepts such liability will increase his price—in effect providing self-insurance. However, this risk is never accepted in the chemical industry. Any single manufacturer's equipment represents a fairly small proportion of the cost and material of an entire facility, and it may be difficult to establish exactly which piece of equipment is preventing the plant from attaining its rated capacity. More important, the profit made on the sale of a single piece of equipment is far less than the losses that a plant may sustain if it is forced to shut down; and an equipment manufacturer may find it very difficult to pay such charges.

Therefore, sales conditions, and usually purchase orders as well, disclaim all such liability. This is done either by a specific clause or phrase to this effect, or by limiting all liability to a maximum amount that usually is the purchase price of the equipment or of defective parts.

A word about the personal-injury aspects: In recent years, there has been an increasing tendency for the courts to allow "third party" suits by individuals against both the user and the manufacturer of equipment, on a claim of injury resulting from improper design or functioning of a piece of equipment. The time interval during which this implied warranty is held to be applicable is much longer than the stated warranty period that we discussed earlier. Allowable grounds for action, and duration of the applicable statute of limitations, varies widely from state to state.

Ownership, Local Laws, Patent Infringement

Because there is an implied warranty that a seller has clear title to the material he is offering, it is not usual for a written warranty to this effect to appear, although certainly there could be no objection to it.

The purchaser also may require a warranty that the equipment complies with all federal and local laws and regulations, and that such portions of the design as are under the manufacturer's exclusive jurisdiction are free of all patent infringement. Many sellers' standard terms make no mention of any of these items.

It is generally agreed that the equipment supplier can offer no protection with regard to the use of his equipment in a chemical process, but only against lawsuits pertaining to the mechanical design of his equipment and its use for the specified purpose for which it had been bought. In those instances in which the terms of the guaranty against patent infringement have been detailed, it will be found that the purchaser invariably is required to cooperate with the seller, but that the latter assumes all costs of the defense.

Should the patent suit be lost, the seller may have the option of obtaining a patent license, of modifying or replacing the equipment to remove the infringement, or of refunding the purchase price. Usually the contract does not say which party bears the cost of all necessary changes, or how costs are to be allocated if they are shared, or how the owner is to be recompensed if plant capacity or efficiency is reduced as a result of these changes.

CPI Practice in Perspective

In comparing the general-industry survey with the one for the chemical process industries, one can see that many warranty provisions are common to both industrial groups. Nevertheless, there are certain dissimilarities.

For example, although the modal length of the warranty period is one year for both groups, half the contracts in the CPI specify a longer period. Similarly, while most warranties state that the material should be free of defects in material and workmanship, over half the terms-of-sale or purchase orders in the CPI also contain various provisions relating to performance specifications.

These differences come about because major equipment in the CPI is bought in accordance with individual performance specifications and must fit into a unique plant with a unique set of conditions; such equipment cannot be readily replaced from off the shelf. In general industry, however, most sales involve standardized equipment, and therefore failure of a single part or single unit may be no more than a temporary inconvenience, rather than being catastrophic to the entire complex.

There are further warranty-related differences between the CPI and other industries in regard to contractual clauses and the philosophy of vendor/user relations, as well as in the practical treatment of claims under the warranties. These differences will be explored in the second article (next issue). We will also cite some typical case histories, and make recommendations relating to the use of service contracts. ■

Meet the Authors

S. A. Bresler

M. J. Hertz

Sidney A. Bresler is a chemical engineering consultant with offices at 500 Fifth Ave., New York City. He has been occupied with the design and economic evaluation of petrochemical and fertilizer plants for many years, and is an experienced project manager. Mr. Bresler holds an M.Ch.E. degree from Brooklyn Polytechnic Institute, and an M.B.A. from Columbia University. He is known to CHEMICAL ENGINEERING readers for his articles on prime movers, ammonia plants, gas turbines, and negotiation techniques. The latter article ("Negotiating With Engineering Contractors"), co-authored with Martin J. Hertz, appeared in the issue of Oct. 11, 1965.

Martin J. Hertz is an attorney with the law firm of House, Grossman, Vorhaus & Hemley, 521 Fifth Ave., New York City. He specializes in the negotiation of agreements for the construction and purchase of chemical and other industrial plants. Mr. Hertz conducts seminars in negotiating techniques for corporate executives, and is the author of "Making Agreements That Are Most Favorable For You," as well as a more comprehensive book on contracts and negotiation techniques to be published in the near future. He obtained his L.L.B. from Harvard Law School, and his L.L.M. from New York University.

Does Your Equipment Warranty Really Protect You?

Here is an analysis—replete with illustrative examples—of the ways warranty claims are generally settled, the factors that enter into the negotiations, and the advantages of a certain kind of equipment service contract.

MARTIN J. HERTZ, *House, Grossman, Vorhaus & Hemley* and SIDNEY A. BRESLER, *Consulting Engineer*

In the previous article, we examined the various and sometimes conflicting clauses used by sellers and purchasers in equipment warranties. Now for the key question: What does a purchaser do if the equipment he has bought does not perform as expected? If he is a member of the chemical process industries (CPI), he never goes to court. Well—hardly ever.

True, when there is an equipment malfunction during the warranty period and the purchaser cannot get the relief he believes himself entitled to, a lawsuit may ultimately be threatened. But since the buyer and seller expect to continue their business relationship, negotiation is generally preferred to litigation. The number of actual lawsuits is also kept down by the difficulty of determining the exact cause of many malfunctions—and the even greater difficulty of proving the cause in court. Furthermore, the purchaser's prime concern is not to win a legal round, but rather to keep the plant running. Thus, according to the usual practice, the equipment manufacturer offers whatever assistance he can during the time of crisis, and both parties argue over costs later.

Experienced buyers have estimated that 5% to 10% of the time of purchasing agents (and presumably also of salesmen) is presently spent resolving claims resulting from the supply of unsatisfactory equipment, and the percentages are increasing. In such post-mortem sessions, the relative strength of the two companies, and the skill of the negotiators, are vitally important.

One might assume that the buyer would be in the most advantageous position. If he is not satisfied, he can hurt the seller by taking his future business elsewhere. But in the CPI, the situation is seldom this one-sided. If the buyer goes as far as to really antagonize the seller, the seller may decline to consider later invitations to bid, and may be very reluctant to assist in the preparation of engineering cost studies. The next time the chemical company needs a similar piece of equipment, it may be forced to pay a higher price to another supplier. The problem, although perhaps minor for standard items that are readily available, is very significant if one is buying special equipment made only by a few fabricators.

The engineering contractor who sells plants on a lump-sum basis must be particularly careful. The outcome of a number of negotiating sessions between him and his suppliers may determine whether or not a job has been profitable. However, if he presses too hard, and too many vendors become antagonized, the contractor will be unable to obtain low equipment quotations in the future, and may find himself in a most disadvantageous position when preparing to bid on the next job.

The Growing Cost of Downtime

In recent years, the amount of time devoted to settling back charges, and the amount of money involved in such claims, has been increasing steadily. As plant capacities grow, and equipment for single-train units

EXAMPLE 1: The equipment has not been delivered when promised. The purchaser makes the claim that the "guaranteed" delivery date has not been met.

Comment: A promised shipping date is not a guarantee; it is a representation. Unless the sales contract also contains a lateness clause (predetermined liquidated damages), the purchaser has no legal recourse except to cancel if delivery is not made within a reasonable time after the date promised. In addition, the terms of sale usually release the manufacturer from his moral commitment if the delay is caused by specifically enumerated causes, or by events beyond his control. (Many contracts use the phrase "reasonable control.")

Many purchase orders expressly reserve to the buyer the right to cancel if the order is not shipped as promised. Some purchase orders state the order will not be canceled if the delay was caused by events beyond the control of the supplier; other purchase orders state that the buyer may not only cancel his order, but may buy the equivalent material elsewhere and bill the original supplier for the difference in cost. This latter provision, if accepted by a vendor, might be considered a form of guarantee of delivery. However, while a vendor may accept bonus-penalty delivery clauses, it may prove difficult to get him to pay a bill for higher charges paid to a competitor for replacement equipment unless a suit is commenced.

What, then, can the purchaser do if specially designed long-delivery equipment is not delivered when promised? Generally, nothing, except to help the manufacturer, and to look into the inadequacies of his own inspection and expediting department.

continually increase in size, the financial losses the operating company may suffer from failure of a single piece of equipment become more and more serious. The equipment suppliers' exposure to loss of capital and loss of reputation grows proportionately.

Ironically, it is the ability of equipment to perform so well in the past that has led to this situation. For example, years ago when ammonia plants were built, many single-train plants erected in this country were provided with two 60% reciprocating compressors, and frequently a spare compressor as well. Multitrain plants contained a spare full size compressor. As good experience with compressors was accumulated, the requirement for a spare compressor and then for half-sized compressors was dropped, and large single units were installed. Formerly, when a compressor in a 200 ton/day plant broke down, there was either no loss of production or, at most, losses might equal $16,000 per day (with ammonia selling at $80/ton). Today, technological and economic pressures have made the use of a single centrifugal compressor in a 1,000 ton/day plant practically standard design. When something goes wrong with this machine, sales losses may amount to $45,000 daily (based on an ammonia price of $45/ton), resulting in increased pressure on the plant operator and his maintenance people to keep the plant running.

This pressure inevitably is transmitted to the equipment supplier. However, his profit bears no relation to the gross sales of an entire chemical facility, and he cannot be expected to accept more of a financial liability than he now has. Thus, the basic problem is not assignment of financial penalties but rather of furthering the common effort to keep equipment in operation.

Another Look at the Warranty Problem

While items sold in other industries may or may not be self-contained units, items sold in the CPI are more likely to be links in a process. The emphasis in warranties should therefore be on performance and continuity of operation.

When one talks to purchasers, they say, "We expect the manufacturers to stand behind their equipment and the contractors to stand behind their plants. We have enough economic strength to persuade these parties it's in their own best interest to do so. If there is a malfunction and they don't fix it, they lose a customer. So what need do we have of closely worded contracts for each major piece of equipment, and all the extra paperwork and negotiation this would involve?"

But the suppliers say, "If they don't buy from us, from whom else can they buy? The number of vendors is limited. We'll do what is reasonable, but not much more." The contractors may think along similar lines. Thus the circle is complete. Each member of the group needs the others in aggregate, and each is convinced that his money or products or know-how is the most essential and most wanted element. There appears to be a deliberate policy of vagueness, of not forcing the issues—of saying "when we have a problem, we can sit down and work it out."

As already indicated, the general practice is to fix the breakdown first, and to argue about who foots the bill later. Generally, the purchaser pays a percentage of the cost. The vendor may claim that it is not his process or equipment that is at fault, but another piece of equipment, either earlier or later in the process. He may claim his equipment met specifications, and that environmental factors that weren't originally specified are at fault. There may be a dispute as to what the specifications called for. And even if there is agreement on what they called for, the conditions of performance or operating conditions specified are rarely the same as those assumed and set forth in the original bid or contract.

Some facts just may not be available—either because

log sheets were not filled out, or instruments may have been unavailable or improperly calibrated, or because the pressure of startup work limited the testing that could be done. In addition, both parties may attempt to withhold relevant facts—the operator not wanting to call attention to possible abuse of equipment, and the manufacturer not anxious to talk about breakdown of the same or similar equipment in other installations.

How, then, can disputes in this delicate area be minimized? The two most obvious approaches involve:

1. Using warranty clauses that meet the real and reasonable needs of the buyer and seller. The party that can point to the warranty and show that the terms have or have not been met is in far better position to resist pressure. Some clauses that are important to the buyer and seller are discussed in the box on p. 223.

2. When possible, doing business with firms that have a good reputation, and that can be expected to make settlements in a spirit of reasonableness. No warranty, contract or specification can cover every possible technical or economic contingency; a certain degree of good faith must be assumed, and this is normal among reputable companies in the industry.

We would now like to propose an additional and perhaps less obvious approach—one involving the use of service contracts.

What Service Contracts Have to Offer

The service contract may be considered a form of a "major maintenance" insurance policy, similar to "major medical" insurance. Because much greater sums now depend upon the continuous operation of equipment, and because the number of man-hours per ton of product has been reduced drastically, operating companies may find that now they are able to afford the necessary insurance premium.

The servicing organization would most logically stem from the suppliers of such major equipment as compressors, pumps, boilers and furnaces. These suppliers would expand their staffs and train the skilled maintenance craftsmen who would visit operating facilities at regular intervals to inspect the equipment and to talk with the plant operating staff. Increased inventories of spare components of long-delivery equipment would be manufactured and kept in central warehouses; this should be feasible with multi-use or readily adaptable parts.

It has been estimated that between 50% and 75% of problems that result in delayed plant startup are caused by defects in equipment resulting from either improper design, fabrication or installation. The use of a service contract would allow an adequate staff of manufacturers' service representatives to review the installation and check the equipment—with less of the cost and time pressures that tend to limit this sort of service under existing policies.

A service contract might commence with equipment installation and startup (when the requirement for supervision would be intensive) and continue for at least one year; thereafter, it might be renewed. Plant or contract personnel would continue to perform such

EXAMPLE 2: A steel valve containing a copper seat has been welded into a pipeline carrying aqueous ammonia. The seat corrodes two weeks after operation starts, and ammonia spills into an operating area. The plant has to be shut down for the valve to be replaced.

Situation A: The valve had been specified by catalogue number.
Comment: Obviously, the supplier has no liability.

Situation B: The bid invitation had stated no copper alloys were to be used.
Comment: The manufacturer might offer to repair or replace the valve if it is shipped back to him. However, the labor for replacing the valve may cost more than the valve itself, particularly if the valve is not readily accessible and scaffolding must be installed to permit a welder to work. In this instance, the owner or contractor will press the valve manufacturer to accept the labor charges. The actual terms of the purchase are not followed strictly, and the settlement is negotiated.

Situation C: The valve fails after having been in service for two years, because the wrong valve was supplied by the manufacturer.
Comment: If the warranty period has expired, no legal claim will be accepted. Nevertheless, the supplier may accept some material and labor costs of a good customer.

Situation D: Failure occurs well within the warranty period. Because it required two hours to flush the line and change the valve, during which time the plant was shut down, the owner suffered losses of $10,000. To satisfy previously accepted orders, the owner had to make open-market purchases of chemicals that he normally manufactures.
Comment: This is a consequential damage, which is expressly disallowed in almost all terms of sale. The only recourse of the owner is to buy business interruption insurance.

Situation E: An operator walks below the leaking pipe and is injured by the stream of ammonia.
Comment: Responsibility is not clear-cut. The operating company and/or the contractor may be liable. The manufacturer's liability may be limited if he was not told of the nature of the process fluid (see **Business Week,** Sept. 16, 1967).

EXAMPLE 3: After one month of operation, the air compressor does not deliver the specified quantity of air, and the owner's maintenance people cannot determine the cause of the trouble.

Comment: Undoubtedly the manufacturer's representative will be asked to visit the jobsite. The manufacturer will want to be paid for the time of his service engineer.

The first check of the service engineer will be to try to measure the air flow when the compressor inlet and outlet conditions are as specified in the purchase order. Frequently this is impossible—perhaps because summer ambient conditions were specified and the test is being made in the winter.

The second check is of the air-flow measuring instrument. If calculations based on this review prove that the compressor is indeed operating satisfactorily, then there is no question but that the cost of the service engineer's time and the use of any special measuring instruments are the owner's responsibility.

Complications arise if it is found that lowered capacity has been caused by a defective seal. In accordance with some terms of sale, the vendor's responsibility would be limited to replacing or repairing the assembly after it has been delivered to his shop—if it can be proved that the seal was defective when shipped. One must now determine what really caused the seal to fail. If it was indeed defective when shipped, which may be hard to demonstrate, the manufacturer may be urged to accept all, or part, of the cost of repairing the compressor as well as all costs of his service man. However, if the seal failed because its components were not aligned properly, and the assembly of the compressor was done entirely by the owner, then the manufacturer may disclaim liability.

Alternate Situation: The capacity decreases after four months, well within the warranty period.

Comment: The procedures are the same as above, but the manufacturer's representative may want to review log sheets to be sure that the compressor has not been overloaded during its months of operation in the past.

Alternate Situation: Because the mechanical efficiency of the compressor is not as high as had been anticipated, it is found impossible to increase its capacity without also increasing its power requirements and hence the operating costs of the chemical facility.

Comment: This is a difficult matter. Repair work must be done under the supervision of the manufacturer's representative, or by authorized personnel, if one is not to run the risk of having the guaranty voided. Regardless of whether the sales terms specifically exclude any warranty with respect to capacity, or specifically guarantee performance in accordance with specifications, the adverse publicity that would saddle a manufacturer trying to escape liability would be most unwelcome.

One alternative open to the supplier is to furnish a more efficient machine. But there probably would be a long delay until the new compressor is manufactured and delivered, and then additional time lost while the old compressor is removed and the new one installed. It might be necessary to have piping and instrument changes made. Therefore, the owner may not consider this to be a satisfactory solution.

The seller's other alternative is to offer a fixed amount of money as liquidated damages to compensate the user for the added power costs.

normal maintenance as changing the pump packing, replacing compressor valves, plugging defective boiler tubes, and so on. Plant or contract maintenance personnel would also be used for most of the work of a plant turnaround, including disassembly and cleaning of equipment and most repairs—certainly all those of a minor nature. But special service personnel would be expected to visit the plant periodically to review performance of equipment, point out hazardous operations, suggest improvements, and describe the manufacturer's latest innovations. Such personnel would be available at times of emergencies and also during plant overhaul and other critical periods.

Such contracts are now offered by some manufacturers, particularly those selling gas turbines and, to a lesser extent, compressors and other equipment.

Finally, as more trained and qualified personnel become available, and as more extensive experience with this kind of service contract is obtained, a single independent organization might be established that would be staffed by people who had been former employees of equipment manufacturers, and who would continue to go back to these manufacturers for training and reindoctrination courses. Such an organization would offer several advantages. First, there is the elimination of divided responsibility; communications would be simpler and more rapid if the man servicing a pump and the man servicing its steam drive both work for the same company and have had the opportunity to get to know each other. Second, having a larger staff upon which to draw would make it easier for the service organization to send a well-qualified man to an operating company as soon as emergency arises. Third, it would be easier for an operating company to deal with a single service organization than with many.

A service company of this kind could evolve as a division of an engineering contracting company that has well-qualified people who now do contract maintenance and turnaround work. However, for security reasons, operating companies and contractors may both be reluctant to have other contractors' employees enter their plant. For the present, and until more experience has been gained, it is likely that the service contract of the type being suggested will be made available only by equipment manufacturers.

Advantages to the Operating Company

All plants have maintenance crews that take care of normal equipment repairs and must take immediate action in case of emergency. It is extremely doubtful that there would be any reduction of this staff if a service contract of the type contemplated were signed. However, each member of the plant maintenance crew must be somewhat of a generalist—particularly in the smaller or medium-sized facility. He may or may not have had his experience supplemented by a training course in the shops of a major equipment supplier, but certainly he is not as aware of all the intricacies of a major piece of equipment as would be the service man who has undergone a long and continuing training period with the equipment manufacturer. It is this very specific and very up-to-date information and capability that would be made available by means of a service contract or a service organization.

At national AIChE meetings, most recently at the one held in St. Louis in Feb. 1967,* insurance underwriters have stressed the importance of preventive maintenance, of more appropriate training, of the use of sophisticated instrumentation to check equipment reliability (i.e., ultrasonic detectors to determine the presence of fatigue cracking), and the advisability of using service engineers during major equipment inspection and overhaul. Instances were cited where preventable losses amounted to millions of dollars.

Although the operating company now can, and does, call in a manufacturer's representative whenever emergencies occur, the use of a continuous servicing and inspection policy should help forestall emergencies. Furthermore, as the servicer becomes familiar with each installation and its problems he can suggest improvements, as they are introduced by the equipment manufacturer, which would serve to keep the plant up-to-date, and more competitive with newer plants.

Advantages to the Vendor/Servicer

An efficient service organization—one that could suggest new methods and improvements to the operator—would be an ideal sales adjunct and marketing asset to its parent, the equipment manufacturer. The

*See paper by M. Telesmanic: "Accident Experience With Axial Flow Centrifugal Compressors"; and paper by D. A. Naughton: "Preventable Accidents to Turbines and Speed-Increasing Gear Sets."

The Basic CPI Warranty: Statements That Are Important to Buyers and Sellers of Process Equipment

1. Equipment contains no defects of material, workmanship or design.
2. All equipment will perform according to physical and performance specifications under normal use, when properly installed and maintained (**optional** . . . and is fit for the purpose intended).
3. Supplier will repair or replace parts (**optional** . . . and provide labor); notice must be given as soon as trouble occurs.
4. Warranty terminates upon any of the following events:
   - Misuse, negligence, accident, alteration, tampering or unauthorized repair.
   - Expiration of a certain period that starts from date of operation.
5. Warranty is extended to ultimate user.
6. Supplier disclaims liability for:
   - Consequential, special or indirect damages.
   - Other warranties, implied or in fact.
   - Direct damages that exceed cost of repair or replacement of the equipment, whichever is less.
7. Equipment is of sufficient size and capacity to fulfill the operating conditions specified in all respects.
8. Supplier has complied with all applicable laws (e.g., Fair Labor Standards Act).
9. Purchaser is protected against patent liability.

Comments on Terms of Particular Interest to the Purchaser:

- If a clause terminates the warranty in the event of unauthorized repair, check to see that the vendor has a readily accessible repair service, or arrangements with local dealers or other convenient organizations to provide authorized repairs.
- If the vendor does not warrant certain components of his equipment because they are supplied to him by other manufacturers, find out whether these manufacturers provide warranties that extend to the ultimate user, and whether the commencement time for these warranties is shipment to the user (rather than to the equipment fabricator).
- If a clause says that the vendor will supply relief only if, **in the vendor's opinion,** the equipment does not meet specifications, you must recognize that reasonable men might differ as to whether specifications are being met, although these differences can usually be reconciled in a negotiated settlement.
- Suppose a subcontractor or component manufacturer has a clause that says "we will make repairs provided we get immediate written notice of the defect." How soon is immediate? If the warranty says "subject to the laws of the State of . . . ," it may be wise to check what these laws are. Otherwise, even though your foreman calls ten minutes after the breakdown, you may not have an effective warranty if he has failed to give formal written notice, for instance.
- Because the relief obtained under the terms of even the best warranty is a poor substitute for equipment that performs without trouble in the first place, someone who is about to buy equipment that is not entirely standard or that is going to be used in a novel manner should not neglect to ask such questions as: Where has such equipment been used under similar process conditions? How long has it been running? How has it performed? May the user be contacted?

service organization can be the first wedge in a sales campaign for new installations. All other things being equal, the salesman who is present when the need arises has the best chance of getting the order; he is a constant reminder and symbol of his company. Furthermore, he will notice new or changed requirements within the plants he is servicing. He can then furnish valuable sales leads to his marketing people, so that they may be the first to make a proposal or to receive a bid invitation on a particular program.

A proper service-guaranty program can also be a sales tool when bids are being evaluated, and vendors selected. The vendor that can cite such a program may have a better chance of getting an equipment order, because the operators will know that there is a service organization ready to take up any problems that may arise after the legal warranty liability of the vendor has ended.

Incidentally, service contracts would minimize the unpaid or marginally-paid consulting and equipment troubleshooting that vendors sometimes have to provide for goodwill purposes even after the warranty has ended. This reduction could be reflected in the price charged for the service contract.

How Much of a Cost Problem?

Returning to the general topic of guaranties for a moment, we have already seen that a guaranty is a form of liquidated-damage clause. If the manufacturer fails to live up to the terms of a contract with regard to workmanship, process, performance, or cost of utilities per unit manufactured, he must make the proper repairs, modification or replacement, so that the purchaser gets 100% of what he paid for. The vendor must base the duration and extent of a guaranty on the likely cost of such a guaranty as a percentage of the contract. If the guaranty is unrealistic, a theoretically profitable contract can become a losing proposition. Therefore, the vendor must calculate what percent of his profit he is willing to risk on an actuarial basis so that he can induce someone to buy his equipment.

As for service contracts, the rates charged to individual users must be sufficient to render the over-all operation profitable on an actuarial basis, while being low enough to be attractive to users. Whether we speak of the cost of a guaranty or the profitability of a service-contract operation, both depend on the accuracy with which one can predict the cost of keeping a facility running. This introduces some problems, but they are not insoluble.

One such problem involves new processes or novel applications, where there is no history of repair or other servicing that can be used as a guide. However, it would be possible to establish the cost of the service contract with regard to those pieces of equipment that are neither novel nor being used in a novel way; then, given a service contract over a number of years, the servicer might be able to arrive at a reasonable and profitable annual average figure for all equipment.

For instance, let us consider a manufacturer who sells centrifugal machines for compressing air, ammonia and synthesis gas for an ammonia plant. As a result of increased plant sizes, it becomes possible to increase the discharge pressure of the synthesis-gas compressor. When this is done, the last stages of the compressor may be subjected to increased hazards; more particles of metal might be carried into the compressors as a result of the added lifting power of the higher-density gas. The compressor manufacturer may be able to directly use his prior experience with air and ammonia compression to establish service rates for these machines, but he may not be able to estimate increased hazards to the synthesis-gas compressor created by the new operating conditions. He may therefore want to treat the latter on a cost-plus basis for a number of years, until a fair yearly service rate can be set.

Probably one of the reasons there has not as yet been a general trend toward service contracts in the CPI is the difficulty and cost of staffing a service organization. It must be acknowledged that the ability level required of someone who is to service sophisticated processing equipment would have to be far greater than that required of oil-burner maintenance men. But the amount of money his services could save would also be greater, as already discussed. Therefore, from an actuarial viewpoint, the fact that the servicer of ammonia compressors might have to be paid twice the $7,000 salary of a home oil-burner mechanic should not be a bar to the formation of an appropriate service organization; neither should the recruiting and training expenses.

As for the risk element, a small servicing company might be financially incapable of taking the chance that many of the serviced installations may develop major problems at the same time. Therefore, the way must be led by major companies in the field, as is now being done to some degree with gas turbines and compressors.

Setting the Price

The unit price of a service contract need not be the same for all participants. A single price would benefit an operating company with poor operational control and high maintenance costs, and be less attractive to a well-run company. Therefore, a dual charge system would be more equitable. The basic charge would apply to all operating companies that accept a service contract, and would be related to capacity and number of units. Additional charges would be based on predicted and actual use of the service organization's equipment and personnel by the operating company. In this way, charges will tend to be in proportion to actual expenses, and the operating company will have an incentive to follow efficient maintenance practices.

So much for service contracts. We hope we have demonstrated their usefulness and feasibility as a supplement to equipment warranties—and have shed some light on the other aspects of these warranties that are of particular concern to the chemical process industries. ■

Engineer as a Witness Must Learn His Lesson

Product-liability suits are multiplying, and they now involve more money. A decisive factor in who wins such cases may be the engineer witness who can explain technology to laymen

JOHN KOLB, Product Engineering (A McGraw-Hill Publication)

When an accident happens today, a newly aware public looks for fault in a product—and for someone to sue. Proof of fault or of sound design and manufacture usually involves expert technical testimony, and more and more engineers find themselves in the witness box.

Such engineers are often the key witnesses on whose testimony cases are won or lost. Thay may be the designers of products charged with having caused death, injury, or property damage, or they may be expert witnesses paid by either party to study the design and engineering of a product and the circumstances of an accident and to submit reports and/or testify.

It is not uncommon in the legal arena to see engineering witnesses testifying to diametrically opposed "truths" for the two sides.

Even though the testimony of such witnesses is based on sophisticated technology, it must—to be effective —be expressed in terms that are clear to lay judges and juries. Not many engineers are articulate enough to translate technical jargon into everyday language, under pressure.

Bitter lesson. Trends in the courts as well as in Congress are forcing new standards of care on manufacturers, causing them to scrutinize their engineering designs more intensely. A few manufacturers already have learned a bitter lesson from product litigation and are applying up-to-date technology and engineering techniques to eliminate litigation-prone defects before production.

In a very few companies, such efforts already borrow from reliability, safety, and other system-effectiveness programs that aerospace engineers still are developing.

Much of today's litigation consciousness has developed in the less than three years since PRODUCT ENGINEERING published its last major article on product liability (*PE*—Aug. 1 '66, p 34). At that time, courts in only a handful of states had imposed strict liability on manufacturers of products demonstrated to have been defective and responsible for death, injury, or property loss. Today, strict liability applies in all but about 10 of the 50 states.

Personal-injury litigation, including auto cases, is a multi-billion-dollar business. Insurance carriers collect an estimated $57 million a year in premiums for product-liability policies alone. And such coverage is becoming ever more difficult to obtain, especially if it is to include defects of design and engineering.

Professional responsibility. In such a setting, it is critically necessary for product engineers—detail men as well as project leaders and administrators—to know how to handle themselves, not only in the courtroom but also in pretrial proceedings (most cases are settled out of court).

In the eyes of judges, juries, government officials and legislators, and even the general public, such tribunals are forcing engineers more than ever to face up to their professional responsibilities, not only to their employers but to users of their designs as well.

Before trial. The roles that engineers may be called upon to play in litigation are varied, and they may be played largely backstage. It may take months or even years for a newly filed suit to have its day in court. Meanwhile the engineer witness is kept busy, sometimes in baffling and frustrating ways.

First, the plaintiff's counsel may require him as an engineer for the defendant manufacturer to submit to so-called "discovery procedures," one of which is known as an "interrogatory." Under the legal power of subpoena, he must answer a sort of questionnaire, usually in his own office, that requires him to search through his records and provide all pertinent data, in writing.

This pretrial procedure is perhaps

analogous to an open-book quiz, so it may not seem formidable. But the engineer who thinks he can pick and choose the facts he will divulge may be in for a surprise. Chances are, the lawyer had an expert's analysis of the product's engineering design and manufacturing methods at his elbow when he drew up the interrogatory. And he is sure to have expert help in evaluating the product engineer's answers.

Thus, if the first answers don't satisfy him, he will come right back with an even more detailed and searching interrogatory.

Still more grueling. Another pretrial discovery procedure known as taking a deposition can be even more trying to a designer. The engineer is required to submit to questioning by opposing counsel at a time and place of the lawyer's choosing—perhaps in the engineer's own office, perhaps at the law office on the other side of the country.

The engineer must appear with all his records and testify on the basis of them. A skilled lawyer can make the taking of a deposition a grueling experience for an engineer, probing not merely into the written record of an engineering project but also into the designer's command of current technology and the motivations and philosophy he brings to his work.

Such sessions to take a deposition may go on for days, with recesses for the engineer to dig deeper into records that he may or may not have brought with him the first time.

In all discovery procedures, as in the courtroom, the engineer is entitled to have his own or his company's counsel with him, to protect and advise him. He can even have his own expert scientific or engineering consultant.

On his own side. Pretrial preparation always involves the defendant company's counsel. Lawyers must learn from the engineers all technical information that's needed for a successful defense of a product. Here the engineer must be willing to tell all, good and bad, and trust the attorneys to make a case.

John Tarpey, a New York attorney who has successfully defended a variety of engineered products, finds that sometimes his clients' engineers don't fully trust him. That can be a handicap.

"If I am to devise a suitable defense," Tarpey says, "I must know the worst aspects of a product's design before litigation begins. However, I've had men responsible for the engineering of a product at issue in a suit who couldn't bring themselves to admit to a non-colleague that there might be questionable factors in the engineering."

Tarpey has found, though, that once he can gain an engineer's confidence, the man will usually work hard to prepare what is needed for a satisfactory defense.

Taking the objective view

Engineers responsible for design of a product that's in litigation react quite differently under examination by the plaintiff's counsel. Lee S. Kreindler, who has represented victims (or their families) of several headline-making air crashes, recalls contrasting attitudes of two engineers responsible for designing components of an American jet airliner that crashed in Europe.

One engineer, Kreindler says, was frank about his own work and his company's policies and engineering practices. He gave his employer the benefit of the doubt whenever he could, but he conceded inadequacies in component design when they could be scientifically substantiated. The other engineer, who was on an administrative level, fought every question Kreindler asked, admitting as little as possible.

"I consider these men were equally loyal to their company's interests," says Kreindler, "and believed equally in the policies and practices by which its engineering staff operated. Also, as it turned out, the man who was so recalcitrant ended up having to yield as significant evidence as his more reasonable colleague did."

Odds and ends of hardware provide analogy

A propeller newly installed on a private plane tore loose in flight, and passengers were injured. They sued the propeller manufacturer, charging that the washers used had not been strong enough to protect the bolt assemblies. Admittedly, the washers did not meet Federal Aviation Agency specs, but it looked to investigators as though the bolts had bent back and forth, a condition not likely to develop merely from failure of washers.

A theoretical analysis by Frank Arnold, expert witness who was formerly an engineering professor at Stanford, supported the conclusion that the fault lay elsewhere than in the washers. Sure enough, under questioning, the mechanic who had installed the propeller admitted he had tightened the bolts by "feel" rather than using a torque wrench as specified.

Show and tell. The problem then was to convince the jury that even carefully hand-tightened bolts might not provide secure enough connections. The effect of various torque levels can be a rather abstract concept for laymen to grasp. To demonstrate the difference between tightening bolts by "feel" and by torque wrench, Arnold constructed a working model (photo and drawing) that he made up of common hardware items: eyebolts, nuts, sections of I-beam and angle iron, and a couple of ordinary spring scales of the kind you might find around the house or farm.

Using the model, Arnold then told the jury, "The flange with the wire handle welded to it represents the propeller flange on the engine crankshaft, and the 15-lb. pressure you see registered on the center spring scale is analogous

Value of candor. Craig Spangenberg (cover), a plaintiff counsel from Cleveland, says: "What we look for are signs that the truth is being hidden or distorted. So long as a witness persists in keeping back some of the facts, just so long is he open to continued attack."

Tarpey, who usually represents manufacturers, prefers to have the product designer testify—if he can express himself effectively.

"I have found," says Tarpey, "that when it appears that the engineer has tried to make a product as safe and dependable as he could, it pays to have him describe the alternative ways he might have designed the product, and then explain why he made the choice he did—complete with results of any tests he made. Judge and jury are more likely to believe the designer did his best."

Ethics. What happens, though, when an engineer is asked to testify about a product that he helped to design but that he believes to have been faulty—especially if he said so but was overruled during development? For the good of his employer, he should if possible refuse to testify, according to most experts.

The chances of his being able to hide his qualms about the design are slim, says William I. Stieglitz, veteran expert witness and safety consultant who resigned as head of the National Highway Safety Agency after a dispute over auto design standards in its early days.

According to the code of the National Society of Professional Engineers, an engineer cannot ethically perform any professional function for which his remuneration depends on the acceptability of the information he develops. Stieglitz goes even beyond this code:

"The only possible role for the engineer in litigation is as a technical expert. He must remain objective even about his own work when testifying. As soon as he starts thinking in terms of 'their' side and 'our' side, he makes himself a party to the adversary system on which our courts are based."

Coming across. Most attorneys and expert witnesses interviewed by PRODUCT ENGINEERING agree with the ideal behind Stieglitz' thinking but don't think it works out that way in practice.

Arnold Elkind, New York attorney who is chairman of the National Commission on Product Safety, observes that every engineer is the sum of his education, experience, attitudes, opinions, and associations, so he is unlikely to be totally objective. "It is important to know who pays his salary," Elkind says.

Lawyers on both sides are concerned about the impression an engineer witness will make on the jury or the judge (many product-liability cases that reach trial are heard by a judge without a jury).

Stanley Mosk, associate justice of the California Supreme Court and former state attorney general, compares the role of the engineer in the courtroom to that of the doctor. But he thinks the engineer's job may be harder. "At the present time, people seem to be more informed about medicine than they are about engineering. Also, doctors work with people all the time, so they can more easily relate to a jury."

John J. Francis, associate justice of the New Jersey Supreme Court and the author of several historic product-liability decisions, tends to agree more with Stieglitz. "The engineering witness should function as objectively in court as he would in a discussion of a technological matter with his colleagues. There can be only one truth, although the inferences may be different, depending on which side you're on."

How to behave in court

About the only thing that everyone—lawyers, judges, experienced witnesses—agrees on is that the product designer who is to testify should be neat, but not overdressed; personable, but not necessarily handsome; believable, but not overbearing; convinced he is in the right, but

to prove importance of torque wrenching

to the amount of torque that should have been put on the bolts holding the propeller in place."

Hooking a second spring scale through the handled flange, Arnold then pulled back and forth on it, maintaining readings between 0 and about 10 lb. Arnold then told the jury, "Note that the pointer on the central scale has not moved, even though the amount of pressure with which I've been pulling on this second scale is analogous to that put on the propeller bolts during normal in-flight engine power fluctuations."

QED. Next, Arnold backed off one of the eyebolts so the central spring scale read only about 7 lb., and then again pulled back and forth on the outer scale, keeping the pressures within the same range of 0 to 10 lb. Now, of course, the pointer of the central scale moved back and forth, as the two scales' springs exerted pressure on the eyebolts in alternating directions.

"This variation in pressure which you now see being imposed on the model's eyebolts is analogous to what happened to the propeller," said Arnold, "because the mechanic tightened the bolts by 'feel' rather than with a torque wrench, as he was supposed to have done."

Arnold and the manufacturer also produced the results of ground tests to back up his demonstration. These tests had been conducted with a propeller, an engine, washers, and other components identical with those involved in the crash. The jury decided in favor of the defendant, ruling that the fault lay with improper maintenance rather than with inadequate washers.

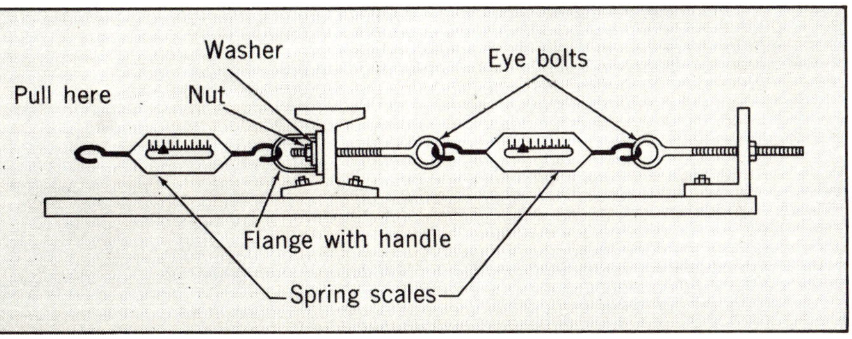

not locked in by bias; patient and calm, not easily riled by opposing counsel.

As a lawyer in the San Francisco area expresses it, "The effective engineer witness must come across with the honesty and sincerity of a clergyman and the confidence of a good doctor."

Attorneys disagree on how an engineer witness, with his expert knowledge, should comport himself.

Some lawyers recommend sticking to "yes" and "no" answers whenever possible. "Never volunteer information that isn't asked for," Michael R. Gallagher (cover), Cleveland defense attorney, says. "You may be opening doors to the opposing lawyer that he didn't know existed."

Alfred S. Julien, a New York plaintiff counsel, recalls a case in point. An engineer who persisted in ending his replies with "becauses" inadvertently revealed that he had no practical knowledge of the subject on which he was testifying but had been coached by colleagues who had done the actual design work.

Can be helpful. On the other hand, Jack Campeau, a mechanical engineer with O. Edward Kurt & Assoc., Detroit specialists in reconstructing motor vehicle accidents, suggests an expert witness can be helpful in some situations. Once, he says, the attorney who had called him to testify in behalf of his client forgot to include a vital part of a hypothetical question they had prepared before trial. "I was able to guide the lawyer back to the point where he could rephrase his question," Campeau says.

Indeed, still more help from the engineer witness is welcomed by Ronald R. Pawlak, a defense attorney who handles much of the product-liability litigation against Chrysler Corp. In the October, 1968, issue of the Defense Research Institute newsletter, he suggests that the engineer witness should watch for opportunities during cross-examination to catch the opposing attorney off-base on technical matters.

However, most attorneys and expert witnesses interviewed by PRODUCT ENGINEERING warn the engineer not to try to assume the functions of the lawyer—on either side.

Be prepared. Fortunately, the areas of agreement as to the engineer witness's responsibilities and functions exceed the areas of disagreement. Everyone experienced in product litigation stresses the importance of being prepared to discuss all aspects of the technology involved, making sure the information is consistent with latest developments.

Above all, the experts agree, the man who designed the product in question should take part in investigating its role in any accident.

The engineer witness should also remember that, under court rules, he can testify only to what he himself knows—not what he is told. The chief product-liability counsel for one of the largest U.S. manufacturers urges: "Please realize what you do know and please realize what you don't know, and don't be reluctant to make these limits known."

Picture of poise. An engineer who has major responsibilities in product litigation for a leading industrial-equipment maker describes the best engineering witness he has ever seen in action:

"He is slow, deliberate, gives off an aura of confidence. He is consistently polite to both attorneys, and no matter how long he is on the stand or how much the opposition is trying to incite him, he never forgets his manners. He even excuses himself when he steps down to make use of a visual aid. And when he shows such pictures to the jury, he displays them to each member, not just those at his end of the box. Best of all, his answers are always brief and always in terms understandable to the layman."

What can go wrong. In contrast, the same man recalls another engineer he once saw opposing his favorite witness:

"He was very verbose and never waited for the completion of the question. Then, in a mistaken attempt at thoroughness, he would explain over and over. Also, he was argumentative and tried to control the situation. When objections were made by opposing counsel, he would talk right through them.

"But what really shot him off the

If a component can be reassembled wrong,

The sports car had been having persistent brake trouble ever since its purchase, less than 11,000 miles ago. Attempted remedies had included replacement of the friction pads in the disk brakes. Then the car failed to stop one day and was involved in a collision that damaged chiefly its left front.

Underneath the car was found a piece of metal that proved to be part of the brake disk (photo A) on the left front wheel. An engineer was called in to help determine if the owner had cause for suit against either the car manufacturer or the service garage.

Telltale groove. The expert witness found that a deep groove had been cut in both faces of the disk, around the periphery of the hub (photo B). Plate thickness of the disk was originally 0.378 or 0.379 in. Thickness at the bottom of the groove ranged from 0.265 to 0.270 in. The engineer-witness concluded that the fracture of the metal had been caused by side impact.

Disassembling the brake that had the broken disk, he noted that the support plates that keep the pistons from rotating were severely worn down on the edges facing the disk (photo C). The worn edges matched the grooves.

Examining the piston assembly further, the investigator then noted that when the pistons were installed after replacement of the new friction pad, they had been rotated 90 deg. from their proper position (photo D). This maladjustment had caused them to push the support plates against the disk as well as against the friction pad, which showed a ridge on its inner face where it had been riding in the groove cut into the disk.

The expert witness found that the front right brake had also been incorrectly reassembled, and its disk had also been grooved by the support plates. The rear brakes, apparently properly assembled, were free from grooves.

The engineer's written report concludes, "In my opinion, the front wheel brake units of the subject automobile were incorrectly assembled, thereby causing defective braking action." It will be up to the attorney who engaged the expert witness to decide whether to file a suit against the manufacturer as well as the garage. Grounds might be that a design fault opened the way to misassembly by the garage that replaced the brake's friction pads.

its design may be at fault

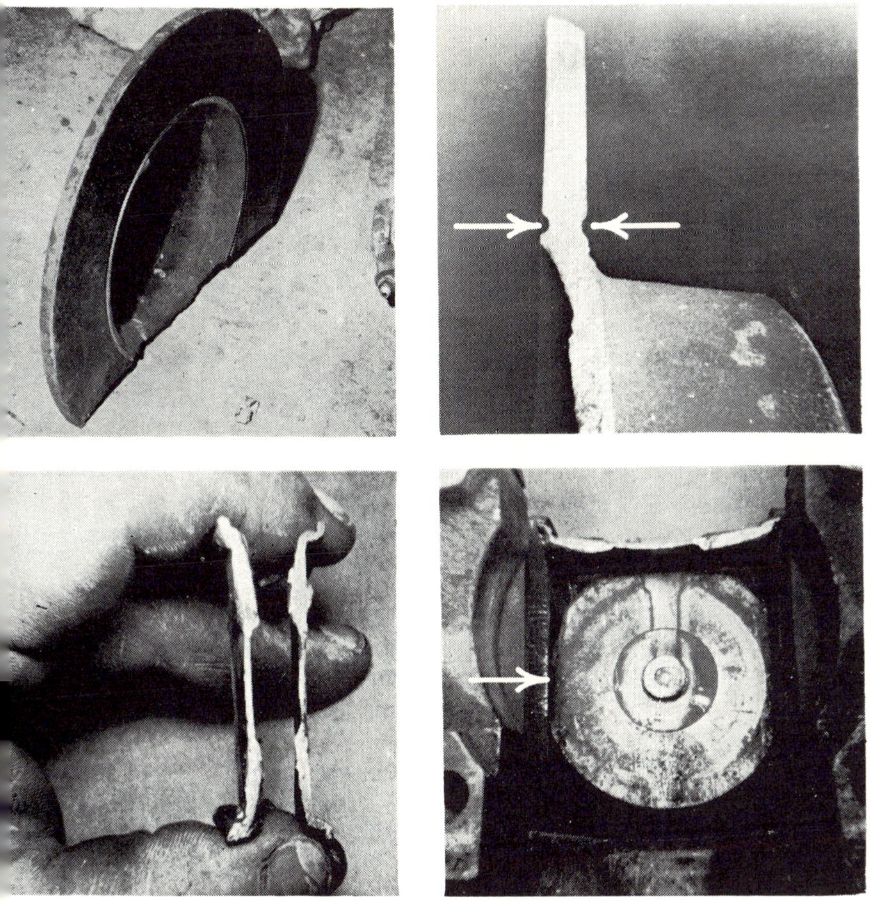

stand was that he had exaggerated his qualifications—he didn't have a degree. But don't get me wrong—I've seen a lift-truck mechanic testify as effectively as an expert."

According to Donald Dodson, engineering-department manager for Aetna Life & Casualty Co., Chicago, the very first thing for the subpoenaed engineer to remember is not to get excited. "There's plenty of time," he says. "It may be years before the case gets into court."

Dodson, like others experienced in product litigation, stresses the importance of reducing all scientific and technical concepts and data to terms that are understandable by lay juries and judges.

Like teaching a class

Paul DeGarmo, an industrial and mechanical engineering professor at the Univ. of California's Berkeley campus who has been testifying as an expert witness for 25 years, says: "When you are talking to a jury, it is much like teaching a class. You have to win their confidence. You have to adjust to them."

Dr. Nicholas Grant of MIT, an experienced expert witness who specializes in problems of brittle fracture, agrees with DeGarmo. "To convince a jury that a structure such as a bridge can stand for years and then go bad, all because certain small changes took place day after day—well, you have to get down to fundamentals; you have to use simple, simple language."

Says George Peters, an expert witness who also is the research director of the Product Safety Research Corporation, Sherman Oaks, California: "You've got to be able to talk engineering to the guy in the street to have a jury understand you."

Visual aids. Most experts agree on the value of showing, as well as telling, the judge or jury. According to Gordon Close, Chicago defense attorney: "Models of key components or sections of an engineering design aren't just nice to have in the courtroom—they're necessary. And they must be able to demonstrate the truth of the statements the engineering witness is making."

Close also notes the need for such witnesses to be skilled in the use of analogies to illuminate difficult technological concepts.

A dramatic example of both models and analogies in practice is the rig devised by Frank Arnold, a Portola Valley (Calif.) expert witness, to demonstrate the importance of a torque wrench in tightening bolts to prescribed limits (box, pp 226 & 227).

What lawyers know. Of course, the engineering witness faces the problem of explaining complex technology long before he gets to the courtroom. He must first explain it to his own side's attorneys.

The wise engineer will not automatically underestimate a lawyer's capacity to absorb and understand technical data. Lawyers have proved over and over again, both in deposition-taking and in the courtroom, that they can learn and retain for short periods a startling amount of unfamiliar and complex knowledge.

Moreover, product engineers should keep in mind that opposing counsel may have had the assistance of some of the best engineering analysts in the business.

Hypothetical questions. The favorite device of lawyers on both sides may well prove the most troublesome problem for engineers to deal with. That is the hypothetical question, often long and rambling.

According to Lawrence R. Keltner, an expert engineer witness for B. F. Goodrich Co.: "The usual form of hypothetical question will be one in which the lawyer describes a set of circumstances and then asks if I have an opinion as to why the product failed. For example, I have been asked, 'If a tire is 50% worn, how many miles could it have been run?'

"A tire expert can answer that in one of two ways: Either that it is impossible to make such an estimate since there are too many unknown factors, such as speed at which the tires are driven, size and make of car, loads carried, inflation pressure, etc., or—if the lawyer insists on an estimate—that it could have been anywhere from 1000 miles to 20,000 miles, obviously meaningless."

Keltner offers the somewhat comforting reminder that even a hypothetical question is limited to those elements already accepted as evidence. But this won't stop lawyers from seizing on one element from a hypothetical question already answered and trying to get the witness to alter his opinion or contradict himself. Under court rules, such methods are a legitimate means of examining witnesses.

Rules of evidence

That brings up the question: What are the courtroom rules under which an engineering expert is allowed to testify? They are something special, because ordinarily such testimony would be classified as "hearsay" and would thus be inadmissible.

Like that of a medical expert, the engineer's testimony is admitted under an exception to the general rule. It is permitted when it seems likely to provide what the courts call "probative" evidence—a means of trying and/or proving a fact. The expert's testimony is welcomed only for its value in helping a judge and jury better to understand the issues—a critically important fact for the witness to remember.

Law of probability. Craig Spangenberg, a trial lawyer reputed to have a superior grasp of technology, thinks it is equally important for engineers to recognize the difference between the meaning of probability in the engineering office and that in the courtroom.

"When an engineer says something is probable," says Spangenberg, "it means he thinks it is about 98½% certain. Not so in court. There, 'probable' means more like 51%, or maybe only a strong possibility."

At least one expert witness who prefers not to be named takes a dim view of the logic system of judges and lawyers in seeking truth without accepting engineering assumptions.

"Some rules of evidence are so arbitrary," he says, "that they cannot fail to frustrate the engineer witness, who will not be allowed to introduce certain proved data—maybe because the way he obtained it is deemed too abstract. For example, an engineer may not be allowed to testify as to how heavy an object was, although without this fact certain energy calculations will be impossible."

This witness warns, though, that not only are lawyers and judges convinced that theirs is the only way to

Bits and pieces prove a bottle case

Anyone who has ever dropped a glass bottle on the floor and then tried to gather up the pieces can appreciate what William Harper, a forensic physicist for the last 35 years, is up against in gathering proof of why a bottle failed. Worse yet, he must put the fragment through a number of painstaking analytical steps before he can draw the conclusions that often decide whether his attorney-client (he works most often for defense counsel) will take the case to trial or settle out of court.

Jig-saw puzzle. First the fragments must be cleaned, so the fracture edges can be clearly seen. Then, like working a jig-saw puzzle, Harper assembles the pieces back into the original shape of the bottle, numbering them as he goes and using Scotch tape to hold them together. This reassembly enables him to identify the fracture pattern drawings, which differs with each cause of bottle failure from impact, thermal shock, or internal pressure.

Next, Harper makes a "crack map," tracing the directions of the cracks and the sequence (identified by color) in which they formed when the bottle broke. The edges of the fragments along the crack lines must be examined to determine whether the leading edges are on the inside or outside. The crack maps then form the basis of the expert witness's discussions with attorneys, as well as of the exhibits he uses in court.

Interpreter of technology. Harper became involved with the interface between technology and jurisprudence while a student at Caltech, working part-time with the Pasadena police department's crime laboratory. He also serves in motor-vehicle and firearms cases as an expert witness—an unfortunate term, in his opinion. "Nobody has a corner on absolute knowledge. I look upon myself as an interpreter of technology, helping to give judges and jurors a clearer, more accurate understanding of engineering and scientific matters."

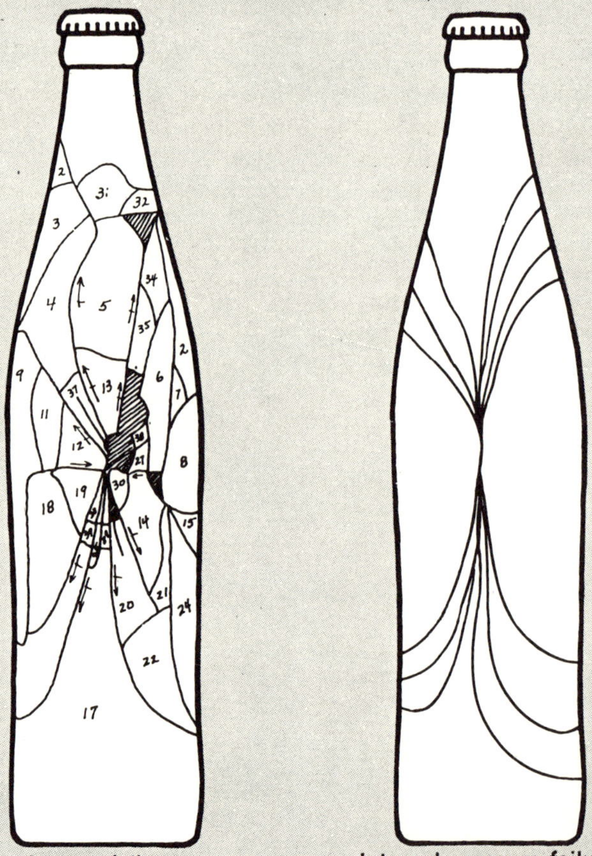

Impact failure   Internal pressure failure

conduct a trial but also that, in court, their way prevails despite what any engineer might argue.

Being an expert witness

Undoubtedly the technologists who best know their way around the courtroom are the full-time or part-time professional expert witnesses. These engineers play many roles in the preparation or trial of a product case. Some of these men, who often call themselves consultants, say a substantial number of their cases go no further than tests and analyses, and the reports they make of their findings.

Very often such reports determine a plaintiff counsel's decision that a case is not worth pursuing or a defense attorney's recommendation that his client settle out of court.

Analytical approach. The investigatory procedure followed by Gordon River, professor of mechanical engineering at Wayne State Univ., Detroit, is typical of methods used by experienced expert witnesses.

"When I get a new case," River says, "I like to tear the product down. Then I look at the drawings to see if specs are correct. Next I examine each part, first to see if any are broken and then to see if all have been made to the applicable specifications.

"Next I look for evidence that the manufacturer has improved his product. Finally I look at the product the way a layman would—I try to operate it as he would. I also study competitive devices after I have found what seems to be the faulty element. And if, in the course of studying such competitive products, I find what seems to be a flaw, I feel obligated to advise the other manufacturer, too.

"This is as important a contribution to safety as my analysis of the product under litigation."

Motivation. River makes it plain that his concern for safety is the only reason why he serves as an expert witness.

"There's not enough money in it, for the time it takes, for money to be the reason," he says. "I guess the big thing is that I am a better teacher because of my experience as an engineering witness, and I think the students who leave here are better engineers for it, too."

Isaac Stewart, of West Orange, New Jersey, who regularly testifies for either side on a wide range of products and technologies, insists that this doesn't make him a jack of all trades. "I remember one attorney who tried to discredit my testimony because I testify about so many kinds of products," he says. "The presiding judge, Irving Ben Cooper of the Southern District of New York federal court, came to my defense. He told the attorney, 'This engineering witness is not claiming to be an expert in design—only in testing.' "

Even John Tarpey, with his preference for putting the product designer himself on the witness stand to defend the innocence of a design, also often backs up such testimony with that of an independent expert

Contradiction on the witness stand

Engineer vs. engineer

When engineer confronts engineer in the courtroom, both often testify to contradictory "truths." From precisely the same facts, they draw opposite conclusions. Not long ago, a suit involved an explosion in a petroleum processing plant, allegedly because of negligence on the part of a maintenance company in performing cleaning chores. The key issue became whether or not a defective light bulb had set off the explosion, and this in turn depended on the wattage of the bulb.

An expert witness for the maintenance company testified that, judging by the leads remaining in one of the broken bulbs, the wattage could not have been more than 100 w. An expert witness for the process plant testified he also had measured the leads and had compared them with leads in new bulbs ranging from below 100 w to above 200 w. He concluded the bulb in question had been 200 w.

Judgment of Solomon. Impasse, until the judge suggested the witness go out and buy a selection of bulbs, break them in court, and measure their leads. The process plant's expert did so—50 bulbs of each wattage. His measurements of the leads confirmed his testimony. But the opposing expert witness, in his turn in front of the judge, read data from his micrometer that supported his own testimony.

The judge settled matters by stepping down from the bench, taking the micrometer, and doing the measuring himself for a third time. His measurements agreed with those of the process plant's expert witness, and the court awarded damages to the plant.

Engineer vs. lawyer

Out of the mouth of one of its own engineers came the admission that held a manufacturer liable for injury of an operator of one of its machines. The man was crushed when a foundry tumbling machine was accidentally actuated during unloading. The unit had no safety interlock to prevent operation while the drum was open. (Until quite recently, no foundry tumbling machines were so equipped.)

Also, the actuating switch, made by another of the defendants, had no ring guard around its pushbutton. An engineer for that company testified that the guard had not been provided because of its possible detrimental effect in another application of the same-model switch in a chemical process. The following exchange then took place:

Lawyer: Was your company more interested in guarding and protecting the process because of possible financial loss rather than the human being?

Engineer: We never forget the human being. In this particular case, that was the lesser of two evils.

Lawyer: Which is the lesser?

Engineer: The personal injury.

Lawyer: The personal injury did not merit the concern that the damage to the chemical process merited?

Engineer: That is correct.

witness. This tactic may overcome any feeling of the judge or jury that the designer is prejudiced in favor of his own work.

Tarpey also recommends having the designer and the expert witness go over the engineering details together. "Observing the interchange between two such men," he says, "is the best way for an attorney to probe for a design's weak points."

Rosters of experts. Tarpey says he prefers practicing engineers as expert witnesses over teachers or men from the sciences. But other attorneys disagree. The 2000 or so expert witnesses available to both plaintiffs and defendants include engineering design consultants, professors in a wide array of technologies and sciences, retired engineers and scientists, and full-time engineering witnesses.

Lengthy lists of such specialists are maintained by the Defense Research Institute, Inc. (1212 W. Wisconsin Ave., Milwaukee, Wis. 53238) and the American Trial Lawyers Assn. (Suite 1005, 6 Beacon St., Boston, Mass. 02108). Both groups are excellent sources of bibliographic data on product litigation.

Counsel for both sides are learning to beware of the very few, but still existing, expert engineering witnesses who will for a fee "arrange" the kind of testimony desired.

Evaluating the risk

Relatively few insurance companies these days provide coverage of the risks in product design and engineering, and fewer still are as effective in counseling their policyholders on how to achieve greater product safety as they are in spotting unsafe conditions in the same companies' plants and offices.

"We don't know how to assess the risk level of a new product," a spokesman for one of the largest carriers told a PRODUCT ENGINEERING editor several months ago.

From user's viewpoint. An exception is the evaluation system used by Dodson and his staff of engineers at Aetna. Says Dodson: "We begin with the finished product and work backward, beginning with a careful examination of the product itself. Then, with a picture of the intended customer in mind, we work our way up the ladder through such factors as how the product will be used, how well it meets existing standards and qualifications, how it was made, how it was designed, what its ultimate purpose is, what warranties are involved."

Dodson says his staff also tries to determine if the product's design invites misuse.

"Too often," he says, "we find there is little thinking by designers about possible problems of the user—or 'misuser.' It may be the best product in the world, but it can be shot full of potential product liability."

Forensic engineering. George Peters, the expert witness from California, is among the founders of the Systems Safety Society. He has coined a term, "forensic engineering" (*PE*—Sept. 9 '68, p 65), for the procedures that assure product safety and required performance.

The society's members have developed, for the armed services, some methods of investigating accidents, analyzing product failures, and determining fault. These methods, Peters believes, can be applied by makers of industrial and consumer goods to reverse the upward trend of product-liability litigation.

"Once such a technology is developed by industry, and the demand for it becomes evident," he suggests, "the engineering schools will develop the necessary specialized courses. Right now, about all that's available is a couple of hours touching on the subject of product liability in a commercial law course."

New awareness. Craig Spangenberg says some manufacturers have already started to meet the challenge posed by the courts—the auto makers, for example. "Tén years ago," he says, "they were doing nothing. Today they are piling up acres of crashed cars as their engineers investigate what happens when vehicles collide at various speeds."

Progress still isn't fast enough to satisfy the small band of plaintiff's counsel, such as Spangenberg, who undertook the early product-liability suits. As late as the '50s, the common law afforded small redress for victims of product defects, but the attorneys hoped financial pressure would force improvements in products and reduce the toll of deaths and injuries. In proportion to that toll, few victims have been helped.

Moreover, says Dean Thomas J. O'Toole of Boston Univ. Law School, financial impact of litigation is blunted by the fact that "the costs of paying judgments against manufacturer or seller for consumers injured by the hazardous product are borne in approximately equal measure by the federal treasury."

Still, as Richard M. Markus, a plaintiff's counsel in Cleveland, put it: "Most reasonable people will agree that product-liability litigation serves a valuable public service. Manufacturers are being forced to adopt some sensible standards."

Stieglitz, however, warns that even though auto makers and other manufacturers are learning to build safer and more effective products, it doesn't necessarily follow that the improvements will find their way to market, especially if they are considered expensive.

"That will take continuing pressure both in and outside the courtroom," Stieglitz says. "It is a struggle from which engineers and their organizations cannot afford to continue to stand aside." ■

Most non-research engineers don't think in terms of patentable inventions. Patents, however, not only improve your company's earning power but also can hasten your personal progress. This article provides guidelines for evaluating the patent potential of the ideas you come up with.

Wanted: More Patentable Ideas

RICHARD C. WILLSON, JR., *Marathon Oil Co.*

A while back, a plant manager developed an improvement in a commonly used piece of chemical apparatus that permitted savings of more than $50,000 a year at each major installation, and there are over 100 such installations in the U. S. and Europe.

A technical salesman thought of an inexpensive substitute for a widely used agricultural chemical, which put his company in a position to license the substitute widely and earn an annual royalty income in six figures.

Recently, two plant engineers became inventors: one discovered a better process for refining a common chemical raw material; the other made an improvement in process control systems, permitting more rapid response with marked increase in product quality.

These chemical process industry employees had two things in common: (1) they were inventors, (2) invention was not a major duty of any of them. Although each of these men assigned his invention to his employer, the chances are excellent that each will personally benefit from his discovery because invention, and methods of encouraging it, is a growing concern to the highest levels of management in many American corporations.

This article deals with this increasing interest in invention, tells how you can personally benefit from patenting, and briefly summarizes in checklist form the U. S. laws governing what is patentable.

Patents, the symbol of invention, are more valuable than most people realize. As an index of their durability, the percentage of patents held valid by the U. S. Court of Appeals has increased sharply in the past five years and is now at a thirty-year high, as shown in Fig. 1.

Chemical industry patents are now being put into commercial practice at a much higher rate than is popularly believed. A survey of major U. S. chemical corporations indicates that over 50% of all chemical process industry patents are put into use before they expire.[2] There is evidence that the income realized from the average U. S. patented invention may exceed $1 million.[3] U. S. corporate revenue from licensing alone is now estimated at more than $300 million per year.[4]

Behind the interest in stimulating invention is the amazing lack of increase in patentable inventions in the face of much greater corporation spending for research. A 3,000% rise in industrial research expenditures during the past 35 years has increased patentable inventions by less than 5%.[5] Fig. 2 shows this vexingly stable rate of invention in the face of skyrocketing research costs.

Mangement's awareness of the problem was indicated by William J. Sparks of Esso Research & Engineering Co., who said early this year, "The declining interest in inventive ingenuity gives me great concern. There's a growing attitude that science for science's sake alone is good and noble. Greek science had little effect on Greek life. It was not until the Industrial Revolution that invention built upon science to make it a major force in our lives. . . . There is no substitute for invention."[7]

Rewarding Invention

The increased value of inventions and the increased difficulty in getting them have recently led some companies to start formal invention award programs. The elaborateness and expense of some of these plans underscore current management concern with the stimulation of creativity.

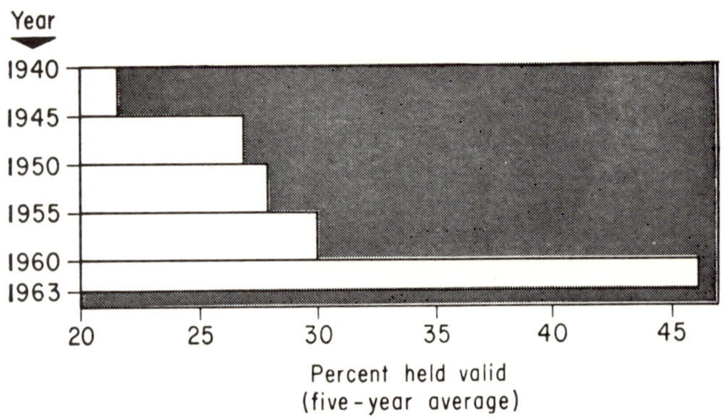

The courts have increasingly tended to uphold patents that have already issued—Fig. 1

Monetary awards (either on disclosure, or on filing of an application, or on issue of patent) have ranged from the nominal $1 to more than $500 per invention, with some companies permitting inventors to participate in licensing revenues. Non-monetary approaches have included bronze scrolls on which the inventors' names are engraved, miniature statues symbolizing invention, annual inventors' dinners with wives and company officers, and even granting of fellowships in the inventor's name to the university of his choice. One U. S. company has recently been giving Hawaiian vacations to certain employee-inventors and their wives, with all additional income tax being paid by the corporation.

IBM, which has awarded as much as $40,000 to a single inventor, reports that its invention incentive program has increased patent filings by nearly 40%.[8] Such success would seem to indicate that more corporations will adopt similar incentive programs.

Even if your company does not have a formal plan for rewarding invention, it can be a definite factor in attaining merit increases. Most corporations are careful to note in the inventor's personnel file when patents are applied for and granted, and some corporate presidents receive copies of all patents so granted. A patent application can also be useful in preventing management's forgetting just who thought of the idea as it moves upward through channels.

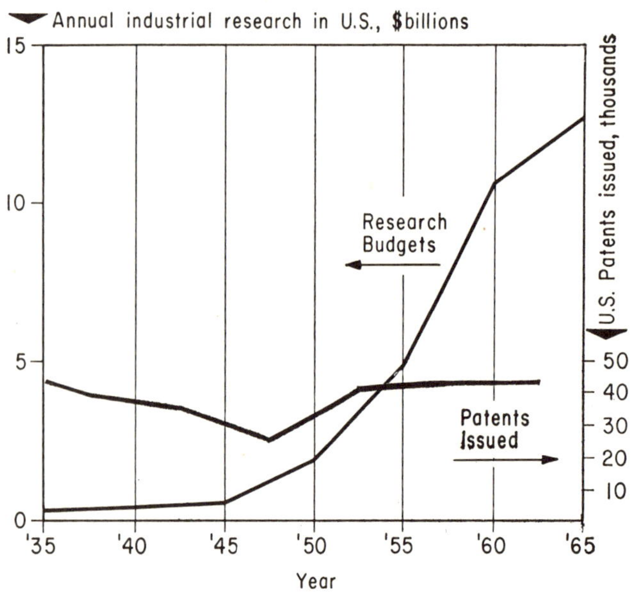

Output of patents has not kept pace with the increase in R&D spending by U.S. industry—Fig. 2

Wanted: Non-Research Inventors

Without speculating on why invention has failed to respond to increased research expenditures, or how a corporation can best foster invention, one solution seems apparent. There is a definite need to enlist the inventive resources of the non-research engineers and scientists who constitute a major portion of America's technically trained personnel. These production, sales and management people whose jobs do not involve invention as a major duty currently contribute less than 2% of the inventions in typical chemical process companies.

While their jobs do not afford as many opportunities for invention as do research positions, such employees have one great advantage over the research scientists or engineers: ideas generated by non-research personnel usually have a high degree of commercial practicality—the factor most frequently lacking in research department inventions.

How's Your Inventive Awareness?

Not only do non-research engineers often neglect to think in terms of inventions, but there is also good evidence that some useful, commercial inventions already worked out are neglected until it's too late to patent them or, worse yet, until someone else patents them. This is a particular danger with inventions that are made by large teams, where the inventive quality may not be seen by the individual contributors.

Solutions arrived at over long periods by discarding unsuccessful approaches can qualify as a patentable invention. (This was not true some years ago, but it is important to remember that standards of invention change gradually and that current law definitely permits invention by repeated trial-and-error.)

The basic requirement for patenting is inventive awareness—a conscious seeking of new solutions to existing problems, coupled with a constant automatic appraisal of one's ideas to determine whether they should be reported to the patent counsel. The importance of this awareness is proved by the fact that once a person has made his first invention disclosure, frequently he makes a second one very shortly thereafter.

The second important requirement for invention is a good knowledge of what has been done by others in the field. While a beginner in an area may occasionally make an off-beat valuable contribution, more often he will tend to "discover America" all over again by suggesting solutions that have been previously tried but of which he is not aware. As dis-

CHECKLIST FOR PATENTABILITY

FOR PRELIMINARY EVALUATION OF INVENTIONS

A. OBVIOUS?[11]

1. Was the invention obvious at the time it was made to a person having ordinary skill in the field and having knowledge of everything published in the field? — YES □ NO ○

Examples of things frequently found to be obvious:

- *a. Mere change of size or shape.*
- *b. Making a process continuous or automatic.*
- *c. Isomers of previously known compounds.*
- *d. Mere optimizing of reaction conditions or reactants.*
- *e. Aggregation of a number of known devices.*

Examples of things frequently found not to be obvious:

- *a. Any of the above where an unexpected result is obtained.*
- *b. Selection of a much better set of conditions or reagents from a large number of possibilities.*
- *c. Combining known devices in an unusual way or to obtain a new result.*

B. SECRET?[12]

1. Was the invention a secret 12 months ago? — YES ○ NO □
   - a. Was the invention in public use in the U.S. more than 12 months ago? — YES □ NO ○
   - b. Was the invention on sale in the U.S. more than 12 months ago? — YES □ NO ○
   - c. Was the invention published in a printed publication anywhere more than 12 months ago? — YES □ NO ○

Examples of things usually found to destroy secrecy:

- *a. Commercial use in a chemical plant within the U.S. without special precautions.*
- *b. Use in presence of one or two members of general public within the U.S.*
- *c. Virtually any printed publication anywhere.*
- *d. Any sale for profit in the U.S.*

Examples of things usually found not to destroy secrecy:

- *a. Experimental use necessary to perfect the invention.*

C. CLASS?[13]

1. Is the invention within one of the following classes protected by patent law? — YES ○ NO □
   - a. Process (method of making or doing something or new use for known chemical or device).
   - b. Machine (device, apparatus, etc.).
   - c. Article of manufacture (virtually any man-made item not covered by the other classes).
   - d. Composition of matter (formulations, new chemical compounds).
   - e. Or an improvement of one of these?

Examples of things usually within the protected classes:

- *a. Control systems.*
- *b. Special tickets, cards, etc.*
- *c. Petroleum derivatives.*
- *d. Most growing plants.*
- *e. Any new chemical compound.*

Examples of things usually not within the protected classes:

- *a. Basically natural products (other than plants), e.g., deveined shrimp.*
- *b. Methods of doing business.*
- *c. Scientific theories.*

D. NEW?[14]

1. Is the invention new? — YES ○ NO □

Examples of things usually found to be new:

- *a. Process of using a known formulation for a different purpose.*
- *b. Processes, formulations, apparatus, etc. that differ appreciably from those previously known.*

Examples of things usually found not to be new:

- *a. Processes or apparatus that have previously been used for an analogous purpose.*
- *b. Compositions that were made previously under non-secret conditions.*
- *c. Anything previously generally known in the U.S.*

E. USEFUL?[15]

1. Is the invention useful? — YES ○ NO □

Examples of things usually found useful:

- *a. Chemical compounds useful as intermediates for producing other compounds.*
- *b. Uneconomic processes.*

Examples of things usually found not useful:

- *a. Things entirely frivolous*
- *b. Devices that clearly will not work.*
- *c. Things detrimental to public welfare.*

F. INVENTOR[16]

1. Is the person filing the application the first inventor? — YES ○ NO □
2. Was the invention known or used by others in the U.S. before the applicant thought of it? — YES □ NO ○
3. Was the invention published in a printed publication anywhere before the applicant thought of it? — YES □ NO ○
4. Was the applicant not the first inventor for some other reason? — YES □ NO ○
   - a. Prior inventors who use secretly or abandon before successful use usually can be disregarded.

G. OTHER TESTS

1. Is the invention not patentable for some other reason? — YES □ NO ○
   - a. Did some other person file patent application in the U.S. previously? — YES □ NO ○
     - (1) Other application may make invention obvious. If identical inventions have been made, interference is declared.
   - b. Did some other person file for foreign patent more than 12 months ago? — YES □ NO ○
     - (1) Foreign patent must have issued; mere filling is not sufficient.

NOTE: Answers falling in circles tend to indicate patentability; those falling in squares should be explained fully to patent counsel.

cussed in two CHEMICAL ENGINEERING articles published in 1964, the files of issued patents can be very helpful in keeping up with what others are doing.[9]

The third most important factor in stimulating invention is a willingness to have one's ideas examined by corporate or outside patent counsel. Free flow of ideas with a minimum of formality between inventor and patent counsel is essential to adequate patent protection. There should never be any embarrassment at having submitted an unpatentable idea.

For the company's part, every inventor is entitled to a reasonably prompt, objective evaluation of the ideas he has submitted. If a corporation has inadvertently built up a cumbersome, overly formal system for reporting disclosures, the inventor should not hesitate to submit ideas to patent counsel informally for a preliminary evaluation. No patent practitioner is ever offended by a request for a preliminary opinion of patentability. When in doubt, disclose—at least informally.

The Ground Rules of Patentability

The checklist on the previous page is intended to increase patent awareness of persons who are not in day-to-day contact with patent matters. It should be used as a quick and very rough review of what is patentable under present-day standards. The check list is not an invitation to make final judgments on patentability and cannot substitute for the opinion of a technically trained lawyer.

The points in the check list are discussed in rough order of their importance. Obviousness comes first—more than half of all unpatentable inventions fail because they are obvious to persons familiar with the subject. The test is simple to state but sometimes very difficult to apply. Take a reasonably well-trained man having ordinary skill in the field of the invention and give him everything that was written or generally known in the field when the invention was made. Then ask yourself the question, "Would the invention be obvious to him without much experimentation on his own?" If not, the chances are that the invention is patentable. Note: something might be fairly obvious to a person having extremely intense knowledge of a field and still not be obvious to persons of ordinary skill and knowledge in that field.

The next most important factor in determining patentability is secrecy. The government grants the patent monopoly in exchange for divulging of new ideas; and if the ideas are already known, there can be no bargain. Secrecy must exist during at least part of the last 12 months prior to the application. This 12 months is a grace period that the U. S. government grants to permit inventors sufficient time to go through the mechanics of applying for a patent. But if the invention was used publicly, or sold, or published in any printed publication more than 12 months previously, chances of patenting are poor.

The invention must also come within the classes of things protected by patent law and generally must not have been successfully applied by others previously. Only minor usefulness is required by the patent statutes; but, in practice, corporations will require some reasonable hope of commercial payoff before approving the cost of patenting.

The remaining questions are reasonably self-explanatory. To use the check list, simply go down the list, answering each question in turn. Inventions that fall within the circles tend to indicate patentability. If all of the answers fall within circles in the check list, chances for patentability will probably be good. Any answers falling within the squares indicate points of possible difficulty, and the circumstances regarding these should be promptly reported to your patent counsel.

One final benefit from invention should be mentioned. The personal satisfaction that comes with inventorship will itself often be worth the effort involved. Vladimir Zworkin, whose electronic inventions have won him the position of honorary vice president of RCA, said last spring, ". . . one's name on a U. S. Letters Patent is an award made under a Constitutional provision and is of a great value to the inventor."

The author expresses his appreciation to Joseph C. Herring, Esq., for his suggestions and comments on this manuscript. ■

References

1. Wood, L. I., general counsel and secretary of General Electric Co., *Va. Law Review*, May 1964.
2. *Pat. Trade & Copyright Jour.*, 4, No. 1, p. 91.
3. Sanders, B. S., "Patterns of Exploitation of Patented Inventions," *Pat. Trade & Copyright Jour.*, Spring 1964.
4. *Pat. Trade & Copyright Digest*, June 1963. (Extrapolated from 1955-59 annual royalty income, estimated between $200 to $300 million.)
5. *Chem. Eng.*, June 8, 1964, p. 260.
6. *U. S. Patent Office Official Gazette* (five-year averages).
7. *Chem. Eng.*, June 8, 1964.
8. *Amer. Engineer*, July 1964, p. 12.
9. Willson, R. C., Jr., "Don't Overlook Patents," *Chem. Eng.*, Feb. 3, 1964; "How to Get Information From Patents," *Chem. Eng.*, Mar. 2, 1964.
10. *Idea*, Vol. 8, p. 33.
11. U. S. Code, Title 35, PATENTS, Sec. 103 (U. S. laws governing patents can be obtained from Supt. of Documents, Washington 25, D. C. for 50¢).
12. 35 U. S. Code, Sec. 102.
13. 35 U. S. Code, Sec. 101.
14. 35 U. S. Code, Sec. 101.
15. 35 U. S. Code, Sec. 101.

Meet the Author

Richard C. Willson, Jr., is a patent counsel for Marathon Oil Co. at its research center in Littleton, Colo. He previously was in the Patent Dept. of Pennsalt Chemicals Corp., Philadelphia. Willson is a chemical engineer, a graduate of the Temple University School of Law, and a member of the American Bar Assn., American Patent Law Assn. and the National Society of Professional Engineers.

5

Exploiting the Techniques of Economic Analysis and Operations Research

One of the strongest trends in modern management, as well as in engineering, involves a more quantitative approach to decision-making. This means translating the advantages and disadvantages of various courses of action into figures or models relating to costs,* revenues, profitability, time factors, effect on other activities, etc. Here is how this is done – via investment analysis, sensitivity charts, operations research (including linear programming), critical-path methods, and R&D project evaluations.

*Readers interested in a more comprehensive coverage of costs and profitability are referred to a companion volume: "Modern Cost Engineering Techniques," H. Popper (ed.), McGraw-Hill, 1970.

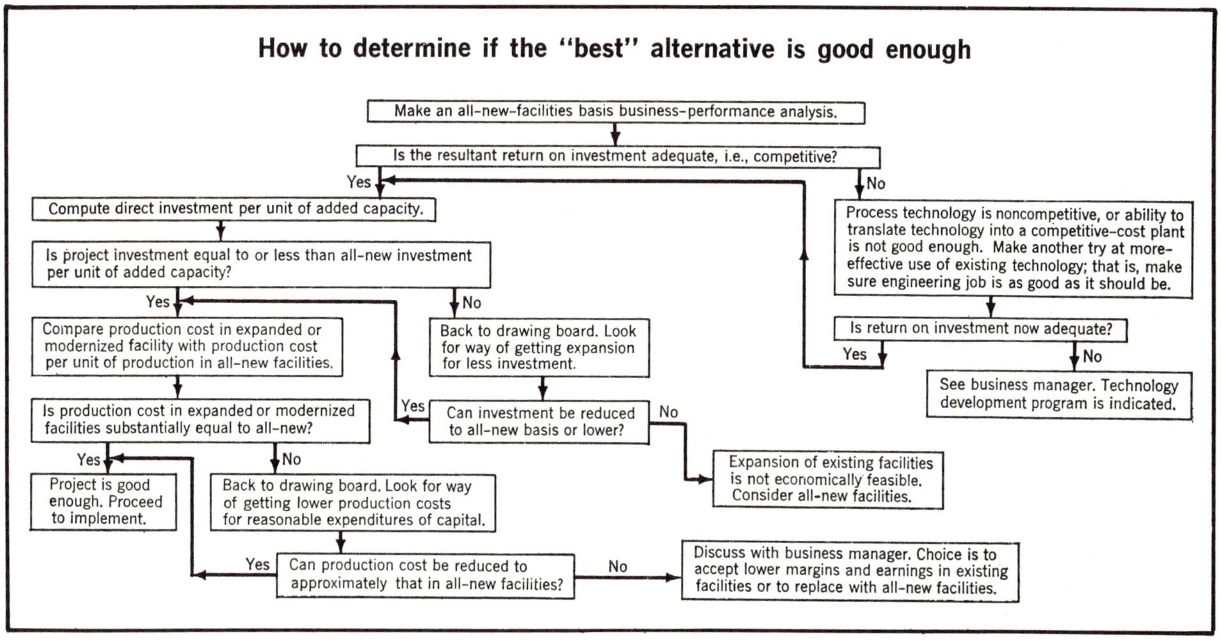

A "Business" Approach to Capital-Budgeting Decisions

You've submitted a project proposal having a high rate of return to management, but it is rejected. Why? Perhaps because the project may be in conflict with business realities. Here is a look at how business management views investment alternatives.

JAMES A. CHAMPLEY, Union Carbide Corp.

Conventional approaches to capital-investment decisions assume that the analyst is selecting a "portfolio" of projects in which to invest limited funds. Various methods are used to rank projects based on some criteria such as rate of return, present worth, profitability index, etc. It often happens, however, that a project with a high discounted-cash-flow rate of return is rejected by management. Why and how can this happen?

Basically, a seemingly promising project can be rejected because of the way management looks at investment opportunities. Management usually views this situation as one of selecting the portfolio of businesses in which to invest resources (as opposed to selecting the portfolio of projects in which to invest funds). Thus, management allocates resources to businesses rather than to projects, and management's prime concern is the overall performance of businesses as they are accounted and reported, rather than of projects as they are normally evaluated.

Adapt Management View

This conflict can be resolved by adapting capital budgeting to a business-oriented environment. The purpose of this article is to present the business approach to capital-budgeting decisions.

Question: When are 30% and 40% return on investment projects not good enough?

Answer: When they do not result in the ability to perform competitively in the business.

Corporations are like baseball players. They compare their batting and fielding averages (growth in earnings, return on investment, market share) against other players in the same league.

If the other players are doing better, our favorite baseball hero is dissatisfied. If other corporations are doing better, our corporation is also dissatisfied.

Our goal, whether we be baseball players or corporations, is to perform at least as well as any other player or corporation in the same league.

Let us see how this desire to perform competitively in a business influences the capital-budgeting process.

Of course, the examples that follow are not intended to be guidelines to the sometimes very complex analyses required for investment decisions. Their purpose is to show how a chosen investment alternative can be evaluated to see whether it is good enough to satisfy the goals of competitive business performance.

An Expansion Project

Sales forecasts for next year exceed production capacity. The production manager believes that his plant output can be increased by about 10% with a moderate investment.

The engineering manager is asked to evaluate methods for increasing capacity. An engineering study confirms the feasibility of an incremental expansion of the present plant.

Financial analyses of the proposed expansion project are made and the results are favorable. A capital-authorization request is drafted and submitted to the division head who has profit responsibility for the business. A summary of the financial analyses supporting the request is shown in Table I.

Let us now look at how the profit-responsible manager, i.e., the business manager, views this draft of the capital-authorization request.

Performance of the business has recently deteriorated because of sales price reductions forced by competition. This has resulted in lower margins, reduced profits, and a lower return on investment. Consequently, the business manager has come under executive pressure to improve the profit performance of his business, as measured by growth in profits before taxes and return on investment, ROI. (Return on investment is measured as the percentage ratio of the profit before tax to the total investment assigned to the business measured at original acquisition cost or gross book value.)

The business manager notes that the project promises a handsome 30% return on investment. The project will be "salable" and there should be little difficulty in obtaining a capital-budget authorization. However, his first concern is understandably his job performance, which is related to the profit performance of his business as measured on the business per-

Capital authorization request for expanding plant facilities—Table I

Annual Items		Thousands of Dollars
Sales increase		5,000
Production cost	3,000	
Freight, warehousing, etc.	500	
Total direct costs		3,500
Margin (sales minus direct costs)		1,500
Overhead costs		400
Operating income (margin minus overhead costs)		1,100
Depreciation (on fixed investment)		200
Profit before tax		900
Investment Items		
New, this project	1,000	
Share of supporting facilities (utilities, laboratory, office, etc.)	1,000	
Working capital increase	1,000	
Total additional investments		3,000
Return on investment (ROI) = (900)(100)/3,000 = 30%		

Ground rules and assumptions for Table I:

1. Evaluation criterion is return on investment (ROI), defined as the average annual increase in profit before taxes as a percentage of total added investment.
2. Official project cutoff rate is 20%.
3. Depreciation is straight-line basis, 10-yr. life.
4. Overhead costs are incremental, based on an assumed rate of overhead increase equal to ⅔ the rate of sales increase.
5. Working capital increase consists of added inventory, increased accounts receivable, and increased cash.

Business performance statement for expanding facilities—Table II

	Now		This Project		After This Project	
	Total[a]	Unit[b]	Total[a]	Unit[b]	Total[a]	Unit[b]
Sales	50,000	1.00	5,000	1.00	55,000	1.00
Production cost	30,000	0.60	3,000	0.60	33,000	0.60
Freight, warehousing, etc.	5,000	0.10	500	0.10	5,500	0.10
Total direct costs	35,000	0.70	3,500	0.70	38,500	0.70
Margin	15,000	0.30	1,500	0.30	16,500	0.30
Overhead	6,000	0.12	400	0.08	6,400	0.116
Operating income	9,000	0.18	1,100	0.22	10,100	0.184
Depreciation	5,000	0.10	200	0.04	5,200	0.095
Profit before tax	4,000	0.08	900	0.18	4,900	0.089
Investment	**Total[c]**	**Unit[d]**	**Total[c]**	**Unit[d]**	**Total[c]**	**Unit[d]**
Direct	40,000	0.80	1,000	0.20	41,000	0.75
Share of supporting	10,000	0.20	1,000	0.20	11,000	0.20
Working capital	10,000	0.20	1,000	0.20	11,000	0.20
Total	60,000	1.20	3,000	0.60	63,000	1.15
Return on investment	6.7%		30%		7.8%	

a. Thousands of dollars/yr.
b. Dollars/unit of production.
c. Thousands of dollars.
d. Dollars per unit of production/yr.

formance statement. His first reaction to the project, therefore, is: "What will this project do for my business or profit center?" The data in Table II give the answer to his question: After the project is completed, the business will show a return on investment of 7.8%.

The business manager judges a 7.8% return to be noncompetitive and, therefore, unacceptable. Overall return for the industry is about 10%, and this business should be higher than the industry average because of its specialized nature. Other businesses in his organization with similar characteristics show returns of 12 to 25%.

The business manager also is aware that other companies are building new production facilities for the same product. His company would not commit itself to major plant expansions at such a return, and likewise, neither would his competitors. Therefore, at least some of the competition must have the ability to get a higher profit margin, or use less investment, or both.

All-New Facilities

The problem becomes one of how to bring the performance of the business up to an acceptable level of return on investment. But before this can be solved, it is essential to know where the business stands today —what are the current capabilities in the business?

The business performance statement shows the historical investment to date in present facilities and today's costs of operating facilities designed and installed in the past. It does not show what today's performance would have been if today's knowhow had been used in the design of the facilities.

The business manager, therefore, asks his financial analyst, with the help of the engineering group, to reconstruct the business performance statement, using production and investment costs that could be attained in all-new facilities based on today's technology. Results are shown in Table III, together with the "after this project" column of Table II for comparison.

The result of the all-new facilities analysis is a pleasant surprise since a 20% return on investment is considered satisfactory by the business manager. It indicates that capabilities in the business are competitive.

What is the reason for the poor performance of the present facilities after the expansion project? A comparison of the two "unit" columns in Table III shows that:

- Unit production cost of the expanded facilities is higher than production cost in the all-new facilities ($0.60 vs. $0.50).
- Unit direct investment in the expanded facilities is higher than all-new-facilities investment ($0.75 vs. $0.60).

Historical investment is a sunk cost and it cannot be changed by anything we do now. Unless there is a reduction in production cost, there is nothing to be gained by scrapping old facilities and building new ones just to reduce book investment so as to increase return on investment.

However, production cost, unlike investment cost, is not a sunk cost. It is something that is continually committed anew by the continuing decision to remain in business with the current facility.

If the all-new facilities were to be operated at a production cost equal to that in the expanded facilities ($0.60/unit), profit before tax would be $0.10/unit rather than $0.20/unit, and the return on investment would be 10% rather than 20%. Thus, the earning power of the business would be halved.

A corporation, if it is to perform competitively, cannot afford to allocate its resources, particularly its scarce management resources, at 50% of the efficiency of which it is capable and at 50% of the rate required to be competitive.

Reducing Costs

Therefore, costs of production in the expanded facilities must be reduced. There are two possible ways:

- Modernize existing facilities.
- Build all-new replacement facilities.

If production costs in the existing facilities can be reduced to near the production costs in all-new facilities with the expenditure of a moderate amount of new investment, then that is the preferred path. All-new facilities and the use of new technology generally

Comparing all-new plant—Table III

	All-New-Facilities Basis		"After This Project" From Table II	
	Total[a]	Unit[b]	Total[a]	Unit[b]
Sales	55,000	1.00	55,000	1.00
Production cost	27,500	0.50	33,000	0.60
Freight, warehousing, etc.	5,500	0.10	5,500	0.10
Total direct costs	33,000	0.60	38,500	0.70
Margin	22,000	0.40	16,500	0.30
Overhead	6,400	0.116	6,400	0.116
Operating income	15,600	0.284	10,100	0.184
Depreciation	4,400	0.08	5,200	0.095
Profit before tax	11,200	0.204	4,900	0.089
Investment Items	**Total[c]**	**Unit[d]**	**Total[c]**	**Unit[d]**
Direct	33,000	0.60	41,000	0.75
Share of supporting	11,000	0.20	11,000	0.20
Working capital	11,000	0.20	11,000	0.20
Total	55,000	1.00	63,000	1.15
Return on investment	20%		7.8%	

a. Thousands of dollars/yr.
b. Dollars/unit of production.
c. Thousands of dollars.
d. Dollars per unit of production/yr.

involve greater startup delays and costs, and increased demands on skilled manpower when compared to modernization. However, if production costs in the existing facilities cannot be reduced, or if the new investment required to do so is excessive, the only viable alternative is to build replacement facilities.

If it happens that modernization is not feasible and replacement is the choice, conventional capital-budgeting practice would be to treat the replacement as a cost-saving project compared to continued operation of existing facilities. However, this conventional approach will lead to conflicting results, as will be seen below.

Normally, replacement would be compared to continued operation of existing facilities; but there is a way of expanding existing facilities. Therefore, replacement is compared to the alternative of continuing to run the existing facilities after they have been expanded. On this basis, a cost-saving project evaluation yields in thousands of dollars:

Increased investment—$33,000 for replacement minus $1,000 for expansion	=	$32,000
Increased operating income—$15,600 minus $10,100 (Table III)	=	$5,500
Increased depreciation—10% of $32,000	=	$3,200
Cost saving	=	$2,300
Incremental ROI—(2,300) (100)/(32,000)	=	7.2%

The 7.2% ROI for replacement vs. continued operation of existing facilities is less than the official cutoff rate of 20% and, therefore, replacement would not seem to be the thing to do.

But there is an apparent dilemma here. Performance of the expanded facility is inadequate and unacceptable, yet the alternative of replacement vs. continued operation of existing facilities shows an inadequate return by the usual standards.

The cause of the apparent dilemma is that the assumptions underlying a differential analysis of replacement vs. continued operation of existing facilities are invalid. These assumptions are:

• Existing facilities can be operated for at least 10 years without any new commitment of funds other than the $1 million for the proposed expansion project.

This is not a realistic assumption. Continued operation of the old facilities will require a series of replacements and improvements. Because competitors are continually improving performance, a continuing upgrading program is required just to stay even. Doing this in obsolete facilities is more costly and less efficient than in modern facilities.

• There will be no difference in performance other than in production costs and investments in direct facilities.

Inadequate margins and low returns have a debilitating effect on all aspects of business performance; the market position of the business will deteriorate because competitors will have more "maneuvering room" with larger margins; organization morale will suffer, management efficiency will be lowered, and the competitive position will deteriorate further.

• Capital funds spent for replacement here at 7.2% ROI will cause other projects with returns of 20% ROI or better to be foregone.

Most companies are not fund limited but people limited. Foregoing replacement but staying in business with the existing facilities does not free people to manage other businesses and investments.

• Staying in a business at a continuing inadequate performance is a viable long-term alternative.

If a business is not to be given the resources to be competitive (which in this case means the investment for replacement), the only viable alternative is to withdraw. Otherwise resources of the organization are used up for lower profits than the competition is getting and the "batting average" becomes poor.*

Realistically, the alternative to replacement is to disengage from the business, and the return on investment for replacement vs. withdrawal from the business is 20%.

Return on cost-saving project—Table IV

Annual Items	Thousands of Dollars
Cost saving	500
Depreciation	100
Profit before tax	400
Investment Items	
New, this project	900
Share of supporting facilities	100
Total additional investment	1,000
Return on investment (ROI) = (400) (100)/1,000 = 40%	

Ground rules and assumptions for Table IV:

1. Evaluation criterion is return on investment (ROI), defined as the average annual increase in profit before taxes as a percentage of the total added investment.
2. Official project cutoff rate is 20%.
3. Depreciation is straight-line basis, 10-yr. life.
4. No change in overhead costs.
5. Share of supporting investment is for added utilities usage.

Conclusions of the Business Manager

• The business is a good one, and the company has the capabilities to perform competitively.

• The expansion project, although yielding 30% ROI, will not result in adequate performance for the business.

• The draft of the proposed capital-appropriation

* It may be necessary, however, to continue operation on a short-term basis, until new technology is developed.

request should be returned with a request to develop a new capital-appropriation request for an economically feasible way of modernizing the facilities in conjunction with the proposed expansion so as to produce lower production costs. If modernization is not feasible, then draft a request for an all-new-facilities project sized to meet market demand.

A Cost-Saving Project

Let us consider another typical problem. The production manager is under pressure to reduce costs. He asks the engineering manager to help him with a cost-saving program.

The process engineer assigned to the job soon sees that the efficiency of using a raw material can be greatly improved by adding a residue-recovery processing unit. The engineer makes design studies and then profitability analyses that show a good cost-saving opportunity. For a new investment of $1 million, the saving will be $500,000/yr.

A capital-authorization request is submitted to the division head who has profit responsibility for the business. A summary of the financial analyses supporting the capital-authorization request is shown in Table IV.

Let us again look at how the business manager views this request for capital expenditure. As before, he is primarily concerned with the profit performance of his business as shown on the business performance statement: "What will this project do for my business?" The data in Table V provide the answer to this question.

After the project is completed, the business will show a return on investment of 7.2%. The business manager judges a 7.2% return to be non-competitive and, therefore, unacceptable. The reasons for this judgment are the same as those outlined before in discussing the expansion project.

All-New Facilities

Again, the problem is how to bring the performance of the business up to an acceptable level of return on investment. But first, what are the current capabilities in the business?

The business manager asks his financial analyst, with the assistance of engineering, to reconstruct the business performance statement, using production and investment costs that could be attained in all-new facilities based on today's technology (see Table VI).

Results of the all-new-facilities analysis are disappointing. Return on investment of 11.4% is not competitive since it is below the expected 20% return. Therefore, it appears that we do not have competitive capabilities in the business.

This judgment is confirmed by the just-announced action of Competitor A in withdrawing Product X from the market–X is a byproduct in these facilities. Competitor A has been the aggressive price leader since his new plant came onstream. It appears that

Business performance statement for cost-saving project—Table V

	Now		This Project		After This Project	
Annual Items	Total[a]	Unit[b]	Total[a]	Unit[b]	Total[a]	Unit[b]
Sales	50,000	1.00	....	.....	50,000	1.00
Production costs	30,000	0.60	−500	−0.01	29,500	0.59
Freight, warehousing, etc.	5,000	0.10	.....	.....	5,000	0.10
Total direct costs	35,000	0.70	−500	−0.01	34,500	0.69
Margin	15,000	0.30	500	0.01	15,500	0.31
Overhead	6,000	0.12	.....	.....	6,000	0.12
Operating income	9,000	0.18	500	0.01	9,500	0.19
Depreciation	5,000	0.10	100	0.002	5,100	0.102
Profit before tax	4,000	0.08	400	0.008	4,400	0.088
Investment Items	Total[c]	Unit[d]	Total[c]	Unit[d]	Total[c]	Unit[d]
Direct	40,000	0.80	900	0.018	40,900	0.818
Share of supporting	10,000	0.20	100	0.002	10,100	0.202
Working capital	10,000	0.20	.....	.....	10,000	0.20
Total	60,000	1.20	1,000	0.02	61,000	1.22
Return on investment	6.7%		40%		7.2%	

a. Thousands of dollars/yr.
b. Dollars/unit of production.
c. Thousands of dollars.
d. Dollars per unit of production/yr.

Return on all-new-facilities unit—Table VI

	All-New-Facilities Basis	
Annual Items	Total[a]	Unit[b]
Sales	50,000	1.00
Production cost	28,000	0.56
Freight, warehousing, etc	5,000	0.10
Total direct costs	33,000	0.66
Margin	17,000	0.34
Overhead	6,000	0.12
Operating income	11,000	0.22
Depreciation	4,600	0.092
Profit before tax	6,400	0.128
Investment Items	Total[c]	Unit[d]
Direct	36,000	0.72
Share of supporting	10,000	0.20
Working capital	10,000	0.20
Total	56,000	1.12
Return on investment	11.4%	

a. Thousands of dollars/yr.
b. Dollars/unit of production.
c. Thousands of dollars.
d. Dollars per unit of production/yr.

he has found a way of improving his business economics with a new process that does not involve X as a byproduct.

This indicates a need for a process-development program to improve process economics. Under such circumstances, a cost-saving project is no longer attractive because it will probably be made obsolete by the new process.

A return of 40% for the cost-saving project assumes a 10-yr. economic life. If the economic life is less than 10 yr., the return is less. Inspection of Table IV shows that the project requires 2 years or more just to break even (2 years of cost saving at $500,000/yr. is required to pay back the initial investment of $1 million). In 2 years, an improved process will probably be ready to operate.

Conclusions of the Business Manager

- The business is a good one for the company, but the company has temporarily lost its competitive capabilities. Competition appears to have developed a superior process technology.
- A research-and-development program to improve process technology should be started at once.
- The capital appropriation request should be returned with a request to stop all work on the cost-saving project.

Lessons for Functional Management

What can line people, and particularly engineers who develop capital authorization requests and make financial analyses of capital projects, learn from these examples? Engineers should remember that:

1. A business manager looks at overall business performance and not just project results. A project is a means to a business end and not an end in itself.

2. A business manager is measured by the realized profits of his business in the way the internal financial scorekeeping system reports them, and not by the promised performance of projects: (a) Project evaluation methods that involve criteria different than the ones used in business performance measurement systems, such as discounted-cash flow returns, net present worth, and profitability indexes, are not too helpful to the business manager. What he wants is the effect of the project on his business as it is accounted and as he is measured. (b) Project evaluations are based on differences between alternatives (one alternative may be to do nothing) while business performance is measured on the overall results of the chosen alternative. Project results are, therefore, not always representative of business results; a 40%-return project does not mean a 40%-return business.

3. The business manager is concerned not only that the project return is good enough to "sell," but also that it results in giving him a competitive overall business performance. This means that each project should:

- Be the best choice among the available alternatives, such as best process, best location, best capacity, etc.
- Promise results above the budget cutoff criterion; for example, a return better than the cutoff return.
- Be good enough to result in competitive business performance; that is, result in a facility that has competitive costs and investments.

4. Evaluations between alternatives, if they are to be meaningful to the business manager, must be between real alternatives and not hypothetical ones.

"Do-nothing" and "maintain status quo" are not real alternatives. A decision to be in and continue in a business implies a commitment to become and to remain competitive in the business. Keeping competitive requires attention to share of the market, matching competitive customer service and product quality, and maintaining a competitive cost and investment position. In a dynamic growing market where competition is continuously improving customer service and process technology, standing still is going backward, and the only real alternative to moving ahead in step with competition is to withdraw.

Recommendations

Specifically, what can the engineer do to enhance his professional stature with the business management of his company? The following practices are recommended:

- Whenever a project proposal is submitted to a business manager for his consideration, translate the project results into business language. Show the effect of going ahead with the project on business performance.
- Determine beforehand whether current business performance is considered acceptable. If it is not, evaluate current business capabilities by making a business-performance analysis on an all-new-facilities basis. This analysis will assist the business manager in selecting the proper development and facility strategies to improve business performance.
- Do not be satisfied with simply selecting the best project alternative. Be sure the "best" alternative to be proposed is good enough to give competitive business performance. The approach outlined in the figure on p. 239 will help answer: Is the best alternative good enough? ■

Meet the Author

James A. Champley is a capital-projects consultant on the corporate staff of Union Carbide Corp., 270 Park Ave., New York, N. Y. 10017. He is responsible for corporate capital-budgeting procedures and is a consultant to operating divisions on business and project planning and analyses. Mr. Champley has also been involved in production, process design and project analysis with the Chemicals Div. of Union Carbide. He holds a B.S. (Ch.E.) from Notre Dame (1941).

RATE OF RETURN

PAYOUT TIME

FULL-LIFE PERFORMANCE

CAPITALIZED COSTS

Comparing Techniques For Appraising Project Alternatives

No single economic appraisal method for capital projects is "best" for all situations. Here is a comparison of important techniques to help you choose the most suitable for a particular project.

HOWARD H. STREET,
E. I. du Pont de Nemours & Co.

The problem of economically appraising alternate methods for completing an engineering project is faced by all firms. How successful a firm is or has been is largely determined by effectiveness in choosing the most profitable alternative.

In an economy where profits tend to be squeezed by costs and competition is keen, a good set of tools for appraising proposals for capital expenditure is an especially valuable asset. Therefore, chemical engineers must develop the ability to recognize the most economic approach to projects, if their firms are to remain highly competitive in today's market.

Recent literature on economic evaluation has been concerned with presenting the merits of a particular appraisal method. However, it is doubtful that there is one method of analysis that serves equally well in all capital investment cases. We will compare some important techniques and leave it up to the reader to draw his own conclusions as to what technique best fits a particular investment situation.

Four accepted yardsticks of economic appraisal of project alternatives used by manufacturing industries are as follows:*

- Rate of return on investment.
- Investment payback period.
- Full-life performance using discounted dollars.
- Capitalized costs.

To determine the order of economic desirability for alternatives to a proposed project, three quantities must be established: (1) amount of investment required, (2) anticipated profit and/or cost schedule resulting from the project, and (3) anticipated length of life. In all the examples that follow, it has been assumed that these quantities have been accurately developed. Accurate forecasting of the above values is imperative for effectively choosing the most economic alternative.

Three Routes to Rate of Return

Each capital project must, in itself, produce profit if the company's success is to be maintained. However, maximization of profit alone is not sufficient as a goal. What is really desired is to pursue some level of profit efficiency from each dollar invested. The efficiency ratio that relates earnings to investment is known as rate of return on capital investment. Return on investment is an important method for measuring the relative desirability of alternative investments or project opportunities.

There are many ways to make such a calculation. Three of these methods are as follows:

- Return on original investment. Computation by this method involves using average income or savings from the project divided by the original, total capital outlay.
- Return on average investment. This form accounts for the fact that the original investment depreciates over the project's life. Average annual income is divided by half the original investment or the mid-point between original cost and salvage value.[1]
- Discounted cash flow rate of return on investment. Calculation by this method is based on the investment unreturned each year over the economic life of the project. This technique requires a trial-and-error solution to find the rate of return at which yearly cash intakes will produce a given profit and simultaneously reduce the original capital investment to zero over the project life. A more detailed explanation will follow.

Return on original investment and return on average investment are "point functions." They only indicate earnings rate at any one time or for a particular year. Neither calculation makes allowance for variation of project income over the life of the project. Each is based on the assumption that savings or income from the proposed project will compose a flat time vs. earnings curve. When this is the situation, a single point on the curve will adequately describe the project alternative being considered. However, most project evaluations are not that simple.

For instance, earnings may rise over the project

*The analytical methods presented in this paper are derived from the cited references and therefore are not necessarily the procedures used by Du Pont for project appraisal.

life due to increased sales, or they may decline due to the burden of heavy maintenance costs or a decrease in product price. In these situations, original return on investment and average return are inadequate measures of the true profitability of project alternatives.

Discounted Cash Flow for Profits

To apply the return on investment concept to cases involving variable income, it's necessary to base calculations on the net investment outstanding at the end of each year. Application of the concept in this manner is called the discounted cash flow rate of return. To understand what this means in terms of calculation technique, consider the following project proposal:

- Capital investment: $100,000.
- Total forecasted earnings: $200,000.
- Life of project: 10 years.

Forecasted total earnings of $200,000 will flow to the project over the ten-year life according to the profit schedule presented in Table I.

For simplicity, assume that the company loans the $100,000 of capital to the project for the duration of the project life at an interest rate of 14% compounded annually. During the first year, cash return to the project is $16,000. Fourteen thousand or 14% of the original capital invested is committed to payment of interest, while the remaining $2,000 will reduce the investment balance outstanding at the beginning of the second year to $98,000. During the second year, the project is forecasted to earn $18,000. Thirteen thousand or 14% of the $98,000 outstanding investment will go the company as interest payment, and the remaining $4,300 will go to amortize the outstanding investment. If these computations are carried out over the ten-year economic life of the project, it will be found that a rate of 14% on the yearly outstanding investment plus the amount amortized each year will just equal the sum of the yearly cash intakes.

Obviously the company does not set or charge an interest rate on the capital allocated to the project. This explanation is merely presented for the purpose of simplification. Actually the calculations must be repeated substituting trial "interest" rates until the one (14% in the example) is found at which the yearly cash intakes will produce a return equal to the trial rate and reduce the capital investment to zero at the end of the economic life. The resulting rate,

Mechanics of discounted cash-flow calculations—Table I

Year	Forecasted* Cash Flow, $	Interest† Returned, $	Capital Returned, $	Balance of Investment Unreturned, $
0	(100,000)			(100,000)
1	16,000	14,000	2,000	98,000
2	18,000	13,700	4,300	93,700
3	18,000	13,070	4,930	88,700
4	20,000	12,375	7,625	81,075
5	18,000	11,300	6,700	74,375
6	20,000	10,375	9,625	64,750
7	21,000	9,050	11,950	52,800
8	20,000	7,400	12,600	40,200
9	24,000	5,700	18,300	21,900
10	25,000	3,100	21,900	0

* Earnings after taxes.
† Based on interest rate of 14%.

Computing discounted cash flow rate of return—Table II

Year	Forecasted* Cash Flow,$	15% Trial Discount Factors	15% Trial Present Value,$	12% Trial Discount Factors	12% Trial Present Value,$	14% Trial Discount Factors	14% Trial Present Value,$
0	(100,000)	1.000	(100,000)	1.000	(100,000)	1.000	(100,000)
1	16,000	0.869	13,900	0.893	14,300	0.877	14,020
2	18,000	0.755	13,600	0.797	14,350	0.769	13,850
3	18,000	0.656	11,800	0.710	12,790	0.675	12,150
4	20,000	0.571	11,400	0.635	12,700	0.593	11,870
5	18,000	0.498	8,950	0.567	10,200	0.520	9,360
6	20,000	0.433	8,650	0.506	10,110	0.456	9,120
7	21,000	0.376	7,900	0.453	9,510	0.400	8,400
8	20,000	0.327	6,550	0.404	8,080	0.351	7,030
9	24,000	0.284	6,820	0.370	8,880	0.308	7,400
10	25,000	0.247	6,180	0.323	8,070	0.270	6,750
Total			95,760		108,990		99,950

* Earnings after taxes.

then, represents the true profitability or profit efficiency of the project proposal.

Use Discount Factors

This method of calculation, needless to say, can be very tedious and time-consuming. Fortunately, there is a short-cut method for determining the discounted cash flow rate of return. This involves applying a discount factor to the annual net income, and summing as shown in Table II. The discount factor is found from the following equation:

$$d_n = \frac{1}{(1+i)^n}$$

i = rate of return.
n = year of project life being considered.
d_n = discount factor.

This relationship represents the present value of one dollar based on an interest rate of i for n years. Actually, if amount d_n were invested for n years at a rate of return equal to i, the investor would have one dollar. The discounted cash flow rate of return, then, is computed by finding the rate whose discount factors will discount the annual net earnings down to a present value equal to the original investment of \$100,000. Table II illustrates this trial-and-error process applied to a proposed project's forcecasted income schedule. The rate that satisfies the trial-and-error process, then, is the discounted cash flow rate of return. In this case, it is 14%.

The major advantage associated with discounted cash flow rate of return is the allowance it makes for the difference in time at which project alternatives would generate their income. This method accounts for the fact that money develops value over a period of time when invested in the economy. Thus, a project alternative that produces a high return in its early years would tend to be more economical than one that generates an initial low return that would later increase. This occurs because the dollars returned early would be available for other investment opportunities at some rate.

Work An Example

Fig. 1 illustrates three alternatives to a single project objective. Alternative B is a steady earner, while alternative C has a declining income, and alternative A increases in earning ability over the economic life of the project. Rate of return on the original investment, and rate of return on average investment, fail to discriminate among the three alternatives. But discounted cash flow rate of return resolves the order of investment desirability. This aspect is very important because alternate approaches to a project goal often have divergent income patterns.

Rate of return is an outstanding tool for project profitability evaluation. Because of its simplicity and versatility, it has gained wide acceptance in the chemical industry. Of the methods presented, the discounted cash flow rate of return is recommended:

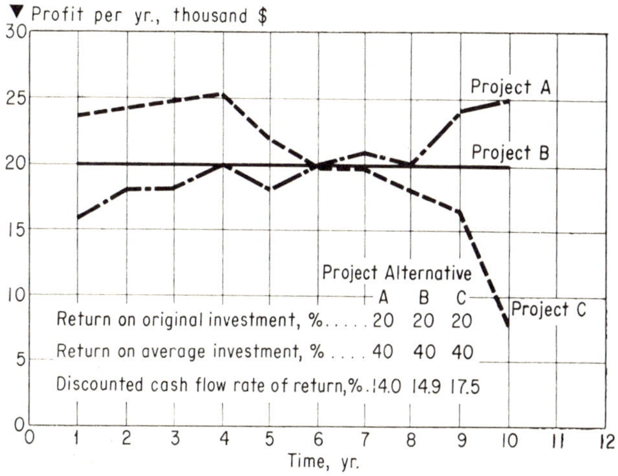

Discount cash-flow calculation pinpoints best project—Fig. 1

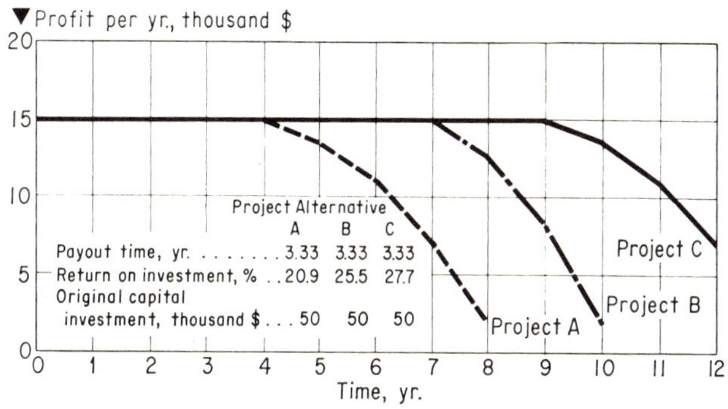

Payout time does not spot most suitable investment—Fig. 2

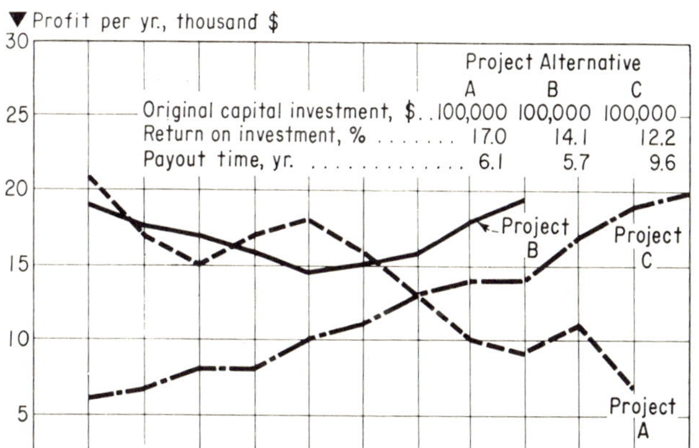

Payout time disagrees with return on investment in ranking projects—Fig. 3

• It accounts for capital amounts recovered over the project's economic life.

• It accounts for the timing of cash outlays and intakes, resulting in proper ranking of the alternatives considered.

Calculating full-life performance of two projects—Table III

	Alternative A	Discount Factors at 5% Interest	Today's Cost for A	Alternative B	Today's Cost for B
Project capital cost (paid today), $	20,000	1.000	20,000	12,000	12,000
Cost to operate:					
Year 1	6,000	0.952	5,770	7,000	6,660
2	6,000	0.907	5,450	7,000	6,350
3	6,000	0.864	5,180	7,000	6,050
4	6,000	0,823	4,940	7,000	5,760
5	6,000	0.784	4,700	7,000	5,480
6	6,000	0.746	4,480	7,000	5,230
7	6,000	0.711	4,270	7,000	4,980
8	6,000	0.677	4,060	7,000	4,740
9	6,000	0.645	3,870	7,000	4,510
10	6,000	0.614	3,780	7,000	4,300
10-yr. total cost, $			66,500		
Average yr., $			6,650		
11	.....	0.585	.....	7,000	4,090
12	.....	0.557	.....	7,000	3,900
12-yr. total cost, $					74,050
Average yr., $					6,170

• The figures obtained are comparable to those used in business finance. The interest on bonds or borrowed funds is computed in the same manner. This allows quick comparison of the project's value with other returns on equity.

Payout Time: Advantages and Limitations

Payback or payout period is the time necessary to recover the original investment in a project.

Payout time is usually calculated by accumulating the year-by-year net project earnings plus depreciation until the sum total equals the original depreciable investment. However, as with return on investment, various alternative computation techniques are employed. In some firms, depreciation is not added to profits. When this is the case, payout time is essentially the reciprocal of the return on original investment.

Another variation in the computation of payout time is the inclusion of working capital. Since working capital* is always assumed to be recoverable, it is not usually included in the investment base for payout time. However, it actually is not recoverable until the project is discontinued, and, for this reason, there is justification for inclusion of working capital in the investment figure.

The major inadequacy of payout time is that it only tells how soon money will be returned and in no way indicates profitability of the investment. Companies, of course, do not make investments merely to get their money back. The true worth of the investment depends on how much income is actually generated after the original outlay has been recovered. Thus, as a yardstick to measuring the worth of a project alternative, payout time is not very good.

The principal user of this method has been the oil industry.[2] However, in recent years, many of the large oil firms, after reappraising their methods for judging capital investment opportunities, have shifted to return-on-investment measurements. This shift was apparently precipitated by the drive for more efficient use of investment dollars when the era of oil-refining prosperity began to fade in the latter years of the last decade. Proper appraisal of project alternatives in this industry is particularly important since it has the highest capital invested per worker.[3]

Payout Time Can Be Misleading

The payout time figure does, generally, give a reliable appraisal of the relative worth of project alternatives if the income producing lives are all the same. However, this is not always the case. Fig. 2 illustrates how misleading payout figures can be. The figure shows three alternate proposals for completing a project, each proposal calling for an investment of $50,000. The annual income of each alternative begins at $15,000 and declines in later years as a result of increasing maintenance costs and/or decreasing price to maintain a competitive position.

The alternatives' incomes decline at different points in the estimated life of the project. Since the early incomes are identical and do not turn down until the original capital outlay is recovered, each project has the same payout time. By this standard, each of the alternatives would be equal from an investment

* Working capital is made up of money and goods that must exist in an organization at any given time to enable it to function. Working capital chargeable to a given project is the net increase or decrease in the working capital of the corporation due to addition of that project.

point of view. However, return on investment for the alternatives ranges from 21% for alternative *A* to 28% for alternative *C,* using discounted cash flow.

It may be implied from this example that the problem can be overcome by considering both estimated economic life of the project alternative and payout time. Unfortunately, this example is not representative of the usual situation. Most investment proposals have different payout times and economic lives, as well as different figures for return on investment. As an example of the more usual case, consider the three investment alternatives illustrated in Fig. 3. The life of each investment is different. Payout time figures favor *B-A-C* as the order of investment desirability, but the discounted cash flow rate of return appraises the order of economic attractiveness as *A-B-C.* If payout times were used for this problem, the ranking of investment desirability would not have been the proper one, and a poor investment decision would have followed.

Payout time can serve as a coarse screen to pick out high profit proposals that are clearly desirable and don't require any refined rate-of-return calculations. However, payout time does not consider the dimension of profitability, an aspect most relevant to making decisions among engineering project alternatives.

Full-Life Performance Weighs Cost

If the future were always accurately predictable, correct economic decisions would be made by everyone. The engineer and business manager would know in advance about the outcome of alternative routes to the fulfillment of a project or venture. In short, decision-making would not be a very difficult task, and profit would be perfectly predictable. However, in real life, this is not the case. Since return on investment and payout time are based upon the forecasted profit picture of the project alternative, determination of desirability of capital investment alternatives using these methods is only as accurate as the ability of the engineering-analyst to predict the future.

There are, of course, degrees of uncertainty involved in business prediction. For some projects, it may be easy to predict profits for each alternative with a reasonable amount of certainty. In other situations, it may be inconvenient or virtually impossible for the engineer to accurately measure expected income of a project alternative. But it is not always so difficult to develop *costs* involved in the various alternatives being considered. This is because costs are more certain than profits and, as a result, elaborate methods of cost estimation have been established in the chemical industry. It is, therefore, important that the chemical engineer understand the principles and basic mechanics of appraisal methods for project alternatives based upon costs. For this reason, the two yardsticks, full-life performance using discounted dollars, and capitalized costs, are worthy of careful consideration.

Comparing Discounted Costs

Appraisal by the full-life performance method involves comparing the discounted or present-value cost of alternatives. The most economical alternative, then, is the one with the least total discounted cost.

In the full-life performance method, the "discounted cost dollar" is computed in essentially the same manner as the "discounted profit dollar" for rate of return. The interest rate is handled the same way—it is the rate that could be earned if the money were put to some other use. Usually this rate is taken as the present interest rate on borrowed funds. Thus, assuming that the worth of money is 5%, $1,000 will return $50 interest at the end of one year. In the second year, interest is earned on the original thousand dollars and on the interest from the previous year. This is merely the process of "compounding" the worth of money annually. In ten years, the original $1,000 would be worth $1,629. This means that $1,000 today equals $1,629 ten years from today. The discounted value of $1,629, then, is $1,000.

The purpose for discounting cost dollars is to evaluate each alternative on a common basis: today's dollars. With this method, all of the project's cost over its entire life are assumed to be paid out today, and in the worth of present-day dollars.

To illustrate the practical applications of the full-life performance method, consider the problem of having to make a decision between the following two alternatives:

	Alternative A	Alternative B
Total investment, $	20,000	12,000
Total project life, yr	10	12
Average annual cost of operation, $	6,000	7,000
Depreciation, $	2,000	1,000
Average cost per year	8,000	8,000

Total average annual cost of each alternative is identical. However, for an additional capital expenditure of $8,000, an operating cost advantage of $1,000 per year can be obtained by purchasing alternative *A.* On the other hand, alternative *B* offers longer service for less capital investment. To make the choice on the basis of the forecasted cost picture alone may lead to a poor decision. The present-value costs based on today's dollar will aid in making the correct decision.

The discount factor applied to forecasted costs converts costs to present-day dollars is given by:

$$d_n = 1/(1 + i)^n$$

where i is the chosen interest rate and n the year of the project life being considered. For application in the above problem, the value of i will be assumed to be 5%. The year-by-year computation and the average annual discounted costs are shown in Table III. For simplicity, it has been assumed that operating cost is paid at the end of each year rather than throughout the year as would be the actual situation.

This method of appraisal favors alternative *B* since the average annual discounted cost is $6,170 compared with $6,650 for alternative *A.* Use of this appraisal

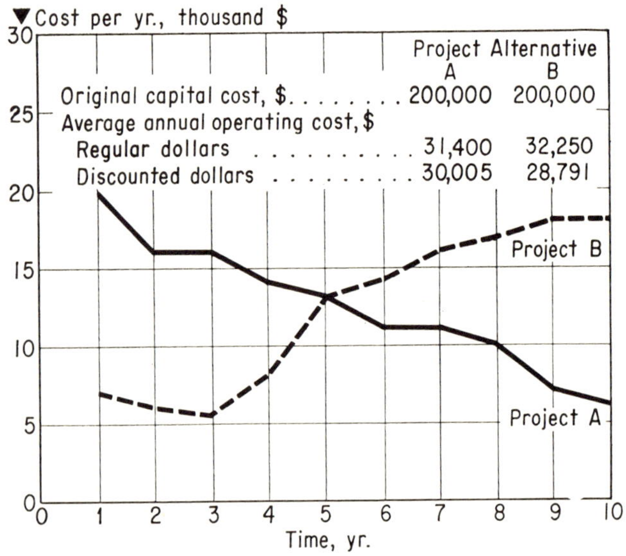

Discounted costs are helpful in evaluation—Fig. 4

method permits a differentiation between alternatives, while consideration of the bare cost may misguide the choice.

Computation using regular dollars gives too little consideration to the fact that an additional $8,000 has to be paid out at the start of the project for alternative *A*. As with the discounted cash-flow rate-of-return method applied to profits, this method applied to costs accounts for the time at which the costs are paid. Time becomes important to economic decisions when forecasted costs are increasing or decreasing with passage of time. It is obvious that two alternatives with the same average cost are not necessarily equally desirable from an economic point of view.

Since money put to work in our economy develops value over time, it is logical that the time schedule of a project alternative's cash outlays be considered when making an economic appraisal. In Table III, the discount factor decreases over the economic life of the project. Thus, the dollars expended early in the project's life are discounted the least. A project alternative that calls for heavy capital outlay or operating costs in the early years of its economic life tends to be less attractive economically than an alternative whose costs start out low and rise over time. Once a dollar is spent, it is not available for use elsewhere in the economy of the firm.

It should not be implied from the previous example that the "discounted"-dollar method of appraisal is useful only when actual dollar costs are equal. In practice, this appraisal method may rate the order of alternative desirability just the opposite of the order based on actual costs. Illustration of such a case is presented in Fig. 4 (here costs include depreciation).

The full-life performance method offers and excellent means of economical appraisal when only the costs of a project alternative are available. However, when a profit schedule can be accurately composed, return on investment should be the major criterion of economic choice. The company is basically interested in the profitability of a particular venture, not its costs. Costs are only half of the profitability picture, which is composed of costs and sales considered simultaneously.

Capitalized Cost Compares Alternatives

Capitalized cost is another method available for choosing the most economical approach to a given project. As in full-life performance using discounted dollars, this method involves development of a mathematical relationship of cost factors that puts the comparison on a sound economical basis. The alternative with the lowest capitalized cost is the most economical approach.

Capitalized cost places project costs on a perpetual basis. This involves computing the first cost of the project alternative plus the present value of an indefinite number of renewals.[4] The following simple example illustrates the basic concept.

A project costing $5,000 will last for ten years. It is assumed that money is worth 5%. The cost to generate the project perpetually is $12,950. This is called the capitalized cost.

To understand how this figure was mathematically developed, trace the following steps: subtract the original cost, $5,000, from the capitalized cost. The remaining, $7,950, compounded annually at an interest rate of 5% for ten years just equals the capitalized cost of $12,950. At the end of ten years, when the original equipment is worn out, $5,000 is available for the replacement. Also, the excess of $7,950 is available to earn at a rate of 5% annually over the next ten years of the project's life. The $12,950 is the exact amount necessary to provide the original cost of the project plus an infinite number of renewals.

From the above example, the capitalized cost factor may be developed. Capitalized cost minus the original capital investment compounded at 5% annually for ten years yields capitalized cost. This may be mathematically stated as: ($7,950) $(1.05)^{10}$ = $12,950. Assigning letters to the above values, the formula is the following:

$$(C - C_o)(1 + i)^n = C$$

C = Capitalized cost.
C_o = Original capital outlay.
i = Interest rate.
n = Economic life of the project.

This may be rearranged as follows:

$$C = \frac{C_o (1 + i)^n}{(1 + i)^n - 1}$$

The capitalized cost factor, then, may be stated as:

$$f = \frac{(1 + i)^n}{(1 + i)^n - 1}$$

For a more-advanced application of this cost-appraisal method, consider the problem discussed under full-life performance. The capital outlay of project alternative *A* is $20,000, while that for project *B* amounts to $12,000. Capitalized cost is computed by multiplying each cost by the capitalized-cost factor. The respective capitalized costs, are $51,800 for alter-

Obtaining capitalized costs—Table IV

	Alternative A	Alternative B
Actual Costs		
Initial capital investment, $.	20,000	12,000
Annual operating costs, $.	6,000	7,000
Capitalized Cost Factor		
Factor for capital investment.	2.59	2.25
Factor for operating costs.	21	21
Capitalized Costs		
Initial capital investment, $.	51,800	27,000
Annual operating costs, $.	126,000	147,000
Total capitalized costs, $	177,800	174,000

native A and \$27,000 for alternative B. These are the capitalized costs of the original capital outlay for each alternative.

Handling Operating Cost

To complete the analysis, however, the yearly operating cost must be included. To do this, the net operating cost is considered equivalent to a piece of equipment that will last for one year. Thus, operating cost for each alternative is multiplied by a capitalized cost factor computed with the value of $n = 1$ (not the life of the project as is the case when considering the original capital outlay). Capitalized operating costs are \$126,000 for alternative A and \$147,000 for alternative B. These figures are added to the original outlay capitalized costs to obtain total capitalized costs for each alternative. Alternative A total capitalized cost is \$177,800 while the total for B is \$174,000. Since project alternative B has the lower total capitalized cost, it is the most economical. The development of these figures is illustrated in Table IV.

The above example illustrates application of the capitalized-cost method when annual operating costs do not vary over the life of the project. In practice, however, annual charges almost always vary, usually upward. Computation of capitalized cost now becomes a little tricky. It is accomplished by first calculating the present value or discounted value of each year's cost as in the full-life performance method. Present values of the annual costs are then summed, and capitalized over the life of the project in the same manner as the original capital outlay. This can be consolidated:

$$C = f \sum_{1}^{n} (d_n) C_n$$

where C is capitalized cost; f capitalized cost factor; d_n discount factor; n year of the project life being considered; and C_n the actual annual cost in year n.

Although capitalized cost is based upon perpetuation of the project indefinitely, it does not mean that the project must or could be used forever. Capitalized cost is merely the common denominator upon which project alternatives with different service lives and cost patterns can be compared. Like the full-life performance method, capitalized cost considers the time value of money, and is an excellent method for comparing alternatives.

Conclusion

In the past, few companies formulated mathematical methods for economic appraisal. However, with passage of time and maturity in the chemical industry, competition will become increasingly vigorous. Under such circumstances, only firms with high efficiency of investment will survive. Therefore, it is imperative that the engineer be able to identify and push to early completion the high-profit projects and weed out those of lesser value. ■

References

1. Hunt, P., Williams, C. M., Donaldson, G., "Basic Business Finance," pp. 486-487, Richard D. Irwin, Inc., Homewood, Ill., 1958.
2. Ruel, R., "Spotting Bargains in Capital Investment," *Chem. Week*, Jan. 20, 1962, pp. 115-116.
3. Alderfer, E. B., Michl, H. E., "Economics of American Industry," p. 271, McGraw-Hill, New York, 1957.
4. Jelen, F. C., "Replacement Problems Using Capitaliized Cost," *Chem. Eng.*, Aug. 1955, p. 181.

Meet the Author

HOWARD H. STREET *is a process engineer in Du Pont's Teflon Div., Parkersburg, W. Va., where his interests are in the areas of process design and economic evaluation. He has a B.S. in chemical engineering from Worcester Polytechnic Institute and a masters degree in business administration from the University of Delaware. He is a member of AIChE. This article is based on a talk by Mr. Street, given at a tri-section meeting of AIChE in Detroit, Mich., last year.*

The Sensitivity Chart—

Any forecast contains assumptions—some near-certainties, others guesses. The sensitivity chart graphs this and indicates how sensitive the forecast is to errors in assumptions.

RICHARD STRAUSS, *Chemical Consultant*

It is seven o'clock on the night of May 28, but the lights are still burning in the executive offices of the Able-Made Chemical Div. of Amalgamated Acme, a large and diversified company. The Able-Made management team is just completing the annual, divisional operating budget for the fiscal year beginning July 1. The budget must be turned in to Amalgamated Acme headquarters by June 1, for consolidation with the other divisional budgets.

Even though the Able-Made management is accustomed to making regular short-range predictions to facilitate continuing divisional planning, the annual budget-preparation time is always a traumatic period. The budget must reconcile two conflicting requirements—the informal Amalgamated Acme "guidelines" for satisfactory divisional performance in terms of growth rate and return on investment, and the desire of the Able-Made Div. management for a budget that is conservative enough to be met or surpassed despite the unforeseeable problems that will inevitably occur in the year ahead.

Predictions Are Made

At last the small Able-Made group breaks up and the managers head for their homes. They have hammered out what appears to be a good budget—the predicted 15% increase in sales and 10% in profit for the coming year more than satisfy the corporate guidelines, and the individual sales predictions and manufacturing-cost figures seem conservative enough for the department heads to live with comfortably.

Disturbing Thoughts

But second thoughts are already beginning to disturb the Able-Made men as they drive home. The sales manager is particularly uneasy. There are many unpredictable factors that will affect his performance during the current year, especially the rather ominous pressure of imports on the Ableox line, an old product but one that still makes sizable profits for the company.

The sales manager is confident that he will be able to hold Able-Made's market share, but he may have to drop prices during the second half of the year before the grumbling by his old customers reaches the stage of open revolt. On the other hand, he is comforted by the budget's modest 15% increase predicted in sales of the profitable new product Ablegard, and he has a pretty good tip from the labs of Activated Artisan (Ablegard's largest potential user) that their new line will start up in a few months, virtually doubling the total sales of Ablegard.

The Able-Made plant manager also has his reservations about the budget. He is just completing the final stages of a major process-improvement job on Ableox production. In his eagerness to obtain approval for the much-needed plant improvement, he had used some rather optimistic cost savings in the project justification. Now that the project is coming into operation and he is obligated to show the cost savings in the manufacturing budget, he wishes that he had been a little more conservative 18 months ago. Although faced with a startup date of December 1 for the new process, he has not yet received a firm delivery date on a major item of equipment. Experience has shown that this equipment supplier can actually be several months late in delivery, despite continuous promises of shipment "next week."

The Able-Made president is happy that a satisfactory budget was prepared this year with only a minimum of horsetrading, but the uncertainties in his department heads' estimates have made him uneasy. He decides to ask the chief accountant to calculate a few "provisional" budgets, showing the effect of using both more-optimistic and more-pessimistic assumptions than those used in the budget.

And, finally, the chief accountant drives wearily home, knowing that for him the annual budget ritual has just begun. In the weeks ahead, the whole job will have to be repeated innumerable times in response to requests to "see what happens if . . ."

The Able-Made example illustrates that any projection—whether it is an operating budget, an invest-

Giving Meaning to Shaky Estimates

ment analysis, or a ten-year plan—is the product of many assumptions about the future. Some of these assumptions are likely to be quite accurate, but some will be sheer guesswork.

The accuracy of the predicted result will be quite sensitive to some of the assumptions and virtually insensitive to others. Yet in most analyses all of the assumptions are taken at face value.

The final result of the assumptions—annual operating profit, payout or discounted cash flow—is usually expressed as a single number which hides the uncertainties contained in the predictions.

Methods have been suggested for expressing the uncertainties in investment analyses. Hertz[2] and Thorngren[3] describe some techniques in which probability distributions of the input variables are specified. The sensitivity of the discounted cash flow is then determined by reiterated computer calculations, using values for each variable within the desired probability limits.

Probability analysis is a most useful tool for management when questions take the form of: "Shall I invest in Project A, Project B, or in no project at all?" However, for the problem of predicting next year's results and then acting on that prediction, this method is inadequate. Unfortunately, management cannot examine next year's predictions and then elect not to have that year at all.

Don't Over-Refine Predictions

The Able-Made experience also illustrates the well-known fact that managers, like other humans, tend to concentrate their attention on factors that are known or knowable, while taking a quick guess and hoping for the best on those that are not readily predictable.

In his excellent article on the profitability of new projects, Johnson[1] shows that the factors affecting the success or failure of a new venture which receive the most attention are those that are most susceptible to analysis—these are not necessarily the factors that can have the greatest effect on the end result. Much effort is expended in refining an investment prediction to ±20% and in obtaining a number (usually Discounted Cash Flow) calculated to 0.1% that is intended to represent the relative value of the project.

However, the success of a new venture is usually far more sensitive to the possible swings in sales price and sales volume over the life of the project, but it is seldom that as much effort is spent in refining these estimates as is spent in the technical development of a new process.

It would be desirable to accompany any projection with the following information: 1) What are the assumptions contained in the project? 2) What is the degree of uncertainty in the assumptions? 3) How sensitive is the projection to variations in the assumptions?

A Graphical Solution

A graphical procedure has been developed that presents the answers to these questions in an easily understood and readily usable form. To illustrate how this chart is constructed, we will return to the problems of the Able-Made Div. Trying to forestall the usual rash of "what if" questions concerning the newly completed budget, the Able-Made accountant has decided to take the initiative in asking questions.

He approaches the department heads in turn and asks each one to list the major uncertainties in his forecast. For example, he discusses with the sales manager the possible effects of imports on the Ableox sales price. It seems inevitable that the domestic price will eventually approach the foreign price level, but the question is, how quickly? Together they guess that there is "less than 10% chance" that the average price will fall below 50¢/lb. during the budget year.

Note that the probability limits are phrased as "less than 10% chance" or "greater than 90% chance" to lend a certain air of mathematical exactitude to the proceedings. (It is likely that the responses would be exactly the same if the questions were based on a "less than 5% chance" or a "less than 15% chance.") As a practical matter, the estimator is asked to express the upper and lower limits between which the values of the variables could reasonably fall. Techniques that require expression of these data as continuous probability curves may have a psychological effect opposite to that intended by the exercise—instead of illustrating areas of uncertainty, the rigorous mathematical treatment may tend to reinforce the false feeling of certainty.

In his discussion with the sales manager, the accountant establishes the two major marketing variables—the sales price for Ableox and the sales volume for Ablegard. The probable lower and upper limits for these variables are agreed upon. In similar fashion the accountant discusses with the plant manager the limits of the manufacturing variables. (These vari-

Tabulation of variables in Able-Made forecast—Table I

Variable	Value Used In Forecast	10% Probability Value			90% Probability Value		
		Value	% Change From Fore-Cast Value	Effect on Annual Manu-facturing Profit	Value	% Change From Fore-Cast Value	Effect on Annual Manu-facturing Profit
Marketing variables:							
Sales price, Ableox.......	$0.55/lb.	$0.50/lb.	−9.1%	−$91,000	$0.57/lb.	+3.6%	+$36,000
Sales volume, Ablegard..	200,000 lb.	150,000 lb	−25%	−$20,000	350,000 lb.	+75%	+$60,000
Manufacturing variables:							
Ableox manufacturing cost after yield improvement....................	$0.39	$0.37	−5.1%	+$18,000	$0.41	+5.1%	−$18,000
Time before startup of new Ableox process.....	5 mo.	4 mo.	−20%	+9,000	8 mo.	+60%	−$27,000
Annual maintenance costs.....................	$150,000/yr.	$125,00	−16.7%	+$25,000	$170,000	+13.3%	−$20,000

ables are tabulated as shown in Table I.) He is now ready to construct the sensitivity chart.

Drawing the Graph

He draws the horizontal axis, which represents the percentage deviation in the variables from the selected most-probable-value, and the vertical axis, which repsents the effect that the possible deviations in the independent variables will have on the dependent variable.

In this example, he is analyzing the sensitivity of the Able-Made operating budget, and the dependent variable is the possible change in annual manufacturing profit.

(When he applies the charting technique to sensitivity analysis of a proposed new investment, he uses either percentage return on investment or discounted cash flow as the dependent variable; and investment, yield, sales prices, and sales volume as typical independent variables.)

The zero-zero point, the intersection of the two axes, represents the most probable, or budgeted, result. Starting with the sales price for Ableox, he indicates the points (−9.1%, −$91,000) and (+3.6%, +$36,-000) on the chart, taken from Table I. He then draws a straight line from each point to the zero-zero point. (If a normal probability distribution were assumed for each variable, the lines on the sensitivity chart would not be straight but slightly S-shaped (concave away from the X-axis). However, using normal probability distributions greatly increases the complexity of preparing the sensitivity chart, and adds little to its usefulness.

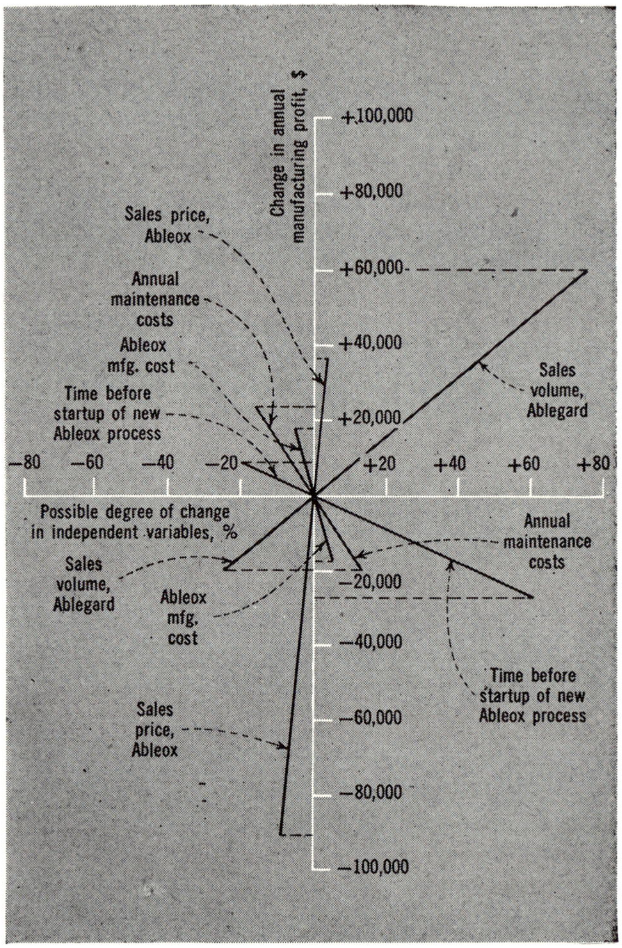

SENSITIVITY CHART for Able-Made forecast—Fig. 1

Tabulation of variables in investment analysis—Table II

Variable	Value Used in Forecast	10% Probability Value			90% Probability Value		
		Value	% Change	Calculated Return on Investment	Value	% Change	Calculated Return on Investment
Selling price	$0.18	$0.21	+16⅔%	48%	$0.13	−28%	20%
Sales expenses	$40.6 MM	....	....	.....	$60.6 MM	+49%	22%
Cost of key ingredient	$2.14/lb.	$1.00/lb.	−53%	47%	$4.00/lb.	+87%	30%
Yield	64.4%	50%	−22½%	26%	80%	+25.7%	51%
Permanent investment	$13 MM	$6.5 MM	−50%	46%	$26 MM	+100%	29%

What the Lines Mean

The line on the chart for the Ableox sales price has a characteristic length and slope, and both characteristics are indicators of the nature of the variable:

The *slope* shows the degree of change in the profit resulting from an assumed percentage change in Ableox sales price. In this case, a small change in the predicted sales price will produce a large effect on the profit. The profit is extremely sensitive to Ableox sales price.

The *length* of the line is the resultant of the sensitivity of the variable and its degree of uncertainty.

Dotted lines are drawn from the ends of the line to the y-axis to aid the eye in estimating the relative effects of the variables. Despite the high sensitivity of profit to Ableox sales price, the low probability of a higher price than that budgeted results only in a moderate ($36,000) possible improvement over budget. However, the high sensitivity combined with a somewhat higher probability for a lower-than-budgeted sales price results in a large ($91,000) possible negative deviation from budget.

The new product, Ablegard, shows characteristics different from the established product, Ableox. As Ablegard has not yet reached substantial sales volume, profits are not sensitive to large percentage changes in sales volume—hence a rather low slope of the sensitivity line. However, with a newly introduced product, purchases by one or two big new customers can make the difference between a modest and a large success.

The Able-Made sales manager maintains that it is virtually impossible to predict when a new product will be accepted by a major new customer, and he attempts to be conservative in his estimates. The resultant line on the chart shows that possible overestimation of Ablegard profits could amount to only $20,000, but the possible underestimation of profits (by underestimation of sales volume) could be as much as $60,000.

The accountant now completes the chart from the manufacturing variables in Table I. Whether the slope of each line is to be positive or negative can be decided arbitrarily when the chart is prepared. In this case, the accountant has assigned positive slopes to the marketing variables and negative slopes to the manufacturing variables. This helps keep the chart from becoming too cluttered and also permits a quick comparison of the relative importance of the two groups of variables.

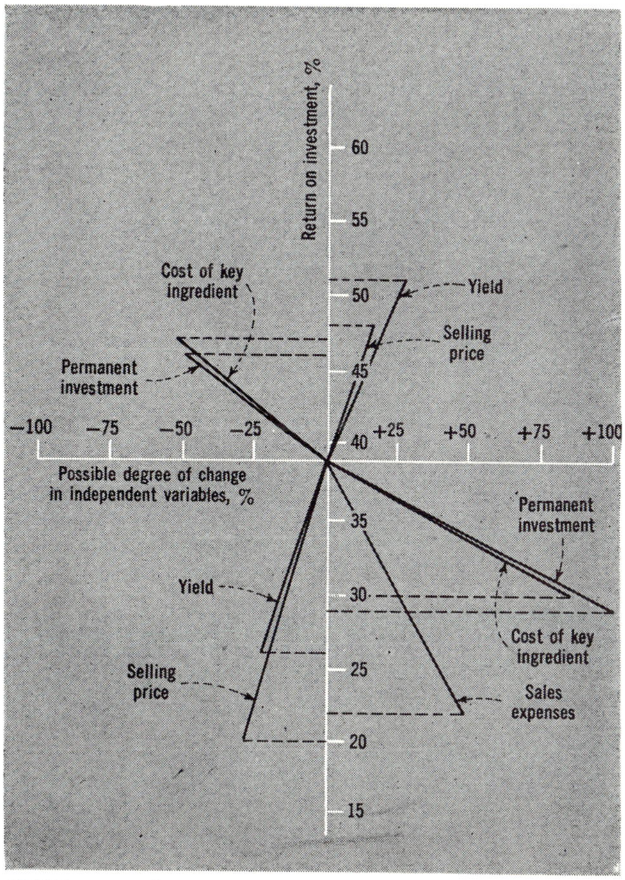

SENSITIVITY CHART for investment analysis—Fig. 2

Value of the Sensitivity Chart

What is the value of the sensitivity chart? To the process engineer, it indicates that he is running into diminishing returns in continuing to sharpen up on the Ableox process improvement in the pilot plant, and that he should be putting more emphasis on expediting the installation of the new process equipment. To the plant manager, it provides a new way of telling the other members of the management team that all the attention they can devote to maintenance can only affect the maintenance costs by ±$20,000, a significant sum but not one which will have a decisive effect on the year's performance.

To the sales manager, it shows graphically the opportunity to build up new product sales to offset the inevitable loss of profitability through price erosion in an old product.

To the president, it provides a "feel" for the nature of the uncertainties in his budget. He can see the quantitative effect of the possible slide in Ableox prices, and he knows that the uncertainty relates only to timing—if it does not occur in the budget year, it will certainly occur in the next year. He resolves to put more of his attention into the development of Ablegard and other new products.

To the central accounting department of the parent company, Amalgamated Acme, the sensitivity chart accompanying each divisional budget gives another insight into the projected performances. It gives a quick method of comparing the division with relatively low profitability but certain performance against the more spectacular division with excellent profit projections but large areas of uncertainty. The sensitivity chart identifies for all levels of management the variables that should be monitored closely during the year.

Characteristic Charts

The chart shows a pattern of an operation that is as characteristic as a fingerprint. Well-established, stable operations will show short lines clustered around the zero-zero point. Operations that depend heavily upon a few products will show very steep lines; operations that are entering new market areas or introducing new products will show wide possible changes in the variables. As a company leaves one phase of operations and enters another, its "fingerprint" will change.

Charting a New Investment

A similar type of sensitivity chart can be prepared for a new investment. For example, we have taken the data from the recent article in CHEMICAL ENGINEERING on evaluation of new projects, by E. A. Ransom,[4] and constructed a sensitivity chart from them. Mr. Ransom mentions that probability limits were set during the study, but as they were not indicated on the curves, we assumed limits from the range of variables shown on the curves. The data for this venture are shown in Table II, and Fig. 2 is the resulting sensitivity chart.

The possible degree of change in the variables we have selected is characteristic of a project in the early stages of investigation. Certainly the investment and the cost of the key ingredient would be known to ±25% before a final decision on the project would be made.

An interesting feature of the Ransom data is that the return on investment is far more sensitive to selling price than to investment. This is a very common finding in the analysis of new investments; yet invariably an enormous effort is spent refining the investment estimate, while the final sales price and sales volume figures more resemble guesstimates than the best possible estimates.

The preparation of the sensitivity chart has some psychological advantages. Everyone who prepares a budget or a new project profitability analysis feels that he should accompany the report to explain personally all the assumptions and uncertainties in the projections. The sensitivity chart gives all estimators the opportunity to express their hopes and fears and know that they will receive the proper attention. Most estimators seize the opportunity eagerly.

Another psychological advantage is that estimates accompanied by the chart more nearly represent best guesses and are thus less likely to have large contingency cushions built into them. Expressing the uncertainties in the sensitivity chart lessens the need (and the opportunity) for incorporating them into the basic projection.

Although in this discussion we have emphasized the use of the sensitivity chart in the analysis of projected operating results, the same techniques may be applied to analysis of investment decisions, research programs, or any other quantitative prediction of results. It is a foolhardy estimator who does not accompany his estimate with a readily understood presentation of the effect of his guesstimates on the final result. It is a foolhardy manager who does not insist on an analysis of the assumptions contained in the estimates on which he must base his decisions. ■

References

1. Johnson, L., *Chem. Week*, July 21, 1962, pp. 61-72.
2. Hertz, D. B., *Harvard Business Rev.*, 42, (1), p. 95 (1964).
3. Thorngren, J. T., *Chem. Eng.*, August 14, 1967, pp. 143-151.
4. Ransom, E. A., *Chem. Eng.*, Nov. 6, 1967, p. 291.

Meet the Author

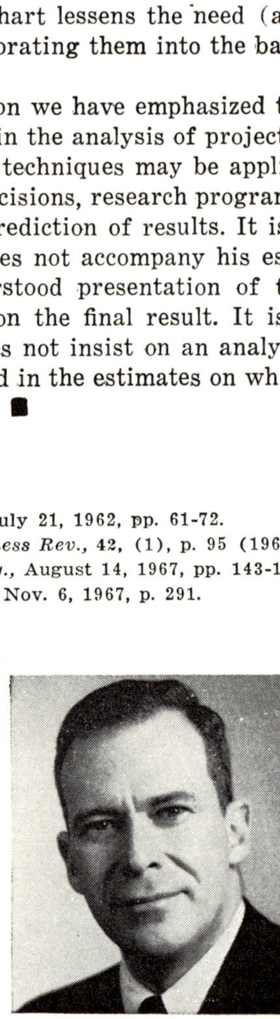

Richard Strauss is a consultant in product and process development, technical economic studies and corporate development problems. His office is at 4 White Terrace, Lexington, Mass. 02173. He has worked in the pulp and paper industry and in specialty organic chemicals with Arthur D. Little, Inc., American Cyanamid and National Polychemicals, Inc. He has B.S. and M.S. degrees in chemical engineering from M.I.T. and is a member of ACS and Sigma Xi.

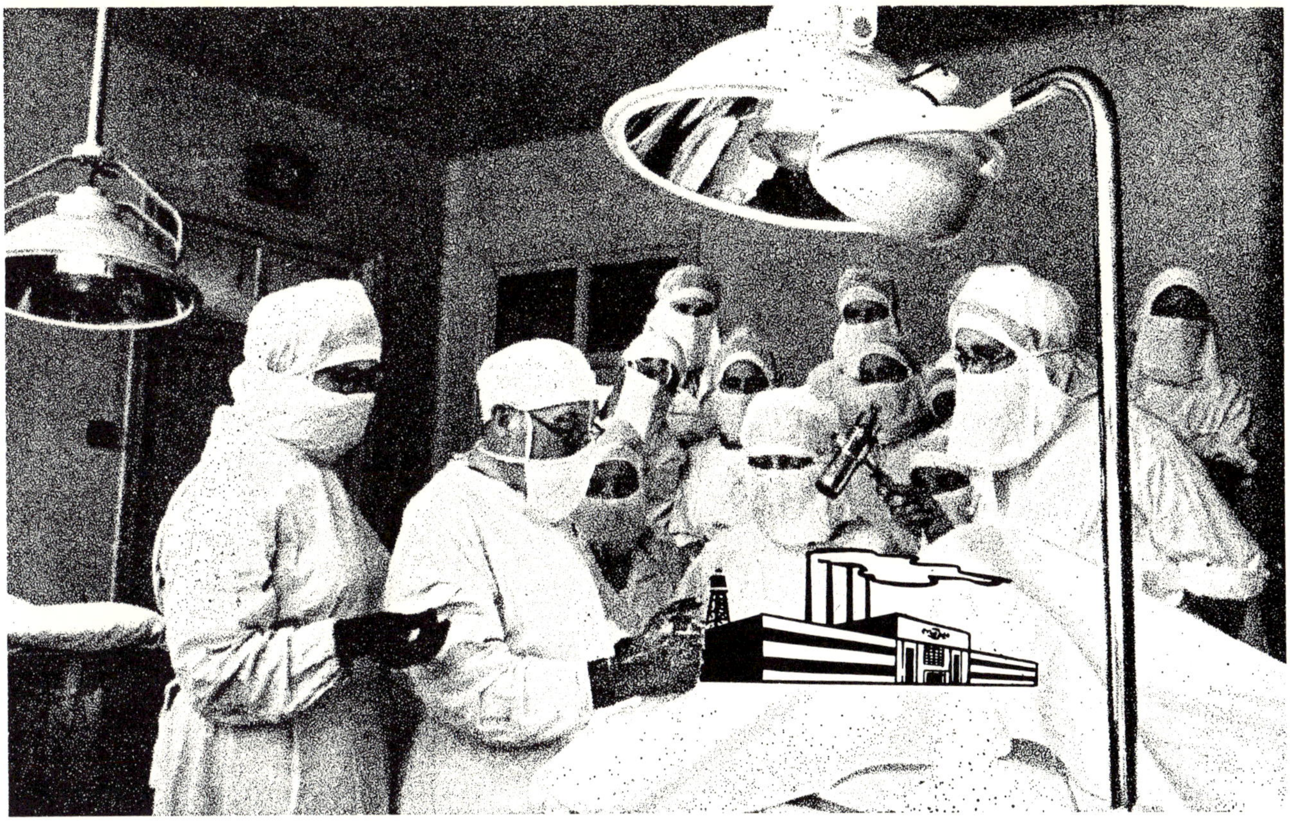

Successful application of Operations Research requires that the man who first identifies the problem follow it through to the end. This follow-through, plus teamwork are the keys to . . .

Resolving Problems With OR

JAMES S. MURPHY, *Allied Chemical Corp.*

Operations research, commonly termed OR, has received a considerable amount of attention over the last few years. The benefits to be gained are: numerical answers to almost any problem, which can then be directly acted upon; and an added insight that eases decisions made subjectively.

The beginning and the end of problem solving consist of identification and final action. OR lies between these two poles. Narrowly viewed, it is the reduction of related operations to mathematical terms as an aid to problem solving. Modeling is the term used to describe such a process of translating a real situation into mathematical symbols and relationships, and models have long been used in business. Cost sheets, budgets, material balances, etc., are common examples. The type of OR models normally used fall into seven categories: inventory, waiting line, allocation of resources, competition, replacement, search, and network (see Table 1, p. 259).

Despite the widespread interest in OR, however, the use of this technique has, for the most part, fallen far short of its promise. Applications have generally been limited to the more obvious, mathematically oriented problems, such as inventory control and distribution, and not enough use of OR has been made in the more general areas of marketing, production, administration, research and engineering.

Much of what has been done has been carried out by OR specialists and has been presented in those terms most readily understood by other practitioners. It is not surprising to find that opinions of the techniques vary across a broad spectrum, ranging from unbounded enthusiasm on the one hand to an equally unlimited disdain on the other.

If industry is to realize the full benefit of OR, the operating personnel, managers, supervisors and engineers must acquire at least some familiarity with it. OR methods must be applied to practical situations, with complete involvement on the part of those responsible.

As a planning specialist, I see opportunities within the planning function to help line management gain some knowledge of OR techniques. I put emphasis on planning, because its overview of all operations permits greater opportunity to spot applications.

In planning, one is concerned with recognizing opportunities and formulating strategies that will exploit the maximum potential of the corporate assets. Carrying out these strategies brings on risks and problems that OR techniques are well suited to solve.

The corporate planner's job is to guide those who are more knowledgeable about specific aspects of the company's business toward the best possible solutions of specific problems. To accomplish this, he should have some knowledge of both the management tools available and the people best qualified to apply these tools, including operations research and those specializing in its use.

The Need for Involvement

Most important, from the planner's overview, is this: the man who has been responsible for initially identifying the problem should remain with it until its solution. The problem thus becomes the planning/OR interface. The OR specialist has the mathematical tools to help solve the problem. But he may not be familiar with all of the circumstances surrounding it or the conditions under which the solution must be implemented.

As previously stated, a start is made by defining the problem. Then an estimate is made of the value of obtaining an answer. Time spent to solve mere annoyances or to satisfy curiosity is pointless. Like any other activity, problem-solving costs time and money; the value of any OR effort must justify its cost.

If it is decided that a solution to the problem is worth the effort, the planner identifying the problem cannot abdicate his responsibility at that point of decision. He must understand and agree with the format for solution, its implementation, and every assumption that is made, before accepting the results of the analysis. Actually, a team approach is the most fruitful. Members of functional groups contribute knowledge of the problem and the conditions under which the solution must be implemented; OR specialists must adapt their models to the limits of application of the specific situation.

Here is the case of a solution that failed because of poor implementation: A production department requested an OR model to cope with an allocation problem that was arising at regular intervals. The OR specialist recommended a ready-made or "canned" linear program. This would have solved the mathematical problem, but the OR specialist failed to recognize the extensive assembling and ordering of information needed both prior to and following the computer solution. The production clerk simply did not, in the course of his normal work, have the time to perform such extra work, and so the problem remained unsolved. This difficulty could have been avoided had a team, including the planner with his knowledge of the organization, an operations manager closely associated with the problem, and an OR specialist with programming background, all worked together to develop a method for ordering the data in their proper format by means of machine, and thus relieve the production clerk of the extra work.

Another example: a few years ago, a PERT-COST program was considered for a problem whose resolution allowed for alternative courses of action. In studying the model, it was discovered that the programmer had assumed that the costs varied directly with time. This information, buried in over six pages of description, made the entire program unsuitable.

Within each of the broad categories of OR techniques, there are numerous specialized mathematical models. A representative sampling of those available is described to provide some familiarity with these methods and the type of situations where they might be used. Table I, p. 259, lists examples of models used in various business functions.

All of these OR models require the same basic steps:

- Think about the problem.
- Define the solution required, in terms of:
  - answer.
  - format.
  - parameters.
- Identify all the factors involved, and express them physically or symbolically.
- Try to express factors algebraically.
- Show the relationships that exist between the factors.
- Screen out the insignificant factors.
- Combine all factors that are dependent on single variables.
- Consider the format of the model.
- Seek parallels to the type of problem that may lead to a format for solution (don't be afraid to get help).
- Obtain estimates of the variables involved.
- Solve the problem.
- Consider the solution in terms of:
  - validity (How good is it?).
  - reduction to action (What do we do with it? Can we make it work? How does solution requirement check against our available resources?)
- Recycle as needed.

Practice makes perfect. It is only by working with OR techniques in problem-solving that we realize how close most of them are to methods we have been using all along. Practice eliminates fears or awe of OR and of those who specialize in its application.

Inventory Models

Inventory models are among the most common of OR techniques. They help solve the problem of maintaining a supply sufficient to satisfy demand at minimum cost. The material involved may be physical items—such as machines or people—or may be more-intangible items such as ideas, plant capacity, research projects, etc. The information inputs to an inventory model consist primarily of cost data and demand forecasts. The availability of the required data varies considerably, depending on the problem being analyzed.

Physical inventory models are usually more easily visualized, and the required forecasts and costs more readily obtained, than models for intangibles. Fore-

How operations research models have been used in functional departments—Table I

O-R model → / Department ↓	Inventory	Waiting-line	Resource Allocation	Search	Competitive	Replacement	Networking
Marketing	Analyze warehousing needs.	Schedule sales calls. Locate sales offices and warehouses.	Determine distribution handling.	Search for customers or competition.	Assess effects of price changes. Evaluate effects of changes in sales policy. Plan competitive bidding.	Replace or rotate personnel for increased efficiency.	Prepare sales strategies.
Production	Schedule operations including batch sizes and reserve stocks. Determine plant capacities.	Schedule production. Staff maintenance crews. Locate plants.	Assign men, materials and machines. Determine distribution patterns for multiplant systems. Control process losses.	Retrieve process information.	Assess process changes.	Replace equipment or parts.	Establish production schedules.
Administration	Stock equipment and supplies. Size the work force. Size liquid assets.	Staff service functions.	Assign clerical help. Set salary and dividend policies.	Determine business strategy. Search for personnel.	Assess personnel recruitment or industrial relations policies.	Replace or rotate personnel.	Control integrated programs.
Research	Stock equipment and supplies. Size staff and the reserve of projects.	Staff and assign priorities to projects.	Allocate manpower and funds to projects.	Search for hypotheses or data.	Plan marketing and acquisition of patents and know-how.	Replace analytical or pilot-plant equipment.	Analyze, formulate, and control projects.
Engineering	Schedule purchasing and staff assignments.	Schedule work. Locate plants and facilities within plant sites.	Balance cost-time considerations. Allocate men, materials and machines.	Retrieve process information.	Select construction contracts.	Replace equipment.	Establish and control design and construction projects.

casting is especially difficult in dealing with a supply of intangibles. To illustrate: How many research projects will be needed in the future? What will be the future loss if such research remains undone? Obviously, inputs to such an inventory of intangibles are difficult to obtain. The first problem is that of projecting "flashes of genius," which will complete specific research projects. An impossible task? Not so. Research projects fall into the following main classifications:

- Developing new technology, with no well-defined goal or approach.
- Developing partially new technology to accomplish a specific result, which implies a well-defined goal, although the approach is not necessarily evident.
- Putting into practice existing technology, which implies a well-defined goal and approach.

The first type of project we cannot do much about because we are not working within any defined parameters. The best we can do is guess at the outcome, based on a probability of success derived from the past performance of the scientist conducting the research. However, this kind of research work is minimal in an industrial organization. It represents risk money bet on the performance of an individual and should be recognized as such. Choose the creative individual with care, then leave him alone.

The bulk of contemporary industrial research funds are spent in the last category; the parameters are defined in terms of incentive, allowable funds and required timing. Based on past experience, projects are scheduled in terms of timing, manpower and costs. The class of research that is intermediate between the two extremes is also intermediate in terms of accuracy of projection. Because the goal is defined, it is possible to calculate the expense that can be justified for success.

Estimates can be made of optimistic, pessimistic, and most likely, probability of success with corresponding costs. Then a comparison of justifiable expense vs. the costs is used to rate the project. Similarly, time estimates can establish reasonably accurate schedules. It is important to remember when making estimates of the value and costs of research efforts that these must be made continually over the term of the project and updated as new information becomes available, always comparing benefits and penalties.

Our inventory model comes into play in these circumstances by helping us to anticipate the number and type of projects that must be kept on the shelf in order to maximize the effectiveness of our organizations. The foregoing analyses provide inputs for solution of this model.

Waiting-Line Models

"Waiting-line" or "queuing" models are special forms of inventory models. Waiting-line models are used in cases where something or someone must be serviced from one of a limited number of facilities. In such instances, costs are associated with delays caused by something or someone having to wait for available facilities. Such problems are not easily solved by intuition alone. To eliminate waiting lines implies increasing capacity to handle peak demands, with attendant idleness in slack periods. Discovering a capacity that will yield acceptable waiting periods requires the use of probabilities and often "Monte Carlo" techniques.

This brings up the problem of how to obtain probabilities. The source of such estimates must be the line person directly responsible for successful performance (e.g. sales from the Sales Manager, operating rates and yields from the Production Supervisor, research costs from the Research Manager, and so on). Sometimes the line personnel must rely heavily on staff assistance in cases where they have no direct experience. Sales may have to base its estimates on Market Research's data. Production may require Research's assistance for technical parameters.

Despite such help, however, line personnel will be held accountable for results, so they must supply the input. The direct approach to probability would be to ask for the limits (most probable, optimistic, and pessimistic) of success. By virtue of a person's role in the organization, biases become apparent. Sales and Research tend to be optimistic, while Production and Engineering are more conservative in making estimates. The most successful way to balance such biases is usually to (1) have a meeting with personnel from the various departments having different points of view, (2) hear their points of view, and (3) estimate limits and probabilities based on the give-and-take among them. This raises the danger of relying heavily on the planner, who may also have his biases. However, experienced planners become reasonably astute at making such judgments.

Waiting-line models are often used in solving staffing and scheduling problems. How many employees are required for optimum effectiveness of a particular function? What is the best way to schedule work for maximum utilization of resources? Such problems must be structured to consider costs associated with possible lost opportunities as well as the direct charges.

Resource Allocation Models

Resource allocation models involve the allotment of resources to various activities in such a way as to obtain maximum profit or minimum costs. These models are most useful in deploying limited resources among several activities. Solution of this type of model depends upon a procedure that evaluates costs of solutions one after another, until no further improvement can be found. Specific resource allocation models include transportation modes, as well as linear and dynamic programming models.

These help to distribute resources and capabilities for maximum effectiveness. Specific examples are listed in Table I. One of the most common examples is in a product distribution system where there are several plant and warehouse locations from which to service customers.

Storage and shipping costs are fairly readily obtained, and an economic balance can be structured and solved. Problems arise with more-sophisticated techniques for estimating the value of improved service and customer relations—a case, for example, in which the total economics of the system make it cheaper to service customer A from Plant 2 even though Plant 1 might be closer.

A close relationship between the customer and the manager of Plant 1 may have developed over the years, whereby the customer received exceptional service and the plant obtained all or most of his business. It may be difficult to switch this customer to Plant 2 without risking the loss of his business. What can be done in such a situation depends on the specifics involved.

Such realistic instances do show that some attempt should be made to assign an economic value to goodwill and the corresponding cost of special customer services. It is in these instances that the second benefit of OR—added insight—becomes most apparent.

Competitive Models

Competitive models are used to analyze situations in which two or more participants strive for the same objective. The solution makes use of games theory. Despite considerable theoretical work in this field, there have been few successful applications to practical problems. Failure in real situations may be due to:

- The use of the "zero-sum" game, which is the most highly developed type of competitive theory. The total payoff of all players is zero: what one gains, the other loses. There are extremely few zero-sum situations in real life. A classical example, showing how the base of what is apparently a "zero-sum game" tends to shift, is the instance in which there are only two companies producing the same product. One increases his advertising. Instead of cutting into the competitor's sales, this advertising expands the total market and both companies profit in some measure.
- The use of zero-sum game theories for more than two competitors. The number of possible alternatives expands considerably, bringing into play a myriad of possible combinations.
- The use of models that require a postulate of the actions of both competitors. While such postulation does provide insights, it does not reveal how a competitor will react under actual conditions.

Despite such deficiencies, competitive models have

been effectively used to provide insights into competitive situations where it is not intuitively obvious which mixed strategies might be useful.

Replacement Model

The replacement model deals with problems in the replacement of parts or components in a complex system. When an item fails, it must be replaced by a similar one. The problem is to establish an optimum policy of replacement. Preventive maintenance models fall into this class. The general characteristics of replacement models include:

- Projections of useful life.
- The length of time required to provide a suitable replacement.
- The extent of the losses due to "downtime."

While the statements above apparently refer to physical items such as parts and machines, the same sort of analysis is equally effective in setting up personnel policies, as is evident from the examples listed in the table.

People tend either to gain or lose efficiency in relation to their tenure, their capabilities and their aspirations. In addition, losses from staff through retirements, illnesses, promotions, resignations and discharges demand continual changes in an organization. While the human element will temper the program, a basic personnel rotation and training program can be statistically devised, based on probabilities, available from past experience, which will help to dramatically improve the system. This type of work can provide real savings to a corporation if adequately analyzed and implemented.

Search Model

Search models refer to broad-type mechanisms for gaining or retrieving information. Using such mechanisms, the OR specialist can define the means for more-effective search, or provide storage facility for later recovery. Because planning relies so heavily on information, it is important for planners to realize that OR specialists can set up data search procedures. Search models are continually being developed but are not yet as completely defined as the other models. However, lack of knowledge should never deter the planner from exploring the possibilities with an OR man. Specific examples of search models include the techniques used by librarians and engineers in conducting literature searches. One of the major faults of technical people is that they do not preplan their literature searches, and thereby lose efficiency. An application of OR can again provide insight, even if no model is developed.

Networking Models

Networking models such as PERT, CPM, RAMPS, etc., are a means of pictorially defining, integrating, coordinating and controlling the individual tasks required to achieve a given objective. The network diagram relates graphically all that is known about how to accomplish a given job. The completeness and accuracy of the diagram is a reflection of the completeness and accuracy of the thinking used in preparing the plan. Thus, the network acts as an excellent check on the plan, because gaps that weren't foreseen during the thinking process become apparent when thoughts are reduced to lines on paper.

Improvements in the network can be made by including costs, manpower, other resources, and probabilities, as well as time. The underlying principles are simple, well known and readily applied to all projects where actions are required and interdependencies exist. While most companies use CPM or PERT, it is well to remember that the same basic networking procedures can be modified to fit almost any particular problem and circumstances. (The tool should always be adjusted to fit the job, not the reverse.)

The main value of networking lies in its attendant thinking and visualization processes. Even if the diagram were destroyed immediately after it was finished, 80 to 90% of its value would still be realized from the effect that structuring it had on those carrying out the assignment. It is the detailed thinking process, not the picture, which is the main product. Many of the cases where networking failed have resulted from slavishly following rigid rules and format rather than adjusting basic principles to the situation at hand. Networking models have perhaps suffered the most from such misuse, since these clearly visualized diagrams seem to attract those who prefer rules.

This article has purposely avoided numerical examples of models. Those readers who would like more specific information stated in very understandable form are referred to: "Executive Decisions and Operations Research," by D. W. Miller and M. K. Starr, Prentice-Hall, Inc. (1961). This is an excellent book on the subject—one that has been used for a number of years.

In conclusion, there is really only one way to learn how OR might be helpful. Use it. The corporate planner need not be a specialist in OR methodology, but he can identify the problem, aid in finding a feasible solution, follow its implementation, collect and assess results, and review the proposed solution for acceptability in practical surroundings. The corporate planner has, if he is willing to accept it, a vital role to play in developing better solutions to business problems by using and promoting Operations Research. ■

Meet the Author

James S. Murphy is Assistant Manager-Planning at the Industrial Chemicals Div. of Allied Chemical Corp. (Morristown, N.J.). He came to this post through prior positions as research chemical engineer, senior research chemical engineer and project engineer, since he joined Allied in 1962. He also worked for Johns-Manville Products Corp., Pitt-Consol Chemical Co., and Allied's Barrett Div. He holds a BS in chemical engineering from Newark College of Engineering and an MBA from Rutgers University, and he is currently working toward a Ph.D.

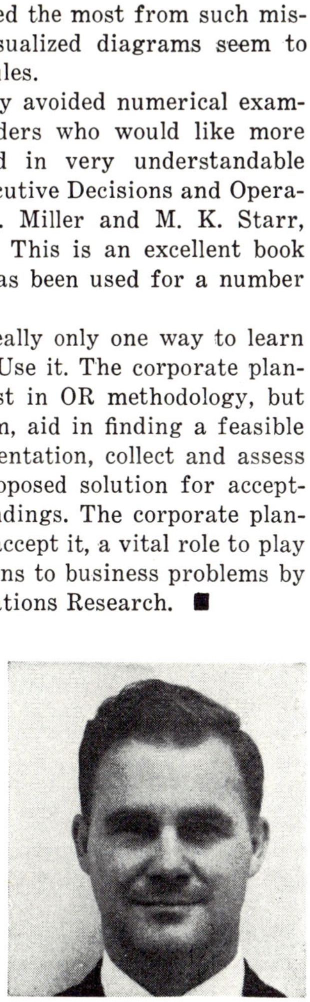

Experience With and Evaluation

An examination of 10 years of experience with critical path methods for scheduling the design, equipment purchase, and construction of major chemical-process plants, with pertinent comments on pitfalls and successes with these methods.

ERIC JENETT, Brown & Root, Inc.

Today's reaction to the Critical Path Method (CPM), first introduced about 1958, is still mixed. CPM can be simple or difficult, useful or useless, cheap or expensive, a tool or a weapon. Few techniques have engendered so much controversy or been around so long without becoming either an unequivocal success or a passing fad.

We will discuss some problems and solutions that we have encountered during 10 years of experience with CPM. The following general subjects will be considered: objectives of CPM, experiences from other areas, CPM and the contract, use of computers, standard CPM logic outlines, irrefutable vs. desirable logic, resources and CPM, cost and CPM level of detail, coding, updating, scheduling to resources, misconstruing and limiting float, and comment on precedence diagramming.

Because my experience is largely with the application of CPM to engineer-constructor projects; because I hope to avoid broad, sweeping generalities; and because I prefer to relate first-hand (rather than second- or third-hand) examples, most of the discussion will seem to be directed toward engineer-constructor projects. This should not by any means discourage or turn away the reader whose principal interest is management, operations, research, sales, maintenance or almost any other phase of chemical engineering practice.

Much of what will be discussed is valid across the board. Where effective practice varies, the closing paragraphs will attempt to point out some such differences.

Objectives of CPM

CPM networks attempt to build the entire project "on paper" at a very early stage of the project—even when the scope is not defined, vaguely defined, or incorrectly defined. In a certain sense, CPM is

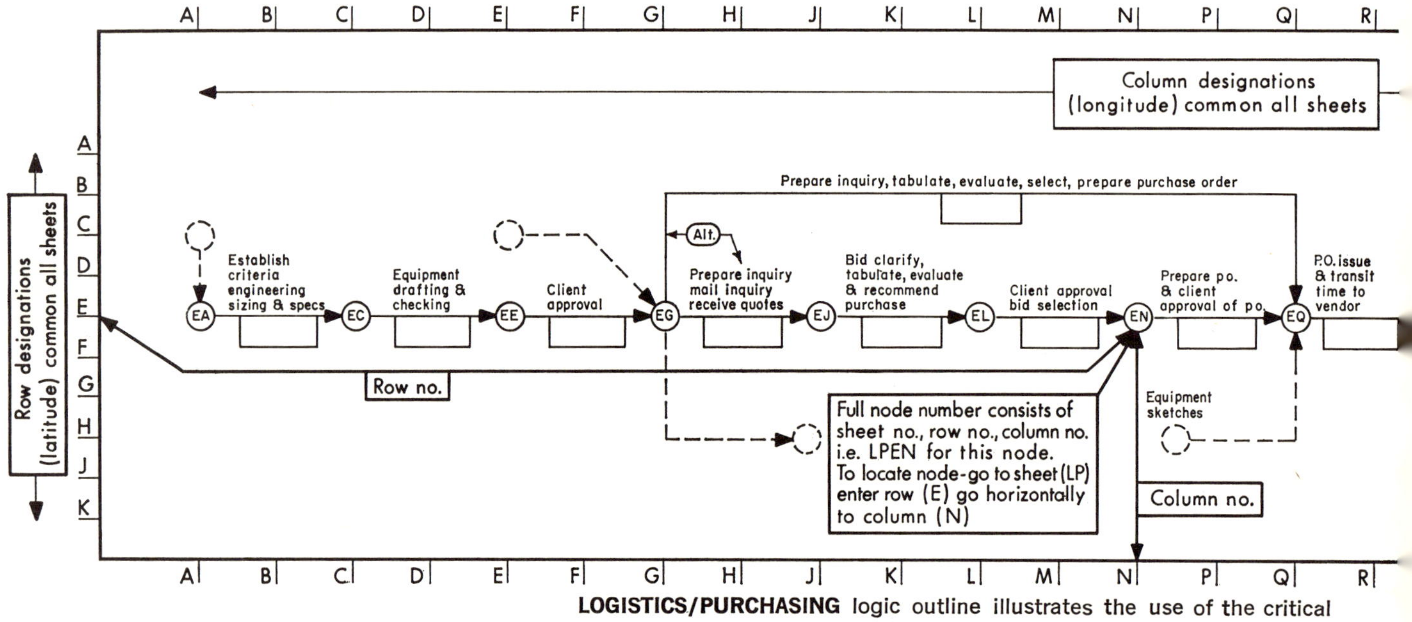

LOGISTICS/PURCHASING logic outline illustrates the use of the critical

# of Critical Path Methods

to project management what modeling or simulation is to economic studies, process control, transportation problems and plant design.

A lot of resistance to CPM occurs because of the planning and thinking through of the entire project that is required to produce the initial data for the diagramming. However, the necessity to enforce early planning for the entire project produces some of the greatest benefits.

Actually, output from the CPM effort is only a tool to be used by the client and the contractor to guide and control the tempo and direction of a project. It must be used as a tool and not misused as a weapon. We have found it effective and beneficial to encourage a brief seminar on the technique and purpose of CPM for all supervisory personnel at the beginning of each project on which the technique is to be used.

Let me emphasize that CPM is a tool; it cannot and will not replace the value judgment of people. CPM networks are pieces of paper that, in themselves, accomplish nothing, and the computer is only a fast calculator that does what you tell it to do. To be a useful tool, the logic networks and durations must reflect reality as closely as possible and be continuously revised to make them currently accurate. CPM must help you control the job you are building, not the one you thought you were going to build.

In utilizing the computer printout for analysis and action, it is important to remember that Early and Late Start and Finish dates represent the earliest and latest time a job can be started and completed *only if* the planning logic is correct. De-

Experience With Critical Path Method

Brown & Root, Inc., received its first exposure to the Critical Path Method (CPM) in 1961 while building a facility in South America. Since that first hand-computed network, we have prepared, run and updated CPM networks on more than 50 engineering and construction projects, ranging from about 250 to more than 5,500 activities, including such diverse projects as a $16-million grassroots power plant, a $28-million chemical plant expansion outside the U.S., a $12-million grassroots chemical plant in Spain, a $28-million papermill addition, utilities and offsites additions and reworks. In an expansion or an addition to an existing plant, turnarounds and tie-in shutdowns are frequently involved; and these have required networks of 260 to 2,600 activities. CPM networks used on actual jobs have been run on six different computer programs, and we have carefully studied and evaluated at least another dozen programs.

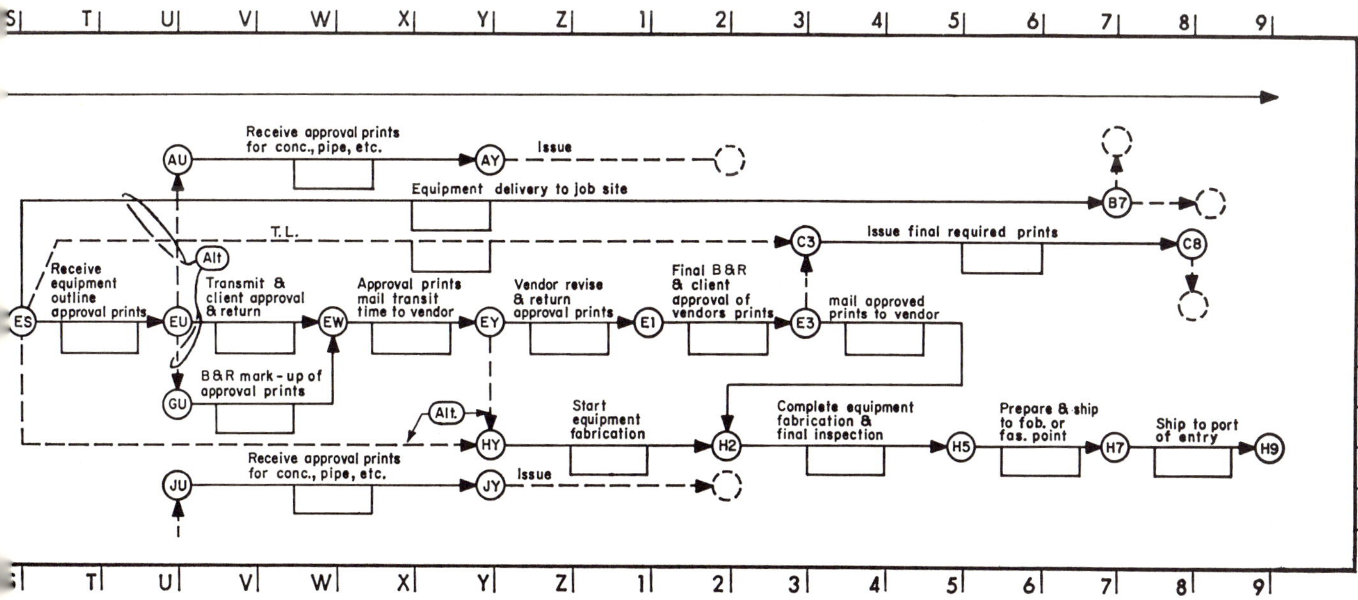

path method for the sizing, purchasing and delivery of equipment—Fig. 1

pending upon whether externally imposed schedule dates are involved on the project or not, the Total Float may run from negative values through zero and on to positive values. We generally consider that anything negative through zero and up to positive (plus) 10 to 20 days float represents a potentially critical chain of activities that should be watched by the project management.

Experience From Other Areas

Application by an engineer-constructor of techniques associated with CPM must be handled differently than in other areas such as commercial construction, maintenance work, prototype manufacture, R&D efforts, state-of-the-art projects such as aerospace or Polaris programs, or sales programs.

These differences: (1) put some limitations on the value of, and approach to, using CPM; (2) invalidate, or at least markedly weaken, some of the literature claims made for the technique; (3) affect the acceptance of the techniques; (4) make extremely difficult, if not impossible, the effective measure of benefits; and (5) yield scheduling costs that can be usually measured more accurately and effectively than when other techniques are used.

CPM and the Contract

I have never yet seen a good set of specifications for the use (or the form) of CPM on a project. I have, however, seen some horrible specifications. In some cases, these were unduly restrictive; in others, they attempted to treat a small project as though it were a several-hundred million-dollar job; in still others, they resulted in considerable cost to the job at no corresponding benefit.

Frankly, I do not know how to write a good set of specifications. Equally frankly, you may be comparing apples and oranges when two organizations just say "The project will employ the critical path method of scheduling." Particularly discouraging is the fact that in the case of CPM, the price ratio of apples to oranges can easily vary 20- to 30-fold, so that it helps to ask for a representative network and output from some previous job.

It is also highly desirable to insist that the project engineer, the project field superintendent, the project manager (if the position exists), and the client's representatives all sign the diagram that is finally used to set the job schedule. As with any good schedule, the networking should start when the job starts but, equally important, must reach a firm state within a useful time—not when the job is 20 to 30% complete.

While CPM can be used to record job progress, it becomes an expensive way to do that if other documents are also kept to record and report progress. Consider making CPM *the* progress document, with provisions in the activity description to show progress in the conventional units used for that particular type of work. Also, if CPM is to be progress and status document, make the entry of actual start and finish dates mandatory. While the use of the "data date" (or date on which the updating information is gathered) can serve the computer very well, since remaining-time-to-complete as of the updating must also be entered, it does however mean that the CPM runs cannot provide a job history that has validity, and therefore other documentation becomes necessary.

Use of Computers

The computer is used only to handle quickly and accurately large quantities of information that you cannot afford to handle manually either because of time or cost. It doesn't do a single thing that could not be done manually. However, CPM on the computer seems to be particularly sensitive to that old data-processing idiom—GIGO (garbage in—garbage out). Brown & Root has studied, tested, utilized and rejected a great number of computer programs for

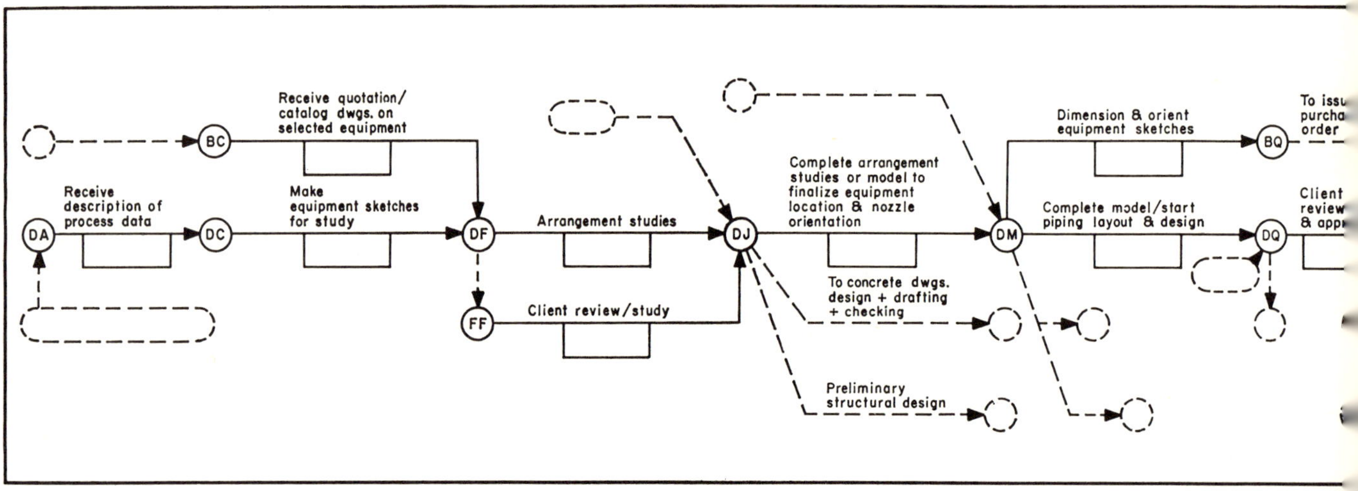

PIPING LOGIC outline is representative of the many networks that are required

CPM; some 10 or more were actually run on test networks of 500 to 2,500 activities.

For runs made outside, we are presently using the MCS (Management Control System) program, which is proprietary with the McDonnell Automation Center of Texas, Inc. This is about the best program we have used for very large networks and where scheduling-to-resources is needed or desired. If you wish to run small networks on a "do it yourself" basis, we believe the RPSM program available from Mauchly Associates is an astoundingly good program for the money.

Brown & Root has developed a proprietary program for handling CPM up to 4,000 activities. It is very flexible on input and output and particularly well suited to allowing easy changes in logic, durations, and updating information. It will accept resource loading on activities, and we plan to add scheduling-to-resources when we have established the most effective way to approach this problem. We generally believe that any project network worth drawing is worth running on the computer. The break point seems to be somewhere in the 50- to 100-activity range.

Standard CPM Logic Outlines

As a result of our CPM efforts over several years, a number of points became evident:

1. An unreasonable amount of lead time and manhours were required drawing and redrawing very similar looking engineering and construction networks.

2. Input listings for keypunching need an inordinate amount of manhours, and are a major source of input error.

3. The description of any activity should be sufficiently complete and detailed so that its intent is clear to one and all when that activity stands alone on an output report.

4. It would be extremely helpful if the physical location of the activity on the network diagrams could be pinpointed by the use of information in the input listing and computer printout.

5. In some cases, limiting the size of the drawing sheet for networks to 24 in. × 36 in. and using multiple sheets make the diagram easier to handle than using a giant sheet showing the entire project network.

All of these considerations helped to point us toward the concept of standardized guide networks. We adopted the grid system of node numbering, which uniquely solves the activity location problem. The example in Fig. 1 shows the application of this method.

The availability of programs permitting additional descriptions, plus judicious choice of wording for these expanded descriptions, produced computer-activity output listings that are meaningful and useful to all project personnel whether they ever look at a diagram or not.

In reviewing the networks that had been drawn, it became apparent that in any area of activity—piping, concrete, electrical, erection, purchasing—the same steps were usually present as a series of parallel paths of sequential activities, with some interpath restraints. The emphasis, in terms of degree of detail in networking, might vary, and the specific status points (nodes) used to establish intereffort restraints might also vary. Thus, by drawing up "generalized" or guide diagrams, 90% of the projects could be covered simply by deleting those alternates not desired, changing some restraints, and adding those restraints peculiar to the project. Typical portions of a "set of strings" designed to cover the design, engineering, procurement and construction activities needed for a petroleum or chemical type installation are shown in Fig. 2.

By appropriately selecting the order of listing of the activity data for generalized nets, and using partially standardized input listings, input preparation manhours can be cut to about 30% of those required to draw and list each network separately.

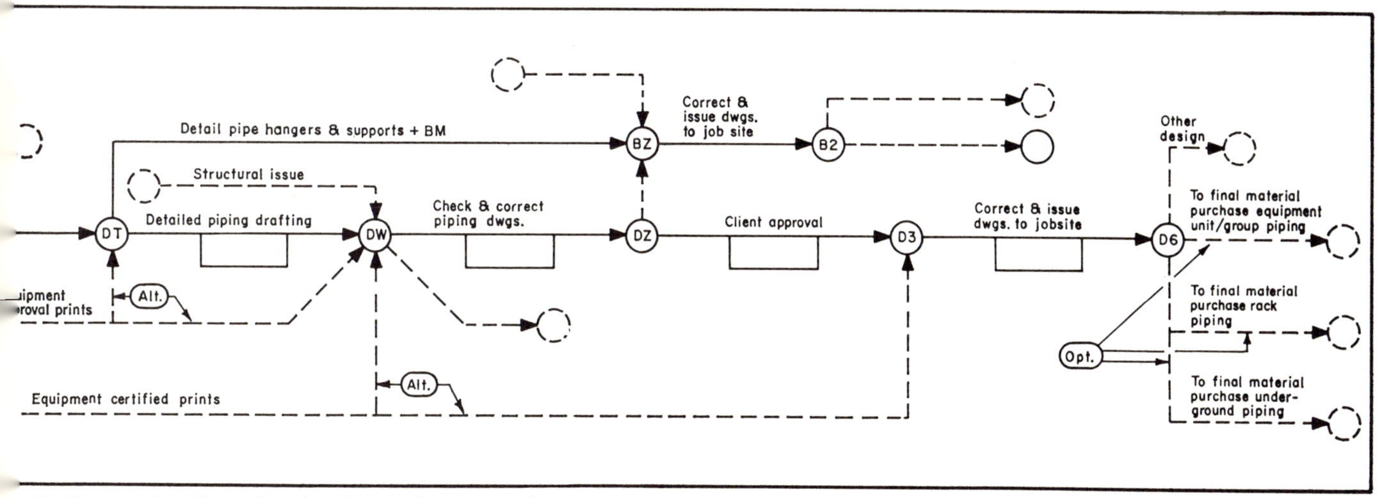

for the construction of major chemical-process plants—Fig. 2

Irrefutable Vs. Desirable Logic

Misleading results, headaches for the scheduler and project manager, or even ridiculous results, occur from the problem of irrefutable logic vs. desirable logic. For example, a draftsman desires to have all necessary information, approved or certified prints, etc., prior to starting a group of drawings. However desirable this logic may be, it is certainly not irrefutable and in most cases is virtually unobtainable.

What does the scheduler show on his diagram as regards information restraints from the engineering design and procurement activities to the construction drawing layout and drafting activities? One approach is to forget it and assume that things will work out as they always have. That may be acceptable, but the result is not a "CPM schedule." This is an example where "parameter float" (discussed below) can help, but it is not a cure-all. In fact, we have not yet found a really satisfactory answer.

Recognizing the two different types of restraints can go a long way toward ensuring realistic diagrams and proper interpretations of computer output. Industrial-plant design and construction abounds with examples of this type of competing logic. Largely leaving out restraints can easily lead to a network that cannot accurately and realistically reflect job status or react properly to updating. Inserting too many "desirable" restraints produces a laughable CPM result, in that the calculated job-completion time is much longer than it ought to be.

Resources and CPM

Since each activity consumes both time and resources, it would seem logical to establish resource needs for it, and use the machine to tell us something about the resource requirements on the project, and about their timings. Typically, you can specify the type of each resource (men or major construction equipment) by using a code, and also the quantity of each resource for each activity. With some CPM programs, this information can only be used to have the computer indicate how many of each resource are required on an Early Start and Late Start basis to meet a project completion date. With other programs, such data can be used to indicate the expected project completion date when certain resources are in limited supply, or to schedule the project activities (to other than Early or Late Start) utilizing "float" in an attempt to fully use available resources while still meeting the specified completion date.

Most programs, or at least the people who use them, approach scheduling-to-resources (also called leveling) by requiring as input the availability levels (pool sizes) for all crafts. When you do this, the computer generates little new information but largely tells the job supervision that its judgment was pretty good.

We believe there is an effective way to involve field and engineering personnel in the "scheduling to resources" when a computer is used. To be effective, we believe the computer must supply them new information for successively better approaches to an acceptable project schedule.

We have seen enough Early and Late Start manpower usage curves to be convinced that those who talk about following a "CPM Schedule" really mean "CPM Guide." Straight CPM produces only a permissive network and some limiting dates that can be used to guide the scheduling of the job except for the near-critical activities (zero to ten days float). The bulk of the results are a long way from being a schedule that a project engineer or a job superintendent can use effectively. For example, the project supervision may be planning to delay the start on the first activities of a chain by not assigning resources until sometime after CPM Early Start but before Late Start. If this is done, it will reduce the float available to the later activities on that chain and make float shown on the CPM output for the

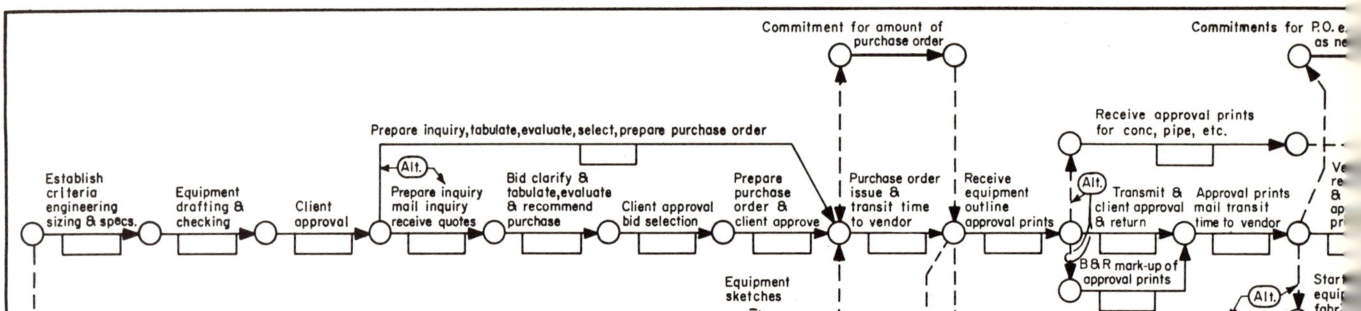

COST CONTROL NETWORKS, using critical path methods, show marked differences

later activities unrealistic with respect to the actual project schedule. Only when resources are considered do you begin to get a realistic, effective schedule for the entire project.

For programs requiring arbitrary, "planner-generated" priorities on each activity, a basic problem occurs because the planner must assign priorities in a context of assumed or visualized work that has been accomplished and competing work that is in progress. When the program attempts to use one of these priorities in scheduling-to-resources, the chances are excellent that the "job status" the machine faces is markedly different from that visualized by the planner at the time he assigned priorities. In fact, could he intervene and face the "job status" the machine faces, he would change priority assignments.

Costs and CPM

There is a great tendency on the part of clients, programmers, consultants and the computer-sales groups to propose and push for cost reporting and even control on the CPM network using the computer program. The argument goes that you are already scheduling, reporting progress, forecasting progress, etc., so why not get the same benefits for costs that you do for time. Additionally, the offhand statement is usually made that "really all you need to do is add the cost of the activity to the input data you've already gotten, and the machine will do the rest." We have attempted some of this and have tentatively concluded that including cost on the scheduling network is misleading. If used in project control, it can lead to serious problems in job management.

The root of the problem lies in the somewhat naive assumption that a network needed and useful for scheduling will serve equally as well for cost reporting and control. This simply is not true. The chain of activities needed for the scheduling effort and the chain needed for cost reporting and control on a piece of equipment are shown in Fig. 3. Note that there is very little, if any, similarity. While it would be possible to add the activities for cost control to a CPM scheduling network, the results would probably not be acceptable either to those interested in cost control or to those concerned with job schedule and progress.

A further complication arises in that most computer programs assume that cost is linear with time over the duration of the activity. While this may be nearly valid for heavily labor-sensitive activities, it certainly is not true, from the purchaser's standpoint, for such activities as fabrication and delivery of equipment.

This method of handling costs on the CPM computer program can produce even more unrealistic and disastrous results when it is used along with activity progress reporting to generate a "value earned" report in order to check constructors' monthly billings. It is suggested that several trial runs be made early in the project life, preferably before this feature becomes a contractual requirement. The ability to use a cost feature on CPM can be useful when studying the cost and time effects of alternate sequences of doing the work; of overtime to shorten project completion, or to keep a project on schedule; or of estimating probable job-fund requirements. At present, cost features of CPM programs are primarily useful as guides and tools but not for cost reporting or control.

Level of Detail

One question that always seems to arise relates to the degree of detail that the networks should reflect. The use of the grid system and the availability of 6,000-node (or larger) computer programs virtually permit the detailing of a design-construct project (where the field work and the preparation of each document and drawing could be shown) almost

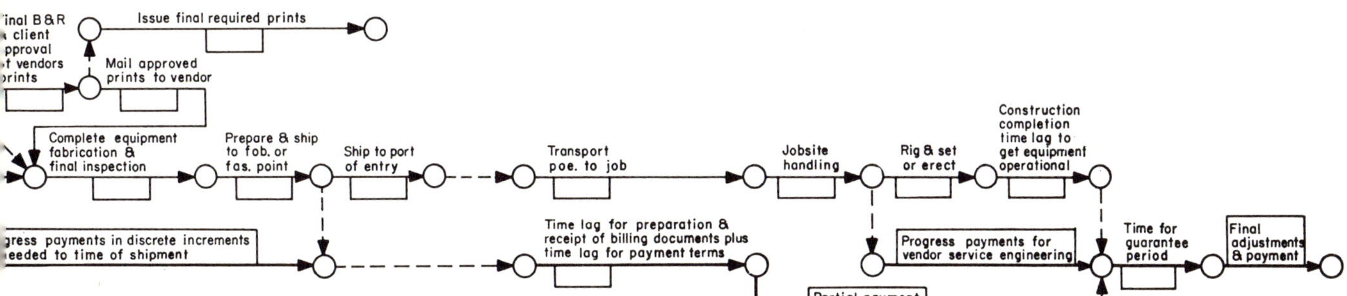

to the corresponding complex scheduling networks—Fig. 3

to the level and elements of a time study. However, the most important thing to remember is that CPM on a design-construct project should be used as a tool by management to set over-all job sequences and timing, and by the working supervision to guide the meshing of their daily activity towards project goals. On such a project, it should not and cannot substitute for the skill, adaptability and experience of the supervisory personnel. Information status and material status, together with the resultant work possibilities, change too rapidly to make a network in great detail usable, unless there is daily updating.

There are some additional guidelines, however:

1. Those interfaces where responsibility (between groups within an organization or between organizations) changes should be represented as clearly and accurately as possible.

2. The activities represented should respect the basis assumed for the CPM mathematics—that once started, the work can and will be carried out continuously and expeditiously until completed in the time shown.

3. If resource loading on activities is provided and scheduling-to-resources is practiced, the activities should reflect accurate resource usage. For example, if cranes are critical, the use of multiple activities (called "erect piping") with durations of 50 or more days may result in excessive crane requirements compared to those that any knowledgeable job superintendent would assign for the total work to be done. Another example would be the structural welders needed to work (but part-time only) with the electricians in running conduit or setting motor-control units.

4. CPM results should establish meaningful target dates (Early Start and Finish) that enable the lowest project-supervisory personnel to draw up their own daily work schedules and assignments. This may sound like a platitude, but you simply cannot do the kind of daily or hourly craft scheduling on a CPM run for industrial design-construct work done for maintenance work or some types of commercial construction work.

There are some tricks that can be used to permit greater detail at an effective level and time. One such is the use of a summary network, covering the total job through which a band of greater detail progresses, representing the work for the next 2 to 3 months.

Uneven degree of detail can be a serious problem in acceptance and usefulness of networks. The scheduler is always influenced by his background and tends to expand the detail of activity breakdown in the areas where he has had experience. The use of generalized logic outlines does much to provide a more uniform, consistent, and useful level of detail.

Typical activity coding for CPM networks—Table I

Coding Structure	Code Significance
Organization	
CC	Client
EA	Engineering consultant or advisor
ED	Engineering designer
FA	Contractor, prime or general
Craft	
DD	Clients department, design
FA	Field forces, insulators
GA	Design & drafting group, architectural
SE	Suppliers' personnel, equipment
Work Done	
BE	Billing, equipment
CB	Construction, building erection or completion
DB	Assembled bid package and specs
EE	Erection, set equipment/components

Coding

It is difficult to overemphasize the many advantages deriving from a sound, logical mnemonic coding structure for activities and resources. Such coding is of no value to the computer in performing its calculations, but is of inestimable value to the planner in assembling the computer output—in readily usable and easily handled form.

A suggested coding for activities is shown in Table I. We believe a four- or five-digit activity code is the minimum to permit the needed controls. A single alpha digit for resource coding can be made to work, though mnemonic significance may suffer.

Updating the Project Schedule

In general, we have performed updating on projects at approximately one-month intervals. We prefer that no engineering or construction activity exceed about 20-days' duration. The major exception is deliveries for which the information is received as a "promised shipping date" rather than as "time to complete." Using a 20-day maximum duration on a monthly update, we are reasonably assured that:

1. No great number of activities will have to be in progress on any one report date; thus we reduce the number of estimates on time-to-complete that must be made.

2. An error in estimating time-to-complete probably will not markedly affect job completion, usually not more than one week.

3. The number of large and loosely defined activities appearing in the project network that can adversely affect job logic, planning and monitoring tend to be minimal.

Scheduling of Resources

In performing scheduling of resources, we have taken a novel and desirable approach to getting people to participate with the computer in the scheduling step. To accomplish this, each activity is appropriately "labor loaded" with direct labor (real crafts, and equipment if desired) plus two fictitious crafts called Total Labor and Crew. Total Labor uses the same resource code for all activities but Crew uses a code that matches the performing nature of the activity such as piping, electrical, millwright, rig and set, concrete, etc. The quantity for each or both of these crafts in any activity is equal to the sum of the direct-labor (real craft) resources assigned to that activity.

The initial step in scheduling-to-resources is to obtain: (1) the time distribution for the fictitious-craft Total Labor on both Early Start and Late Start, and (2) the total resource-time units for each resource needed to perform the project network as drawn and resource-loaded. These data are obtained from a simple CPM run, and they form the basis for all subsequent input of availability levels and timing for resources against which the computer program is asked to schedule the project.

Looking at the Early-Late Total Labor curves from the first computer run, the project engineer and the job superintendent provide their estimated total direct-labor manpower curves for the project. This permits a scheduling-to-resources run against only the Total Labor resource, by using availability values to match these curves while all other resources ("real" and Crew) are considered as unlimited. The output from this run is a time distribution of the Total Labor resource needed to hold the project's scheduled completion, while attempting to hold the desired input availability curves. In addition, Crew usages making up that Total-Labor curve are output.

In the next step, any adjustments desired in the Total Labor pool are made, as well as in the Crew curves for those Crews in which scheduling of the work by the program (with only Total Labor limited) produced unreal or unworkable Crew usages. This step may involve the craft foreman to establish desired crew sizes.

A second run is now made, limiting Total Labor to availability values from the desired curve, and adding the restrictions on Crew availabilities. The output this time is usage of Total Labor and Crew as a time distribution, plus individual craft usages making up the Crew usage.

In the final step, the individual crafts are reviewed, availabilities established or adjusted for those that are needed, and the last computer run made for scheduling to Total Labor, selected Crew and selected Craft availabilities. The output from this third step generally is a thoroughly workable schedule that considers resources; one that all those directing the project can readily accept with confidence.

In this approach, the computer-generated output on labor usage always contains at least one level of greater detail than that which the project personnel were asked to provide as input.

Users should recognize the power of this program capability. For example, a common yardstick for reasonable, effective fieldwork is 100 to 150 sq. ft. of plot area per man. Plot area is ground area plus working area on platforms and structures. If this density limit can be established for each area of the plant, then an estimate of the "plot area" can be used to establish a maximum number of men in the area. By coding this "resource" properly and by assigning it for each activity, plus setting the area "density limit" as a resource pool, you can get the machine to consider personnel density at the same time that it is attempting to schedule work to actual craft pools and for desired completion dates.

The same approach can be taken to evaluate the effect of weather on a job. For example, to establish a resource code of "rain," assign a quantity of one to each activity that could be affected by rain, and set a "rain" resource pool that is infinite (very large) on the days when good weather is expected and zero during rainy days. When the program is presented with these activities and the resource availabilities, it will not schedule rain-affected activities during anticipated rain periods.

When considering programs for handling resources, there is an important difference between industrial-construction activities and maintenance or turnaround activities, or even commercial work. In the latter, something akin to a production-line operation occurs, except that crews flow to and through the work instead of the work going to machines or stations. This difference is reflected in the programs that "schedule to resources" as opposed to those that "level resources."

It has been our experience that to be useful the program must allow incrementing and decrementing available resource pools in steps that are multiples of one resource unit and one time unit. Programs that "level" resources by attempting to make a rectangle of the labor usage histogram are of little value in industrial plant-construction projects.

Misconstruing and Limiting Float

One of the most insidious problems in the effective use of CPM appears to lie in the definition and use of Total Float. While everyone can follow the definition of Total Float, it becomes very difficult for each group having activities on a chain of float with other groups to continually remember that the float is shared, and is really a "communal chain" property. The problem is aggravated when you break up output according to area of interest or responsibility in order to get the paper volume down to manageable size. Everybody tends to feel the float is his alone, to use without regarding others on the chain. Suppressing the printing of float and Late Start and Finish dates eliminates one of the potentially useful and effective results of a CPM calculation.

We have found no good solution to the problem except to continually remind people that float is communal, to urge them to look at the diagram to see who they're going affect by using float, and to consult with and advise others of any action.

On engineering construction projects, we have found that the actual expected logic or sequence of the work in many areas cannot be accurately diagrammed at reasonable time and cost. There is also the problem (inherent in CPM algorithms) of considering essentially unlimited resources available to the project as a whole, so that only duration and logic considerations are handled. The net result is that some misleading Total Float can appear on a number of chains. It is false and will never exist on a project, even assuming the logic and duration are correct, because it is not economically feasible. Programs permitting a so-called "Parameter Float" adjust the Late Finish for specified milestone nodes, or events, such that each has no more than the specified Parameter Float.

Once established, this limiting float works on back through all entering chains. Despite the potential of confused CPM output if used indiscriminately, Parameter Float can markedly improve the acceptability of a CPM computer output. This is one of the two devices that can be used to establish the Late Finish for activities.

The other device is the schedule (or "not later than") finish that must be entered as a date, and can only be used if it is a contract date or if the "right" date is known beforehand. Parameter Float can be used when no dates are known, and will continually adjust late finishes as project updating and progress reporting are accomplished.

Miscellaneous

One of the most glaring deficiencies of CPM when applied to industrial-plant design-construct projects is that CPM cannot produce, expand or replace the real working schedule. However, it can make such a schedule more realistic and easier to generate. It can and does produce permissive or limiting dates and valuable information about the interaction of project components that enables effective scheduling (or rescheduling) to be done.

CPM considers only time, which is not the real world of the engineer constructor. It requires accurate representation of continuous work segments at essentially constant production rates, which are not technically and realistically feasible for the engineer-constructor. No CPM program or network can produce a really viable schedule without resorting to scheduling-to-resources; gimmicking and gadgetry such as time lags, Parameter Float, scheduled dates; or manual extraction and manipulation of the output.

CPM is an excellent tool for setting target dates and for calling attention to areas threatening orderly and timely job completion. Because of the origins of CPM, its mathematical base and the algorithms used, there is not much that can be done about this deficiency except to recognize it.

One of the most elusive problems in connection with CPM is in the effective presentation of results. We still are not satisfied in this area, and believe it is a very fruitful one for further research, since it has a healthy effect on the reception accorded the CPM technique. Graphic output either on a plotter or on the line printer can help considerably. Condensation can also help through such techniques as (1) a Hot Line Report showing only those activities starting or finishing in the next 30 days, or (2) a Limited Float report showing only those activities with float less than, say, ten days. A computer program generating over-all pseudo-activities to summarize the total time span over which similar activities are occurring can be helpful too. Coding has already been mentioned as a way to assemble the output in meaningful packages.

We have looked at computer-drawn networks. Except for being pretty and perhaps more legible than hand-drawn ones, we see no presentation advantage and have incurred some respectable costs. We do feel plotting of early or late (or leveled, if used) resources is very helpful in communicating with project personnel.

Time-scaled network diagrams always seem to come up in talking about output. Let me say first that we do not practice time scaling of network diagrams and we do not believe it is justified even once. Properly sorted, the bar chart can serve the same purpose more effectively. We have done some experimentation and do not presently feel that computer-plotted time-scaled network diagrams are feasible costwise, or even make a very effective presentation. If the diagrams are to be drawn time-scaled, draw them with chains of increasing float, moving down from the critical path at the top of the sheet. Drawing chains with more than 10 to 20 days float does not serve much purpose. If only CPM has been run, there is a problem in deciding whether to plot each chain on Early Start or Late Start, and how to properly and effectively show float. There doesn't seem to be any good answer.

Establishing an effective, real, duration within the context of how the program will handle that duration can pose a problem and a potential trap—depending on the activity and how the network is drawn. For instance, if all major restraints are not realistically drawn, a duration estimate in lapsed job-days can lead to a much better result than the more conventional working-days estimate emphasized in the CPM literature. The classic example is the piping drawing that represents 120 man-hours of applied effort (15 working days) but may take 60 lapsed job-days (3 months) to reach the issue-for-construction stage because of the need for vendor's information, client review, checker availability, etc.

Note that resource usages, summaries and leveling can go badly astray if such lapsed-time duration activities are assigned resources. Entry of the drawing issue as a dummy restraint on field activity having a start-not-sooner-than date may be preferable. We try to select the handling of the activity to match the type of network being generated, and to get those people who are estimating duration to visualize how the computer will handle the duration value.

One facet in using CPM on design-construct projects that has received no mention in the voluminous literature relates to the ability to fake or purposely tell stories with CPM. Make no mistake about it—IT CAN BE DONE! I would hasten to add that it takes considerably more effort, skill and experimentation on the part of those scheduling to fake with CPM than to fake with the more conventional bar chart. However, CPM can be used to show that almost any progress, up to about the 40 to 50% point of the job, is indeed on schedule.

Sooner or later, of course, your sins catch up with you if you can't make up progress. If other job-control tools are being used with experienced judgment, you will be caught in your lie when the job is somewhere between 10 and 20% complete. This faking is particularly attractive and easy to do when only CPM is being used to monitor the project, and when activities are described as "Start . . .", "Continue . . .", or as ". . .% finished" in three or four increments. Generally, when faking is done, "dummies" and time lags, as well as not-sooner-than dates and Parameter Float options, are changed on successive updates. This may be hard to distinguish from a printout only (since all such changes can be easily suppressed) and a close scrutiny of the network diagram may be required to uncover them. I do not mean to imply that networks cannot, will not or should not change during the course of a job, but there should be a sound, logical explanation for each change.

One last point relates to the mountain of literature that CPM and allied techniques have generated. The author has compiled a 41-page bibliography available upon request.

Comments on Precedence Diagramming

Precedence Diagramming is a newer version of CPM designed to more nearly fit (in terms of diagramming technique, activity coding and permissible representation) the manner in which design engineers and field supervisors plan and execute their work. It has almost universally received acceptance from the people who should use (and benefit from) the planning and scheduling effort. From the standpoint of the planner, it permits simpler techniques for showing certain relationships, and by eliminating dummies as a computer requirement avoids unintentional logic errors. Scheduling to resources can get sticky, and lead-lag factors that are permitted can sometimes make the interpretation of output more difficult. Depending on the networking technique adopted, the degree of detail needed and the computer program selected, there can be some savings in computer costs for running a precedence diagram over a CPM network. All that has been said for CPM applies to Precedence Diagramming. The difference is largely one of technique.

Precedence Diagramming has not yet had the wide acceptance nor received nearly the amount of publicity accorded CPM. I feel it has numerous distinct advantages, and the coming years should see much wider acceptance of it, both by present users of CPM and nonusers of networking.

Techniques for Other Projects

The discussion and comments on the use of computers, standard CPM logic outlines, level of detail, coding, misconstruing and Limiting Float, miscellaneous and Precedence Diagramming are of a general nature, and should be usable by individuals in any field of endeavor. However, the discussion has been oriented toward the use of CPM on engineer-construct projects. There are some variances and even contradictions when you talk about other types of projects. The following paragraphs will attempt to relate our experiences and developments as an engineer-constructor to other areas. In so doing, I must pass from the realm of first-hand experience to the position of "It seems to me that . . ."

1. *Irrefutable vs. desirable logic*—In maintenance or turnaround scheduling, the use of irrefutable logic might be desirable to get an idea of what the maximum time could be. Likewise, the cost of downtime for maintenance or of the repeated cycles for commercial construction could provide enough economic incentive to warrant diagramming so detailed that the work is accurately represented. Our problem basically stems from the complexity of our work. Other fields may not have the myriad, almost daily interacting, restraints found on an engineering-construction project, so that a concise, very accurate arrow diagram can be economically drawn.

2. *Resources and* CPM—I feel that scheduling to resources is one of the most underused features of CPM programs that allow the use of resources—even if they are tabulated on an Early and Late basis. For instance, an "arrival activity" having a negligibly short duration with respect to the total job duration and the proper coding of "resources" and "resource" quantity are useful. Therefore, the resource-plot program will draw a histogram of "arrivals," be they drawings, parts, equipment or materials.

For certain types of endeavor, such as maintenance scheduling, commercial construction, production planning and perhaps R&D work, I can visualize that planner-assigned priorities for activities will be very helpful. They could result in acceptable schedules that would consider resources even if only some activities were assigned a priority scale and the others treated uniformly below them. However, priorities must be realistically assigned, and the output from such a scheduling run needs careful review to assure that these assigned priorities are indeed valid in the context of competing activities as seen by the computer program.

3. *Costs and* CPM—To those knowledgeable in the field, the most glaring omission in this article has

been any mention of those techniques and programs that consider costs in compressing (by machine calculation) the project schedule and generating costs for alternate schedules. They are known by various names, though the generic names most commonly used are probably Least-Cost Expediting, or Time-Cost Tradeoff. Central to their entire operation is the concept that you can and will establish for each activity a so-called normal cost with its associated time, and a higher crash cost with its associated shorter duration.

Existing programs can treat a fairly complex network (1,000 to 1,500 activities) with simple, rigid cost models, or treat insignificant networks (50 to 100 activities) with fairly elaborate and flexible cost models. The real problem is one of developing input cost data that are realistic. If you have a project for which data are available and if your project fits the premises and limitations of the programs, you can obtain with relatively little effort some very intriguing and effective results. But you must be sure your data and project features fit the program requirements. We have never felt we were that fortunate on any of our engineer-construct projects.

I understand these programs have been run with considerable success on turnaround maintenance projects and saved some handsome sums in terms of over-all net cost.

An interesting development that is somewhat analogous, and is known as the Decision Tree has been used with outstanding results in R&D. This approach defines a project as a series of decision points occurring in time. At each decision point, there are only two branches, and you must put a money value and a probability of success on each branch. You then carry each branch out through further decision points until all routes reach the desired end. From such a model, you can obtain valuable investment return data to assist in R&D decisions.

In any resource-time tradeoff or expedited cost program, the machine must follow your logic. It never changes logic but only looks at the time-cost slope of activities to shorten duration. It has been our experience that revisions in logic, which only human beings can introduce, frequently make a much more significant contribution to shortening project time and even reducing project cost than multiple computer runs that massage the same old non-inevitable or infallible logic cranked in the first time.

4. *Updating*—Frequency of updating must depend on the nature of the project and the value of the updated information. I have heard of one turnaround maintenance operation that updated twice a day by remote teleprocessing. They updated at noon, had reports back by 3:30 to 4:00 p.m. and planned the next shift's work. A subsequent update produced a new set of schedules and control documents for an early morning briefing. The basic criteria expressed above for updating frequency are really valid regardless of the time scale. The more frequent the update, the more desirable an update report and simplified "file maintenance" procedures for the program become.

5. *Scheduling to Resources*—Much of what has been said above is valid in principle. It is making or finding the analogy that is difficult. In maintenance or plant revamp work, for example, the "rain" resource could become items of production equipment needed for modification, relocation or rework. Instead of a weather forcast, then, to establish resource availability, the forecast for that equipment is used.

For production planning, it could become sales commitments from a seasonal market standpoint or from a committed schedule basis. In R&D, it could become vacation schedules for critical personnel or the previous commitment of essential and scarce facilities, instruments, etc. Fundamentally, resource availabilities used this way enable you to establish on a dated time-scale the ambient against which the project is to be run. You do not have to establish where in that time-scale the project will be interacting with that ambient, until the computer is running and making the Early, Late calculations.

With respect to difference between "leveling resources" vs. "scheduling-to-resources," the former may actually be preferable for a maintenance foreman or an R&D director. ∎

Acknowledgements

The author wishes to express his appreciation to the large number of people who have helped to educate, enlighten, and encourage him on CPM. In particular, thanks are due to M. Datz and the members of his scheduling group at Brown & Root; Russ Archibald, now of Booz-Allen-Hamilton; and to U. F. O'Brien, a former Brown & Root employee. E. Engman, of Mactex and J. D. Craig of Brown & Root were particularly helpful on computer techniques and applications in project control. Thanks go to the management of Brown & Root, Inc., for permission to publish this material.

Meet the Author

Eric Jenett is coordinator for the process group at Brown & Root, Inc, Houston, Tex. 77001. Following pilot-plant work at Tennessee Eastman Co., he joined Brown & Root in 1951 as a process engineer. His later assignments have included liaison and startup, in the U.S. and as a resident on construction for ANIC at Ravenna, Italy. He has an M.S. in chemical engineering from Columbia University and is a registered professional engineer in Texas. His society memberships include AIChE, Phi Lambda Upsilon and Sigma Xi.

Linear Programming: Advantages and Shortcomings

Linear programming is a useful management tool. But the user should be aware of its pitfalls as well as its virtues.

W. E. NORRIS, Union Oil Co. of California

Linear programming is a mathematical method for solving problems concerned with decision making. It is essentially a means of making systematic selections.

Linear programming (LP) is an optimizing technique. That is, it not only gives a workable solution to a problem, but also the best, or highest profit, solution from a large number of possible alternatives.

LP is used to plan operations, blend products, make production schedules, calculate prices, design units, control inventories, evaluate raw materials, assess sales opportunities, and perform many other functions.

It has both good features and shortcomings. This article explains what LP can do, without delving into how it works. The weaknesses of LP are described, as well as the strengths.

What Is Linear Programming?

As the term "linear programming" is generally used, it includes the formulation of the problem, finding the solution, and exploring or analyzing the effects of change. Practical problems tend to become rather large, so it is common to use a computer to calculate the answers. This is not because the mathematics are difficult, but because of the extremely large number of calculations involved.

Excellent computer systems are available to assist the user of LP. Hence, the user does not need to write the intricate computer program required to solve LP problems. He merely needs to arrange his problems in a prescribed manner and furnish the required input data. The computer solves the problem and prints the answers.

Generally, the user need not have a detailed knowledge of how the computer finds the optimum solution, or how it measures the effect of changes. He should, however, understand the elements of these methods so that he can both formulate his problem effectively and interpret the results intelligently.

Linear Programming Requirements

Linear programming can be applied to a problem only if certain criteria are met. These are:

- The relationships between the variables must be expressed by a set of linear equations. In two-dimensional systems, the equations would be straight lines—hence the term "linear." In multidimensional systems, this means that variables appear only in the numerator with an exponent of plus one.*
- The problem must contain more variables than equations. (If the number of equations equals or exceeds the number of variables, a unique solution can be obtained by ordinary algebra.) When there are more variables than equations, a very large number of solutions is possible. This is the realm of linear programming—the situation where it is

This article is a slightly shortened version of a paper presented at a meeting of the 1969 National Petroleum Refiners Assn. at San Antonio, Tex.

* A linear equation is one of the form $aX + bY + cZ = \text{RHS}$, where a, b, c, and RHS (right-hand side) are constants and X, Y, and Z are variables. No terms of the type X^2, XY or a/X may appear.

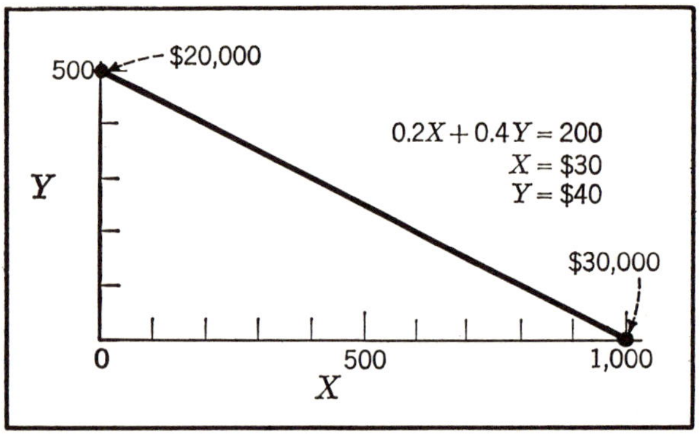

ECONOMICS plus equation yields LP solution—Fig. 1

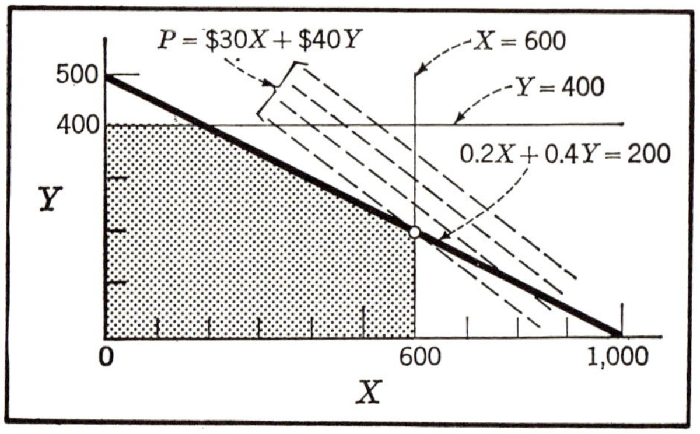

GRAPHICAL solution of wax production problem—Fig. 2

desired to find the best solution from many feasible alternatives.

• No quantity can be negative. This ensures that all answers are significant. Thus, an LP solution will not prescribe a meanlingless operation such as a negative charge to a process, a negative yield, or a negative transfer from one operation to another.

• There must be a directing force or economic objective to be maximized or minimized during the solution. This is what guides the calculations to select the one best solution from the many possible alternatives.

A Simple Example

The equation $0.2X + 0.4Y = 200$ may be used to explain the preceding requirements. As shown in black in Fig. 1, this is a straight line that intersects the Y axis at 500 and the X axis at 1,000. It meets the first three of the LP criteria: it is linear, has more variables than equations, and positive quantities. It has no unique solutions—a feasible solution lies anywhere along the line.

However, if we assign economic values to X and Y ($X = \$30$ and $Y = \$40$), the situation changes. We can now find the most profitable solution. The solution $X = 0$, $Y = 500$ shows a profit of $20,000 while $X = 1{,}000$, $Y = 0$ shows $30,000. The latter is obviously the best, or optimum, solution provided neither X nor Y is allowed to have negative values.

Linear programming functions much in this manner. It operates in a stepwise fashion such that the profit is continuously improved until the computer program detects that no further improvement is possible. The value of the variables at this point is the optimum solution.

Formation of Matrixes

LP involves a matrix-type solution. This means that the problem must be formulated in a row-column format, with the rows representing equations or restrictions, and the columns representing variables. The matrix elements are the coefficients of the variables in the equations, with the sign intact. Thus the equation $0.2X + 0.4Y = 200$ would be represented in matrix form as:

X	Y	RHS
0.2	0.4	200

where RHS means right-hand side.

Inequalities

Few practical problems can—or should—be expressed in terms of true mathematical equalities because of the need to provide freedom of action. A specification limiting the benzene content of cyclohexane, for example, would not be set equal to 0.05 volume %, but to a *maximum* of 0.05. Mathematically, this would be expressed by an inequality:

Benzene content $\leqq$ 0.05 volume %

This means that 0.05 or any value below this will satisfy the specification. Likewise, cyclohexane purity could be assured by controlling the cyclohexane content to a minimum of 99.8 volume %:

Cyclohexane content $\geqq$ 99.8 volume %

Thus 99.8 or any higher limit will satisfy this specification.

Also, a unit-capacity limit or a product requirement would seldom be set equal to a fixed amount because factors such as quantity, quality, economics or other considerations would make it impractical or impossible to exactly meet the prescribed figure.

Consequently, a unit-capacity limit could be set to a maximum of 32,000 lb./hr., and a product requirement might be set to a maximum of 20,000 gal./day:

Unit capacity $\leqq$ 32,000
Product requirement $\leqq$ 20,000

Slack Variables

An LP system cannot handle inequalities directly, but it can handle them if the "slack variables" concept is used to convert an inequality into an equality. Take the unit capacity used above:

(Inequality) Unit capacity $\leqq$ 32,000
(Equality) Unit capacity used +
Unit capacity not used = 32,000

Thus, if 32,000 lb./hr. are made, the slack will be

zero. If 30,000 lb./hr. are made, the slack will be 2,000 lb./hr. If none of the capacity is used, the slack will be 32,000 lb./hr. (In an LP solution, neither the slack nor any other quantity can be negative.)

Since a slack variable appears in nearly every LP equation, it is common practice to identify each equation by the name of its slack. Thus slack variable *XYZ* would appear in Eq. *XYZ*, and slack variable *XSCAP* would appear in Eq. *XSCAP*.

A Two-Dimensional Sample Problem

Let us formulate a simple LP problem. It will be two-dimensional, so that we can draw a picture of it.

A wax press produces two grades of wax. Raw material restrictions limit monthly production to not more than 600 tons of Grade X and 400 tons of Grade Y. The press takes 0.2 hr. to produce a ton of X and 0.4 hr. per ton of Y; there are 200 available production hr./mo. Profit realized is \$30/ton X and \$40/ton Y.

(Availabilities) Wax $X = 600$ tons
Wax $Y = 400$ tons
(Time) $0.2X + 0.4Y = 200$ hr.
(Profit, P) $P = \$30X + \$40Y$

Graphical Solution—Fig. 2 shows that the solution must lie within a polygon bounded by Eq. $X = 0$, $Y = 0$, $X = 600$, $Y = 400$, and $0.2X + 0.4Y = 200$. The solution occurs where the family of parallel profit lines $P = \$30X + \$40Y$ touches this polygon at its farthest point from the origin. The point $X = 600$, $Y = 200$ meets this requirement; the maximum profit is:

$$P = (600 \times \$30) + (200 \times \$40) = \$26{,}000$$

LP Solution—This same problem can be solved by LP if we use slack variables to remove the inequalities. The equations then become:

(Wax X) $X + S_X = 600$
(Wax Y) $Y + S_Y = 400$
(Time, hr.) $0.2X + 0.4Y + S_T = 200$

We now have three equations in five variables. These may be arranged in matrix form as shown in Fig. 3.

As there are three equations and five variables in the matrix, only three of the variables can have non-zero right-hand sides or activities. These variables are called "basis" variables. The other two variables are called "nonbasis" variables. They always have zero activities.

In LP, all nonbasis variables have cost sensitivity indicators associated with them. These indicators are given many names, but are most commonly called "reduced costs" or D_j values. We will call them D_j's in the balance of this article. They are an important part of the LP solution because they indicate the profit change that will result if a change is made in the current status of the solution.

The D_j's reflect the monetary incentive or cost of an incremental quantity. They indicate the incremental profit opportunity if another unit is sold at the same price as the last. Also, the algebraic sign of the D_j is important in that it indicates whether optimality has been reached. In a profit-maximization problem, such as the one above, optimality is reached when all D_j's have positive signs. (This is what tells the computer that the solution has been reached.)

The negative D_j values shown in Fig. 3 for variables X and Y indicate that this is not an optimum solution. Also, it shows that the overall profit could be improved by \$30 for each ton of X and \$40 for each ton of Y that can be brought into the solution.

Without going into the intricacies of LP calculations, the final solution is shown in Fig. 4. The positive D_j values indicate that no further profit improvement is possible, and that the solution is optimal.

An LP solution gives a simultaneous material balance and cost balance. The Fig. 4 RHS (right-hand side) values show material balance or activity values: $X = 600$, $Y = 200$ and $S_Y = 200$. These values are identical to answers from the graphical solution in Fig. 2. The cost balance is given by the D_j values. Here the profit of \$26,000 is the same as that calculated from the graphical solution. The D_j values for the nonbasis variables indicate the economic penalty for deviating from the optimum solution. Thus the D_j for S_X shows there is a \$10 profit loss

	X	Y	S_X	S_Y	S_T	RHS
Cost	\$30	\$40	0	0	0	0
S_Y	0	0	0.5	1	−2.5	200
Y	0	1	−0.5	0	2.5	200
X	1	0	1	0	0	600
D_j	0	0	10	0	100	26,000

LP MATRIX; starting basis—Fig. 3

	X	Y	S_X	S_Y	S_T	RHS
Cost	\$30	\$40				
S_X	1		1			600
S_Y		1		1		400
S_T	0.2	0.4			1	200
D_j	−30	−40	0	0	0	0

LP MATRIX; optimum solution—Fig. 4

for every ton of X not produced. Also the D_j for S_T shows that the penalty for not using all available production time is \$100/hr. Additional information may be obtained regarding the cost and volume ranges over which the optimum solution applies.

A comparison of the graphical and LP solutions indicates that LP solutions always occur at the point where two or more equations or restrictions intersect. If the cost equation should happen to be parallel to one of the bounding equations, alternative optimum solutions would exist, but most LP systems would report only one corner solution.

Model Formulation

Linear programming problems are formulated by describing the process (or operation) to be simulated in a series of independent linear equations. Such a formulation is often called an LP model.

In a typical petroleum industry model (Fig. 5), the equations or rows would represent such things as component availabilities, product inventories, unit capacities, blend requirements, quality specifications, and the like. The variables, or columns, would represent components to blends, unit charges, transfers, recipe blends, purchases, sales, unused components, unused inventories, unused unit capacities, unfilled product requirements, quality giveaway, etc. Appropriate costs or prices would be placed on the variables.

In this way, an LP model of an entire production plant, or a corporate model of an entire company may be put together. Such models may contain hundreds or thousands of equations. Computing time may be large, even on high-speed computers.

One model of a process will seldom satisfy all requirements. One designed for long-range planning may not suffice for production scheduling, for example. Thus, it is common to develop several models, each for some specific need.

LP WEAKNESSES

LP is not a panacea for all things. It has many shortcomings, the chief of which are:

Linearity

By definition, LP's can handle only linear problems, yet many process industry problems are highly nonlinear. With ingenuity and effort, the engineer can often develop linear approximations for his data. But the effect of such approximations on the results is sometimes poorly known and, in extremes, may cause serious inaccuracies.

Computing Time

Practical LP problems tend to require large amounts of computer time. A typical 460-equation LP model, with which I am familiar, requires about 30 min. from a slack basis for matrix generation, LP solution and report writing, using a 256K, IBM 360/65 computer. And computing time increases approximately with the square of the number of equations.

LP vectors may be combined to reduce matrix size, but this results in a loss of flexibility and control. Substantial savings in computing time can be made by using previous solutions as a basis for new problems, rather than starting them from a slack basis. With the 460-equation model, computer time was thus reduced from 30 to about 10 minutes, and more-dramatic savings are frequently obtained.

Milli-Cent Optimization

The term milli-cent optimization describes the tendency of an LP program to go to any extreme to obtain a fraction of a cent in calculated profit, even though the accuracy of input data does not warrant such precision. The engineer must use the

Rows \ Columns	Components to blends	Unit charges	Transfers	Recipe blends	Unused components	Unused capacities	Unfilled requirements	Quality giveaway	Purchases	Etc.	RHS
Objective function											
Component availabilities											
Product requirements											
Quality specifications											
Capacities											
Limits											
Etc.											

GUIDE such as this one can be used for model formulation—Fig. 5

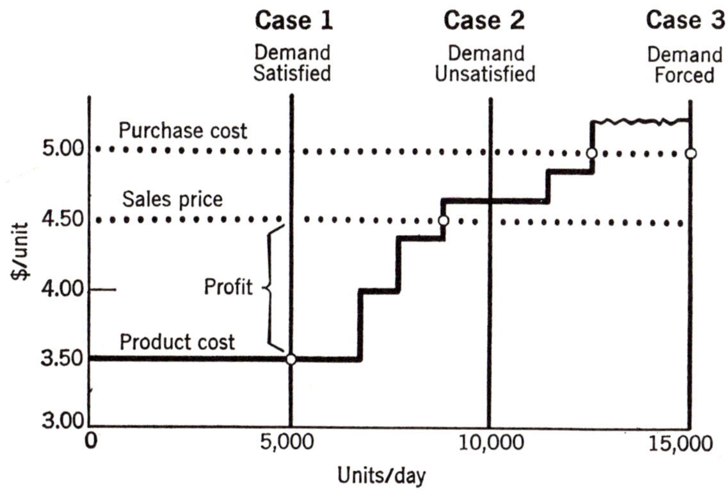

D_j on slack of:	Case 1	Case 2	Case 3
Product equation	\$3.50	\$4.50	\$5.00
Demand equation	\$1.00	0	\$0.50

LP PARAMETRIC profile (typical) of a product—Fig. 6

best available data, even though there are known inconsistencies. LP cannot cope with errors, and will find and use inconsistencies to generate fictitious profits or unwarranted yields or product qualities.

LP models are hard to debug, because the optimum solutions are far from obvious. A model will run on bad data or wrongly punched cards just as well as on good data. All data must be closely checked, and results analyzed and cross-checked for reasonableness.

Costs

The LP system can handle only those costs that vary directly with throughput, such as raw materials, utilities, catalysts, royalties and transportation costs. Capital costs, if included, must be expressed as dollars per unit of throughput.

LP cannot handle fixed costs, or those that do not vary consistently with throughput—such as labor, maintenance materials, insurance, and taxes. Note this when analyzing or interpreting LP results.

The D_j's, or cost sensitivity values, for variables in an LP solution indicate the cost penalties for deviating from the optimum activities. The D_j values are "last unit produced" costs in that they apply only for a limited range around the optimum activity level. Some people erroneously ignore this and assume that the D_j value applies to the entire amount produced. The D_j value is actually the cost of an incremental unit after all fixed costs have been written off. It indicates the incremental profit opportunity if another unit is sold at the same price as the last, but has no relationship to the total cost of producing the last unit or any previous unit.

D_j values are commonly used in calculating product values or break-even costs for individual products. In analyzing LP cost sensitivities, it is important to determine whether a variable is restricted in the final solution. The sensitivity interpretation may change if a variable is so restricted. Three situations generally exist (see Fig. 6):

1. In a typical LP solution, where the product demand is filled, the D_j on the slack variable of the product equation indicates the incremental break-even or production cost for the product. The D_j on the slack of the demand equation indicates the incremental profit. The sum of the two D_j's is equal to the sales price.

2. If the demand volume is not filled, or if production is on economic incentive, the same generalization holds, but here the D_j on the product equation slack becomes equal to the sales price and the D_j on the demand equation slack is zero. Here the LP analyzer must ascertain the nature of the restriction that limits production before he can judge whether the indicated D_j's are valid for his problem.

3. If a purchase option for meeting demands in excess of production becomes active, the D_j on the product equation slack becomes equal to the purchase cost. The D_j on the demand equation becomes the difference between the purchase cost and the sales price. Here again, the analyzer must determine whether these values have meaning for his problem.

Incomprehensibility of LP

LP is not understood by many who come in contact with it, and some people distrust answers they do not understand, particularly if they are not intuitively obvious. Many people prefer to live with a problem they cannot solve rather than use a solution they do not understand. LP solutions are presented in a coded abbreviated form and are often difficult to interpret by anyone other than the problem formulator.

The way the problem is posed can have considerable effect on the answers. Even the formulator sometimes has difficulty in deciding on the amount of freedom to permit the LP to have. He must use his intimate knowledge of the problem, the process and the LP model to judge whether to use maximum, minimum, or equality equation to most effectively express the requirements of the problem. Frequently he must employ a trial-and-error approach to force the LP to do what he wants it to do.

Time Element Cannot Be Handled

The time element cannot be handled by LP. In a continuous process this can be ignored if availabilities, capacities, demands, inventories, etc., are all expressed in the same time-rate units—bbl./day or 1,000 gal./mo. or 10^6 tons/yr., etc. The LP "sees" only the numerical value—the time element is apparent only in the mind of the user. LP input and answers represent averages over the time period selected.

The time problem becomes difficult in a batch process or blocked-time operation where several different products are produced in sequence from the same piece of equipment. Here the LP model would intimate that all products are made simultaneously, and that each would consume appropriate proportions of the equipment's capacity.

LP STRENGTHS

Despite the weaknesses discussed, LP has proved to be an excellent and versatile tool for predicting, evaluating and analyzing many short- and long-term process industry problems. Knowing its weaknesses will guide the user to capitalize on its many assets. LP is only a tool; judgment is needed to interpret the answers. Even though an LP solution is mathematically correct, the answers can be no better than the data used or the correctness of the model.

Now let us look at some of LP's strengths:

Simplicity

Despite the great detail required, an LP model is relatively simple to devise, to use and to maintain. Certainly, LP is much less demanding than the process-flow types of simulation model. The availability of good general-purpose LP computer programs, complete with matrix generators and report writers, plays a major part in this. The user need not instruct the computer in how to solve an LP problem. He need only set up his problem properly and let the computer perform the details of generating the matrix, solving the problem and writing the report. The LP system is very flexible and requires no reprogramming even to accommodate major changes in problem configuration.

Simultaneous Solutions

LP makes a simultaneous solution for all equations. Thus, the order in which equations appear in the model is immaterial. This greatly simplifies the development and maintenance of the model. It also simplifies the handling of such items as recycle operations or transfer of products from a final operation to an earlier one.

Optimum Solution

The LP solution always reflects the economic optimum, within the limits provided in the formulation of the model—the highest profit or lowest cost solution. It opens up a new horizon by making feasible the comparison of alternative cases on a consistent optimum basis. It is particularly well suited for comparative evaluations of objective function values between two similar cases for studying the effect of change.

Nonlinearities

Techniques have been developed for handling many nonlinear relationships in an LP system. A curve, for example, may be approximated by a series of straight-line segments. In some cases, slight nonlinearities may be ignored if it is felt that inaccuracies in the data do not warrant a rigorous approach.

Computing Time

While computing time is quite long for large LP problems, there are techniques available for easing this burden. Equations and variables may be eliminated by compositing two or more processes in the LP model. A process may be modeled by simulating the strategy of the operation rather than all the details of that operation. As previously noted, dramatic savings in computer time may be obtained by saving previous solutions and starting similar new problems from them rather than from a slack basis. ■

Meet the Author

W. E. Norris is Manager, Technical Services, Union 76 Div., Union Oil Co. of California, 200 East Golf Rd., Palatine, Ill. 60067. He received the degree of B.S. in chemical engineering from the University of Alabama.

Much of the published information helpful for economic evaluation of a project at its early stage of development is scattered in the literature. And in many cases, the material is not focused towards use for research evaluation. To remedy this situation, we are presenting a coordinated series of articles—a collection and interpretation of significant evaluation techniques as reported in the literature, with suitable references for those interested in more details. Emphasis in this series will be on how to go about reaching conclusions as to whether a proposed research project will be "profitable."—Editor

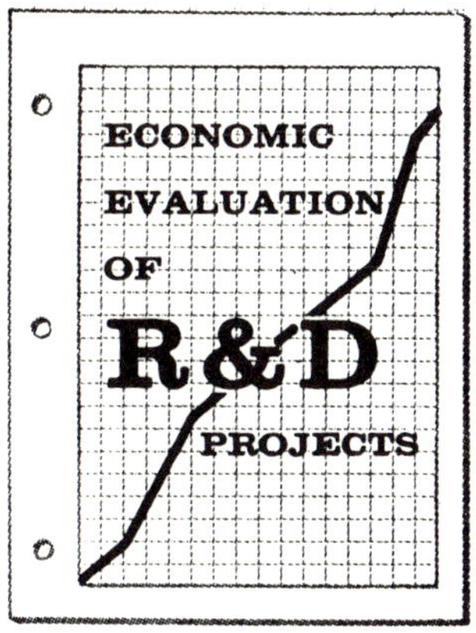

Part I

Improving R & D's Batting Average

Economic evaluation techniques will show which promising research projects should be expedited, and flash caution signals for those that look less attractive. Judicious use of these methods can result in important contributions to corporate profits.

A. J. WEINBERGER, *American Cyanamid Co.*

Suppose your boss dropped by your desk and said, "Paul Black, over in Research, has a new product he's rather excited about. Sales has had a quick look at it and thinks it may be attractive. I'd like you to examine the project and see whether we could make money from it." Would you know how to go about such an assignment? This article is the first in a series intended to give you guidance in the steps required to come up with an answer.

If you did get such an assignment, you would be among a growing number of chemical engineers involved in these considerations. Research and development is itself a "growth" area; the chemical industry alone spends about a billion dollars a year on it. Managements have been increasingly willing to spend money on R & D for new products because they see it as a way of raising profit margins and assuring growth. New products are more likely to be profitable than the older, more competitive ones (Fig. 1), and the volume of sales that is added by a succession of new products is one key to increased growth for a company whose older products have reached relative maturity (Fig. 2).

And yet, with all the emphasis on research and development, an increasing number of managements are wondering whether they are getting their money's worth from it. No over-all figures are available for the chemical industry, but one management consulting firm[1] has estimated that only one in eight research and development people working on new products is working on one that will be a commercial success. The

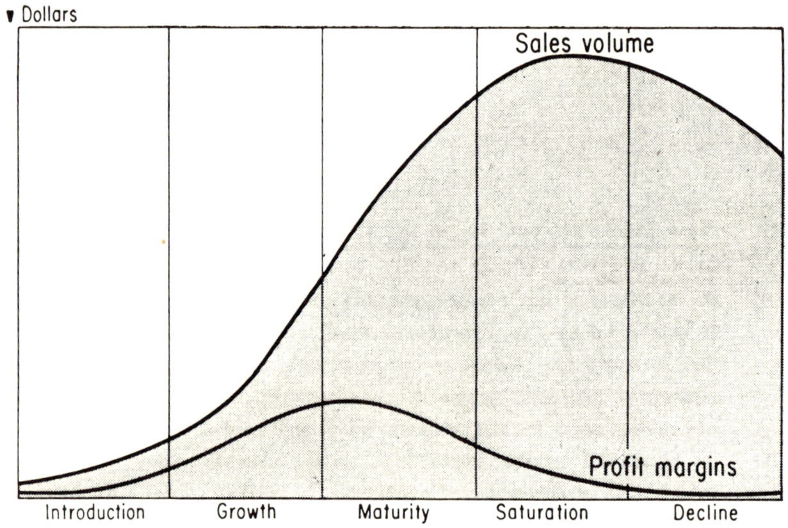

Product life cycle for typical competitive product Fig. 1

firm also found that lack of commercial success could have been predicted in many cases if the projects had been analyzed at the research and development stage. That is why more engineers are getting involved in economic analysis of R&D projects.

R & D Evaluation Is Relatively New

Economic analysis is an assessment of all the costs involved in carrying through a project compared with the dollar returns, to see whether a project would return a profit and, if so, how much. The methods used have, of course, been developed over a considerable period of time and applied to a variety of investment decisions, such as whether or not to build a new plant, modernize or replace equipment, etc. A widely used text that gives background on these concepts and methods is "Principles of Engineering Economy."[2]

Economic analysis in research and development projects is a more recent development that is considered controversial by many who feel that it may channel research into "safe" areas, where gains may be modest, and discourage work in more speculative areas. In these more daring areas the return could be higher but would be more difficult to judge. However, the fact remains that an increasing number of companies are applying economic evaluation methods to research and development projects to improve research profitability.

The evaluation can logically be applied at a series of stages:

Screening Suggestions for New Projects—This determines whether projects are worthwhile. At this stage, relatively little may be known about costs, yields, and the size or even type of market for the product, and so assumptions will have to be made. This can be done by using a range for each important, but uncertain, factor—one limit being conservative, and the other optimistic. If even the most optimistic set of assumptions shows the project would not be profitable, why undertake it at all?

Project Selection—The most promising projects are chosen and given proper emphasis depending on their relative value to the company. Many rating schemes have been proposed[3 to 7] most depending on a ratio of profit to cost and a factor considering the risk or uncertainties involved. The review, at this point can also show up factors that are most significant in reducing costs, and the areas of uncertainty likely to have most effect on the project's outcome.

Selecting Alternatives—This is selection either in methods or materials, to assure best over-all economics. Some alternatives are relatively simple, such as choices between two purchased raw materials. However, when there is a choice between a purchased material and one available from within the company, or where there is a balance between process steps whose capital, labor, and material costs will all vary, a more complete review may be required. This has been discussed previously in some detail in articles by Gilford and Rasmaussen[8], and Bates.[9]

Review of Projects—This is either on a scheduled basis or as critical information becomes available. Frequently a project is started and continues under its own momentum, fed by the interest and any empire-building tendencies of the proponents. A periodic review, taking into account the latest information developed through the project, can show whether the economics remain favorable.

This review should also take into account competitive developments. New materials now or shortly to be available, price changes in older materials, and other external factors could affect expected market and price for the developing product.

Before Transfer to Manufacturing—Though this stage is late for review to evaluate the project as research and development, it may be the point at which purchased know-how will become available, and serves as a last chance for research and development management to examine the project internally before it is turned over to others. Results of an economic analysis made at this time can be compared

Company growth based on new products
Fig. 2

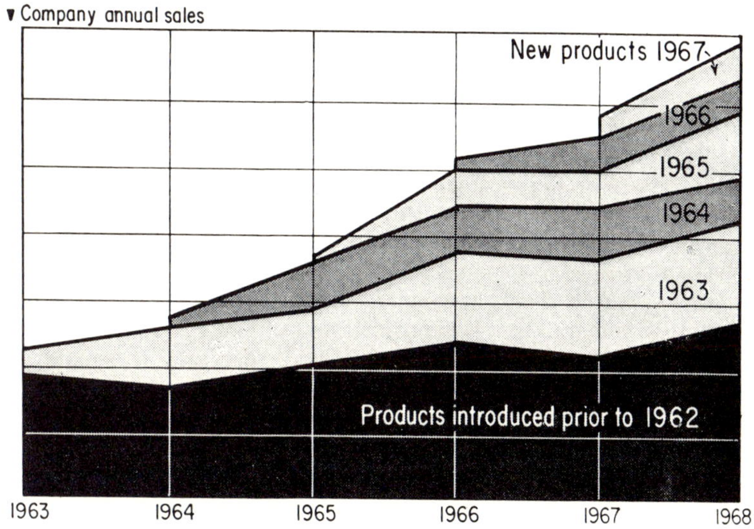

with earlier ones and help to sharpen future preliminary evaluations.

Why at R & D Stage?

The "firm" decision as to whether a project should be commercialized will be made on the basis of relatively accurate and complete knowledge. Such data will not be available early in the research and development phases. The purpose of evaluating at the research or development phase is to decide whether the project should be continued, dropped, or perhaps shelved for reconsideration later as circumstances change.[10]

This simplifies the problem of economic evaluation of R & D projects. Early in the research stage of a project, costs are usually relatively low, and we risk little by continuing the investigation. We also are likely to have little information on which to base a firm evaluation, so only a brief review is warranted. This may consist of assuring ourselves that a market for the product exists at the required price, based on reasonable yields and processing costs; that the market is large enough to justify the proposed expenditures for R & D and plant investment; and that we can develop a satisfactory product with the proposed research and development program. The "and" is important; but if any of these conditions is not fulfilled, the project cannot succeed commercially.

As the project advances through research and possibly pilot-plant and semicommercial stages, we will be increasing the rate of expenditures, as well as getting better information on which to base evaluations. Thus more thorough studies will be warranted.

For convenience, four classifications of evaluations are proposed. These are:

- **Exploratory**—This would be appropriate before appreciable experimental work has been done. Processes may be based on textbook reactions, literature (including patent) references, or a few brief laboratory experiments.
- **Preliminary Range**—At this stage, laboratory work is well advanced. Process steps, yields, and process conditions are fairly well worked out, and raw-material grades have been explored.
- **Probable Range**—An evaluation at this stage would be done after a considerable amount of pilot-plant work has been completed. Raw-material grades, processing conditions, yields, and product quality will be established. Materials of construction, treatment of recycle streams and wastes will have been considered even if not established definitely, and samples of product will have been subjected to the intended uses.
- **Pre-Production**—This would be the final evaluation after research and development has been completed. The project is now ready for a decision on commercialization.

No hard and fast rules can be established as to when these classifications should be used; they are offered as a guide to be adapted to the needs of a company, as well as to the circumstances inherent in the specific project. For example, an exploratory estimate would probably not be justified where the R & D project involves a modification of a single step in an existing process for a product on which considerable experience has been accumulated.

Important Factors

Regardless of the extent of an evaluation, there are certain factors that have to be considered, though the methods of gathering information and the extent and sophistication of calculation of results will vary as the project approaches commercialization. The factors include cost, sales and marketing aspects, and timing throughout the project. Let's look at each of these in more detail:

Research and Development Costs—This is the total dollar expenditure required to bring the project far enough along so that information is sufficiently complete to design and build any new plant required, and to start commercial production. Costs should include, in addition to the direct technical work on the project, costs of evaluations, development of special analytical methods, corrosion tests, overhead charges, and market research and preliminary market development if these are not charged as company overhead. The question arises whether costs of research and development prior to the evaluation should be included. Money already spent is not recoverable unless the "know-how" is sold or licensed, but this is a matter for company policy to determine.

Plant Investment—Total cost of new or modified facilities required to commercialize the project may have to be estimated for several levels of production. It may be desirable to consider several levels of sales volumes and prices, particularly during early evaluations. The method of estimating plant investment will depend on the type of evaluation to be made and the available information.

Manufacturing and Other Expenses—Total cost of making, selling and distributing a new product for commercial sale is required. During early years, costs may include heavy commercial development, plant startup and modification, and other expenses that must be accounted for to give a realistic appraisal of the project. In addition to direct project costs, provision must be made for a fair share of company overhead.

Sales Volume and Price—As a general rule, the sales department tells the engineer how much of a product can be sold, and the sale price. In novel and highly technical products, sales personnel may require assistance from the research and development department in visualizing uses for this new product. Usually, however, responsibility for the accuracy of sales price and volume estimates remains with sales.

When the evaluator reviews the entire project, he should consider whether projected sales volumes and prices are realistic. It would not be reasonable to expect that a new plastic with only marginal improvement in properties over similar ones could enjoy double the sales of all competitive materials when sold at a premium price.

Timing—The time schedule for a project is important for a variety of reasons. Top management wants to know whether it is proceeding as planned, financial management needs to know when capital expenditures will be required, and sales wants to know when a product will be available to sell. Timing of the project is important to the engineer doing an economic evaluation if he is using discounted methods, in which money is considered as having a time value.[11]

The engineer will have to assure himself that the timing is realistic, based both on his own experiences and those of others with comparable projects.

Be Consistent

In gathering information for an evaluation, it cannot be emphasized too strongly that cost and other data, as well as methods of calculations, be consistent with company practices. If, for example, plant overhead is allocated based on capital investment in the process unit, it is foolish for the engineer to base plant overhead charges on direct labor costs. It is in the best interest of the project, and of the engineer, to become familiar with and follow his company's accounting practices. These are the standards on which any cost analysis will be made once the project is in operation, and will probably also be used in presenting figures to top management for approval of the project. So it is well to make all evaluations on a comparable basis.

As mentioned previously, the effort justified in preparing an evaluation will depend on the status of the project. It should also be related to the quality of information available. Frequently, the sales aspects, especially for a new product, are the most uncertain data of all. This does not reflect on the integrity of the sales forecaster. It is exceedingly difficult to prejudge the extent of acceptance of new products in a competitive market; knowledge in this area is limited. Errors in sales estimates and prices can have a drastic effect on profitability, especially since the tendency is to overestimate sales potentials.[12] For this reason, the evaluator must exercise discretion as to how much effort is justified in working out other elements of the evaluation with which he is more familiar, such as capital and manufacturing cost estimates. ■

References

1. *Chem. Eng. News*, Dec. 5, 1960, pp. 42-44.
2. Grant, E. L., Ireson, W. G., "Principles of Engineering Economy," 4th ed., Ronald Press, New York, 1960.
3. Miller, T. T., *Chem. Eng. Progr.*, June 1958, p. 56.
4. Rosen, B. H., Regnier, A. L., *Chem. Eng. Progr.*, Dec. 1956, p. 500.
5. Harris, J. S., *Chem. Eng. News*, Apr. 4, 1961, p. 110.
6. Ansoff, H. I., "Evaluation of Applied Research in the Business Firm," p. 209, from "Technological Planning on the Corporate Level," J. R. Bright, editor, Harvard Business School, 1962.
7. Disman, S., *Chem. Eng.*, Dec. 24, 1962, p. 87.
8. Gilford, W. S., Rasmussen, L. E., *Ind. Eng. Chem.*, Sept. 1958, p. 62A.
9. Bates, A. G., *Ind. Eng. Chem.*, May 1959, p. 97A.
10. Boyer, C. D., Krayer, R. S., *Chem. Eng. Progr.*, Mar. 1963, p. 24.
11. Kroeger, H. E., *Chem. Eng.*, May 16, 1960, p. 143.
12. Tyler, C., *Chem. Eng.*, July 23, 1962, p. 123.

Meet the Author

Arthur J. Weinberger is a senior chemical engineer in the Central Research Div. of American Cyanamid Co. at Stamford, Conn. He has specialized in technical and economic evaluation of research and development projects, and has also handled surveys of potential research topics. Mr. Weinberger was graduated from M. I. T. (S.B. and S.M. degrees), and has had experience in production supervision, plant startup and troubleshooting, and process development at Polaroid Corp., E. R. Squibb and Heyden Chemical. Mr. Weinberger has been active in the AIChE as a founder, and chairman of the Fairfield County section. He is currently a candidate for Director of AIChE.

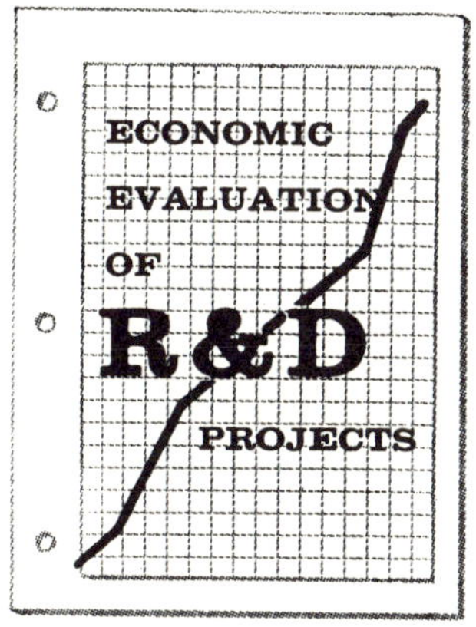

Ratio, flowsheet and equipment-list methods are available for estimating the investment required at the research and development stage—but each has limitations as well as advantages.

Part II

How to Estimate Required Investment

ARTHUR J. WEINBERGER
American Cyanamid Co.

The first article in this series[1] outlined why it is important that research and development projects be evaluated economically from the time of their inception. This article and succeeding ones will outline methods for gathering data and doing the calculations needed to perform an evaluation.

Since the investment required for a project enters into calculations of other items in an evaluation, it will be considered first. And since timing of these investments may be important in a large project of long duration, we will also cover it.

Strictly speaking, from the viewpoint of the accountant, investment in a project should include only fixed capital (funds required to construct any necessary facilities) and working capital (funds required to cover expenses such as raw materials, wages and salaries, supplies) until payment is received for the product sold. However, since we are considering the over-all economics of projects before or during research and development, we should also include all money still to be spent on the project before it reaches commercialization, and this must necessarily include research and development. Because these latter costs are treated as expenses, they are kept separate and handled in a different manner, as will be discussed in a later article in this series.

Estimating Research Costs

R & D and other pre-production costs are generally estimated for small projects by a (presumably) informed guess on the part of the manager or project leader as to how much time and how many people will be required for the effort. A usual method is to estimate R & D costs based on the man-years or man-months of technically trained personnel required for the job, and use a flat rate (frequently \$25,000-\$30,000 per man-yr.)[2] to include technical and supporting personnel, and overhead. When an unusual ratio of support to technical personnel, or unusually high materials costs are expected (as in a pilot plant), they should be figured separately. When special facilities will be required (laboratory or pilot plant), a rough cost estimate for this should be included.

For a comprehensive project, it is well to set up a detailed schedule and timetable for each segment of the entire project from the current stage through sustained operation of the plant. Such a schedule helps the estimator in detailing the steps required, and minimizes omissions.

Table I is a checklist for typical R & D cost items that might apply in a comprehensive project. In unusually complex projects where time or manpower may be limiting in the scheduling or where multiple alternatives exist, PERT or other sophisticated scheduling methods have been suggested,[3] but this is not generally required in R & D projects in the chemical process industries.

The first schedules drawn up for a project are likely to be crude and inaccurate, but through periodic review they can serve as a check on progress of present projects and as a guide in planning future projects. The tendency is generally to underestimate time required for research and development. For a while, it was fashionable to talk about shortening the traditional "seven years from test tube to tank car," but at least one study showed that this figure may still have merit[4] for new products. However, each case must be judged on its own peculiar circumstances;

Checklist of some pre-production project costs —Table 1

Literature survey	Process evaluation
Bench-scale synthetic work	Market development and product field testing
Preliminary market research and product evaluation	Physical property and engineering studies
Patent search	Patent review and filing
Preliminary process evaluation	Preliminary plant design
Bench-scale process development	Final project evaluation
Development of analytical methods	Detailed plant design
Corrosion studies	Plant construction
Toxicology studies	Plant startup
Pilot-plant design and construction	Initial sales promotion and marketing
Pilot-plant startup and operation	Technical service and trouble-shooting
Sample and trial lot distribution	
Byproduct and waste-disposal studies	

modifications of an existing product do not require the market development time required of a completely new product.

Capital Cost Estimating

Capital-cost-estimating methods may be classified as:

- Ratio methods, based on sales value or production capacity.
- Flowsheet-plus-factor methods, based on process flowsheets.
- Equipment-list-plus-factor methods based on equipment lists as well as flowsheets.

In addition, it is sometimes possible to find in company records, or in published literature, cost estimates for the process but at a different production rate. Methods for converting these will also be discussed.

Ratio methods are the quickest, and can be completed in a few minutes. Factor methods, including a rough process flowsheet, may require several man-hours. And detailed estimates would involve man-days for all but the simplest projects.

Ratio Methods

One early method for quick estimates of capital costs is based on a ratio of dollars invested per annual ton of capacity. Published ratios for chemical plants cover a wide range—from about $5 to over $1,000/annual ton of capacity. These ratios can be grouped based on the type of process—heavy inorganic chemicals, large-volume simple organics, specialty organics, etc.[5, 6, 7] Even when no ratio has been reported for the process being considered, an approximate figure can be estimated by analogy from the processes for which data are available. Table II gives some typical capital investment ratios.

Additional data for process plants (sizes and costs) may also be found in the Plant Inventory summaries of *Chemical Engineering* and *Chemical and Engineering News*.

Literature values for plant investment ratios—Table II

(Corrected to 1963)

	Plant Cost, $/Annual Ton	$ Plant Investment Per $ Annual Sales	Plant Capacity, Tons/Yr.
Inorganic chemicals			
Lime	5–8	0.35–0.56	....
Sodium silicate	12–17	....	2,000
Sulfuric acid, from sulfur	24–27	1.1–1.2	80,000
Disodium phosphate	60–65	0.37–0.4	....
Phosphoric acid, wet process	80–90	0.7–0.8	35,000–45,000
Soda ash, Solvay process	90–105	2.9–3.4	300,000
Organic chemicals			
Acetaldehyde, from acetylene	70	0.35	25,000
Carbon disulfide	75–105	0.9–1.2	20,000–30,000
Methanol, from synthesis gas	200–340	2.2–3.7	30,000–120,000
Aniline via nitrobenzene	410–460	1.4–1.5	5,000
Ethylene dichloride	570	3.0	10,000
	510	....	....
Phenol	570	2.5	20,000
Salicylic acid	1,150	1.5	1,000
	1,000	....	....
Hexamethylene tetramine	1,730	3.7	2,500
	1,550	....	....

A second ratio method is based on capital investment per sales dollar or its reciprocal. Turnover ratios (annual $ sales/$ of investment) for various types of plants have been reported,[8] and capital ratios ($ of plant investment/annual $ sales) have been reported for specific process plants[9] as well as for types of products.[10] An analysis of capital ratios[11] showed the majority to be within the range of $0.20 to $5.00 of investment per annual dollar of sales, with $1.00 as a median.

In general, these capital ratios tend to fall into groups that relate to type of process and plant size. It is often possible to pick a value for an unreported process by comparing it with available figures for a comparable operation. Some selected ratios are shown in Table II.

Ratio methods are admittedly crude generalizations. They are suitable only where a rough approximation is needed, as in an exploratory evaluation of an R & D project at the stage where little or no laboratory work has been done and the objective is to screen the proposal quickly to determine whether any further work is warranted. Accuracy of ratio methods has been reported to be in the range −70% to +90%.[12] While this range is wide, it must also be realized that at the exploratory stage many other factors—yields, process steps, sales prices, and potential sales volume—are likely to be uncertain, and more elaborate estimating methods may not be justified.

For a new product, the selling price would not be known until costs have been worked out. This means that capital/sales ratios would not be available to use in determining capital investments unless a good guess

Some unit values for installed equipment[17]—Table III

One Unit Each	Two Units Each
Column, carbon steel	Column, stainless steel
Reactor, carbon steel	Furnace
Evaporator, carbon steel	Centrifuge
Blower, carbon steel	Compressor
Storage, feed or product (liquid)	Refrigeration unit
	Product storage (solid or gas)

These apply for pressures of 100 psig. or lower. At higher pressures, multiply units by ratio of working pressure (psig.)/100

Factors for estimating total plant cost[18]—Table IV

Each of the following should be added to installed process equipment cost:	
Process Piping	
Low (solids processing plant)	7–10%
Average (mixed processing plant)	10–30%
High (fluids processing plant)	30–60%
Instrumentation	
Low (little automatic control)	2–5%
Average (some automatic control)	5–10%
High (complex, centralized control)	10–15%
Manufacturing Buildings	
Low (outdoor construction)	5–20%
Average (mixed, indoor & outdoor construction)	20–60%
High (indoor construction)	60–100%
Auxiliary Facilities	
Low (minor additions at existing site)	0–5%
Average (major additions at existing site)	5–25%
High (complete at new site)	25–100%
Outside Lines	
Low (among existing facilities)	0–5%
Average (separate processing units)	5–15%
High (scattered processing units)	15–25%
Each of the following should be added to total physical cost (installed process equipment plus factors above):	
Engineering and Construction	
Low (straightforward engineering)	20–35%
High (complex engineering)	35–50%
Size Factor	
Low (large commercial unit)	0–5%
Average (small commercial unit)	5–15%
High (experimental unit)	15–35%
Contingencies	
Low (firm process)	10–20%
Average (subject to change)	20–30%
High (speculative processes)	30–50%

Select appropriate percentage factors for process piping, instrumentation, outside lines, manufacturing buildings auxiliary lines. The sum of these factors is applied to and added to "installed equipment cost" to obtain "total physical cost." Factors for engineering, size and contingencies are applied to and added to total physica cost to get total project cost.

were made of the sale price level. For this reason, such ratios are most useful as a check on the reasonableness of estimates for capital investment obtained by other methods.

More Realistic Estimate

Ratio methods have several shortcomings. For instance, they are based on the assumption that capital investment varies directly with size, and this may lead to serious errors. A more realistic estimate can be made if the plant investment is assumed to vary as the 0.6 power of the plant size; as proposed by Chilton:[13]

$$\$\ \text{Investment}_2 = \$\ \text{Investment}_1 \times (\text{Capacity}_2/\text{Capacity}_1)^{0.6}$$

Frequently, reported data do not specify whether the cost is for the process unit alone, or for a complete plant including site development and utilities, or for some unit between these two extremes. This contributes to the inaccuracy of the ratio method, but will probably not lead to serious problems if the evaluator recognizes the accuracy limitations inherent in the method.

An easy way to correct plant investment figures to current costs is to multiply the old cost by the ratio of construction cost indexes:

$$\$\ \text{Investment (now)} = \$\ \text{Investment (previous)} \times \left(\frac{\text{Cost index (now)}}{\text{Cost index (previous)}} \right)$$

For the chemical industry, the Marshall and Stevens[14] or CE[15] construction cost indexes are frequently used, as well as the ENR and Nelson indexes.

Capital/sales ratios cannot be corrected so readily, but probably fluctuate less with the construction cost index. Capital investment and sales prices are both likely to be affected by cost index changes. The median capital ratio of 1.0 was based on 1950 data, and is the same as a 1938 approximation[16] for typical capital/sales ratios.

Flowsheet-Plus-Factor Methods

Flowsheet-and-factor methods for capital cost estimating are based on a rough process flowsheet for the plant. The number of pieces of major process equipment (reactors, columns, filters) is added up, then multiplied by one or more factors to obtain total investment.

In the method proposed by Hill[17] for petrochemical processes, individual items of process equipment are classified as being equivalent to one or two "units," depending on their type and construction. Simple equipment such as carbon-steel columns and low-pressure reactors would be classified as one unit; more-complex and expensive equipment, such as stainless steel columns and furnaces would be classified as two units (see Table III). For equipment with pressures over 100 psi., the number of units would be multiplied by working pressure (psi.)/100 (i.e., a 300-psi. carbon steel reactor would be rated as three units.) Each unit for a 10-million lb./yr. plant would be considered to have an installed price of $30,000 (based on M & S cost index of 185), and total installed equipment cost

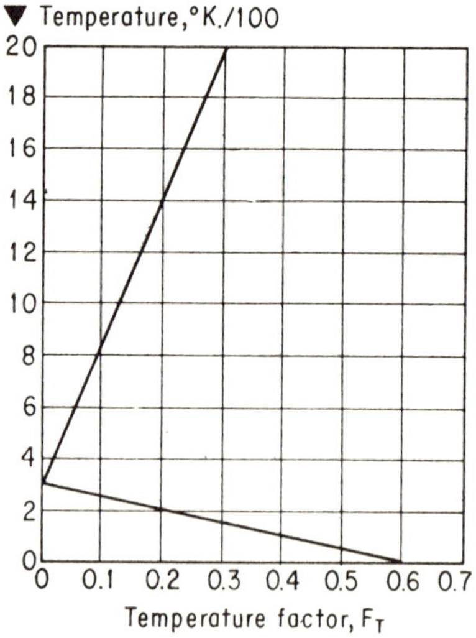

Select temperature factor based on operating temperature—Fig. 1 . . .

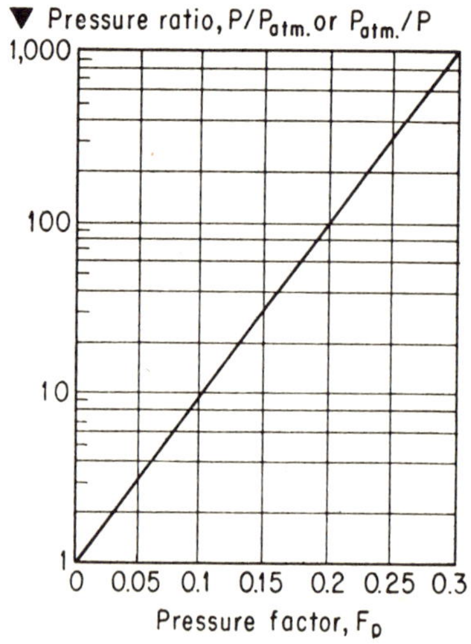

. . . And pressure factor based on plant operating pressure—Fig. 2

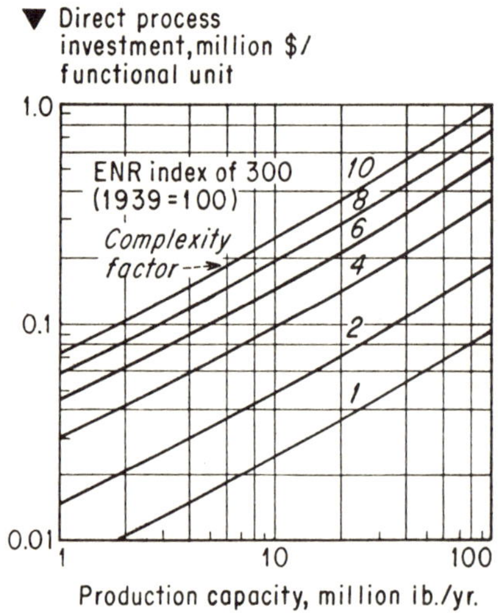

Investment is a function of plant capacity and project complexity—Fig. 3

would be the sum of units from the flowsheet times $30,000. The equipment cost for plants of other sizes would be calculated assuming cost to vary with the 0.6 power of capacity as noted previously.

Total plant investment can be estimated from installed equipment costs by adding percentages as proposed by Chilton[18] to include piping, buildings, electrical insulations, and other costs (Table IV). The author reported that this method gave ±25% accuracy in 90% of over 20 low-pressure petrochemical processes estimated, with a maximum deviation of 40%.

Functional Units

In a more recent estimating method developed by Zevnik and Buchanan,[19] the capital investment per functional (process) unit is obtained from a graph relating cost, plant size, and a process complexity factor that allows for cost variations due to materials of construction and operating conditions.

The process complexity factor (*CF*) is defined as:

$$CF = 2 \times 10^{(F_t+F_p+F_a)}$$

Where: F_t = Process temperature factor (from Fig. 1)
F_p = Process pressure factor (from Fig. 2)
F_a = Equipment alloy factor (0 for cast iron, carbon steel, wood; 0.1 for aluminum, copper, brass, Type 400 stainless steels; 0.2 for Type 300 stainless steels, nickel, Monel and Inconel alloys; 0.3 for Hastelloy alloys; 0.4 for precious metals).

Once a complexity factor has been established, the direct process investment per functional unit can be read from the graph of process investment vs. production capacity (Fig. 3).

The number of functional units (major items of process equipment, including storage and waste disposal facilities) is totaled from the process flowsheet to give:

Plant investment = 1.33 × number of functional units × cost per functional unit × current ENR Index/300. The *Engineering News-Record* cost index of 300 was used as the basis for the correlations.

The authors suggest that potential users try the system on cases for which they have their own data, as a check, and, if necessary, to modify the factors. The authors cite no accuracy figures for the method, but the six examples shown ranged from 75% to 125% of costs reported in the literature.

Some Limitations

One limitation of the factor methods: they make no allowance for major variations in yield or processing under dilute conditions where significantly larger equipment and hence higher costs would be expected. Although it is not specified, if the plant operates less than 24 hours a day, 7 days a week, plant investment should be based on full-time capacity.

Another limitation of the flowsheet-plus-factor methods is the possible confusion in the counting of process units by an engineer not familiar with plant costs. Hill[17] spelled out the particular types of process units, some of which were more expensive and, therefore, were counted as two units for purposes of arriving at a total cost. Zevnik and Buchanan[19] provide examples of what they call functional units but the user is left to his own discretion as to what to include.

The flowsheet-plus-factor methods of estimating capital costs are most applicable to exploratory estimates, where sufficient information is available to define the process. However, it must be remembered that when such an estimate is based only on proposed reactions or meager laboratory data, the results obtained may be no more reliable than the ratio methods. It is not uncommon to find, as more definite information becomes available from experimental work or if there is a change in product-purity requirements, that additional process steps will be necessary. This can increase investment considerably.

Equipment-List-Plus-Factor Methods

Equipment-and-factor estimating methods require preparation of process flowsheets and a listing of major process equipment. Equipment sizes and materials of construction are specified, so a cost can be obtained for each major item. Over-all plant investment is then calculated by multiplying the total of major equipment costs by a single over-all factor, or by the sum of factors.

For estimating investment costs of projects at the research and development stage, it is not generally necessary or desirable to obtain quotations from vendors—except for special and expensive units. There is considerable literature available on costs of many standard process equipment units. Of particular value are the compilations by Aries and Newton,[20] Chilton,[21] Zimmerman and Lavine[22] and the bibliographies by Katell, et al.[23]

However, the user who does considerable purchasing of equipment may choose to maintain files of recent prices actually paid for pieces of equipment, and use this information where available. Equipment costs given in these references are dated, and should be brought up to date using an appropriate cost index.

Lang Factors

Some articles quote equipment on a purchased basis, others on the basis of installed price. While installation costs for individual types of equipment vary widely, installed cost is usually assumed to average 50% more than the purchase price.[24] On the average, delivered equipment cost is only a few percent higher than purchase price. Single factors have been proposed for converting total major chemical process equipment cost to total investment. Lang[25] recommended:

Total plant cost = 3.10 × Delivered equipment costs (for solid process plants)
= 3.63 × Delivered equipment costs (for solid-fluid plants)
= 4.74 × Delivered equipment costs (for fluid process plants)

and cited accuracies within 10-15% of those obtained by more-detailed estimating methods. A more-recent study by Haselbarth and Berk[26] showed nearly the same ratios, but a wider range of error. A modification has been suggested[27] where an individual factor is used for each type of equipment to reflect the fact that some equipment—such as pumps, processing tanks, and heat exchangers—should have higher factors than furnaces, field storage tanks, and centrifuges. However, no figures were given to show how much this improved the accuracy of the over-all estimate.

A variety of composite factors have been proposed for converting equipment cost to total investment. These are usually made up of allowances for equipment installation, piping, electrical instrumentation, etc., the magnitude of each varying with the type of plant and the circumstances.

One useful set of composite factors for estimating projects at the research and development stage is that proposed by Chilton,[18] referred to previously in this article (Table IV). Other systems have been described but are more appropriate to preliminary project engineering, when more is known about the specific installation site, existing facilities, etc.

No Better Than Data

While equipment-list-and-factor estimating methods are more accurate and overcome the limitations cited for the more preliminary techniques, they also pose problems. The most obvious is that they will be no better than the flowsheet and the equipment list. Since the usual tendency is to underestimate the complexity of a process during its early stages, the equipment-list-and-factor methods are usually applied to projects at the "probable range" evaluation stage referred to in the previous article in this series,[1] in which an extensive amount of experimental work has been done on the process.

If the equipment lists are made up by research and development personnel with little exposure to plant operations, the evaluator must check for their tendency to omit or minimize in-process as well as raw-material and finished-product storage capacity, materials handling equipment (such as conveyors and feeders) packaging facilities, special utilities, and waste disposal. For continuous processes with recycle, it is also easy to overlook the need for purification of impurities that build up in the recycle streams.

Choosing Methods

As might be expected, the more detailed methods of cost estimating would be expected to give greater accuracy. However, the more detailed methods are more time-consuming and, if the process description and yields are not sufficiently well worked out, it is pointless to prepare the more "accurate" estimate. A more detailed method will not compensate for the unknowns or errors in the data, and the estimator may succeed only in convincing himself of an unwarranted accuracy of his figures.

Because preliminary estimating methods have only limited accuracy, it is best to present the resulting figures as a range, rather than as a single number. The evaluation should be based on both the high and low figures to emphasize that a range of investments is involved.

Detailed estimating methods would be appropriate as the project approaches commercialization and other factors, especially market volumes and sales prices, are known with some reliability. Detailed estimating methods have not been covered here, since they do not generally apply to projects in research and development; but the reader is referred to Weaver, Bauman, and Heneghan[28] for an up-to-date discussion of the topic.

Estimating Working Capital

Working capital for a project is, in essence, the amount of money necessary to pay out-of-pocket expenses as they arise. These expenses will include raw materials, wages and salaries, fuel, and supplies, which must be paid for before payment is received for the products sold.

Working capital will, presumably, be recovered when the project is shut down—by selling the last of the product and any remaining raw materials—but it must be included in the capital required to make the project a going business. The amount of working capital required is usually an appreciable percentage of the plant investment, and is worth considering with a degree of care in keeping with its importance.

Company standards often prescribe the amount of working capital to be allocated to a project. If your company does have such standards, find out what they are, and use them; it is pointless to do an evaluation on a basis that is not consistent with your own company's standards. Where company standards do exist, they may be expressed as (1) percentage of annual sales, or (2) a combination of:

- X months supply of raw materials.
- Y months supply of product, at full manufacturing cost.
- Z months sales at full cost of sales.

While apparently similar, these two methods represent a difference in viewpoint—the first is based on classical accounting practices, and the second on actual out-of-pocket costs. Discussion of this point is beyond the scope of this article, but was well presented by Weaver and Lyndall.[29] It is important that the evaluator understand the distinction, and which viewpoint is represented by his own company's definition, because cases arise in which the usual standards do not apply and modifications must be made. Among these are products or raw materials that are:

- Supplied by pipeline (generally very low inventory).
- Produced or used in the same location (also low inventory).
- From remote locations, such as overseas (high inventory).
- Seasonal (high inventory).
- Shipped in very large volume, such as tanker (high inventory).
- Of unusually long processing time, such as aging or curing (high inventory).

Where no company standards exist, the evaluator must make his own judgment of working capital requirements. For exploratory estimates, a figure of **three months sales is suggested as a rule of thumb.**

For more detailed estimates, as the project is further along in R & D, the following ranges can be used, subject to the conditions of the specific case:

- Raw materials (including delivery and handling, when significant) ½ to 2 months.
- Out-of-pocket cost of sales (sale price, less pretax profit and depreciation) 1½ to 4 months.

As a rough check on the total, the ratio of working capital to annual sales for the project can be compared with the company's ratio of inventory and accounts receivable to total sales, as shown by the latest annual balance sheet. Unless the project is unusual for the company, the two ratios should be roughly the same.

In this discussion, working capital has been expressed in terms of production rates. This means that if plant output varies, as it may during the early years of a product just gaining acceptance, the working capital—unlike capital investment—will also vary for each year. This may be important in evaluating the profitability of the over-all project, even at the R & D stage. Also worth noting is the fact that working capital is generally based on steady operating conditions. Startup expenses are generally higher, but are not considered in calculating working capital. ∎

References

1. Weinberger, A. J., *Chem. Eng.*, Oct. 28, 1963, p. 123.
2. Gambs, G. C., *Research Management*, **2**, No. 1, 34 (Spring 1959).
3. Livingston, J. S., *Ind. Research*, **5**, No. 6, 22-25 (June, 1963).
4. Tyler, C., *Chem. Eng.*, July 23, 1962, p. 128.
5. Wilcoxon, B. H., *Chem. Eng.*, May 1947, p. 112.
6. Anon., *Chem. Eng.*, May 1951, p. 164.
7. Faith, W. L., Keyes, D. B., and Clark, R. L., "Industrial Chemicals," 1st ed. (1950), and 2nd ed. (1957), Wiley.
8. Kiddoo, G., *Chem. Eng.*, Oct. 1951, p. 145.
9. Kistin, H. R., Cameron, C. S., and Carter, A. P., *Chem. Eng.*, Nov. 1953, p. 191.
10. Lynn, L., *Chem. Eng.*, Apr. 1954, p. 159.
11. Schweyer, H. E., *Chem. Eng.*, Feb. 1952, p. 164.
12. Nichols, W. T., *Ind. Eng. Chem.*, **43**, 2,293 (Oct. 1951).
13. Chilton, C. H., *Chem. Eng.*, Apr. 1950, p. 112.
14. Anon., *Chem. Eng.*, Apr. 4, 1960, p. 149.
15. Arnold, T. H. and Chilton, C. H., *Chem. Eng.*, Feb. 18, 1963, p. 143.
16. Tyler, C., "Chemical Engineering Economics," p. 51, 2nd ed., McGraw-Hill, 1938.
17. Hill, R. D., *Pet. Ref.*, Aug. 1956, p. 106. Also, see Hand, W. E., *Pet Ref.*, Sept. 1958.
18. Chilton, C. H., *Chem. Eng.*, June 1949, p. 106.
19. Zevnik, F. C., Buchanan, R. L., *Chem. Eng. Progr.*, Feb. 1963, p. 70.
20. Aries, R. S., Newton, R. D., "Chemical Engineering Cost Estimation," McGraw-Hill, 1955.
21. Chilton, C. H. (ed.), "Cost Engineering in the Process Industries," McGraw-Hill, 1960.
22. Zimmerman, O. T., and Lavine, I. (editors), "Cost Engineering" cumulative volume 1951-59 and quarterly issues since, Industrial Research Service, Inc., Dover, N. H.
23. Katell, S., et al., "Bibliography on Investment and Operating Cost for Chemical & Petroleum Plants," U. S. Bureau of Mines Circulars 7705, 7751, 7847, 7884, 7916, 7966, 8035, 8117.
24. Zimmerman and Lavine, *Cost Eng.*, Apr. 1962, p. 11.
25. Lang, H. J., *Chem. Eng.*, June 1948, p. 112.
26. Haselbarth, J. E., and Berk, J. M., *Chem. Eng.*, May 16, 1960, p. 158.
27. Wroth, W. F., *Chem. Eng.*, Oct. 7, 1960, p. 204.
28. Weaver, J. B., Bauman, H. C., and Heneghan, W. F., "Cost and Profitability Estimation" in 4th ed. of Chemical Engineers' Handbook, McGraw-Hill, 1963.
29. Weaver, J. B., Lyndall, F. S., *Ind. Eng. Chem.*, Aug. 1959, p. 49a.

There are a number of techniques and correlations available for estimating the cost of various elements that make up manufacturing costs. Some give order-of-magnitude answers, others can produce fairly accurate estimates. Choice depends on how far along the R&D project has progressed and the accuracy of data used in the calculations.

Part III

Calculating Manufacturing Costs

Arthur J. Weinberger
American Cyanamid Co.

The preceding article[1] in this series reviewed methods for estimating the capital required to convert R & D projects into commercial ventures. This article will outline methods that are useful for predicting manufacturing costs for projects still in research and development.

A distinction should be made at this point between manufacturing costs and operating costs. We will assume that manufacturing costs include all costs for making a projected product and preparing it for shipment. Operating costs include manufacturing costs and other elements such as costs of selling and distribution, indirect costs of company administration, company staff services and other items included in overhead (see Table I).

We will cover estimation and evaluation of manufacturing costs here, leaving sales and administrative costs to later articles in this series.

Allocated vs. Incremental Costs

There are two different philosophies on which costs can be based. The traditional, or "allocated basis," assumes that the project should cover all direct expenses connected with it, and, in addition, assume a portion of all general expenses. These are usually allocated to each product or project based on a percentage of total sales, manpower, utilities required, etc. For example, a new project that would use 5% of the steam generated in a plant (assuming sufficient excess capacity exists) would be charged 5% of the total cost of operating the steam plant. Or, if the new project required a raw material already used and the larger volume would permit more economical purchasing, the new product would pay the new average costs for the material.

The other, and newer, basis of "incremental costs" assumes the project should be charged only for the added costs due to the project. If the new project obtained steam from existing excess capacity in the boiler plant, it would be charged only for the extra fuel and any increase in maintenance, labor, etc., required to generate the additional steam.

If additional quantities of a raw material were involved and this reduced the cost per unit purchased, the new project would be charged at a lower rate corresponding to the incremental quantity purchased. This has been discussed more fully by Ross, et al.,[2] but is important in any cost analysis, even at the R & D stage, since it will determine which elements are to be included. Both "allocated cost" and "incremental cost" methods are in use, and it is well to find out which your company uses and apply it consistently in R&D evaluation.

The estimation of manufacturing costs may be as far as an exploratory R & D evaluation will be carried. If manufacturing cost appears unfavorable by comparison with existing processes, or if the manufacturing cost is not attractive as compared with the selling price of comparable products already on the market, the evaluation need proceed no further—at least until improvements and cost reductions can be made.

Cost of Materials

Materials cost is usually the largest item in manufacturing cost. For this reason, it is given the most attention in cost estimates and evaluations. There are

Important elements of operating costs—Table I

Manufacturing or factory costs	Utilities
Materials	Maintenance
Raw materials and processing chemicals	Miscellaneous supplies
Waste-disposal chemicals	Waste disposal
Packaging materials	Factory overheads
Processing	Depreciation
Direct labor	Plant protection, taxes and insurance
Supervision	Royalties
Process control	Less byproduct credits
	Sales and distribution costs
	Administrative and general costs

three stages of R & D, and each affects the choice of estimating methods. The stages are:

- Exploratory (pre-laboratory stage).*
- Preliminary range (based on laboratory work).*
- Probable range (after pilot-plant work).*

Materials costs are frequently developed and presented in tabular form (see Table II), spelling out the materials used, grade and form, unit cost, usage per unit (generally per pound) of product, and cost per unit of product. For estimation, it is desirable to include usages of all materials consumed in the processing, whether or not they are strictly raw materials. This would include catalysts, filter aids, solvents, and waste-treatment materials (if any). Packaging materials, if required, are generally neglected in exploratory estimates.

Exploratory Estimates: Materials

In the exploratory stage, usages may be only rough estimates based on assumed yields. And it may be desirable to indicate ranges for the materials that contribute most to total cost. Grades of materials specified in an exploratory estimate should, in the absence of definite knowledge, be the lowest priced grades consistent with the type of product to be produced. Generally, technical grades are suitable, except for some food and pharmaceutical uses where N.F. and C.P. grades may be required.

Materials costs for exploratory estimates are generally based on current published prices. The most complete lists of chemical prices are published in weekly issues of *Oil, Paint and Drug Reporter,* and the "Quarterly Report on Current Prices" included in *Chemical and Engineering News.* This latter magazine also contains special annual reports on synthetic resins and plastics, on chemical reagents, and on surface-active reagents and detergents. *Rubber Age* lists prices of various types of rubber, plasticizers, pigments, fillers, synthetic fibers, and other chemical materials used in rubber processing, and *Oil and Gas Journal* includes prices of crude and refined petroleum products.

Since listed prices of purchased materials vary with volume and type of shipping container, the appropriate purchase volume must be considered in obtaining prices. As an example, purchases in tank-car quantities would be preferable for a unit cost where volume justifies it, but would be unsuitable for a material to be used at a rate of 10-20 gal. per week; such a quantity would be obtained more conveniently from drums or carboys.

Table III lists typical shipping quantities obtainable in tank cars, carload lots, and truck loads.

Transportation costs may or may not be included in quotations for purchased materials. If transportation costs are not included, they are generally ignored in exploratory estimates, unless large quantities are to be shipped over long distances, or the product is of low unit value. As a check on the importance of shipping costs, it is frequently assumed that railway shipment costs 1-3¢/ton-mile for distances over 500 miles.[4]

Prices for captive materials (those made within the company) will generally be charged at less than market price. However, this is a matter of company policy and should be determined accordingly. Some

* See the first article in this series[3] for more details.

Table shows details on all materials used in process —Table II

Material, Grade	Unit of Material	Price $/Unit*	Price Reference	Quantity Used, Units/Lb. Product	Cost, $/Lb. Product
X, solid	lb.	0.20(P)	Chem.Eng. News	3	0.600
Y, liquid	gal.	0.08(M)	Company	1.2	0.096
Z, solid	lb.	1.02(P)	OP & DR	0.3	0.306

Total material cost ($/lb. product) = 1.002

*Indicate (P) if purchased, (M) if manufactured.

alternatives used for pricing captive materials for purposes of exploratory estimates are:

- Market price.
- Market price less 10-20% sales costs.
- Manufacturing costs, with or without allowances for overhead.

Since the project being estimated may be years from commercialization, it is also well to determine whether there are plans for expanding production of the captive material, with possibly lower costs. Unless such plans are firm, and have a major influence on materials costs, they should only be mentioned in the estimate rather than used as a basis for estimating.

Preliminary Estimates: Materials

For preliminary-range estimates, materials costs are generally handled the same way as for exploratory evaluations. Usages of major materials will be better defined, although recycled catalysts and solvents, and filter aid requirements, may still be vague. The annual requirements of each material should be reviewed to assure that adequate supply will be available.

Cost of packaging materials should be included if the product is not to be shipped in bulk. Dowling has given a "rule of thumb" of about 0.4¢/lb. for bags[5] and has given data[6] on packaging materials costs (Table IV).

For a preliminary-range estimate, prices for purchased materials would probably still be based on published figures, though the advice of the company purchasing department should be sought on the possibility of lower prices based on contract purchases of major materials. Costs of transportation of major raw materials should be included if prices are not quoted on a delivered basis, and if these are likely to be significant. Von Allmen[7] has reported transportation costs with an accuracy sufficient for preliminary range.

Captive materials prices will depend, as mentioned above, on company policies and the likelihood of expanded production with possibly lower costs at the time of commercializing the project. In addition, the price to be charged may also be determined by projecting demands for the captive material in relation to production capacity. Some alternatives include:

A. If full output of material can be sold:

1. Assume market price (why should the new project penalize the existing one?).
2. Market price less 10-20% (since no sales cost is involved).

B. If excess capacity for material will be available:

1. Average unit cost (if "allocated costs" are used).
2. Incremental cost of the added volume (if costs are figured on an incremental basis).

Probable Range: Materials

For probable-range estimates, where the process has been piloted and a thorough review is desirable, material usages and specifications should be available.

Typical shipping quantities—Table III

	Range	Most Typical
Tank cars	4,000 to 30,000* gal.	6,000 to 10,000 gal.
Railroad cars	2,000 to 4,000* cu. ft.	2,000 to 2,600 cu.ft.
Truckloads		15 to 20 tons
Tank trucks	2,000 to 5,000* gal.	2,000 to 4,000 gal.

* Weight limits for high-density products.

Prices can be based on vendors quotations, and transportation costs should be included. The general practice is to assume current prices (any inflation or deflation will affect both costs and prices). However, if it is anticipated that the project is still some time away from commercialization, possible price changes in the more-significant raw materials should be considered—particularly if new production facilities for these raw materials are planned or being built. The trend of prices has been discussed by Copulsky and Cziner.[8]

For captive materials that will be a significant part of total cost, projected costs or prices (depending on basis used) should be obtained. These figures should reflect not only prices when the project is commercialized, but any price "breaks" expected from expanded production facilities.

Even after the pilot-plant stage has been completed, estimated usages of some materials such as volatile solvents, recycled materials, and waste-treatment chemicals may still only be rough approximations. However, unless these are major elements in the materials cost, a generous allowance should be adequate.

If the product is to be packaged in nonreturnable containers (such as steel or fiber drums, or bags), these costs should be included.

Costs of packaging materials[6]—Table IV

	Unit Cost, $
Steel drums	
55-gal., unlined	6.00
55-gal., 2 coats baked resin lining	7.30
30-gal., unlined	3.10
Fiber drums	
55-gal., removable metal cover (various barriers)	2.50–7.00
30-gal., removable metal cover (various barriers)	1.90–2.20
15-gal., fiber cover	1.00–1.20
5-gal., fiber cover	0.65–0.80
Multi-wall paper shipping sacks	
5-ply, sewn valve, PE moisture barriers (38-in. x 19-in. x 5-in.)	0.20
2-ply, partial open-mouth, extensible kraft (40-in. x 20-in. x 9-in.)	0.08
Burlap bags	
Waterproof, paper-lined bag, cemented seams (28-in x 22-in.)	0.30

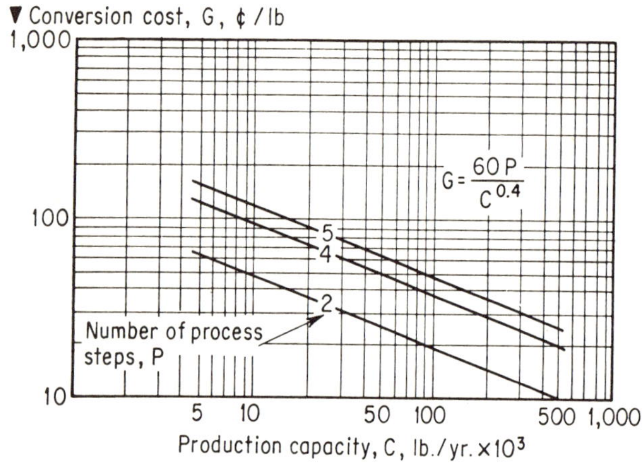

Conversion costs for multi-purpose plant[9]—Fig. 1

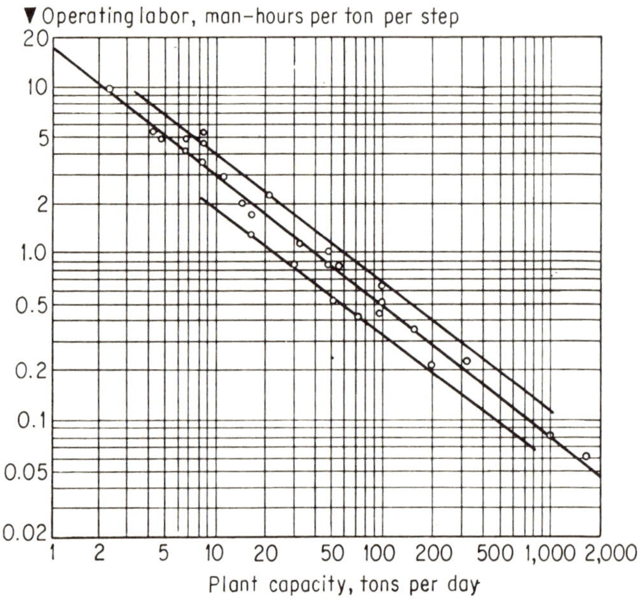

Direct-labor: a function of capacity[10]—Fig. 2

Processing-Cost Considerations

Processing costs are all those elements that, in addition to materials, would be included in converting the materials into the desired product. The individual items are given in Table I. Labor and supervision, depreciation, maintenance and utilities are generally the most important cost items.

Using Multi-Purpose Facilities

The scale of manufacture of the new product and the kind of facility to be used will influence the method of estimating processing costs. If the product is to be produced in relatively small quantities (to perhaps 100,000-1,000,000 lb./yr., depending on process requirements), and is amenable to batch or semi-continuous operation, it may be desirable to consider making it in multi-purpose or "job shop" facilities, some of which may already be available within the company. Berry[9] has described a method for estimating processing cost for complex organics based on the number of process steps involved, and production capacity. He found that conversion costs could be expressed in the form of:

$$\text{Conversion cost (\$/lb.)} = \frac{\text{Constant} \times \text{number of processing steps}}{\text{Production rate}^{0.4}}$$

and presented the results in graphical form (Fig. 1), with the stipulation that the equation could only be used for the over-all process when throughput is roughly the same (within 30%) for each of the process steps. This method would be of interest mainly for exploratory estimates.

It is common to use general-purpose facilities for a number of products, and allocate processing costs on a proportionate share of the over-all costs of operating the facility. In this case, all charges (direct labor, supervision, utilities, maintenance, overhead, etc.) are totaled and divided by total man-hours of labor or equipment-hours. Total processing charges for each project would be allocated to each project depending on the man-hours or equipment-hours required. The cost estimator would then only have to determine the batch size and labor or equipment time requirements to estimate processing cost. Preliminary and probable-range cost estimates for this type of facility are suitable provided they are made in consultation with the supervisor to assure that they are realistic. The tendency of the outside estimator is to underestimate time required for cleaning equipment, and revision of piping when converting from one product to another.

Using Special Facilities

When special facilities are built for the project, processing costs have to be handled in a different manner. For exploratory estimates, processing costs can usually be assumed to be made up of three elements: labor and labor-related costs, investment-related costs, and utilities.

Direct-labor requirements for exploratory estimates can be estimated from number of process steps and the production rate, as outlined by Wessel[10] (Fig. 2). The upper line in this graph would be more suitable to batch operations requiring a maximum of labor, and the lower one for highly instrumented continuous processes. While the method is a rough approximation, it is generally adequate since labor and related costs are generally small as compared with materials costs. If labor costs are significant (over 10-15% of materials costs), the more detailed method of estimating labor costs (discussed under preliminary range estimates) should be used.

Labor-related costs are all items (other than direct labor) proportional to personnel required to operate the project. These include supervision, payroll taxes, personnel services (medical, employment, cafeteria, etc.), fringe benefits (vacation, sick leave, overtime, insurance on personnel, and the like), and plant administration and overheads if these are allocated on the basis of labor charges.* Charges for each of these labor-related items will probably be shown separately in your company's accounting records, but should be totaled to obtain a ratio of labor-related to direct labor costs.

While the ratio or factor for labor-related costs is best obtained from your own plant or company records, Holland and Brinkerhoff[11] have reported a ratio of 1.25 for labor-related costs that may be used if other records are not available. Using this factor, direct labor costs for the project would be multiplied by 2.25 to give total labor and labor-related charges. This method assumes that average supervision, laboratory control, and other factors will be adequate for the new project, and this is usually accurate enough for exploratory estimates. The factor for labor-related costs should be fairly stable for a given plant or company but should be verified periodically.

Investment-related costs include: depreciation, maintenance, property taxes and insurance on plant.

Since all of these are proportional to plant investment, they may be composited. For exploratory estimates, a single factor may be used, and this should be based on company procedures. Holland and Brinkerhoff[11] cite a factor of 0.156 x plant investment for annual capital-related costs.

Utilities costs are minor for most products, and for exploratory estimates an allowance of 0.5-1.0¢/lb. product is frequently made. However, if this would be an excessive cost item, or if there is reason to believe this would be inadequate, there is no substitute for a preliminary energy balance to determine heat, power and cooling-water requirements.

Preliminary-Range Evaluations

For preliminary-range estimates, the three elements of processing costs (labor-related, investment-related, and utility) are considered in more detail. Since the process flowsheet used for estimating investment will be available, the direct labor can be estimated for each of the process units used. Perry[12] and Aries and Newton[13] have published tables of manpower requirements for various kinds of process equipment, and these can be used as a guide (Table V). If special packaging and waste treatment facilities are to be added for the project, labor requirements for these should be included. No special provision need be made for analytical control personnel unless the project will require an unusual amount of analytical work on a routine basis. Labor rates for preliminary-range estimates may be based on averages for the company or plant.

The over-all factor used to convert direct labor costs to total of labor plus labor-related charges may be the same as described for preliminary estimates, but it, and the labor rates, should be checked to assure that they are up to date.

Investment-related charges may have to be examined more closely in preliminary-range estimates. If the life of the project is expected to be shorter than the useful life of equipment (usually assumed as 10-20 years for tax purposes), the depreciation charges should be adequate to repay the fixed investment over the life of the project. For example, a project with an assumed life of 8 years would have a depreciation cost of $100\%/8 = 12\frac{1}{2}\%$ per yr. Uniform yearly depreciation charges are usually assumed for purposes of preliminary-range evaluations, even though the company may use another basis for depreciation of its actual investments.

Maintenance costs will vary depending on the severity of operating conditions (temperature, pressure, corrosion, abrasion, etc.) and the novelty of the process. The factors given by Newton and Aries[14] (Table VI) are given as guides.

Property taxes and insurance should be based on company or plant experience; or the factor of 3% of plant investment per year cited by Holland and Brinkerhoff[11] may be used.

Major utilities should be calculated roughly, and an allowance made for the less significant ones. Typical values given in Table VII may be used as a guide, but actual company experience should be used if utility cost is significant.

Typical labor requirements for equipment[13]—Table V

	Man Per Unit		Man Per Unit
Rotary dryer	1/2	Vacuum filter	1/6–1/4
Spray dryer	1	Evaporator	1/4
Tray dryer	1/2	Size-reduction equipment	1/4–1
Centrifugal separator	1/4–1/2	Tower	1/4
Vacuum crystallizer	1/4	Batch still	1
Mechanical crystallizer	1/6	Reactor, continuous	1/2
		Reactor, batch	1

Maintenance costs for process plants[14]—Table VI

Service	Yearly Maintenance Costs as % of Plant Investment
Light utilization in simple equipment	2 to 4%
Average usage	6%
Severe usage in complex operations	8 to 10%

Costs of some utilities[15]—Table VII (Jan. 1960)

	Large Plant Gulf Coast	Small Plant Northeast
Electricity, $/kwh	0.008–0.012	0.010–0.015
Steam, $/1,000 lb	0.30–0.50	1.00–2.00
Cooling water, $/1,000 gal	0.01–0.02	0.015–0.03
Process water, $/1,000 gal	0.10–0.25	0.15–0.30
Natural gas, $/1-million Btu	0.15–0.25	0.60–1.25

* If incremental costs are used, only those items of overhead added specifically for the project would be included.

Probable-Range Evaluations

For probable-range estimates, processing costs will require closer scrutiny. Direct-labor requirements should be based on a chart for the proposed plant giving the number of workers of each labor classification, analysts (if any) required for in-plant process control, foremen or shift supervisors, etc., with a full complement to provide the coverage required on all shifts during which the plant will be operated. Actual pay rates should be used for each category of labor in computing direct-labor costs.

The factor to be used in calculating labor-related costs will be lower in probable-range than in exploratory or preliminary-range estimates since supervision and possibly analytical services will be included in the direct labor figures of a probable-range estimate.

Capital-related costs for probable range may be calculated in the same manner as for preliminary-range estimates. It may be desirable to check whether the allowance for maintenance is reasonable based on the average amount of maintenance labor required for the project.

Major utility costs should be calculated based on plant accounting figures for unit costs, although a rough estimate will be adequate for minor utilities. It is well to use a check list (electricity, steam, gas, cooling water, process water, refrigeration, compressed air, inert gas, waste disposal) to make sure all utilities have been considered. At the preliminary-range stage, utilities at the proposed site should be checked to determine whether new utilities installations will be required, for this will increase capital cost for the project.

Other Operating Costs

Byproduct credits are generally neglected in exploratory estimates. However, if they are large, an allowance may be made at a fraction (not over 20-30%) of market price. For a preliminary and probable-range estimate, the annual production of byproducts should be compared with current market volume. If it appears reasonable that all or even a significant quantity can be sold, a cost estimate should be made for byproduct recovery, and the capital, as well as operating costs, included as part of the over-all project.

Royalties are not generally considered in exploratory estimates since the processes and intermediates are only tentatively established, and a patent search will not generally have been made. The only exception would be in cases where the product itself is patented.

For preliminary-range estimates, an allowance of 1-5% of sale price may be assumed in the absence of specific knowledge if it is known that licensing of patents held by others will be required.[13]

For probable-range evaluations, the existence of patents should have been explored thoroughly, and any royalties agreed on with the licensor should be used for this item.

Once manufacturing costs have been developed, the unit cost should be compared with expected or possible selling prices to see whether the project is likely to be profitable.

One rough rule is to examine the ratio of manufacturing costs to sale price as compared to the total of all manufacturing costs to sales income for the company (or division in a larger company).* If the new project has a ratio no higher than the average for company experience, and the plant investment per dollar of sales is comparable or below the company average, it is probably a reasonable candidate for commercialization. This is a crude generalization that should be used only when no other company criterion is available for judging the promise of research projects. The point to be made here is that manufacturing costs should be examined and compared with possible sales prices as soon as they are available to determine whether further work on the evaluation is warranted.

This is also the point at which the relative magnitude of materials, and labor-related and investment-related costs should be compared. Such a comparison provides an evaluation on possible directions for further estimating, and can guide research or development towards reducing projected costs for the product. The evaluation should be done after each estimate has been made, but it's likely to be more effective if done while the project is still on a bench-scale and before appreciable time and money have been spent on pilot-plant operations.

The next article in this series will discuss sales and market aspects of R&D evaluation. ■

* Company ratio of manufacturing cost to sales price may be obtained, if no other figures are available, by comparing the "cost of goods sold" item on the annual financial statement with "net sales."

References

1. Weinberger, A. J., *Chem. Eng.*, Nov. 25, 1963, pp. 113-118.
2. Ross, J., Marsik, F. V., Douglas, F. R. and Wagner, R. L., *Chem. Eng.*, Aug. 19, 1963, pp. 145-150.
3. Weinberger, A. J., *Chem. Eng.*, Oct. 28, 1963, p. 125.
4. Scheyer, R. H., *Chem. Eng.*, Sept. 2, 1962, pp. 158-159.
5. Dowling, T. E., *Chem. Week*, Nov. 12, 1960, p. 5.
6. Dowling, T. E., *Chem. Eng.*, Oct. 3, 1960, pp. 85-96.
7. von Allmen, E., *Chem. Eng. Progr.*, **56**, p. 37 (Nov. 1960).
8. Copulsky, W., Cziner, R., *Chem. Eng. Progr.*, **56**, p. 46 (Feb. 1960).
9. Berry, R. M., *Chem. Eng.*, June 27, 1960, pp. 123-126.
10. Wessel, H. E., *Chem. Eng.*, July 1952, pp. 209-210.
11. Holland, F. A., Brinkerhoff, R., *Chem. Eng.*, Feb. 4, 1963, pp. 97-98.
12. Perry, J. H. (Editor), "Chemical Business Handbook," pp. 2-68, McGraw-Hill, 1954.
13. Aries, R. S., Newton, R. D., "Chemical Engineering Cost Estimation," p. 162, McGraw-Hill, 1955.
14. Newton, R. D., Aries, R. S., *Ind. Eng. Chem.*, **43**, p. 2,306 (1951).
15. Weaver, J. B., Bauman, H. C., Heneghan, W. F., section 26, "Chemical Engineers' Handbook," 4th ed., McGraw-Hill, 1963.

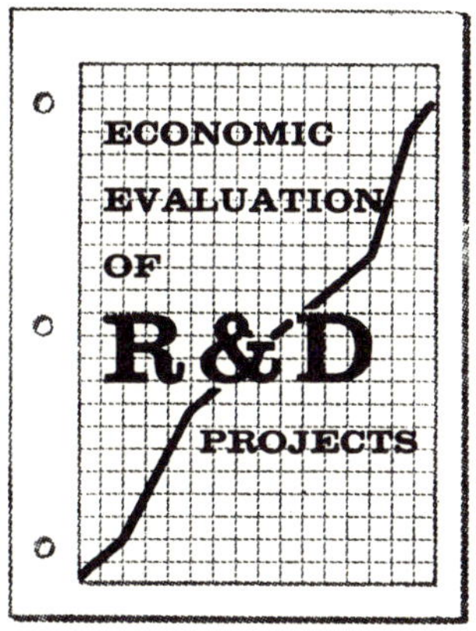

Acceptance or rejection of an R&D project may hinge on the accuracy of estimates of potential sales volume and price. This is the most difficult part of an evaluation, and the evaluator should know where to go for help and guidance.

Part IV

Estimating Sales and Markets

ARTHUR J. WEINBERGER
American Cyanamid Co.

In three previous articles, we discussed the estimation of costs related to manufacture of a new product. In Part IV, we will be concerned with the cost of selling and distributing the product, with general overhead costs, and with an evaluation of proposed sales price and sales volume.

Sales and distribution costs, as well as price, will be required for an economic evaluation, and the evaluation of potential markets will enter into the over-all assessment of the project.

Sales and distribution costs can vary widely, depending on type of product, total sales volume, method of distribution, and the kind of industry served. Sales and distribution costs ranging from 3 to 15% of total sales income have been published.

At one extreme would be large-volume basic chemicals sold in bulk and shipped to a relatively few customers directly from the producing plant. The customers would probably be large organizations with well-developed technical staffs, requiring relatively little technical service. At the other extreme would be sale of specialty chemicals to a large number of small customers—many dependent on technical service from the manufacturer. In the latter case, it might be necessary for the manufacturer to establish a system of local sales offices and warehouses in various parts of the country to provide adequate distribution and service, or the product might be handled through distributors.

Sales and Distribution

It is beyond the scope of this article to discuss details of sales and distribution practices for the chemical and related industries. But it is important that the project evaluator recognize that various techniques are used and that adequate allowance should be made for the particular project even while it is in research and development.

Sales costs include the following major elements:

- Direct sales expense (salesmen's salaries and/or commissions and expenses, sales office expenses, and sales administrative costs).
- Advertising and promotion (including literature sent to prospective customers).
- Technical service to customers (either by helping them in use of the product, or modifications in the product to fit their needs).
- Market development—the search for new markets or applications for the product.

Distribution costs include:

- Cost of factory warehousing, loading material for shipment (packaging costs assumed included in manufacturing costs).
- Transportation charges for product (unless paid by customer).
- Cost of operating branch warehouses or distribution centers.

Sales expenses are likely to be higher for new products than for established ones since the customer must be sought out and convinced of the product's merits, and technical service and market development are required.

Detailed estimates of sales and distribution expenses are not generally warranted in evaluating R & D proj-

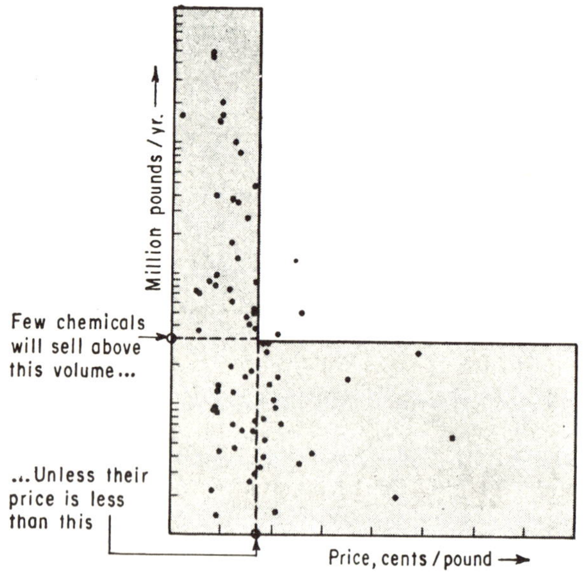

Exclusion chart gives pricing information—Fig. 1

ects, particularly during the early stages, but the outline cited above provides a basis for comparing sales and distribution costs for the new product with those already being sold by the company. If the new product will be sold by the existing sales organization, and technical service and distribution channels are also expected to be similar, it may generally be assumed that sales and distribution expenses (generally expressed as a percentage of sales income) will be comparable to the average percentages for existing products.

For a more sophisticated analysis of a project further along in development, it would be well to add, for the early years, a dollar allowance equivalent to the extra market development, technical service, and preliminary sales efforts required until acceptance of the product is achieved. The evaluator should not attempt to forecast the magnitude of these allowances, but rather see that such allowances (preferably estimated by the sales department) have been added.

Several methods can be used for products that would be distributed in a different manner from those already being marketed by the company. One procedure, useful for preliminary purposes, is to obtain from published financial records, such as company annual reports or industry surveys, sales and distribution costs as a percentage of sales for companies already serving the particular industry or market. For evaluation as the project approaches commercialization, a more definitive estimate of actual sales and distribution expenses involved should be obtained from company marketing management.

Distribution costs can be handled without difficulty, unless the new product is a low-priced bulk material of considerably lower value than products usually sold by the company. If shipping charges for the material are to be absorbed by the manufacturer, they should be estimated based on average distances the product will have to be shipped. For low-value bulk materials, this can be an appreciable part of the sales price, and it tends to limit the radius within which the product can be sold profitably. Some transportation costs have been published that should be useful for a preliminary estimate.[1]

General Overhead Costs

One cost item that must not be overlooked is overall company overhead. This includes salaries and expenses of company officers, administrative staff, and those general company expenses that are not charged directly to manufacturing units or products. These latter expenses, for a large company, might include such items as legal and public-relations staffs, corporate personnel department, and corporate research.

Company overhead is generally calculated as a percentage of sales income, to be charged to all products. It may be argued that the addition of a new product would have negligible effects on company overhead (see incremental cost discussed in Part III). Since this is a policy decision that the company makes, the evaluator will need to know the system used in his own company.

Evaluating Projected Volume and Price

Predicting sales volume and probable prices is the most difficult and inexact part of any evaluation, and yet these factors are usually the most important ones in judging profit potential.

Studies have shown that more R & D projects become economic failures because anticipated markets did not materialize than for any other single reason. Thus, it is desirable that all but very preliminary evaluations include an analysis for the effect of change in sales volume and price on profitability.[2]

The evaluator is not generally called upon to provide estimates of projected sales price and volume, but he should evaluate the information supplied. This is particularly important when the estimates are made by R & D people who are proponents of the project and are perhaps naive about market considerations. If the company has a market research group, these people should provide market figures on which to evaluate the project. However, if this assistance is not available, the evaluator should satisfy himself that the projections given him are reasonable. This is worth an appreciable amount of time, since much depends on the market estimate.

Existing, Improved or New Products

R&D projects may involve new processes for making existing products, or processes for making improved or new products. From the market point of view, we would have for sale:

- Existing products (perhaps at lower price).
- Improved products.
- New products.

Each of these is handled in a somewhat different manner, so we will discuss the cases individually.

Existing products, as discussed here, will be those available now, but not presently being made by the evaluator's own company. (For products that one's own company is now producing, the sales department should be the source of sales forecasts.) The R&D project may be based on a unique position in process, materials supply, or other factors that would give the company a competitive advantage. Therefore, it is worth evaluating the possibility of entering an established market.

Market volumes for many established chemical and related products are published regularly by government agencies. Organic chemicals, plastics, dyes, and some pharmaceutical products are compiled by the U. S. Tariff Commission. Data on minerals and many inorganic chemicals are compiled by the Bureau of Mines and the Bureau of the Census. A general publication giving production information on a variety of chemical products over a period of years is the "Chemical Economics Handbook" published by Stanford Research Institute. In addition, there are periodic surveys of markets and end-uses for individual chemicals by *Chemical Week, Chemical and Engineering News,* and *Oil, Paint and Drug Reporter.* These surveys include estimates of future production and list producers and their production capacities.

Estimating Share of the Market

From this information, the evaluator can form some idea of what might be a reasonable share of the projected market for his company to capture. One method would be to assume a proportionate share. If there were three producers, a fourth might be able to capture $100/(3 + 1) = 25\%$ of the total; half this much might be a conservative estimate. If the projected sales estimates are above this range, there should be a good reason for optimism. The possibility of a lower price based on a new process or an advantageous raw-materials position might be good reason for an optimistic estimate—but it is also well to make an estimate of competitors' probable costs before any decision is made on a price.[3]

If it appears likely that the product can be sold profitably at less than present market prices, it may be possible to not only increase the company share of the market, but also enlarge the total market by selling the product for applications where it is now too expensive to be considered. For example, a lower-cost route to soda ash might permit it to displace ammonia from applications where it is used as a soluble alkali only because it costs less on an equivalent basis. While it is generally beyond the scope of the evaluator's job to make such an investigation, suitable questions should be asked: an important part of the evaluation is to assure that the project is being assessed realistically, and not just to uncover over-optimistic assumptions.

The sizes of competitive plants, sometimes listed in the product surveys mentioned above, or listed among the new plant inventories in *Chemical Engineering* and *Chemical and Engineering News* have another value. They give some insight into competitors' ideas of what constitutes an economical plant size using their own processes.

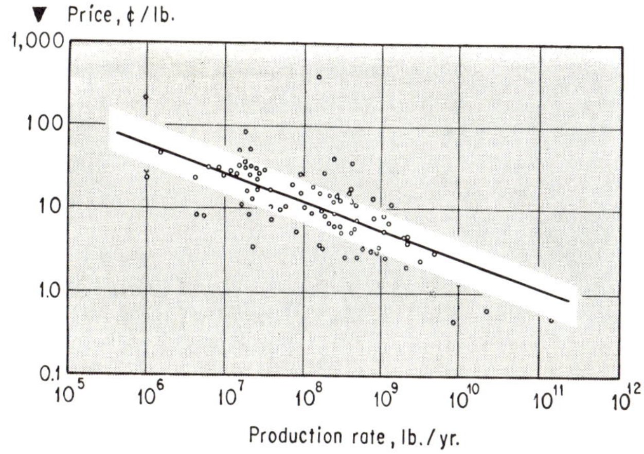

Price vs. production rate: the trend is down—Fig. 2

Evaluating Improved Products

Improved products usually offer markedly better performance over those currently available. They not only permit the product to compete against existing ones serving the same purpose, but may make it possible to compete in new end-uses where previous products were unsatisfactory. For example, improved impact strength has made it possible to use plastics in business machine housings, where previously only metals were considered adequate.

The greater the improvement in the product, the more likely it will find applications where previous products could not be used. In situations of this kind, it becomes necessary to closely examine end-uses and the alternative materials the product may displace.

A considerable volume of market information is available for such a study. The Census of Manufacturers, Current Industrial Reports, and the various publications of the Business and Defense Service Administration are among the most useful U. S. government sources. In addition, the various industry associations (Society of the Plastics Industry and National Electrical Manufacturers Assn. are examples) and industry trade publications, and their editors, can be most helpful. The marketing and technical people of potential customers are not only even more valuable sources of information but are good contacts for later work when you want outsiders' evaluation of the material. Generally, users are quite willing to talk about their material needs, and this feedback can be helpful in guiding future R & D as well as in the evaluation itself.

In cases where estimates of markets are based on applications and end-uses, it is frequently not possible to obtain direct figures on the actual usages of the desired material (as, for example, electrical insulation if end-use is in a motor, or enamel in a refrigerator), but "rules of thumb" often exist for either the amount of such material per unit or the percentage of total value represented by the specific material being investigated. While these are crude measures, they are frequently the only ones available.

Where the improved product is intended to sell at a premium price (the usual situation), caution must be used in judging whether the new material will be worth the premium. When the product is competing on a price-performance basis, this is easily checked. If the improved product is, for example, a thickening agent that would sell for 40% more than an existing product but would only require half the usage to give comparable performance, the premium-price product would represent a real saving to the user and he would have incentive to use it. If, however, the new product cannot be compared on such a definite price-performance basis, care must be taken that the improvement has real meaning for the potential purchaser and is not just a more impressive number on the product specifications list.

Handling the New Product

Markets are very difficult to predict for new products. If a product's characteristics are such that it can displace existing materials—for example, a high-temperature synthetic surface coating that could replace the porcelain glazes used in heaters and oven interiors—the markets for the end-products can be used as guides. However, a great amount of testing is required to establish that the performance of the new material is suitable. While the evaluator would not actually do such checking, he should make sure that adequate allowance of both time and expense have been included for extensive tests when required.

Truly novel materials may find their applications in new products that were not possible before the material became available. In cases of this sort, a considerable amount of market research and market development will be required, and, until this is done, any market figures are likely to be guesses. Under such circumstances, it may be desirable to calculate costs and possible prices for several levels of sales volume to determine how realistic the various combinations of volume (and profitable price) would be. For a truly novel product, one level of production might be a semi-works plant intended more to supply sizable amounts of material than to realize the maximum profit possible from the project.

The introduction of new materials or products involves time for testing in the intended applications and for commercial acceptance, as well as acceptance of the final end-product that will use the new material if it is also new. For this reason, it should be assumed that the market for a new material will develop gradually over a period of years—the full capacity of the plant will not be sold as soon as it is on stream. Even in the case of existing or improved materials, it is usual to assume that there will be a gradual buildup in sales (at a faster pace than for new materials).

The Price Question

While the evaluator of an R&D project will not be setting the price for the final product, he should assure himself that it is reasonable. Therefore, he should be aware of factors that affect price levels.

Usually the price for a material is dictated by the end-use and the prices of other materials that could be used as alternates. For example, the selling price for rigid plastic sheetings for sheathing commercial buildings is limited by the price of corrugated iron sheet. If the cost of corrugated plastic sheeting were too high, builders would not consider it and would use iron instead. In cases where the price-performance relationship can be measured quantitatively, the new material must be at least competitive to be of interest to potential customers. This concept is inherent in the "exclusion charts" prepared by Zabel[4] that outline price-volume relationships (see Fig. 1).

Price information on a variety of chemical and related products were discussed in a previous article,[5] but as was pointed out, these are list prices. While list prices are adequate for preliminary analysis, the evaluations should take account of any price changes due to quantity purchases, contracts, etc., for more detailed analyses.

It is not usual to assess changes in the general price level (either of purchases or sales) with time in analyzing R & D projects. But one important trend is frequently ignored: the reduction in prices, especially of newer materials, as volume increases and as competition becomes more intense. For this reason, it is well for the evaluator to review the projected price for the product and see whether it does make allowances for future reductions. Correlations of sale price level with increasing volume of production have been presented by Bartels[6] and Schuman[7] (see Fig. 2).

Now that we have discussed the methods for preparing or reviewing the individual factors in an economic evaluation, we are ready to put them together. This will be done in the next article in the series. ■

References

1. *Chem. Eng.*, Sept. 3, 1962, p. 158.
2. Douglas, F. R., Marsik, F. V., Tilquist, G. E., Wagner, R. L., paper "Effect of Key Marketing Variables on Profitability," given at Amer. Assn. Cost Engineers annual meeting, San Francisco, July 24-26, 1963. Condensed in *Chem. Eng.*, Oct. 28, 1963, p. 154.
3. *Chem. Eng.*, Nov. 26, 1962, p. 109.
4. Zabel, H., *Chem. Eng. Progr.*, May 1956, p. 183. Also see *Chem. Eng.*, Oct. 19, 1959, p. 112.
5. Weinberger, A. J., *Chem. Eng.*, Dec. 23, 1963, pp. 81-86.
6. Bartels, C. R., *Chem. Eng. Progr.*, Mar. 1958, pp. 59-61.
7. Schuman, S. C., *Chem. Eng.*, Mar. 1955, pp. 180-182.

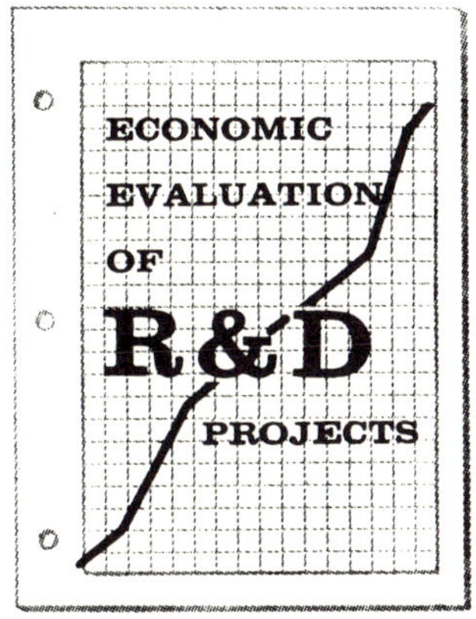

Having estimated the manufacturing and capital costs, and appraised the sales and market aspects, we are now ready to determine just how profitable a particular R & D project is likely to be.

Part V

Methods for Estimating Profitability

ARTHUR J. WEINBERGER, *American Cyanamid Co.**

Previous articles in this series have described methods for obtaining the basic information required to determine the profitability of R & D projects. The present one will describe the methods used for the profitability determination itself once this basic information has been obtained.

In general, relatively rough and approximate methods for estimating profitability are used early in the R & D project—when the data are less reliable—whereas more-sophisticated and accurate methods are used at later stages.

Profitability methods fall into three classes:

- Return on investment, in which the profit after all costs is compared with the investment in the project.
- Payout time, or the number of years between the start of commercial operation of the project and the time when total income from the project equals all expenditures, including plant investment.
- Project value or worth, which is a dollar figure that represents the potential value of the entire project to the company.

There are a number of variations in each of these categories, so only a few of the most widely used ones will be considered here.

Methods for Preliminary Estimates

Unlike the more detailed methods, payout time and the simpler return on investment formulas are generally calculated on the assumption that money has no "time value;" in other words, a dollar spent now will be equal to, and have the same value as, one returned several years later.[1]

For preliminary estimates made at an early stage of R & D, these methods would probably be sufficiently accurate for the rough screening required to determine whether a project appears worthwhile. In these cases, it is usually sufficient to consider the potential return based on figures for a single typical year once the plant is operating at design capacity:

$$\text{Return on investment} = \frac{\text{Profit}}{\text{Investment}} \times 100$$

In the above formula, the investment includes both fixed capital (plant cost) and working capital. Depending on individual preference, the profit may be either before or after corporate income tax, and may be based on original or average (depreciated) fixed capital investment. Although, logically, R & D cost should be accounted for as part of the money invested in the project, this is generally omitted. Table I on the following page shows a sample calculation of return on investment.

The formula for payout time is as follows:

$$\text{Payout time} = \frac{\text{Fixed capital investment}}{\text{Annual net profit} + \text{Annual depreciation}}$$

Working capital is not included in calculating capital investment, since it is held as cash, raw materials, finished-product inventories, etc., which presumably can be sold and converted to cash when the project is discontinued. Depreciation is generally assumed to be uniform for the life of the project in preliminary evaluations.

Project value methods can be used in preliminary evaluation of projects to estimate the amount of R & D justified for a given ultimate profit.[2,3]

Manley[4] has suggested a formula for calculating a maximum research expenditure, *R*, that should not be

Calculating the payout time and return on investment—Table I

DEF Chemical Company
Proposed Organic Intermediate GHI

Data:		
Investment required:		
New-plant investment*	$300,000	
Working capital	200,000	
Income and expense statement (per year)		
Predicted sales—1,000,000 lb./yr.@$0.75/lb.		$750,000
Manufacturing costs	$400,000	
Sales+distribution costs	100,000	
General overhead expenses	150,000	
Total costs		650,000
Net (pre-tax)		$100,000
Net (after 52% tax)		48,000

Return on original investment (R.O.I.):

$$\text{R.O.I.} = \frac{\$48{,}000}{\$300{,}000 + 200{,}000} = 9.6\%/\text{yr.}$$

Payout time (P.O.T.):

$$\text{P.O.T.} = \frac{\$300{,}000}{\$48{,}000 + 30{,}000} = 3.8 \text{ yr.}$$

* Depreciated at 10%/yr. based on 10-yr. plant life and no salvage value.

exceeded if the project as a whole is to be profitable:

$$R \leqq (YSN - P - W)/r$$

where Y is the maximum allowable payout time, in years; S is the annual net sales income; N is the minimum acceptable percentage of net profit after taxes, based on sales; P represents plant investment, and W the working capital; r is the fraction of R & D cost that is not tax-deductible.

If R & D costs are allowed to equal the net profits from one year of commercial operation at full plant capacity (an arbitrary choice), then this figure may be used as a yardstick for measuring the justifiable R & D project. For projects that will require little or negligible capital expenditures—such as revision of existing processes, or manufacture in existing facilities—the justifiable R & D might reasonably be taken as an increased number of years profit at full production.

In one company, values of successful research projects are calculated as: the savings for one year on a process improvement; 3% of net sales for one year on an improved product; and 3% of net sales for three years on a new product.[5] The same reference also cites similar practices used by other companies for assessing the value of research projects. While these methods are generally used for an "after the fact" evaluation of the over-all effectiveness of an R & D department, they can also be used to assess the value of proposed projects.

Since profitability evaluations made at an early stage of R & D, or even before, are likely to be crude, it may be better to express the evaluation results as ranges rather than single numbers. In general, the sales price or volume, or both, will be less accurate than production costs; and so a range in sales estimates (one relatively conservative, the other optimistic) may be advisable. While ranges could be used for costs as well as sales, the evaluator would soon find that his range of results would be so wide as to make them meaningless.

Year-by-Year Analysis

As R & D projects progress beyond the exploratory stage, and more-reliable manufacturing and marketing information are made available, more-accurate measures of profitability become justified.

A first step is to analyze the project on a year-by-year basis, rather than at the full capacity condition mentioned above. Yearly sales (physical volume, unit price, and sales income) are tabulated along with the yearly items of expense, and a yearly profit is obtained (see Table II). Proper allowance is made for the rate of buildup of sales volume, possible reduction in price, initial market development, and plant startup costs, as discussed in previous articles.

Return on investment is calculated as indicated above, based on average profits for five or ten years, or whatever time period is normally used by the company in appraising projects.

Payout time based on a year-by-year analysis is still determined by finding the time at which the net cash returned to the company is equal to the investment in the project. This procedure, called a cash flow analysis, counts as "net returns" the net profit (or loss) on the project each year, after allowing for income taxes plus depreciation on the fixed capital invested in plant facilities.

The investment may be taken as either the fixed

How cash flow is affected by changes in: Sales price—Fig. 1

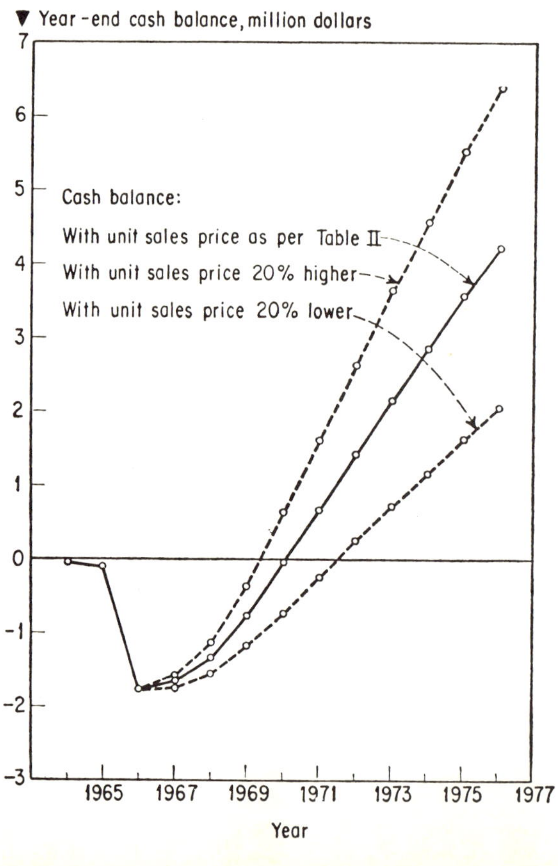

Year-by-year analysis of cash flow for sample project—Table II[1]

Year	Sales Price, $/Lb.	Annual Production, Lb.	Sales Income, $	Unit Mfg. Cost, $/Lb.	Total Mfg. Exp., $	Sales Exp., $	Overhead, Etc., $	Net After 52% Taxes, $	Depreciation,[2] $	Year-End Cash Balance, $
1964							75[3]	−36	...	−36
1965							125[3]	−60	...	−96
1966							175[4]	−84	...	−1,780[5]
1967	1.75	500	875	1.40[6]	700	87.5	162.5	−36	160	−1,656
1968	1.60	1,000	1,600	0.90	900	160	190	168	160	−1,328
1969	1.55	1,500	2,325	0.70	1,050	232.5	232.5	388.8	160	−779
1970	1.50	2,000	3,000	0.60	1,200	300	300	576	160	−43
1971	1.50	2,000	3,000	0.60	1,200	300	300	576	160	+693
1972	1.50	2,000	3,000	0.60	1,200	300	300	576	160	1,429
1973	1.50	2,000	3,000	0.60	1,200	300	300	576	160	2,165
1974	1.45	2,000	2,900	0.58	1,160	290	290	556.8	160	2,882
1975	1.40	2,000	2,800	0.57	1,140	280	280	528	160	3,570
1976	1.35	2,000	2,700	0.57	1,140	270	270	489.6	160	4,219

1. All figures in body of table are in thousands, except as noted.
2. $1,600,000 plant built during 1966 and depreciated at 10% per year.
3. R & D expenses.
4. R & D and commercial development expenses.
5. Includes $1,600,000 capital expenditure for plant.
6. Includes all startup expenses.

capital alone, or it may also include R & D and other pre-production costs. If the latter are included, an adjustment must be made for their being treated as expenses in calculating company income taxes. As a result, the net cost to the company is reduced by the amount of tax credit on these expenses, as follows:

Assume $100,000/yr. expenditures for R & D on a new project not yet commercialized, for a company paying a tax rate of 52% on incremental income. Since this $100,000/yr. in expenditures is subtracted from company net income in calculating tax payable, the actual net cost is $100,000 − 0.52(100,000) = $48,000.

Methods That Allow for Interest

In the methods just described, there is no allowance for the interest that money spent on the project could have been earning had it been invested in bonds or other securities. A simple means of including interest is the "venture profit" described by Happel.[6] If i is the minimum acceptable interest rate, then venture profit equals:

$$\text{Net income} - i\,(\text{Working Capital} + \text{Fixed Capital})$$

Venture profit may be based on typical or average

How cash flow is affected by changes in: Annual production—Fig. 2

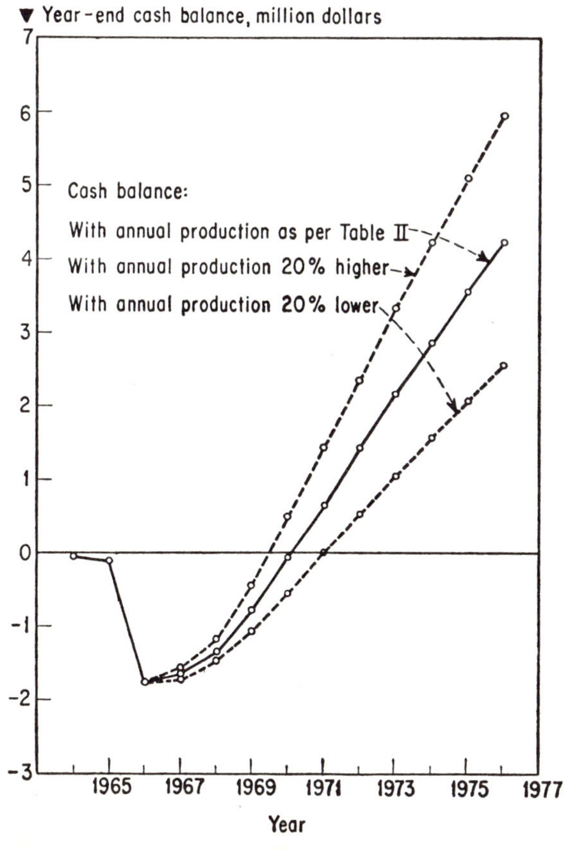

How cash flow is affected by changes in: Manufacturing cost—Fig. 3

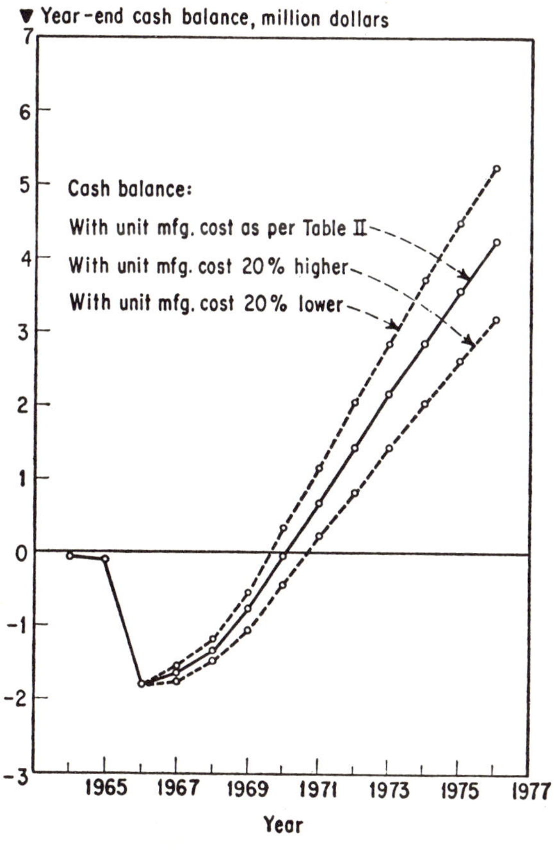

yearly figures, or on a year-by-year basis over the life of the project. Other methods including interest in calculating return on investment have been summarized by Bates and Weaver.[7]

More widely used are the evaluation methods based on the concept of present worth. The present worth of a sum of money to be spent (or received) at a future time is the amount that would have to be set aside, at the specified interest rate, to yield that future sum when the specified time arrived. For example, at 5% interest, the present worth of $105 to be repaid in one year would be $100.

One of the most common methods based on present worth is the Profitability Index described by Reul.[1] Profitability Index is the interest rate that money put into the project would have to earn in order to equal the returns expected from the project itself. Venture worth, proposed by Happel,[8] is also based on present worth, and is the value of the project over and above a minimum acceptable rate of return.

These present-worth methods are somewhat time-consuming to calculate by hand, but can be worked readily with a computer. In general, they are warranted only at later stages of R & D, when the basic information on the project is relatively reliable.

The Sensitivity of the Profitability Estimate

So far, we have discussed only the mechanics of calculating profitability. An important factor in the analysis is to determine just how much effect a change in the major assumptions (such as sale price, volume and cost) will have on profitability. For projects at an early stage of R & D, it is usually sufficient to indicate a range of profitability based on optimistic and conservative estimates of the more important factors. For projects at a more advanced stage, profitability calculations can be made while changing one assumption at a time. Examples of this have been reported previously.[9, 10, 11] A more elaborate method based on a series of alternatives has been described by Anderson.[12]

The purpose in calculating the sensitivity of profitability to changes in input data is to get a measure of the risks involved if the data and assumptions made are incorrect. Frequently, profitability of a new product is affected drastically by modest changes in selling price, output volume, unit manufacturing cost, or some other factor.

For illustrative purposes, a 20% change in one variable (sales price, sales volume, and manufacturing cost, respectively) has been applied to the data shown in Table II, and the results plotted in Figs. 1, 2 and 3. In this case, since Fig. 3 shows the least separation of the year-end cash balances, the profitability is evidently less sensitive to changes in manufacturing cost than to the equivalent percentage change in unit sales price or volume.

The date at which the cash-balance curve for the project turns positive, i.e., crosses the zero line, represents the end of the payout time; note how this point varies with each assumption.

If the project is nearing commercialization, and further market research or more-detailed estimating can supply better data, it may be worthwhile to await these data before reaching a final decision. On the other hand, if the project is long delayed because of this wait, competitive advantage may be lost. The proceed-or-not decision must be made by management, but it is the evaluator's responsibility to provide the facts and alternatives.

Once the estimated profitability of the project has been established, it can then be compared with other R & D projects as well as with alternative uses of company funds, to see which uses would give the best returns. Comparisons will take into account not only the potential profitability but also an assessment of the risks involved. Below are some types of R & D projects arranged in order of increasing risk:

- Improvement of existing processes and products.
- New products related to existing lines and using similar technology.
- Methods for producing materials used within the company.
- Products new to the company, but already in commercial production elsewhere.
- Novel products unrelated to present company activities.

It is obvious that to be attractive the projects involving greater risk should yield higher return. Numerical rating schemes have been proposed to assess the relative desirability of projects, taking into account both profitability and the chances of technical and commercial success; the rating methods described by Rosen[2] and Miller[13] are typical.

A less quantitative but broader checklist approach has been presented by Harris.[14]

These methods can serve only as guides to what others have suggested. In the last analysis, the selection of projects will be based on the criteria of the evaluator's own company. ■

References

1. Reul, R. I., *Factory Management and Maintenance*, Nov. 1955, p. 92; also *Harvard Business Review*, July-Aug. 1957, p. 116.
2. Rosen, B. H., Regnier, A. L., *Chem. Eng. Progr.*, Nov. 1956, p. 500.
3. Disman, S., *Chem. Eng.*, Dec. 24, 1962, p. 87.
4. *Chem. Eng*, Jan. 1955, p. 130.
5. Perry, J. H., (editor), "Chemical Business Handbook," Sect. 4, p. 43, McGraw-Hill, New York, 1954.
6. Happel, J., "Chemical Process Economics," Wiley, New York, 1958.
7. Bates, A. G., Weaver, J. B., *Chem. Week*, June 15, 1957, p. 116.
8. Happel, J., *Chem. Eng. Progr.*, 1955, p. 533.
9. Boyer, C. D., Krayer, R. S., *Chem. Eng. Progr.*, Mar. 1963, p. 24.
10. Thorne, H. C., Jr., Twaddle, W. W., Grahl, E. R., *Petrol. Refiner*, Mar. 1961 p. 171.
11. Johnson, L., *Chem. Week*, July 21, 1962, p. 61.
12. Anderson, S. L., *Chem. Eng. Progr.*, Mar. 1961, p. 80.
13. Miller, T. T., *Chem. Eng. Progr.*, June 1958, p. 56.
14. Harris, J. S., *Chem. Eng. News*, Apr. 15, 1961, p. 110.

ECONOMIC EVALUATION OF R&D PROJECTS

"Auditing" a completed economic evaluation can yield valuable information for future studies. And there are many non-quantitative factors management must consider that might cause a promising project to be rejected.

Part VI

Post Audits and Qualitative Factors

ARTHUR J. WEINBERGER
American Cyanamid Co.

In this last article of the series, we will cover two important aspects of an economic evaluation: (1) learning from a post-investment appraisal of the evaluation, and (2) reasons why management may reject a project that looks good to an evaluator.

What Went Wrong?

As we have seen, evaluation can and should be a continuing part of the R&D project until it reaches commercial operation or has been dropped. But the evaluation should not stop there. The approved project should be subjected to a post-investment appraisal to determine whether it worked out as predicted, and whether there are any lessons to be learned for improving future evaluations. This type of audit is not the usual accounting audit. The probing discussed here is beyond the scope of the accounting function.

These post-investment appraisals require long-term planning to be effective, and a considerable period must elapse between the time of commercialization and the time a project audit can be considered complete. Records of successive evaluations made as the project progresses through R&D stages must be kept in sufficient detail to provide the auditor with all the information available at any particular time on the technical, economic and market aspects. The most obvious items would include:

Technical

- Process outline, requirements for special equipment, severe processing conditions, materials of construction.
- Raw materials (quantities required and specifications).
- Utilities.
- Status of any byproduct recovery steps.
- Status of waste-treatment problems.

Economic

- Capital-cost estimates.
- Manufacturing, distribution and overhead cost estimates.
- Profitability estimates.
- Byproduct disposal or recovery costs.

(All of the above to include substantiating data.)

Market

- Projected sales volume and price.
- Raw-materials procurement.
- Product specifications.

Of course, all this information would have to be collected and dated for future reference. The amount and extent of data to be retained would be a matter of balance, since too much detail could be time-consuming to review, and too little might make it difficult for the auditor to follow. This can best be handled by using a standard format specifying the type and extent of material to be retained in the files.

Getting "Objective" Audit

For an audit to be of real value, it must be objective. Ideally it should be done by a person who has not had previous contact with the project. This is not always possible, but it is good practice to ask for written comment on the auditor's findings. The original evaluator's comments should be a part of the final audit.

It is also valuable to consult as many as possible of

Some questions to ask when appraising an evaluation

- **Did the commercial plant perform as projected from the R & D project?**

  Did product meet specifications?

  Were anticipated yields achieved?

  Did unforeseen process problems occur?

  How about usages of materials, catalysts, solvents, utilities, etc.?

  Was waste treatment adequate?

- **Were sales and market predictions substantiated?**

  Were predicted sales volume and price reached?

  Did the sales volume build up as fast as anticipated?

  Did the expected uses materialize, and did additional ones appear?

  Was it necessary to change product specifications to fulfill expected uses?

- **Were anticipated time schedules and costs met?**

  Do manufacturing costs match predictions?

  How did plant investment and startup expense and time tables compare with predictions?

  Were sales, commercial development and technical service allowances adequate?

  How realistic were facilities and cost estimates for raw materials, byproduct, and waste disposal?

---

the key people in R&D, engineering, manufacturing, and marketing who worked on the project for their views on how the process (or product) worked out in commercial application. Their comments can be very valuable in terms of developing information and procedures that can be applied on future projects. However, it must be clearly understood that the purpose of the audit is to discover possible improvements that can be made in future projects and evaluations, and *not* to place blame for any inadequacies in the performance of the project.

Preliminary Appraisal

Since a period of years will necessarily elapse between the completion of a R&D project and the time the commercial venture is in reasonably "steady state" operation, it is helpful to conduct a preliminary audit after the R&D has been completed. In reviewing the project evaluations made at various steps of R&D, we can seek answers to such questions as:

- Were the R&D cost projections realistic?
- How well was the project time-table maintained?
- How much did the process or product change?
- Did the process become more or less complex? How much?
- Did product specifications change? Did this make for greater or fewer processing problems?
- How much did manufacturing cost and capital investment change in succeeding evaluations?
- How and how much did sales projections change; also did numbers of applications, and price levels, change as R&D progressed?

Post-Commercialization Audit

These and some additional questions would be asked at the time of a post-commercialization audit: the most important questions would start with whether the project realized the anticipated return on investment, but others of greatest value in assessing the R&D evaluations would include those shown in the table.

Little has been published on after-the-fact audits on R&D projects, but one review by Tyler* concluded that overoptimism in forecasts arise from:

- Underestimating the time required for process development, market evaluation, product development, and plant startup.
- Overestimating price and volume.
- Underestimating competitive strengths.

Qualitative Factors

Having completed an economic evaluation of the project, you may now believe that you can tell management whether or not the project will be profitable and whether it should be pursued. However, there are nonquantitative factors, generally outside the province of the engineer, that also enter into the decision of management and may help to explain why the fate of a project is not based solely upon economic evaluations. This is no apology for management's actions, but it does show why we, as engineers dealing in quantitative terms, are sometimes baffled by the decisions made at higher levels.

To get a better understanding of these qualitative factors, let us examine a few.

The Problem of Risk

Most economic evaluations are deficient in that the methods generally used do not assess risk adequately. As mentioned before, a company may require higher profitability before pursuing projects of higher risk; this is at best a rough approximation. Management must apply other judgment factors.

Having a better over-all view of the company, management should decide whether the project would make good use of the strong points of the organization and not be too demanding of "weak" areas. If, for example, the company has an adaptable and above-average production and quality-control organization, it would be in a better position to supply a new product that requires complex processing to rigid product specifications. Similarly, a company with an excellent technical service group need be less hesitant about introducing a radically new intermediate requiring considerable customer education.

Management will usually be more reluctant to authorize projects that would take the company into areas in which it has had no previous experience. These areas may, for example, involve type of market (such as a heavy chemical producer diversifying into consumer products), or kind of manufacturing know-how (a pharmaceutical outfit considering petrochemicals). Management's reluctance to proceed with such projects is an honest recognition that it would require development of know-how and this may be more costly than the economic evaluation would indicate.

The willingness of management to accept the risks inherent in a new project will also depend on the amount of investment required, and the seriousness of any loss were it to prove unsuccessful. This is particularly true of projects requiring an expenditure involving a significant part of the total assets of the company. After all, a primary function of management is to assure continued solvency.

Company Objectives

The management of a company will have a concept of the kind of company it is operating, and what kind of company it wants its firm to be or become. When these are clearly defined, and preferably reduced to writing, they become corporate objectives that can be very valuable in guiding R & D in the selection and evaluation of existing as well as potential projects.

As a result of their long-range planning efforts, forward-looking managements will have information on the growth potential and profitability of not only the markets now being served by the company, but others that might be desirable either now or in the future. Obviously, management will be more receptive to projects in the more profitable, growing areas than those that are uncertain or static.

Looking at Competitive Position

If the project would yield a new or improved product, management wants to know how much competition it would face. For a unique product, this would include asking:

- Does the company have, or would it be likely to obtain, a patent position strong enough to preclude or minimize competition?
- Is the product in a field likely to gain customer acceptance readily (possibly determined by analyzing case histories of other novel products previously offered in the field)?
- Would the product involve governmental regulation or possible legal liabilities in its use (such as potentially toxic materials)?

For any product—unique or not—management will have to weigh: The number and quality of competitors, and the likelihood of their introducing a comparable (or superior) product; whether raw materials are readily available from a number of sources, or only a very few suppliers (one of whom may become a competitor); whether the company has special circumstances that make the product particularly appropriate (supply of raw materials, captive uses, existing manufacturing facilities that could be used, a seasonal product to complement existing ones, etc.); the diversity of end-uses, number of potential customers (product should not be too dependent on any single one), and the stability of the market (so that it is not tied to end-uses that will shrink or be displaced); the effect of the new or improved products on others now being produced by the company; likely trends of future costs in relation to sale price.

The size of the project will also have to be considered in relation to the company. The R&D project may, if successful, require such huge expenditures to commercialize on an adequate scale that it is too large for the financial resources of the company. Similarly a new-product project may be so small that it would not bring in enough net profit to be worthwhile for the effort required if the company were relatively large.

Management may also be reluctant to approve projects that, when commercialized, could pose unusual problems in safety, pollution, and/or waste disposal.

Management's Job

As you can see from this quick rundown of the qualitative and largely subjective considerations, there are many factors that must be considered in evaluating R&D projects. Although many of these are not quantitative, the evaluator should not be discouraged from analyzing aspects that can be examined quantitatively. It is his job to see that management does not have to concern itself with obviously unprofitable projects. Management can then concentrate on the aspects of decision that have not yet been reduced to numbers. ■

ACKNOWLEDGEMENT

The author gratefully acknowledges the contributions of colleagues at the American Cyanamid Co. and elsewhere for background information that made this series possible.

Meet the Author

Arthur J. Weinberger is a senior chemical engineer in the Central Research Div. of American Cyanamid Co. at Stamford, Conn. He has specialized in technical and economic evaluation of research and development projects, and has also handled surveys of potential research topics. Mr. Weinberger was graduated from M. I. T. (B.S. and M.S. degrees), and has had experience in production supervision, plant startup and troubleshooting, and process development at Polaroid Corp., E. R. Squibb and Heyden Chemical. Mr. Weinberger has been active in the AIChE as a founder and chairman of the Fairfield County section.

6

Tools Specific to Plant and Project Management

The last part of the previous section is rather specific to R&D management; other managerial career fields are treated equally in this section.

For actual or aspiring plant managers, there are guidelines on plant organization, maintenance, quality control, foremen relations, disciplinary problems, and automation.

For those interested in project work, there are comprehensive reports on the management of U.S. and European projects. (Also particularly pertinent to project work is the earlier article on critical-path methods, on pp. 262-272).

What a Plant Manager Does And How He Does It Part 1

Engineers who want to be plant manager would do well to look into the job's requirements and responsibilities. This, the first of three articles on the subject, examines how the job varies from small to large companies.

M. K. GOSS, J. T. Baker Chemical Co.

Along with the prestige of being the manager of a plant goes the complete responsibility for its operation. Your success or failure will depend on you.

You will be judged by what you accomplish, and your ability to obtain from management the manpower and facilities necessary to carry out your goals will in no small way affect your performance.

Responsibilities Vary With Company Size

The responsibilities of plant managers vary significantly with the size of companies. Even the job of a plant manager in a small plant of a small company can considerably differ from that in a small plant of a large company.

Although many functions and the number of demands for services are similar in the plants of very small and very large companies, managers in small companies will usually operate with little or no staff, whereas those in large companies will have available to them many staff specialists. Therefore, the primary difference between your responsibilities as a plant manager in a small or large company will be in the diversified knowledge that you must possess.

The sales volume or potential of your company will normally determine the character of your plant, and in turn your duties and responsibilities. Obviously, the smaller your company in terms of sales, the less will be your responsibilities in terms of control of capital. For the purpose of relating these responsibilities to company size, companies may be categorized according to sales in the following manner: (1) very large—sales of \$100 million and up; (2) large—\$51 to \$99 million; (3) medium—\$20 to \$50 million; (4) small—\$6 to \$19 million; and (5) very small—sales to \$5 million.

A second factor that will determine how you will function as a plant manager is that of type of plant, primarily whether your plant is continuous or batch. A continuous plant normally contains expensive, automated equipment, and the sales-dollars-per-dollar-invested in it is usually quite high. A batch plant, on the other hand, typically consists of simple equipment, normally with unsophisticated instrumentation, although there are many exceptions, such as one-product batch plants, in which expensive autoclaves are highly instrumented.

The scheduling of batch equipment that is used for the production of many different products can be quite complex. The problems of handling many different materials, with the consequent complications of raw-material and product storage and materials of construction, get involved. Also, a plant manager can become more quickly familiar with the products and processes of a continuous plant or of a batch plant producing one or few products than with a multiproduct batch plant.

The nature of the process can also complicate a plant manager's job. When a process is hazardous, the amount of time he must spend on safety will be considerable, because safety is essential to continued, profitable operation. Insurance and downtime costs and low morale can generate excessive expenses.

Managing in a Very Small Company

Without specialists to handle quality control, labor and community relations, air and water pollution problems, cost control, inventory, scheduling, distribution, maintenance, engineering, technical service, research and development, accounting, safety and security, wage and salary administration, packaging and traffic, a plant manager of a small plant in a very small com-

This article describes the job of the plant manager in the chemical process industry. The two following articles will discuss how he handles several of his more important functions: dealing with cost control, budgeting, conducting a successful value-engineering program, organizing a maintenance department and reviewing its performance, setting up quality controls and realistic product specifications, and keeping an older plant competitive.

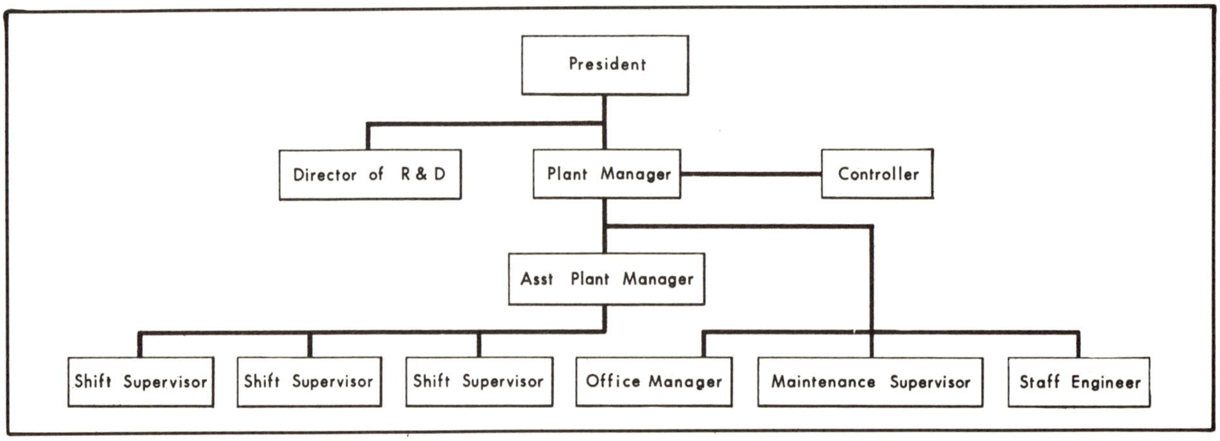

MANAGERS in very small companies, having little or no staff assistance, must own diverse skills—Fig. 1

pany must be a "shirt sleeve" executive who enjoys getting out into the plant to personally solve problems and give direction. He must, therefore, possess a high degree of capability in all of these activities. Much of his time will be spent in the plant solving problems.

Operations in such a plant are usually conducted on the basis of a profit center. The plant manager's ability to direct personnel–laborers, operators and mechanics–will directly reflect the profitability of the plant. Reports will be minimal, and sophisticated operating charts will mean very little. For him, the important thing will be to get jobs done as directly and quickly as possible.

The organization of a small plant typically includes a plant manager, an assistant plant manager, an engineer for special projects and troubleshooting, a maintenance supervisor, shift supervisors and an office manager (Fig. 1). The operating staff should be supported by a quality control group, which would carry out process development. It should include at least one or two Ph.D. holders and one or more chemists, who should in turn be supported by technicians, their number depending on the nature of the process.

Small-Company Operations

Operations in a plant of a small company (sales from $6 to $15 million) are conducted similarly to those in a plant of a very small company, but with greater sophistication and by at least a few staff specialists. Such a company may be listed on a stock exchange or be affiliated with one that is.

Whereas communications in the very small company are informal, being largely oral, reporting in the small company plant is likely to be fairly well developed. As plant manager, you will need reports about the day-to-day status of cost control, production schedules, new projects and special problems. Control by objective is the desired approach, but you personally will have to be aware of the details of plans and objectives to be achieved.

Try to Get a Piece of the Action

Unless you are an entrepreneur, as plant manager you will have to deal effectively with some level of management. If your company is very small, and you are not operating with your own capital, you will most likely report to a president, and your company will probably be closely controlled by its stockholders.

Most of the stock in such a company is frequently owned by the president, his family or close associates, on whom expenditures, profits and losses will have a direct and immediate personal financial impact.

As long as your plant is making money, the president will be easy to get along with. But for when things are not rosy, there is one thing you will need —but which will be hard to get—and that is a share of the business, i.e., some stock.

Unless your credentials are impressive enough so that you can get stock as part of compensation before you accept their job, getting your hands on even a few shares will be difficult. If you can manage it, however, being a part owner of the business will afford you considerable protection. With stock, you might eventually achieve a voice in the direction of the business. Also, stock ownership will offer some protection against the owner's selling out.

Of course, a small business can also disappear quickly as a result of labor troubles, loss of a key customer, or the demise of the owner or a key member of management. Stock would obviously be of no value if a disaster completely wiped out your company's single plant.

At the least, you will need the following key people for plant operations: an assistant plant manager, a chief process engineer, an industrial relations supervisor, maintenance supervisors and a quality-control supervisor (Fig. 2).

As a plant manager in a small company, you should have time for planning and directing an organization. You will probably have a private office without car-

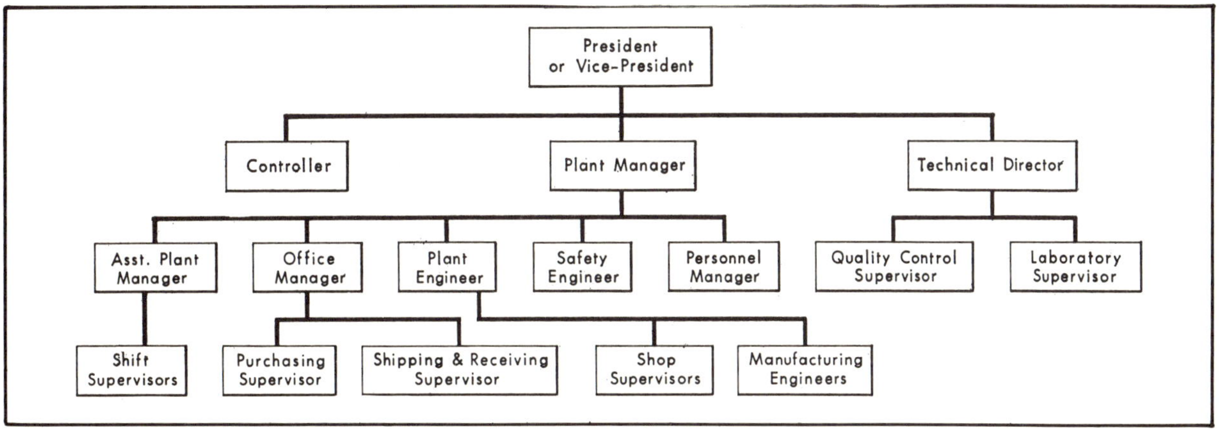

SMALL COMPANY plant managers will have a little more time for planning and guiding an organization—Fig. 2

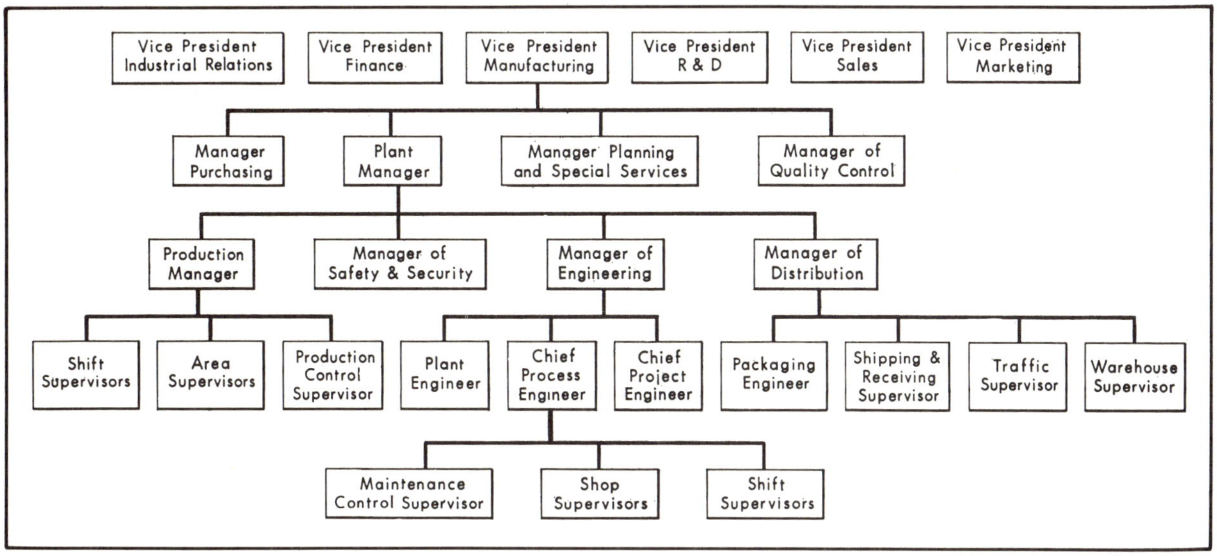

OPERATIONS in medium-sized companies begin to approach the level of complexity in large plants—Fig. 3

peting or other luxuries. If you have a product line that gives your company some basis on which to expand, your future prospects may be quite favorable.

Unlike the very small company, the small company will probably have other plants, in which case you may not report directly to the president, but most probably to a vice-president of manufacturing.

Managing in a Medium-Sized Company

Running a plant in a medium-sized company (sales from $20 to $50 million) involves managerial sophistication substantially different from that in small and very small companies. Operations in this sized company approach the level of complexity found in large and very large plants.

The plant manager will be more concerned with overall operations, organization, objectives and the development of personnel rather than with the actual mechanics of plant operation. He increasingly keeps abreast of plant operations by means of reports from his managers. Functions such as production scheduling and preventive maintenance are likely to be computerized.

Normally, this plant manager will report to the vice-president of production. To a large extent, however, his responsibilities will depend on whether he is the ranking executive at the plant site. (If the corporate offices are located at the plant site, his responsibilities will be substantially different.)

If he indeed is the ranking executive, he will be in charge of all the plant organizations, as well as community relations. His day-to-day responsibilities will include handling administration, production, engineering, maintenance, distribution, warehousing, inventory, labor relations, relations with state and federal governments, quality control, process development, and possibly research and development. More normally, however, research and development personnel report directly to the vice-president in charge

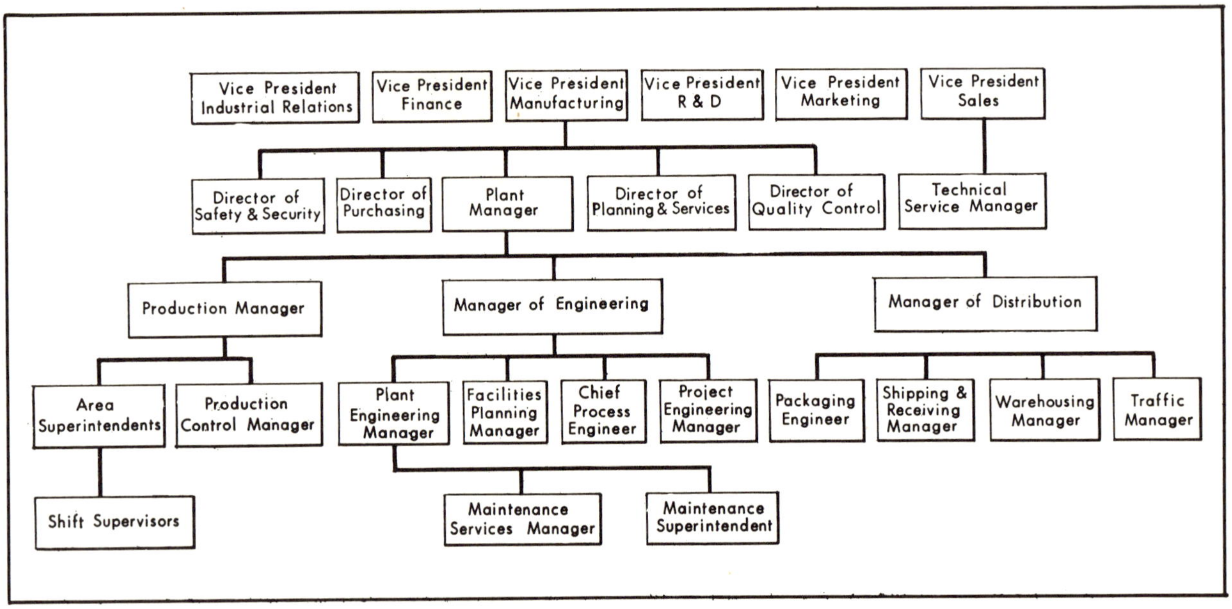

LARGE STAFFS in big companies will assist the plant manager with organization and planning—Fig. 4

of research, who may be either at the plant site or at a distant corporate headquarters.

If, on the other hand, the plant manager is not the highest officer at the plant site, his responsibilities will usually be limited to the actual operation of the plant, and to a specific range of other functions (Fig. 3). Sometimes, the extent of his responsibilities in such a situation may depend on his abilities and background.

Large and Very Large Companies

The responsibilities of a plant manager in a large company (sales from $51 to $99 million) and a very large company (sales over $100 million) will be similar in many ways. Both are likely to function in accordance with well-defined policies and be guided by clearly established precedent. Both will be assisted by large staffs that will concentrate on organization, cost control, and planning for the future. The information that they will rely on for directing plant operations will come to them in the form of comprehensive written reports, supplemented periodically by oral communication, from their various managers. Both will depend on their experience and training to evaluate these reports and make their decisions.

Community relations will occupy a substantial part of their time. They will be called on to address local groups, and act as representatives of their company in handling requests for and granting donations in the name of the company to community organizations.

They will represent their companies in business relations with other companies and with engineering and construction firms. The continued success of their companies will depend on their ability to devise innovations that will boost the output of their plants and reduce operating costs. A substantial amount of their time will also be taken up with planning new facilities, personnel development, organization, and long-range corporate objectives. The plant manager will usually be given the assignment of executing the objectives established by the corporate engineering group.

The chief difference in the job of the plant manager in the large company from that in the very large company may be in the slightly narrower scope of the latter's organizational responsibility. In the large company, managers of manufacturing, distribution and engineering are likely to report to him (Fig. 4). In the very large company, he will probably not be responsible for engineering (Fig. 5). It is also more likely that labor relations will be handled by a corporate group in the very large company.

You and Your Organization

Regardless of his capabilities, a plant manager cannot do everything himself, nor should he want to try. His staff should be an extension of his philosophies, interests and ideas. He should give his subordinates all the freedom they will accept, but he should also set up controls to ensure that they understand and are executing the policies that he has established. He must establish checks and balances to ensure that projects will be completed on time and within estimated costs.

A company that is not expanding will tend to lose its employees of high potential and attract and retain those more security conscious. Therefore, a major problem of a plant manager will be how to attract and retain those of high potential.

A plant manager must impose his control on his organization to ensure compliance. Yet, he should not be overly concerned about subordinates overstepping

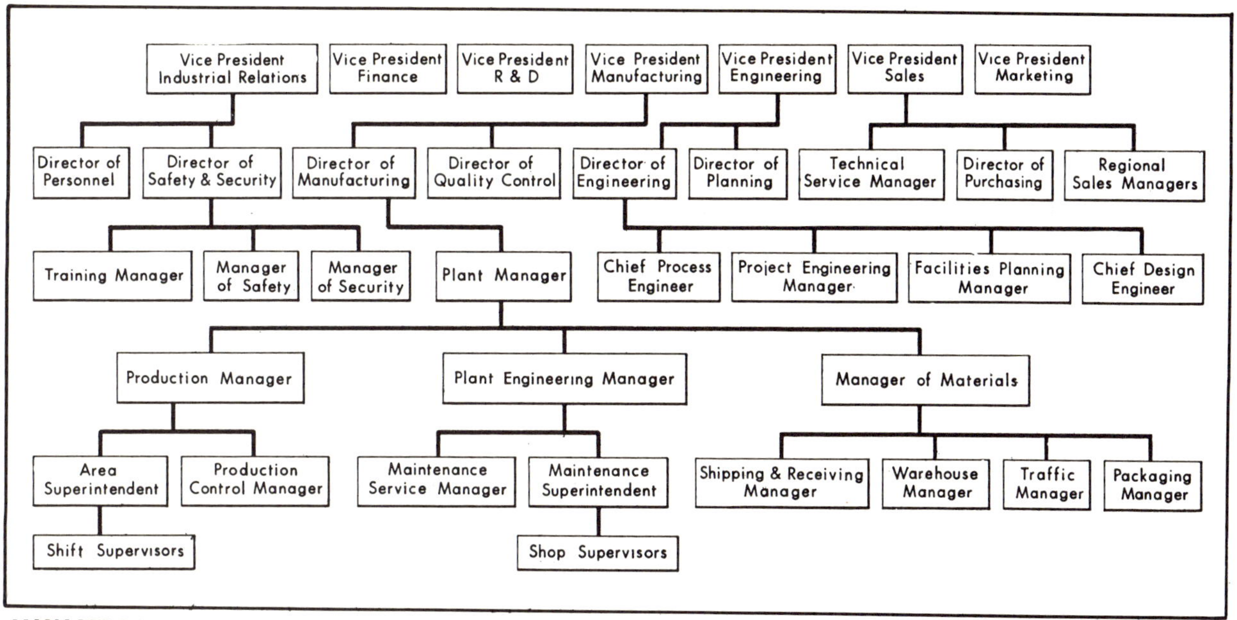

MANAGERS in very large companies will be guided by well-defined policies and established precedent—Fig. 5

Financial Stability: A Prerequisite

Capital is essential for successful operation, and must be assumed to be available. Before you accept a position as a plant manager, you should carefully check the company's financial resources and credit standing (especially if it is a small company).

Money will always be available for a company from some source if it presents the potential for making a profit. You should be sure that your company will have reasonable working capital on hand for its operations, and that it can reasonably be sure of drawing on additional resources according to its requirements and its ability to use the funds profitably.

Returns on investment of projects will usually range from 10% to 20%, with some considerably above 20%. Projects should be grouped in order of importance to the company, and resources then directed toward those that will achieve the greatest return. It is often the plant manager's responsibility to translate projects into profitable installations.

their authority. Normally, his problem will be to motivate and develop more-aggressive performance. When he finds a person who is constantly pressuring for additional responsibilities and new problems to attack, he should give these to him, and then watch other key employees begin to contribute more effectively. Those who do not are either going to drop out or fall into line at some functional level. Not everyone can be a high performer, and the level of plant activity and growth will determine how many of them the company can afford to keep.

Without competent people to extend his capabilities, a plant manager will soon become bogged down in a morass of details that will distract him from planning and anticipating problems—in which case, he will be constantly floundering between one crisis and another.

In a very small company, a plant manager will necessarily operate with a minimal staff. He will be on call 24 hours a day, weekends and holidays. A telephone must always be close at hand, because he will receive middle-of-the-night calls involving equipment failures, process problems, fires, accidents and other disasters.

A plant manager of a very small plant will need an assistant who can fill in for him when he is away on business. But even with an assistant, he will not have much time for vacations. Most of the time, he will also be the one who will check the plant on holidays and weekends. In addition, he will spend many nights in the plant coping with emergencies and solving technical problems that are handled by staff in larger plants.

He will not receive much sympathy, for his president will point out that the plant managers of *his* executive friends in larger plants are not having such problems. While the plant manager will know why, he would have trouble explaining. However, for his troubles, he will be well paid, and there may be a substantial bonus for him at the end of the year if the company makes money.

One thing he must not do is add staff to help with the problems. Staff is overhead, which eats into profits. His boss will be sure to remind him about Parkinson's Law (which states that work expands in direct proportion to the amount of people available to do it).

In a very small company, a plant manager's well-being will depend considerably on the day-to-day fluctuations in the personal fortunes of one or two key company officers. In which case, it will take every ounce of diplomacy and will-to-survive if he is unfortunate enough to have a boss with idiosyncrasies. To prevail in such circumstances, he must be his own man and face his problems squarely. If, in addition, his boss is a dynamic executive who excels in business but is somewhat less expert in plant management, the plant manager had better develop a thick skin. One plant manager needed a double shot with lunch, an early lunch at that, to fortify himself for his daily telephone report to his president.

A plant manager's best strategy will be to take the offensive when he is right, and to stand up and admit it when he is wrong and then offer a positive program for remedying the problem. He will not get sympathy, but he will usually get respect, which may save his head in the future.

Although the plant manager of a larger plant will spend much more of his time away from the operating areas of his plant, he must still remain fully aware of its products, personalities and problems. Although budgeting will occupy much of his time and he will be constantly reviewing new ideas, equipment and techniques, he cannot neglect such things as personnel relations, safety, quality control, scheduling and engineering.

The larger the company, the greater will be the need for controls and procedures. The easy way of meeting these requirements will be to add staff, but this route must be rigorously resisted unless it offers clear-cut benefits. Of course, although staff adds to overhead, it also means support in planning and developing the organization to its maximum potential. A wise manager carefully controls this balance.

It is in the large and very large companies that proclivities toward excessive personnel must be especially controlled. One sign of bad organization is when new groups are set up to take over the functions of older ones. The appearance of the coordinator is usually one sure sign of a deteriorating organization.

In any company, it is a manager's responsibility to turn out more than the sum of the resources put at his disposal.

New Manager in an Established Organization

When a plant manager is brought in from another company or plant, it naturally means that in the opinion of the company's executives there is no one within the organization capable of assuming the management of the plant. Often, of course, there will be personnel within the organization who will not have agreed with this decision.

Although there may not actually be anyone within the organization capable of handling the job, the new plant manager will have to prepare himself for the fact that not everyone will be willing to accept him with open arms. Generally, one or more individuals will be surreptitiously, if not outwardly, working for his demise. Because companies frown on resistance to organizational changes, this challenge will more likely be hidden rather than overt.

This compounds the problems of the new manager. It would be better for him if those who oppose the change would leave and provide him with the opportunity to replace them with people who would be loyal and who would conscientiously work to further his programs.

The ability to hire people offers a substantial advantage to a new manager. It is through the development and placement of personnel for maximum efficiency that he obtains best results. To achieve this, however, he must have a thorough understanding of the company, the organization, and its people.

Some plant managers make no changes for at least one year after assuming their job, believing that at least this much time must pass before they will have gained sufficient insight into the organization to modify it properly. Others, pressured by the urgency that forced the management to hire them in the first place, feel that they must produce tangible evidence of their activity sooner.

In general, the specific situation should dictate a new manager's action. A point in favor of an early shakeup is that management will generally be willing to go along with changes during the first phase of a new administration. This is logical, being based on the need to support a new manager it considers its great hope for salvation. Nevertheless, any manager should be sure he thoroughly understands the problems and personalities involved before he makes changes.

Whether sooner or later, he will have to alter the organizational structure. Good organization is essentially having the best man in the right spot, which depends on the manager's understanding of the job requirements and his insight into the capabilities of the available personnel.

Once he has moved the right people into the key job, he will be ready to move forward. However, this will not mean that the people selected will remain in the same jobs forever, or that even the jobs will be permanent. A manager must constantly revise his organization to cope with changes in products, sales and expansions. His own superior will furnish him with a certain amount of direction, which will depend on his interests and the directives that he receives. Other than that, however, a manager will have wide latitude within the framework of company objectives in which to operate. ■

Meet the Author

M. K. Goss is Manager of Facilities Planning at J. T. Baker Chemical Co. (Phillipsburg, N.J. 08865) responsible for planning, engineering, cost reduction, air and water resources, and construction. Previously, he was plant manager with Gamma Chemical Co. (now part of Ashland Oil Co.) and head of the Engine Manufacturing Dept. of Thiokol Chemical Corp.

He is a graduate of Purdue University with a B.S. in chemical engineering, and a registered engineer in the State of New Jersey.

What a Plant Manager Does And How He Does It Part 2

How the plant manager handles some critical responsibilities—planning, budgeting, controlling costs, and setting product quality— is the subject of this, the second of three articles on the job of the plant manager.

M. K. GOSS, J. T. Baker Chemical Co.

A significant amount of a plant manager's time will be devoted to making presentations to management outlining his plans and, more importantly, his requirements in terms of men and facilities for carrying out his programs. Planning, therefore, will be an important element in his success (Fig. 1).

It is essential that all company projects be reduced to a plan that includes a timetable and that lists the costs and facilities necessary to complete them.

There will be many influences working at cross-purposes to hinder the completion of his plans, and justifiable pressure to finish projects quickly and at the lowest cost. The further a manager or executive is away from a task, the easier it will appear and, conversely, the closer, the more difficult. Somewhere in between these two viewpoints is the middle ground that usually reflects the true status of the situation.

Management will expect a plant manager to run his plant in accordance with company policies and with a minimum of direction. If he is fortunate, he will have short- and long-range plans to guide him. If there are none, he must prepare them himself. In any case, plans should be flexible, subject to changes demanded by exigencies.

Budget and Cost Control

Budgeting occupies a considerable part of a plant manager's time. In addition to annual budgets, he must prepare status reports, as well as special reports on proposed projects, all of which require the collecting of backup data that must be reviewed and edited so as to be transmitted to management as concisely as possible.

He will have to prepare budgets for all aspects of plant operation. Most budgets include the following items: raw materials, direct and indirect labor, salaries, maintenance materials, equipment replacement, distribution, travel and entertainment, education, taxes, depreciation, insurance, maintenance and repairs, supplies, heat, light and power.

Whether or not budgets are highly developed, expenditures will generally be based on expenses of previous years, adjusted for sales forecasts and projected labor costs. Although, in most cases, sales and marketing are not among the responsibilities of a plant manager, he should nevertheless be familiar with the details of the programs of the sales and marketing departments, and work closely with their personnel.

To prepare budgets, a plant mnager must have detailed knowledge of past performance and projections for the future. Yearly improvements in performance will always be demanded of him, and he must factor these into his budgets. Regressions in performance must be offset by cost improvements if his company is to remain competitive. If he does not

PLANNING is critical to managerial success—Fig. 1

The first article of this series (p. 309 of this volume) provided a general insight into the plant manager's responsibilities in all sizes of companies and plants, describing the different types of organizations he might have at his disposal and how he could use them most effectively. The third, last, article will discuss how to set up production controls, organize a maintenance department and review its performance, and keep older plants competitive.

hold costs in line, he will be inviting an aggressive competitor to drive his company out of the market.

Costs of labor and materials (other than raw materials) can be expected to rise at least 5%/yr. A good cost-improvement program will more than offset this increase. And a really effective program will also provide capital for new facilities, as well as support new-product development and increased activity in sales and marketing.

Although cost-reduction programs are generally directed at those operations that are not contributing to profits, cost reduction is all-encompassing in a progressive company. No operation should be immune from scrutiny.

Major possibilities for cost reduction include: raising process yields, switching to cheaper raw materials, trying alternate packaging materials, purchasing in bulk quantities, employing personnel more efficiently, boosting process throughout, lowering utilities consumption, improving materials handling, increasing maintenance effectiveness, discovering uses for by-products, and reclaiming waste materials. Although a plant manager should direct his program primarily at areas offering possibilities for large reductions, he should not overlook those that might result in even the smallest saving, such as expenditures for telephones and periodicals.

Vital to the operation of a successful company is product replacement. In the field of organic chemicals, for example, at least three-fourths of existing products must be replaced every 10 years to assure continued growth. At some point, the failure to introduce new products will seriously hamper the upward progression of a well-planned cost-reduction program because, after a while, those projects most amenable to cost reduction will have become exhausted. A continuing flow of new products, in addition to boosting profits, is necessary for a cost-reduction program to generate—as it should—savings equal to from 2½ to 3% of total annual sales.

A plant manager must take personal interest in cost improvement, and continually search out high-cost area to attack. The fallout from such a program, which must be supported by all levels of supervision, will be a greater awareness and concern on the part of all employees for controlling costs.

To generate and screen cost-reduction projects, as well as to maintain a high level of interest in cost cutting, a plant manager should create a cost-reduction team consisting of key representatives from engineering, operations, maintenance, accounting, sales, research, and purchasing.

Monthly reports on cost-reduction programs are essential. Each departmental manager must report critical cost areas so that the plant manager can maintain a running check on performance. This will eliminate unexpected siphoning off of funds, and time wasted in rectifying mistakes that might have been more easily handled earlier.

Manufacturing and maintenance costs and production scheduling are vital to keeping a competitive stance. Reports from these departments must spotlight existing and potentially costly problems so that they can be rapidly resolved. In this respect, lack of communications is often an obstacle. People who can resolve such problems may not be aware of their existence unless the plant manager's communications network brings them to their attention.

Normally, it is not desirable to set up a formal employee suggestion program until after most of the ideas obtained through normal channels (many of which will have been at hand for several years) have been cleared away. Only after a cost-reduction program reaches the stage at which most of the available suggestions have been acted upon will it be time to set up a system to obtain broad employee participation. Even after this point, a cost-reduction program should be able to net relatively easily up to 3% of sales for several years.

Cost Reduction by Value Engineering

Basically, value engineering, or value analysis (the terms are synonymous), was designed to provide value for customers' dollars at a fair profit to manufacturers by investigating what a product does in relation to the amount of money spent on it.

Interpreted broadly, value engineering involves the critical analysis of each facet of an operation, product or problem. It is concerned with separating every phase of an operation into increments that can be easily analyzed. Nothing is excluded from meticulous examination. The increments are minutely examined either to find justification for not making changes or to come up with alternatives. Those segments or steps of an operation that remain must be shown to be economically necessary.

Value engineering can result in already profitable products being made even more profitable, unprofitable ones being made profitable, and unprofitable ones being eliminated. A product's quality may be reduced so it can efficiently serve a lower-class market, or a superior product can be manufactured at additional cost so it can command a still higher price and be more readily marketable. A product must be geared to the customer. If he wants quality, he must be given it—at a price, of course. If, on the other hand, he is primarily concerned with price, a company cannot hope to compete by having a high-quality product.

The best place to control product price is in the initial layout of a process, in the design and engineering steps leading up to the manufacture of the products. The size of the equipment chosen, the raw materials selected, and the process developed—all will directly affect the largest components of operating cost.

The analysts should examine preliminary drawings and specifications even before they have been formally approved. This, along with an analysis of project objectives, is certain to uncover unnecessary costs. When this is done, the final installation frequently results in substantial savings. A tradeoff must be made between the cost of an installation and that of the product. The initial cost, although important, is not as critical as the cost of manufacturing the product throughout the life of the installation. A 1,000-gal. reactor system may cost only 40% more than a 500-gal. one; however, product from the larger vessel could cost 10% to 30% less, so that, depending on throughput, the larger investment could be recovered in only a few months.

During the information-gathering phase of a project, it is essential to determine the adequacy of the data collected. The preliminary data must be appraised, screened for personal bias, constraints examined—and all this must be done with complete objectivity.

In the creative phase of a project, the first requirement is to define function: What is the product or process supposed to do? Answering involves breaking down the project into basic elements, such as: movement of materials, processing, packaging and shipping. It must also be determined if the product is already being provided by someone else, at what prices and quality levels, at what price competitors might ultimately sell the product, and what are the advantages in the positions of competitors (such as proximity to markets, rapid distribution, or cheap raw materials).

The purpose of this phase is to find less-costly ways of producing an existing or new product, such as by lower-cost raw materials, greater throughput, less material handling, and exploitation of waste products. High-priced, high-volume products whose specifications have been set up by the seller usually offer rich potential for cost cutting.

An examination of production records and sales forecasts for products in the development stage often will reveal opportunities for applying value engineering.

Basis of a Successful Program

Without a favorable working climate, it is impossible to conduct an effective value-engineering program. It is essential that the following types of attitudes first be changed: "They will never buy it;" "It isn't the way things are done around here;" "I'm not authorized to make changes;" "You're ahead of your time;" or "It is not in the budget." Care must also be taken to avoid the implication that because something is being changed, harsh criticism is being directed at those who designed or created the original product or process.

Value analysis contributes immeasurably to a successful cost-reduction program because it is precisely the approach that leads to the critical analysis necessary to generate viable cost-saving projects.

The first step in a value-engineered cost-reduction program is asking such a question as: Why not switch to less-expensive raw materials? Other questions follow logically from such a start: What would be the effect on product quality? Why buy in 100-lb. containers instead of 500-lb., 1,000-lb. or carload quantities (Fig. 2)? Is the quality of the product right for the market it now serves? Should it be of higher quality to command a premium price, or lower to reach a broader market? Is the container too expensive, and would a cheaper one serve as well, assuming ICC regulations would allow a cheaper one?

BUYING in bulk quantities can reduce costs—Fig. 2

Forming a Value-Engineering Team

A value-engineering team must include representatives from engineering, purchasing, research and development, manufacturing, marketing, technical service, and sales. All of the representatives should have had training in value engineering. Programs fail most often because managers do not have personnel trained by professionals.

A program should be started by someone who has received formal training in value engineering and is keyed to cost awareness. He should also have a broad engineering background with a depth of experience related to product lines. He should possess initiative, being a person who seeks out ideas, develops programs and follows projects to completion. His job crosses many functional lines. He will constantly find himself seeking information and cooperation from all the departments in the plant.

Team members must also be widely experienced. They must know products, materials, processes, costs, and most importantly the organization, how it functions and how to get things done.

Important to success is to assign specific objectives to managers, supervisors and foremen. Value analysis must be made a part of the daily function of operating management. At an early stage, supervisory personnel must be told that the program is not being implemented because of shortcomings on their part. They should understand that value engineering involves the bringing to bear of collective knowledge to generate ideas and solve problems under conditions somewhat removed from the pressures of daily operation. To be successful, the program must have the unqualified support of every member of management.

In the selection phase of the program, the team must develop a substantial number of cost-cutting ideas and appraise the possible approaches for implementing them. The economic values of the various ideas should be worked out in terms of positive numbers, and the approaches developed into a matrix. The matrix will reveal previously unperceived economic variables. This will make possible judgments based on facts.

Implementation, the last phase, will follow automatically if the preceding work has been done properly. Innovations requiring changes in procedures, techniques and materials can usually be accomplished quickly. Those that call for major expenditures, such as changing equipment layout and process variables, will take longer.

Typical Plant Projects

Using standard reactors, pumps and centrifuges instead of custom equipment will not only save a lot of money but also cut down installation time. Substitutions of materials of construction can reduce maintenance costs. Often, the savings related to operating a particular piece of equipment over its life span will be much larger than its initial cost.

In the case of a new product, quality level must be considered, because it will determine to a great extent what process will be used, as well as manufacturing and selling costs. Lower manufacturing costs that can be obtained through superior plant layout offer an important competitive edge.

SHIPPING offers cost-cutting possibilities—Fig. 3

Operating and maintenance costs cannot be overlooked during design. The initial expense of an installation may, over its life, be insignificant, if initial price considerations are allowed to adversely affect the final cost of the product. Higher production rates boost efficiency. When throughput is doubled, the cost of manufacturing a product can be reduced by as much as 20%.

New products and processes in either development or design stage are prime targets for value engineering. Still other projects will come to light as a result of analyses of high-cost operations, technical service experiences and sales problems. Studying competing products and the actions of competitors may also offer new approaches to investigate.

How to Implement Value Engineering

The following application of value engineering to a pelletizing process illustrates the approach to reducing costs and improving quality. Involved was an investigation of raw materials, equipment, process yields, maintenance, packaging, rejections, labor, product modifications, and layout.

Raw Materials—The possibility of buying liquid caustic soda at a concentration higher than the 50% solution now being used was investigated. One supplier offered it at 73%, at a premium price.

Freight savings would amount to $834/yr., unloading savings to $350/yr., fuel savings (eliminating concentrating the caustic) to $2,400/yr., for a total of $3,584/yr. However, this reduction would be offset by the investment for heating and insulating the tank, because a 73% solution solidifies below 155 F. Also, the existing tank would be unsatisfactory for maintaining purity at the elevated temperature, and a new alloy tank would have to be installed. Feed lines would also have to be insulated and traced.

Because the 73% caustic would have to be diluted to 50% for other plant applications, there would be an additional cost of $535. A cooler to remove the heat of dilution would cost $5,000.

Switching to a 73% solution was rejected.

Tank-Car Size—Larger tank cars were found to offer a 15% freight advantage but to present problems because of the limited capability of the existing siding. However, modifications to the siding for another project would make possible the use of jumbo cars at a savings of $1,500 (Fig. 3).

Competitive Products—An investigation of competitive products led to the consideration of marketing a flaked rather than a pellet product. This was rejected because the pellet could not be replaced in many applications. Also, marketing both a pellet and flaked product, which would result in additional handling, processing and inventory costs, was considered uneconomical.

Product Composition—An ingredient that prevented the formation of undesirable impurities and added whiteness to the product was found necessary in the quantity now specified.

Reduction of Wastage—Product was being scrapped for poor composition, shape and color, and was being lost through spillage. About 46% of this loss was at the screening operation, 50% of which could, however, be reclaimed by a second screening or by screening more slowly at the first pass, when the pellets were hot. Reducing scrap losses 50% would save $4,500/yr.

Drum Size—Packaging 350 lb. in a 35-gal. drum, instead of 100 lb. in an 11-gal. container would save about $6,000/yr., if customers would accept shipment in larger drums. Shipping in bags rather than in fiber drums was also examined but was rejected as being more expensive.

Liner Modification—Substituting a 2½-mil polyethylene bag costing 6.7¢, for a 4-mil one costing 9.8¢ would not be possible—because the pellets are still hot when packaged—until new equipment could be installed that would increase cooling capacity. The potential savings amounted to $600/yr.

Container Sealing—Sealing drums with a 1½-in., rather than a 2-in., wide tape, and installing a machine to control tape length could save $1.000/yr.

Labels — Because pressure-sensitive labels were needed by customers who reprocessed the material, this type of label had to be retained. However, labels costing 0.03¢ could be replaced with smaller ones costing 0.005¢, at a potential savings of $500/yr.

Pallets—A sales policy had not been developed that included the cost of the pallet in the cost-of-sales price. The Sales Dept. was requested to develop such a policy. Savings could amount to $800/yr.

Maintenance—A review revealed that Dowtherm usage, which was related to frequency of pump maintenance, was excessive. Corrective action saved $3,000/yr.

Gross savings totaling $17,900/yr. were thus identified and programmed for implementation.

Decisions regarding what actions should be taken to realize the potential savings identified in value-engineering programs should be made by those members most familiar with the specialties involved: packaging, laboratory, transportation, engineering, research, purchasing, sales, etc.

Purchasing Techniques

In value engineering, it is not enough to know that a particular product is currently listed by a supplier at a certain price. Far too often, price information is accepted at face value—including by value-engineering analysts.

Value represents a combination of factors that yield the greatest economy to a purchaser, who should be aware of the fact that initial cost is only one of many factors that he must consider. For example, a purchasing agent may establish certain combinations or requirements to take advantage of quantity pricing, or set up relations with a particular supplier who provides services, such as special packaging, transportation arrangements that reduce costs, or a system that reduces inventory requirements. All such factors must be taken into account when materials are purchased.

The field of materials purchasing is a particularly inviting target for value engineering. A typical company spends from 30 to 50% of its sales dollars on materials.

A plant manager must recognize that personnel responsible for selecting materials, evaluating suppliers and other functions related to materials purchasing will not be eager to double or triple their workload by taking on value-engineering projects. Yet, roadblocks will rise up if a plant manager tries to establish a plantwide value-engineering program without involving them.

If it becomes desirable for a value analyst to contact a vendor directly, this must be done with the knowledge of the purchasing agent, who must understand why it is necessary.

Measuring Effectiveness

A measurement system must be established to determine the effectiveness of value-engineering programs. The system should ensure that savings targets are established and that the budget reflects the savings realized under the program.

Measuring the effectiveness of a value-engineering program by conventional cost-accounting procedures may be difficult, and at times even impossible. For example, a mechanical part not ordered because of improvements will lower transportation charges; one part not ordered could also signifiy greater process reliability and, hence, lower maintenance costs. The results in total, however, will certainly tend to improve customer loyalty and good will, which is very difficult to measure in terms of dollars and cents.

Production Specifications and Quality Control

Product specifications determine costs. A product's quality level must be defined by supplier and customer. Once established, specifications become dominant in setting the quality of raw materials, nature of the process, and quality-control techniques.

Specifications must reflect real needs if a product is to compete effectively. Products range from ultrahigh purity through reagent, purified, commercial and technical grades. Within the grades, there are also appreciable differences.

The process and raw materials used will establish what impurities will be present in the product. For some products, impurities may not be important; for others, they may be critical.

Fixing Specifications

Product specifications are designed to serve either the general market or a single, key customer. Naturally, broad specifications favor the manufacturer. Some customers at first may not request specifications, but they regularly become more demanding as they become more familiar with the vagaries of their own processes and with those properties of a product that are more critical.

Eventually, a standoff develops between the manufacturer, who desires broad specifications, and the customer, who wants closely set specifications and little or no fluctuation in properties from lot to lot. Of course, the customer knows that here at some point he will have to pay more for tighter specifications, and this will tend to limit his demands.

Whether a product is new or already available, its specifications will eventually migrate to a range similar to that of competing materials. A manufacturer will not object to turning out a better product if it will not raise his production costs. If, however, the product's impurities impair a customer's catalytic operations, for example, it may become necessary that the process be tailored so as to turn out a product having fewer impurities. If, as well, a competitor is producing a better product—by a different process or with other raw materials—his choice will narrow to either duplicating the competitor's quality or dropping out of the market. A plant manager must always be striving to produce a better product than his competitors at the same or lower cost. There will be no respite for him from this demand.

Keep Specifications Broad

Narrowly drawn specifications will often mean that a manufacturer will incur excessive costs for reworking product, or suffer substantial product losses, or both. He also runs the risk of having his product rejected when customers rely on different analytical methods and when the product changes between the time it is manufactured and used.

A continuous processing plant will normally turn out products within a much narrower range of specification than a batch plant. In a batch plant, regardless of how well controlled, product quality will vary from batch to batch, and blending will frequently be necessary. Color, particle size and other properties will fluctuate critically with slight changes in operating cycles, temperatures and pressures.

Although operations has the responsibility for running the process in accordance with operating procedures, the procedures must accurately reflect the process. In actual practice, this is seldom the case, regardless of how detailed and explicit the procedures. The problem is further complicated by the fact that the procedures must also be updated to take into account new operating techniques.

A periodic review of product specifications by both the manufacturer and customer is a good practice. Most companies have specification teams (or something similar) that review rejected materials and make decisions regarding disposal. Frequently, teams also recommend changes in specifications.

Specification reviews are especially desirable when the rejection rate for a product is excessive. If the process cannot be made to produce material within specifications, a plant manager may have no choice but to relax them. Frequently, specifications are established because of a particular customer's requirements, and these tend to become the model for other customers. Often, a customer is satisfied with a particular product and assumes that its specifications are all vital to his operations, when this is actually not the case.

Some customers may accept a product not within all the specifications. Often, however, they will do so for a price reduction.

Watch for Variant Analytical Methods

Analytical methods for specifications should be detailed. Different techniques can result in variant results, which can be critical. Occasionally, a situation arises in which the laboratory results of the manufacturer differ importantly from those of a customer. If the disagreements cannot be resolved, it may be necessary to resort to a commercial laboratory for results that will be acceptable to both parties. One way of avoiding such a confrontation is to have key laboratory personnel visit customer plants and review their techniques and procedures.

Any analysis of a product is only as good as the sampling technique. Samples taken with inadequate care, contaminated sample bottles, improper handing and unrepresentative sampling can result in the rejection of first-class material and unnecessary, costly reworking.

When analytical results are unfavorable, a second sampling and analysis should be standard procedure. Frequently, a second, or even a third, analysis will contradict the first analysis.

Control Vs. Production

There is usually a continuing conflict between quality control and operations departments in plants. Each side tends to be overly jealous of its prerogatives. Operational personnel often take the stance, "We made this batch like all the others; why should it be rejected?" To which the quality-control personnel reply, "We analyzed this batch like all the others, and we're rejecting it."

It is often up to the plant manager to get both sides to work together, to show that both serve an equally important function, and that one cannot exist without the other. Both must recognize dual responsibilities: to the customer, on the one hand, by providing products that meet necessary specifications; and to the company, on the other, by seeing that material is processed as quickly as possible, without unnecessary costs for reworking the product.

Operating costs must be held in line, but it is also essential that the plant manager back up quality-control standards and provide the necessary enforcement to ensure the reputation of his company's products. ■

What a Plant Manager Does And How He Does It Part 3

To a large extent, a plant manager's success will depend on how well he organizes maintenance and controls its costs. This article, the last of three, also discusses how he can keep an older plant competitive.

M.K. GOSS, J. T. Baker Chemical Co.

A production control group should be a part of any plant's organization. In a small plant, it is likely to consist solely of the plant manager. In larger plants, scheduling of production will depend on sophisticated techniques. In a medium-sized plant, the group may contain from three to six employees.

Production scheduling is intimately tied in with sales and inventory control. On it depend meeting customer requirements, lengths of production runs, and size of inventories. The last is a major cost factor that will directly affect a plant manager's performance.

Too little inventory will penalize operations in the form of unscheduled runs, short runs, interrupted schedules, overtime premium, and excessive transportation costs. It could even mean the loss of an important customer. Too much inventory, on the other hand, represents lost capital. Based on product value, inventory costs should run about 10% annually.

Spoilage and damage add to warehousing costs. Direct costs for warehousing can be expected to run \$2-5/sq.ft./yr. Of course, full warehouses will necessitate capital expenditures for additional storage space. If a plant manager is to hold the reins on costs, he must set up effective controls on inventory and constantly be improving turnover.

Maintenance for Cost Control

Maintenance is costly, as any plant manager knows who must try to explain to higher management why he is spending so much on it. It is also overhead, and as such does not contribute to profit, at least not directly. The lack of it, however, is readily discernible in higher production costs because of equipment downtime, in increased rates of product rejection, and in reduced throughput.

Although maintenance bears directly on operating efficiency and costs, this does not mean that a plant need be maintained in perfect condition. Operations superintendents have a tendency to believe that all of their problems can be solved with more operators, supervisors and maintenance. An economic balance must be struck between no maintenance at all and a plant in perfect working order.

Expenditures for maintenance must be related to production, specifically to production downtime. The probability and the cost of having a particular piece of equipment not function at some time in the future must be weighed against the cost of various levels of maintenance. In other words, how much preventive maintenance is justified, and where should it be directed, to prevent an unscheduled shutdown?

Maintenance must be highly organized in large plants. Its cost should be directly reflected in manufacturing costs and not be allowed to become lost in overhead. Work must be scheduled by a job-order control section that is responsible for categorizing jobs, estimating job time and cost, and planning.

Reviewing Goals and Performance

Depending on the nature of the plant, the process and the corrosiveness of the materials handled, maintenance costs will fluctuate with plant throughput. Many plant managers budget maintenance costs in relation to sales. Ideally, the ratio of maintenance costs to sales should drop from year to year because of improvements to equipment, experience in maintaining and repairing it, replacement of old with new and better equipment, and standardization.

Because the age of the plant and many other factors affect the cost of maintenance, an accurate projection of the ratio of material and labor costs to sales is difficult. A maintenance cost percentage frequently used for new plants is about 2% of the value of the installed equipment. In older plants,

---

The first article of this series (p. 309 of this volume) provided a general insight into the plant manager's responsibilities in all sizes of companies and plants, describing different types of organizations and how he might use them most effectively. The second article (p. 315 of this volume) discussed how he handles some critical responsibilities: planning, budgeting, controlling costs, and setting product quality.

maintenance costs vary from 2.5-5% of sales, depending on the products.

One of the first objectives of a new plant manager should be to determine how maintenance costs have been running in relation to sales. To do this, he will have to first ascertain what is being included in the category of maintenance. Sometimes, costs for plant improvements are lumped together with those strictly for maintenance.

Backlog Is Key to Efficiency

An important indication of the effectiveness of maintenance in a plant—in addition to visual inspection and incidence of downtime—is the size of its backlog of work orders. A backlog makes planning possible. Generally, a 30-day backlog is considered realistic. Of course, it should not represent urgent jobs, but those that must be completed to prevent a shutdown in the not too distant future.

When a backlog is skillfully planned, its total number of manhours is always known. This permits it to be manipulated. It can be kept from rising beyond its upper limit by resorting to overtime or hiring additional manpower. Another means of handling a temporary rise in the workload is with contract maintenance. If, on the other hand, it is shrinking, plant improvement projects can be added to it. If it begins to expand again, these can be temporarily dropped.

The number of maintenance personnel required to support operations will vary directly with the type of process. A ratio of about 15 operators for each maintenance mechanic is not uncommon in many plants having widely varied equipment. This ratio usually drops rapidly as instrumentation and automation are added.

Establishing a Control System

The first step in establishing control of maintenance is to determine how much time is actually spent at work and how much in supportive activities. Once this information has been obtained, it is then possible to set up work-order planning and scheduling.

The percentage of actual work performed by maintenance organization without formal planning normally varies between 25% and 35%. With effective planning and scheduling, this can be brought up to about 55%. Work is often conducted ineffectually because of poor supervision, lack of materials and tools, and excessive traveling to and from the jobsite. More often than not, a low rate of work is not the fault of those who do it.

After the reasons why the work rate is not as high as it should be have been identified, the next step is to establish an organization that consists of at least the following three members: a clerk to process and file work orders; one or more planners to plan the jobs and schedule materials; and an estimator to estimate job costs and arrange for delivery of materials.

The estimator should maintain files on completed jobs. Although the manhours assigned to a particular job may differ from plant to plant, the average time to accomplish it will remain essentially the same within a particular plant. Although circumstances will vary slightly for the same job done at different places and times, these will balance out; or allowances can be made by an experienced estimator.

Blanket orders should be established for all routine jobs, such as safety checks, repacking of pumps and reactors, and lubrication. The maintenance force required to accomplish these jobs can be easily projected, and costs charged to the appropriate departments. Also, major equipment should be suitably identified for cost-charging purposes.

If records are kept on the failures of specific pieces of equipment, a preventive maintenance program can be based on a statistical failure rate. With such information, an economic balance can be established between the cost and frequency of maintenance and the possibility of a failure that could result in a shutdown.

How preventive maintenance can cut costs is illustrated by the problem of replacing lighting elements. Labor constitutes the major part of the cost of replacing burned-out bulbs. Based on statistics, it is uneconomical to replace bulbs a few at a time as they burn out. It makes economic sense, therefore, and is the practice in many plants, to replace all the bulbs—burned out or not—at regular intervals. The same applies to lubrication. Minor adjustments and visual inspections can also be made by lubricators as they make their rounds.

Regulating Maintenance Expenditures

Relating maintenance budgets to sales is one way of ensuring that adequate funds are being spent on maintenance. This practice guards against a plant's becoming run down when it is being pushed to meet rising sales. However, it may not save the plant from being neglected when sales are in a slump. A certain minimum level of maintenance is always necessary to ensure that the plant will be able to operate at open throttle when sales pick up again. Neither should maintenance costs be allowed to advance strictly in proportion to a rise in sales. They should, instead, be allowed to go up at a reduced percentage of the gain

> **Productivity Goals for Maintenance**
>
> **Productivity increases in maintenance can be achieved through the use of better equipment and special tools, along with training, improved materials of construction, standardization and planning.**
>
> **It is first necessary to determine where efficiency now stands in relation to the industry. Once efficiency has been established, the plant manager can then turn to setting goals, and analyzing performance as to whether or not goals are being achieved, and if they are not, why not.**

in sales, and should also reflect at least a 5%/yr. improvement in operating efficiency.

Information about cost goals should be passed as far down the organization as the operating foremen. Charging maintenance costs to specific operating departments will make it more likely that operating supervisors and foremen will be as concerned with the care and cost of maintaining equipment as the engineering and maintenance departments. When costs are not charged in this manner, complaints usually arise from maintenance personnel that those in operations have little regard for plant equipment. Operations personnel are generally inclined to view equipment failures as the fault of the maintenance department, and maintenance personnel tend to be unduly critical of the care that those in operations give to equipment.

Every plant has areas in which a disproportionate amount of money is spent repairing and replacing equipment. When a process causes a high rate of maintenance due to corrosion or mechanical failures, it should be closely examined to determine whether the actual profits are sufficient to pay for the burdensome maintenance. A plant manager can get an erroneous view of profitability when maintenance costs are not charged directly to the process. If the product is worth keeping, however, he should consider new equipment of superior design.

Files on the replacement, materials of construction and spare-parts history of key pieces of equipment can save time in obtaining replacement parts and restoring equipment to serviceable condition. Such histories may also be a flag that the cost of maintaining a particular piece is so high as to warrant replacing it.

Many choices of materials of construction are available today, even for difficult, corrosive applications. Programs to test new materials in actual operating environments should be conducted continuously in the plant. Vendors are always interested in providing samples of materials (such as new plastics, clad materials and alloys) for testing. One of the more important tasks of maintenance engineers is keeping their knowledge of new materials up to date.

Manpower Staffing and Training

Well-trained manpower is the key to efficient maintenance. Therefore, an effective training program is essential if a plant manager is to achieve his goal of boosting operating efficiency and reducing maintenance costs. On-the-job training—always the foundation of any training program—should be augmented by specialized courses, such as the type offered by many equipment manufacturers.

Smaller plants particularly tend to use maintenance crews for both regular maintenance and construction projects. This is sometimes a necessary arrangement because it permits the employment of staff when there is not much regular maintenance. But is also presents difficulties because, when construction work slackens and employees are shifted to regular maintenance work, maintenance costs will rise precipitously. This can only be prevented by rigorously adhering to manpower budgets for construction and maintenance and by proper scheduling. The problem in doing this is that construction schedules are difficult to project because personnel will constantly be pulled away for critical maintenance jobs.

One way of coping with the last problem is to resort to contract maintenance. This course may offer two advantages: if the particular job is contracted for a fixed fee, the cost of the job will be set, and the construction work can proceed without interruptions.

How Much to Spend on Maintenance

How much should be spent on maintenance will depend on the product and the process. In special circumstances, "running equipment until it stops"—that is, without any maintenance—has merit.

When a safety hazard or loss of production is not involved, it may make sense to run equipment until it fails. This will be so in cases where equipment is not critical to continuing operation at the required output rate. For instance, agitators on batch vessels, when there are a number of them, may be "neglected." What must be determined is whether the cost of repairing the equipment will be more or less than that for maintaining it.

Other Ways of Boosting Efficiency

Special equipment, such as vibration analyzers, can greatly enhance the efficiency of maintenance staffs. A vibration analyzer, for instance, can provide advance warning of an impending breakdown of a key piece of equipment. It can also prevent unnecessary overhauling, and thus extend the equipment's normal maintenance cycle.

Labor-saving equipment of all types will especially pay for itself in this era of high labor costs. Equipment such as ladders, power saws, drills, cutters and hoists should be made readily available. One problem with smaller items is that they will tend to disappear unless security measures are enforced. However, a certain amount of loss must be considered inevitable and worth the price of greater efficiency.

Costs can also be reduced considerably through equipment standardization. Standard equipment is less costly to purchase, its delivery time is shorter, and it can be replaced without extensive changes to piping and other accessories. Standardization also makes it easier for mechanics to become familiar with the equipment, and reduces the quantity of spare parts that must be inventoried.

Keeping Older Plants Competitive

Older plants usually have many deficiencies, including poor layout, obsolete equipment and inadequate lighting and ventilation, all of which frequently result in high operating costs. Most often, older plants will have grown haphazardly, with sections added to the original installation without the benefit of long-range planning.

However, older plants do have the advantage of low-capital cost, the result of equipment having been depreciated over many years. If a plant's product line is marketable, this advantage enhances its profitability. To exploit this advantage fully, the plant's weaknesses —primarily its poor layout and obsolete equipment— will have to be remedied.

Installing process equipment in a new plant costs about as much as putting it into an existing plant. However, the cost of installing auxiliary equipment at a new site will often shift the scales in favor of the existing plant. Services, utilities and auxiliary buildings already installed represent an important asset.

Operating an Older Plant Profitably

Certain specific factors determine whether an older plant can be operated profitably. Minor, continuing expenses, such as for coveralls, tools, tape and shipping drums, should be examined for excessiveness. Such recurring costs should be eliminated or at least controlled stringently. Small lump-sum expenditures to eliminate them are frequently preferable.

In an older plant, work orders should be checked to determine areas in which maintenance is repetitive. Many of the problems can be alleviated by changing materials, equipment or processing steps. Sometimes simple changes, such as replacing glass panes with plastic ones, can save many hours. Mechanical seals are usually easier to maintain than packed glands, although higher skill is required to install them initially In highly corrosive operations, plastic materials of construction will save the time required to periodically replace corroded metals. Although the initial cost for new materials may be high, it will be outweighed by the savings in maintenance labor, and the reduction in downtime.

Whether maintenance should be organized departmentally or centrally will depend on the peculiarities of the plant. Centralization offers flexibility. When mechanics are not busy in one area, they can be transferred to another. Although personnel assigned to a specific area will understand its problems better, they frequently will not be working 100% of their time. In practice, some combination of both types of organization may be desirable.

Upgrading Files and Reports

Operating procedures in an older plant must be reviewed, standardized and brought up to date, especially when much of the operating knowhow exists only in the minds of key employees. Otherwise, the loss of such an employee through retirement or an accident could undermine cost control, product quality and the plant's ability to meet schedules. In older plants, product costs often show increases that are more than coincidentally related to the retirement of a key operator who had years of experience with a particular operation.

Various reports are necessary if costs are to be controlled. Yield data, throughput and raw material usage must be available on a daily basis to provide a regular check on cost trends. Equipment records must be updated to simplify ordering spare parts, and cross-referenced to eliminate excess inventory. Frequently, obsolete parts will occupy valuable shelf space.

Review reports should be checked for distribution, and recipients questioned as to their real need for them. Reports that duplicate each other should be combined.

Daily operating reports are essential to eliminate the problems that occur on shifts. An informative logbook not only provides for the efficient transfer of responsibility from shift to shift, but also can be an effective means for firstline supervision to keep management abreast of operating problems. Daily reports to operating personnel should outline the work that must be accomplished, and review accomplishments against yesterday's work schedule.

Rejuvenating an Older Plant

Many jobs in older plants evolve over a period of years and, even in times of fast changes, these somehow remain. In such circumstances, the plant manager must critically assess jobs to see which can be combined or eliminated. Although layoffs are undesirable, except as a last resort, some provision can be made for early retirement, transfers and retraining.

Layouts of older plants frequently are not suitable when movement of materials represents a substantial part of the cost of production. Reactors and other major equipment must often be relocated to facilitate operations with a minimum number of personnel. Drum handling, for instance, can often be replaced by pipeline transfer of raw materials. Excess piping and old buildings may have to be removed to provide space for efficient material movement.

A master plan should be made of the existing plant, and future installations outlined on it, to provide for the orderly addition of new facilities and the elimination of those too obsolete to be operated profitably.

Warehouses should be planned for access to raw materials and shipment of products close to the operating area. Materials movement must be minimized and automated, with bulk materials moved by conveyors. Warehouse ramps should be expanded for larger and longer trucks, and storage tanks installed for bulk handling. The latter should ideally be located near the operating area and on railsidings.

Larger equipment reduces operating costs. In older plants, operating equipment is often small. Replacing obsolete, low-capacity equipment with larger equipment is expensive; however, it is frequently possible to find other uses for the old equipment to soften the economic impact. ■

If apple-polishers, buck-passers, back-stabbers, and empire-builders have infiltrated your plant or department, you should consider . . .

How to Squelch Intraplant Politicking

These ten rules should help you curb politicking among your subordinates, and to channel their energies into activities more useful to you, themselves and the company.

HERBERT POPPER, *Assistant Editor, Chemical Engineering*

Politicking can be loosely defined as an attempt to obtain personal advantage not by playing a better game but by talking a better game.

Healthy competition is based on performance; it benefits the company and usually brings out the best in people. Politicking on the other hand, usually harms the company, and brings out the worst. Like jealousy, it "doth mock the meat it feeds on," and may eventually bring about the downfall of the politicker as well as of his victim. Besides wreaking injustices on individuals, politicking can ruin efficiency, erode professional ethics, and cause your best men to leave or to lose part of their usefulness.

Some degree of politicking can be found everywhere. But its seriousness varies greatly. Unfortunately, in the operation and maintenance field it is often quite serious, particularly when promotional channels become clogged. Almost every operating engineer or production supervisor wants to become a plant manager some day; so do most engineers in maintenance, and quite a few engineers from other service departments. The temptation to jockey for position is sometimes strong.

Furthermore, professional ethics in operations and maintenance have not been defined or emphasized quite as much as in some of the purely technical fields of engineering. Even in the latter, of course, there are some gray areas in professional ethics, as brought out in *Chemical Engineering* recently,[1] so the rules about to be suggested are not necessarily confined to any one function.

While some of the rules may be rather self-evident, a quick review of even these may stimulate your thoughts on the causes, manifestations and effects of intracompany politicking, as well as on what can be done to discourage it in subordinates.

1. Don't Appear to be Playing Favorites

It is not enough to be objective about your subordinates, you also have to appear objective. For example, let's say Joe Smith is on the same civic fund-raising committee you're on. You get to know him socially. He starts spending more time in your office than your other subordinates because you and he are more at ease with each other. Although he has less experience than his colleagues, Joe becomes the one you call if you need help or information. Eventually a promotion comes along and you give it to Joe—not because of friendship or his gifts as a conversationalist, but because you feel he has a slight edge over competition.

But try convincing your other subordinates of this. Typical comments are fairly sure to include "Those hours that Joe spent making points with the boss sure paid off." "Performance doesn't mean as much as being on the right fund-raising committee." And so on.

A bunch of soreheads, you say? Not necessarily. Remember, they don't know what you know; they can only go by outward appearances. In any case, a little care on your part to prevent any impression of favoritism, and to make it plain to Joe's colleagues that their work isn't going unnoticed, would have helped.

A promotion often stimulates everyone to better performance. But as it stands, Joe's promotion is unlikely to have that effect; it may merely stimulate his colleagues to spend more time chatting with you, and perhaps to join your fund-raising committee.

There are various ways of dividing your listening time fairly among your subordinates without letting this consume an unduly large portion of the day. One is to keep individual interviews short. A subtle way of doing this may be to develop "egg-timer" eyes. As suggested in a recent article,[2] there is a subtle something in the gaze of egg-timer-eyed executives that makes callers aware of the smooth but relentless progress of time. At a certain moment (in this case, when most of the constructive business has been dispatched and image building begins to sneak in), the visitor becomes aware that he is "done." An invisible pressure—only resistible by hardboiled types—makes your caller aware that the sand has run out and that he no longer has your full attention.

2. Emphasize Performance

Let's say Ed Brown is prone to politicking. He will spend much more time and ingenuity figuring out ways to pass off a low yield or a poor efficiency as a major triumph over cruel handicaps than in trying to improve said yield or efficiency. He calls attention, usually by a multiplicity of memos, to any activity of his that he feels may impress somebody; but he seldom has the perseverence to carry projects through to the payoff stage. Even if he had the perseverence, he would not have the time, since his writing activities take a big chunk out of his day—and out of his weekend.

Can Ed be reformed? Yes, particularly if the cure is applied in the early stages of the sickness. The cure, in this case, consists of making it quite clear that you judge Ed not so much by his mountainous labors as by the molehill results they bring forth. For a start, you might tell him you're not interested in his correspondence, and that you think he spends too much time on writing memos. Ask him if he can't delegate some of his paperwork, and make his own memos more concise, so he can spend more time coming to grips with his problems and following through on his ideas. Eventually he should get the idea that form alone is not enough, and that time spent obtaining results can pay bigger dividends than time spent on diversionary tactics or excuse formulation.

3. Clamp Down on Politicking Memos

Memos lend themselves so well to various forms of politicking that they deserve a section all their own.

In plant *X*, it was customary for the assistant plant manager, the plant engineer, and the various technical supervisors to send the division manager copies of all their correspondence with the plant manager. Since the division manager had considerable say in promotions and raises, politicking was rampant. The plant manager frequently found that a memo addressed to him was really not written for his benefit at all. To make matters worse, more and more memos got written directly to the division manager, with the plant manager getting an information copy only. If the writer felt safe in doing so, the memos would contain digs at other departments, and even unflattering references to the plant manager's leadership. (The plant was old and relatively inefficient; hence various departments were under considerable pressure.)

To clear the air, it was eventually decided that memos and reports were not to be addressed or routed to division headquarters by anyone other than the plant manager. Memos or reports intended for the division and originating from within the plant were to be addressed to the plant manager; the latter had the choice of transmitting them over his own signatrue, summarizing them in a weekly plant report, or handling them entirely on the plant level. He was given new guidelines on what the division headquarters did and did not want to know about.

The results were constructive all around. The plant manager's authority and responsibility were reestablished, and his performance improved. The division manager was glad he had less correspondence to review and that he didn't have to be embarrassed at the contents of some of it. As for the plant manager's subordinates, most of them were rather glad to be relieved of the temptation to go over his head or maneuver for position now that the "everybody is doing it" rationalization couldn't be used any more.

Perhaps this represents a rather extreme case, but it still may pay to review your subordinates' paperwork output occasionally and ask yourself:

- Are some people writing many more memos, letters and reports than necessary? How does their actual performance compare with the ones who write less, or who write more concisely?
- Is correspondence generally aimed at the person it is addressed to, or is it aimed primarily to score points with other people on the distribution list?
- Is the distribution list proper, or does following it sometimes involve shorting the chain of command?
- Have memos become too personal—i.e. do they take too much personal credit or cast too many personal aspersions?
- Is the tone of the correspondence generally constructive, or is some of it likely to be disruptive to the morale and efficiency of one of the recipients?
- Have you become overly dependent on memos and reports to find out how and what your subordinates are doing?

4. Maintain a Performance-Evaluation Program

Professionalism can be encouraged, and politicking discouraged, by a uniform work-evaluation program. Such a program will help your employees to know what is expected of them and how well they are measuring up to the predetermined standards. It can also make you more objective about the relative abilities of your subordinates.

The program suggested by the National Society of Professional Engineers, which was described in a recent article,[3] can be adapted to nontechnical as well as technical employees. Under this program, every engineer is evaluated by his immediate superior at regular intervals, preferably every six months. Evaluation criteria are publicized in advance.

If you make it clear to chronic politickers that cooperation and attitude are among the traits that will be rated, the announcement itself can have a healthy effect.

The completed evaluation must be frankly discussed with the subordinate when the time comes. The natural tendency to base the evaluation on very recent impressions, or on a single outstanding incident, can be avoided by filling out a very simple monthly report (even using a plain index card) wherein you record your impression of the subordinate's performance over the past 30 days. These monthly reports are also a help if you are asked to explain your rating.

Performance evaluation can be a great anti-politicking tool—as long as you make it clear that employees should not call every minor achievement to your attention, and that you will be negatively rather than positively impressed by "one-upmanship." True, some of your employees may do some politicking at the evaluation interview itself, but this is less time consuming and less disruptive to morale than daily image-building.

5. Keep Supplied With Challenges

A man who has a challenging assignment to work on not only may lack the time for politicking, he will probably lack the inclination. He has something to devote his energy and imagination to—something where his own interests and those of his company coincide.

In a lean, growing company or department, there should be no lack of challenging assignments; in a mature, stable company that is topheavy with technical and administrative talent, the challenges may be fewer (and the politicking worse). But even here, challenges can be created by forcing your subordinates to stay abreast of advancing technology, by creating long-term improvement projects, by encouraging your subordinates to attend technical meetings and bring back ideas.

The writing of technical articles or papers can enhance company as well as individual prestige. Furthermore, it provides a challenge, and a healthy outlet for professional zeal. While no one advocates disclosure of proprietary information, clearance procedures shouldn't be so cumbersome that they stifle initiative.[4]

If you can devise legitimate ways of keeping your department, plant or company growing, that, of course, creates many challenges automatically.

6. Be Understanding About Status

Here again, there are fundamental differences between a rapidly growing, fast moving company or department and a mature, relatively stagnant one. In the former case, an engineer or foreman may not care too much how nice his office is—he knows that if he's on the ball, he's not going to stay in it for very long before he moves up. He may not even care whether he's making more or less money than the going rate, because he knows that eventually the rewards will catch up with him, and in the meantime there's exciting work to be done.

But take a company or department that is growing slowly or not at all. The people in it come to realize that competency alone does not assure promotion. For every one of the few promotional opportunities that open up, there may be a dozen people who feel qualified to fill it, and who may very well be qualified. To prevent politics from becoming rampant here, the men must not only have the challenges discussed earlier, but also enough evidence of recognition to enable them to live with themselves if they don't get promoted.

Naturally, going overboard on offices, secretarial help and status symbols can be unhealthy; but at the other extreme, it is hard for one of your subordinates to consider himself a valuable member of the management or technical team, and to think "big," if he knows that no matter how good his performance, his office for the next ten years will probably continue to be a stuffy cubicle next to the big boiler, with a 1920-vintage desk. Frustrations like that can and do provide rationalizations for vicious politicking and infighting. In this case, an acceptable office and tactful treatment may prevent things from reaching that unpleasant stage, and from having the subordinate consider himself a failure if he is not promoted.

7. Fix Responsibility and Authority

By setting up strict standards of responsibility, accountability and authority, much politicking can be avoided. A certain amount of line/staff or interdepartmental conflict is not at all abnormal or perhaps even undesirable; the situation usually becomes destructive only when there are many areas of vague or overlapping responsibility.

If a man is responsible for getting a job done but lacks the authority to go with the responsibility, he has to rely on flattery, intimidation, lobbying, manipulation or reciprocal agreements when normal cooperation fails. Clear lines of responsibility avoid this.

In many companies, the relationship between first-line supervision and staff functions such as personnel and quality control could be much better defined.

Purchasing decisions are often a source of friction because the engineer and purchasing agent have never had their authority and responsibility spelled out, and have never set up consistent step-by-step procedures that will avoid duplication of effort, tangles in communication, and unsound relations with vendors.

Foremen often have so many bosses, with such a long chain of command, that the temptation to bypass somebody in order to get reasonably quick action or approval is often too great to resist. The bypassed person may then retaliate and salve his wounded pride by disapproving the foreman's next three requests as a matter of principle. To get even, the foreman's next move may be to play off one boss against another, and to freeze out the recalcitrant one. And so on.

It is well to look into areas of chronic friction such as these to see if definitive procedures can't be set up, or the organization chart simplified.

8. Watch Your Communications

Knowledge is power, and many people will politick no end to get some piece of information that can be used for further politicking—particularly information concerning promotions, personnel changes, expansions or cutbacks in production, and so on.

You can minimize this situation by communicating as much information as you can to your subordinates, and doing so promptly before the rumor mill goes into high gear. If your plant engineer first hears about an important development or decision through the secretarial pool, he may get a distorted version that you may never be able to completely correct. Further, he may feel so left out that he'll decide to spend less time engineering and more time cultivating the more-talkative secretaries.

It is important that you explain to your subordinates the reasons for some of your decisions—why you took one person's or group's advice rather than another's. Make sure that everyone understands that his views were understood, considered and appreciated. Don't rely on memos or letters; they may be misinterpreted. Besides, they lack the personal touch.[5]

9. Permit an Honest Mistake

It might have had a very unfortunate effect on George Washington's character if his father had said "George, I appreciate your telling me about what you did to the cherry tree, but since you made the decision to chop it down without consulting me or any of your elders, and since it was at variance with standard procedures, you are hereby disowned and must leave my house forthwith." It might have made George wish that he had covered himself beforehand, so that he could have said "While I did chop down the tree, I did it accidentally. I was just planning to prune it. I would have asked you for permission, only you were busy. So I asked little Johnny Adams and he said he thought it was OK to use an axe."

The intended moral to the above is not that any mistake that is forthrightly admitted must necessarily be forgiven on the spot, but rather that if the punishment exceeds the crime, there will be more politicking and jockeying so as to minimize future reprimands.

Specifically, a person who does not permit his subordinates to make any mistakes will find that it will take very long to get any sort of decision from anybody. The person who should by rights be making the decision will try to either foist it on someone else, or at least to get as many persons in on it as possible so that the blame can be shared in case of failure. Buck-passing is a special but very common form of politicking—one that can stifle initiative and play havoc with efficiency.

10. Watch Out for Mismatches

A person whose forte is getting things done through other people may be unhappy and unsuccessful in a job that involves individual effort and no supervision. It is natural for him to start the empire-building brand of politicking so as to get someone to whom he can delegate part of his work.

Conversely, a person who does good technical work himself but is not good at supervising others, may feel a particular kind of frustration if placed in a purely supervisory position; he may start to blame his subordinates for not handling matters as well as he would himself, and may start politicking for a different supervisory assignment although the problems he would face there would be no different.

In both of these cases, the people may not be aware that they are miscast in the roles they are playing. But if their supervisors are perceptive enough to diagnose their frustration, and can provide them with more-suitable assignments, the company will stand to gain considerably, and politicking will diminish. Of course, this assumes that the company has parallel paths of advancement for supervisory and nonsupervisory professional employees, so that an outstanding engineer is not forced to become a mediocre supervisor merely to make more money or to enjoy more prestige.

On a somewhat different level, take the case of the production supervisor who gets poor yields but loves to write and is awfully persuasive. In fact, he writes such good memos that he almost convinces you each month that his 80% yield is more of an achievement than someone else's 95%. His spell-weaving ability may not improve production, but the man may have a tremendous future in your company's sales or public relations department. ∎

References

1. *Chem. Eng.,* Dec. 9, 1963, p. 177; also Sept. 2, 1963, p. 87.
2. Webster, E., *Management Review,* Oct. 1963, p. 4.
3. *Chem. Eng.,* Jan. 6, 1964, p. 103.
4. Cunningham, W. A., *Chem. Eng.,* June 26, 1961, p. 136.
5. Hertan, W., Rice, R., Ruder, W., *Business Management,* Aug. 1963, p. 26.

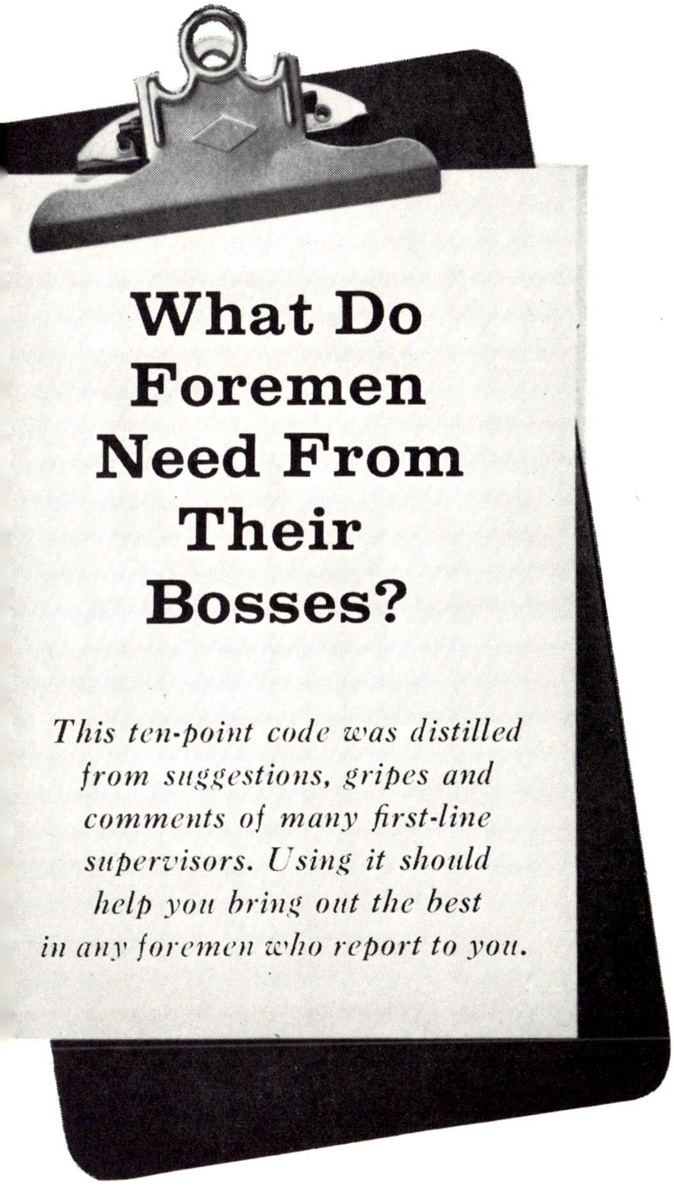

What Do Foremen Need From Their Bosses?

This ten-point code was distilled from suggestions, gripes and comments of many first-line supervisors. Using it should help you bring out the best in any foremen who report to you.

BRUCE VON DER HEYDT
Titanium Metals Corp. of America

A recent cartoon showed a manager answering a new foreman's question by saying "Yes, we do have a very good incentive program here. We fire at the drop of a hat."

At one time, this sort of statement would hardly have been novel or funny, because fear for one's job was regarded as the key to good performance. The old authoritarian managers who relied on this approach certainly had their day when our industry was young. Today, however, with increasing complexities all around us, the motivational type of management is taking over.

Thus the question of how to motivate first-line supervisors is of great importance. Unfortunately, it is often answered by a host of platitudes that don't seem to hit the mark.

To get more-pertinent answers, I decided to go right to the grass roots; hence I interviewed a large number of first-line supervisors and tried to find out why the well-motivated ones, as well as the less-happy ones, felt the way they did. Most of these men were in the maintenance branch, but their comments seem to apply almost equally well to all first-line supervisors.

The results of these talks and of other observations form the basis for the set of rules presented below. I realize that some of these rules may have a familiar ring; but I think that even they can serve as a useful reminder.

1. Delegate liberally

Everyone agreed that the first-line supervisor can only do a good job if he is a definite part of management, and is treated as such by his superiors and by the workers. It is important to delegate enough managerial functions to him to keep him solidly identified with management.

If we haven't already done so, we can start this process by delegating in measured quantities at first, then gradually increasing the amount and extent. This process will require grooming the foreman as progress is made.

Some of us at one time or other might have felt that we were "passing the buck" when we delegated an unpleasant task to a subordinate. However, unless we develop our first-line supervisors by increasing the amount of work delegated to them—so that eventually, on most jobs, we can shoot them the whole works—they aren't going to be convinced that we have complete confidence in their ability as members of the management team.

There are those in middle management who are afraid to delegate more than the essentials. This is short-sighted thinking since, by delegating more liberally, we can make our own jobs easier; help the first-line man gain stature; and furthermore ultimately groom a replacement for our own jobs. How many times has promotion of a man been held up because no suitable replacement had been trained to do his job?

When increasing the tasks that you are delegating to the foreman, try to avoid incessant needling, and be prepared for an occasional goof. Most of us learn by our mistakes, and, unless the foreman is an extremely callous individual, a mistake in itself is going to make him feel badly enough without a lot of second guessing and criticism from his supervisor.

One last precaution: a very important part of delegating is the followup. Check on progress, or have the foreman report on status from time to time. Dropping the ball in the foreman's lap and then completely forgetting the whole matter usually isn't the best technique.

2. Keep your men challenged

Today more than ever, good men shop around for jobs. When, during a hiring interview, I ask a presently-employed applicant why he is considering leaving the job he has, I frequently get the answer, "my present position just doesn't keep me busy enough."

Of course, it's not enough to keep a foreman just busy—there has to be a challenge. An up-and-coming young engineering graduate isn't going to stay favorably motivated very long if he is saddled with a bunch of laborers who are mostly doing housekeeping chores around the plant.

Judicious selection of the right problem assignment for the right man is essential. Putting a man who is long on heavy electrical equipment maintenance in charge of instrument maintenance isn't using good sense.

Another thing—let's not burden our foreman with a profusion of petty tasks that could be performed by clerks or messenger boys. The more innate ability and ambition our foreman has, the more we must keep him challenged.

It may take a little background information to create the feeling in a supervisor that his work is important. It often pays to explain to him how a small contribution or task fits into the big picture. If the foreman knows exactly why something is being done, he will be in a position to get a more favorable response from his work force.

Generally speaking, the higher the caliber of individual we are dealing with, the more concerned he will be with the significance of the work. If our foreman is convinced that he and his men are really working on a prominent contribution to the total effort, he will drive harder and be much more satisfied with his work than the fellow who merely accepts it as another task. It is one thing to go home worn out every day from a treadmill job, but quite another if our foreman is tired but inwardly proud that he has accomplished something.

3. Don't neglect communications

Foremen agree that nothing can be more frustrating than to hear of future management plans via the grapevine or from their own men. This is a particularly acute problem in union shops, because most unions do have good communications with their members and can often jump the gun on management. Since we want our foreman to be a key part of management, let's keep him abreast of new developments—from the dream stage to the arrival of new equipment.

In our plant, resistance to change is evident everywhere. We find that it is worth the time to sit down and not only discuss proposed changes but go into considerable background detail to explain *why* we are considering these changes.

Imagine yourself in a first-line foreman's situation, where one of your men comes up to you and asks, "Is it true that the company is going to install a new furnace that will do the work of two old ones and eliminate two jobs?"

Your stock really hits bottom if you blurt out, "I don't know—they never tell me anything!" You score a little better if you condition yourself to reply, "sounds like a rumor—I'll check and let you know as fast as I can find out." But you are at your very best if you can emphatically state, "that is a false rumor—we are planning to replace a furnace that has been a bottleneck so that we can handle increased business; it won't affect jobs one way or another." This is the answer of an informed key man of management. He may not always be the bearer of glad tidings, but at least he is "in the know."

4. Build a sense of participation

To make the foreman aware that he is on the management team, we should let him have a hand in influencing decisions that affect him, and encourage him to make suggestions. By doing this, we are not relinquishing any decision-making powers, we are merely seeking the participation of our first-line supervisors in determining the best approach to some of our mutual problems.

In my department, much is accomplished in a short weekly maintenance meeting, where plans, progress and suggestions can be taken up. Praise for good work and constructive criticism is also important here. Oddly enough, one supervisor told me that, for him, the severest criticism is not so much when we find some fault with his work, but when we ignore his work completely. That way, he doesn't know where he stands and isn't even sure he is noticed at all.

Besides building a sense of participation, these meetings have a unifying influence, which is perhaps more important in the maintenance field than in production. Most grievances involving maintenance workers are of a jurisdictional nature—pipefitters allegedly doing millwright work, for instance. Much of this can be minimized by continually stressing the collective *we*—to signify the whole maintenance team, not just the electrical gang or the shop crew. Once all the foremen are on the same target, some of this spirit is bound to rub off on their crews.

5. Be consistent

More than one foreman told me that he considered uniform standards, fair dealings, and an objective approach to all problems, as top priority. Foremen who felt strongly about this one attribute—which we can call consistency—went on to explain that they particularly disliked situations where a second-line supervisor let friendship interfere with his day-to-day dealings with his subordinate foreman.

Perhaps some of us are forgetting the very elementary admonishment that we have always given to new foremen—play no favorites. One of the reasons for powerful labor unions today is the favoritism once shown by management in dealing with workers. Impartiality is, at the very least, equally important in dealing with our first-line foremen. We must be especially careful not to be susceptible to subtle apple polishers in our own organization.

Not only must we be sure to base promotions and assignments solely on merit, but we must be consistent in our day-to-day relationships with our foremen. One preoccupied moment, when we fail to render a cheery greeting or miss a chance for a few words about a good job performed, may be interpreted as a deliberate slight.

6. Practice what you preach

There are several aspects to this rule. One foreman I queried scratched his head and insisted he'd be a lot more pleased if his boss didn't always leave the impression of resisting innovations. The latter was always suppressing initiative and holding back everyone, and yet, had few original progressive ideas himself.

This is emphatically a very undesirable commentary for a supervisor to make of his superior. While I promised to avoid platitudes, it goes without saying that the better we can personify qualities like energy, enthusiasm and perseverance, the more likely that these attributes will rub off on our subordinate supervisors.

We must try to keep temperament out of our work. By remaining calm during the rough spots and emergencies, solutions to problems will come easily for all those around us.

In short, we must try to practice what we preach; both in work attributes and personal conduct. One foreman emphasized that although his superior was a hard taskmaster, everyone had to respect him because he lived by the same rules that he laid down for everyone else. In this way, the respect of our subordinates is more easily held.

7. Keep your door and ears open

When we ask our foreman for suggestions, we must be prepared to listen, consider, discuss, and either tactfully reject or try some of them.

One first-line supervisor told me that he would be much happier if his boss were a little easier to approach. "That boss of mine is pretty darn hard to get to. Then when I do reach him, he's too preoccupied to listen carefully."

Being asked to listen doesn't seem to be asking too much. In the union shop, there are very elaborate procedures for the airing of differences. It is entirely feasible for an individual in the bargaining unit to press his grievance right up to the plant manager's door.

What of the foreman? In some establishments, the fact that an employee is on salary seems to mean that the poor fellow can't have a legitimate complaint.

Let's provide for complaints. This can be done in various ways. Usually a very informal system will suffice so long as it is not so informal that it is sidetracked or eventually abandoned. We owe our foremen at least as much recognition as we do their crews.

As we continue to cultivate our supervisors, we shall find that some like to have an occasional performance review. They want to hear a restatement of some of the company policies on selection, promotion, and so on. A foreman's innate curiosity may lead him to insist on answers to questions such as these: "How far can I go?" "What are my prospects?" "How am I doing since the last discussion we had?" "What are my limitations, and is there something that I can do to correct them?" "Does seniority have anything to do with advancement?" "Is the lack of a degree going to hinder my future?" Our first-line supervisor is entitled to truthful answers. He should be informed on where he stands.

8. Stand behind the foreman

Now let's go on to a sure-fire morale breaker for a front-line foreman. This fellow reports, "Every time I stand up to the union and try to enforce the rules or get the workers moving faster, I feel as if the rug is pulled out from under me."

Today we often find the foreman with too many responsibilities and without the requisite authority to carry them out. The poor man can never settle grievances himself. Industrial Relations always handles that. He never decides whether a product is acceptable or not. Quality Control technicians are right there all the time. Scheduling of the work to be done next? No! Production Control tells him what is first priority. Make a field change on a new installation? Not without injuring the dignity of a half dozen engineers or other staff specialists. Work the logically qualified man overtime on a hot job tonight? Never! Seniority or overtime agreement govern this. And so on, ad infinitum. Our key man, the foreman, is harrassed by so many staff specialists, rules and contracts that he gets the justifiable feeling that he isn't directing the work force at all.

We must make a strenuous effort to stand behind our foreman. If he makes a bad decision, let's have him correct it himself, and show him how to avoid a repetition. If there is going to be anything given up by management as a result of a grievance, let's have the first-line foreman (representing the company in the eyes of the workers) do the giving wherever possible.

To help all of us in supervision to be more aware of the union contract, training or discussion sessions are advisable.

Unions almost never send only one or two men in to

a meeting—they use a grievance committee. We should do that too. Get at least two or three other foremen into the first-step meeting, even if they just observe. It will pay off later. They will learn how to act in future similar situations.

9. Follow chain of command

There's nothing surprising about this one. Ideally orders flow from the department head, to general foreman, to foreman who sees that they are carried out. Reports or results flow in reverse order. We've got to keep our key man, the foreman, in the chain. This vital link should seldom be bypassed, and then, only if the situation borders on an emergency. In such a case, the least we can do is to apprise the foreman at the earliest opportunity of any discussion of, or changes in, work in progress.

Few things will discourage a foreman more than having a superior go out and countermand his directions without first telling him. In virtually all cases, it would be better to take the additional time to reinstruct the foreman and let him make the needed changes himself. Indiscriminate jumping of organizational lines right within our own department is a surefire way to undermine the foreman.

10. Don't neglect salary and benefits

You've all heard of the fellow who said, as his boss was praising his work and extolling his virtues, "Don't applaud, boss, send money." While foremen aren't motivated by money and fringe benefits alone, the importance of tangible rewards can't be dismissed lightly. Many foremen told me or implied that it wasn't so much what they themselves earned that they found fault with—it was the small differential between their pay and that of the men in their charge.

In the past twenty years, the hourly pay scale of factory workers may have gone up something like three- or fourfold, while salary schedules for many of our key supervisors have only doubled. The differential between the lowest paid laborer and the machinist has decreased. Similarly, the differential between the skilled craftsman and his foreman has diminished. By the time that shift-differentials, overtime, quick returns, and sixth and seventh days are thrown in, the foreman frequently finds himself on the short end.

It is the manager's job to convince the new foreman that, while his immediate monetary gains may be slim, future salary opportunities resulting from merit raises are worth striving for. We can point out the present advantageous differences in health and accident plans; stock purchase plans, if any; continued salary, in event of sickness; additional vacation benefits, and other tangible gains.

Similarly, it is our duty to provide merit raises promptly when earned, and to see that our salaried foremen get some consideration to compensate for the escalator clauses of the bargaining unit. Furthermore we should do all we can to make sure they take all their vacation time, and that they get occasional compensatory time off when they have worked around the clock to get some piece of equipment back in operation. And we can use our influence to get them a fair shake on monetary and tangible fringe benefits from their employer.

So here we are

I don't think there are any short cuts for these rules. The survey showed that foremen are quite unimpressed with indiscriminate lip praise and backslapping when used as a substitute for challenges, middle-management support, or some of the other points we have discussed.

Not one man in supervision with whom I talked, seriously disliked his work. Neither was anyone wholeheartedly in agreement with everything the company did, or totally satisfied with every facet of his job. (Had I found anybody who was, I would have suspected him of either being abnormal or holding back something.)

Naturally, we can't be expected to keep all supervisors enchanted with our company's policy every minute of the day; but by keeping these ten rules in mind, we should certainly be able to remove a great many causes or excuses for indifferent or merely adequate performance. ■

This article is based on a talk "How to Motivate Supervisors," which the author gave at the 13th Plant Engineering and Maintenance Show at Philadelphia.

Meet the Author

BRUCE VON DER HEYDT *is a graduate of the Drexel Institute of Technology, in Philadelphia, where he obtained a B.S. in Civil Engineering in 1947.*

He began his career as a construction engineer for the New Jersey Zinc Co. An interruption came when the Army recalled him to active duty during the Korean emergency. Since 1953, he has served in various maintenance-supervision and project-engineering capacities for National Lead Co., and later with Titanium Metals Corp. of America, where he is now superintendent of maintenance at the Toronto, Ohio, plant.

Dealing Out Discipline

A labor relations attorney reviews the cornerstones of a fair and effective discipline policy, pointing out some of the pitfalls by drawing on cases that went to arbitration.

JOHN W. WHITTLESEY, *Union Carbide Corp.*

The right to discipline offending employees is one of the most useful and zealously guarded management responsibilities. Its exercise, if firm, fair and consistent, is of enormous benefit to the company.

Moreover, a sound disciplinary policy is of equal benefit to that vast majority of employees in any industrial plant whose main interest in working is to do the job and stay out of trouble. This majority tends to resent those individuals who deviate from accepted standards of behavior and get away with it. In principle, the interest of the plant supervisor in keeping a trouble-free workplace, and thereby boosting production, thus coincides with the desires of most of his subordinates.

While there may be agreement over general principles, knotty problems frequently arise over their application in specific instances. No one disagrees, for example, that the chlorine plant employee who steals mercury for resale should be fired. But how about the luckless individual who pockets a 29¢ light bulb? No one thinks that the employee who slugs his fellow worker on the job in an unprovoked assault should stay on the payroll. But what about an assault that was provoked by persistent heckling, or by a racial epithet, for example? The individual who sneaks off his job early, leaving his equipment running, certainly deserves a penalty; yet, if no damage occurs because other employees take up the slack, what should the penalty be?

Now, most plants have provisions in their union agreements that restrict management's right to discipline, suspend or discharge to situations involving "just" or "proper" cause (the terms are roughly synonomous in this context). Under these provisions, both the problem of whether to discipline, and how big a penalty to mete out, are open to question in the grievance and arbitration procedure. The burden of proof usually rests with the company when discipline cases go to arbitration.

Management is thus placed on its mettle in penalizing misconduct. It cannot indulge supervisory whims, personal likes or dislikes. Management representatives must know what company disciplinary policy is, must apply it consistently and impersonally, and be able to back up what is done with sufficient proof to convince an arbitrator of the merits of the case.

Furthermore, a successful disciplinary policy in any plant must be firmly based on enlightened principles of psychology and motivation. The employee who really should be disciplined is generally the one who is not mature enough to control his antisocial tendencies. Yet, as in child psychology, punishment that either does not fit the crime or is handed out capriciously because of the mood of the moment will hardly serve as an impetus to better behavior. Instead, it is more likely to breed frustration and animosity. It is thus essential for the supervisor to recognize that discipline is not a substitute for training, communication and good human relations, but is essentially a last resort. It must be applied deliberately and without rancor, and there should be an anticipated benefit from it.

Let's Become Specific

None of these goals or principles blazes any new trails of management philosophy. Each has been resummarized primarily to paint a background for the discussion of some specific areas of uncertainty.

The first one concerns the problem of spelling out disciplinary policy to plant employees. Is it best to set forth plant rules detailing the type of conduct that will be penalized, or is a more general statement, or no statement at all, the most desirable solution?

There is, of course, a sharp difference of opinion on this point.

Some believe that employees should be apprised in advance precisely how far the company will go, so they will know exactly what will happen if they ignore the rules.

Others feel that detailed exposition of forbidden categories of conduct is unnecessary, that most employees know right from wrong and will obey instructions.

They envisage such a statement as: "Employees will be penalized, by way of reprimand, suspension or discharge, as the circumstances may warrant, for transgressions against generally accepted standards of industrial conduct and decency, or violations of legal constraints, or for conduct prejudicial to the good order and proper functioning of the organization."

They point out that it is difficult if not impossible to write a code of conduct that will cover all types of conduct subject to discipline. Any statement of company policy that is published, they feel, should therefore be only very general in nature, and they can show support for their views in certain arbitral rulings.

Man Who Was Nabbed Napping

Thus plant rules at a particular location specified (among other things) that "leaving one's post of duty to sleep" was a dischargeable offense. No other rule covered this area, but the company had a strong past practice of applying this one strictly. One of the operators, overcome with late-afternoon heaviness of the eyelids, walked five feet away from his equipment, left it running, sat on a large drum, put his head on his updrawn knees, and fell asleep. He was so discovered by two foremen, who awoke him with some difficulty.

The grievance protesting his discharge eventually went to arbitration. At the hearing, the union argued that (1) going five feet away from his equipment was not leaving his post of duty to sleep, and (2) he had not left his post to sleep in any case; his purpose was to sit down and watch his equipment run (a perfectly permissible act when the equipment needed no attention); his falling asleep was inadvertent. Thus, it was argued that the plant rule was not deliberately violated and no penalty at all should have been handed out.

These technicalities swayed the arbitrator. However, because the specific plant rules were supplemented by a general catch-all clause of the type quoted earlier, no back pay was ordered on reinstatement, and the arbitrator permitted the company to reassign the man to a far more closely supervised job.

The Case of the Untimely Overtime

Another plant suspended two employees for refusal to work overtime. The contract had a standard management-rights clause that contained no prohibition against requiring overtime work; there was a typical overtime distribution provision that had been followed in this case. As for the plant rules, none of them dealt directly with overtime; the one that was applied warned of discipline for "gross insubordination." The rules also contained a general catch-all provision.

The arbitrator in this case ruled that the company could, under its contract, require reasonable overtime; the overtime in question here was deemed reasonable; and the employees were said to have had no excuse for refusing. But, the arbitrator stated, their refusal was mere insubordination, it was not gross. Since the plant rules did not include refusal of overtime in its list of punishable offenses, the employees had a right to regard such refusal as something less than a major offense. The catch-all provision could not be used to thwart this expectation. The suspensions were reversed.

The sequel to the story is equally pointed. After the award, this plant issued a notice rescinding its entire code of plant rules, and substituted a general statement of disciplinary policy along the lines already discussed. Two years later, two more employees refused to work required overtime under closely similar circumstances. (Indeed, since one was a union official, it is possible that the policy statement was being deliberately tested.)

The new policy passed with flying colors. The grievances were denied and the discipline stood. The replacement of a specific code by a general policy, with appropriate notice, was held to be ample. It warned that the company could hereafter rely on whether or not just cause for discipline existed, without limitation to specified types of misconduct. Insubordination of this or any other sort was clearly just cause for action.

Moral: Make It General

From the above two cases, it appears that the best course of action in stating disciplinary policy is to leave it as general as possible, relying on normal employee maturity and experience to minimize the necessity for taking action, but setting forth in clear and precise language disciplinary policy in areas that may require specific treatment.

Thus penalties for violations of safety rules could well be detailed. Similarly, the results of smoking in prohibited areas could be outlined. The company's attitude and procedures for handling absenteeism is an apropriate area for detailed treatment, as is what to do about garnishments or wage attachments—those costly nightmares of the accounting deparment.

Of course, these policies will have to be carefully drawn, to make sure that as many contingencies in the particular field are conveyed as possible.

Consistent but Varied Application

Once these policies are laid down, care should also be taken to be consistent in applying them. This, of course, does not mean that the application cannot be varied to fit different circumstances.

One case involved two employees who were caught walking toward the plant fence, just beyond which was the automobile of one of them. This man was carrying certain plant equipment, to which he alone had access, that could have been useful in the welding shop of the other. On being challenged, they fled, but not before being identified. Both were discharged for misappropriation of company property.

The discharge of the employee actually in possession of the equipment was sustained. That of the other was reduced to a six-month suspension because he was merely accompanying the other, and no motive for his participation in the actual theft could be proved. Possibly this was merely a failure of proof; but consistency here demanded not uniform but diverse treatment on the basis of the available evidence.

The most obvious application of varying degrees of discipline for the same or similar offenses comes in situations where two employees equally in violation of plant standards of conduct have varying past records standing against them. If one of the offenders has a history blemished by disciplinary reprimands or suspensions, and the other has not, it may be entirely appropriate to penalize one man more heavily than the other.

Of course, for serious offenses, discharge may be quite appropriate for the first offense, even without a past record. It should also be kept in mind that if the past record of the employee is unrelated to his present offense, an arbitrator could reject such a record as having no bearing on the case in question.

In other situations, varying penalties for similar offenses may be justified because of the different degree of responsibility each employee bears. Thus, when a wildcat strike in violation of contract has been triggered and led by a union official, companies have frequently discharged the official but inflicted lesser penalties on the other strikers, on the general theory that the union official owes it to his constituents to set an example of proper conduct. Since his is the heaviest responsibility in this area, his offense is thus the greater and deserves what might at first appear to be disparate treatment.

How to Correct Inconsistency

Suppose that a company has actually applied its disciplinary policy inconsistently. It will, of course, find it increasingly difficult to sustain its actions in arbitration unless the violations are quite serious in nature. But, having gotten into such a dilemma, how does one get out of it?

The general rule is that the company must take positive steps to warn employees that it will be consistent in the particular area in the future; and it must then follow through on the promise. Unless the rules of discipline are specifically set forth in the contract and are a matter of mutual agreement between company and union, this course will usually correct the problem.

Normally there is no reason to negotiate such changes with the union if the contract merely states, without specifics, that discipline will be for just cause. This language, coupled with the arbitration procedure, is ample protection for the union; and the company is clearly entitled, under the normal management clause, to specify in particular areas what it considers and will hold to be just cause.

But suppose an employee is suspended for violating an inconsistently applied rule. While there will be a tendency on the part of the arbitrator to reduce or rescind such suspension, even this may well serve to apprise the employee that his future infractions will be more severely dealt with.

At one plant, a crane operator was suspended for two days for failure to follow the signals of his hooker—a violation of safety rules. As a result, he overturned a large container of molten metal on the shop floor. At the hearing, the supervisor admitted that until this accident, he had applied the safety rule in only about half the cases that came to his attention. In consequence, the suspension was reduced to a written warning, but it was allowed to contain the statement that future infractions would be more severely handled.

The Joust With the Janitress

In another case, a janitorial employee suddenly found herself unwilling or unable to complete her job assignments in the time allowed on her eight-hour shift. She had no real problem before; her laxity followed an extensive revamping of janitorial work procedures, which was generally disliked by the affected employees.

The company first warned the employee, then suspended her for three days, then for five days, and finally discharged her; all for the cumulative and continued offense of negligent work performance. Grievances were raised to protest all four actions, and were consolidated for hearing. The company's case was heavily documented by the testimony of several foremen who had tried to cope with the employee.

All four disciplinary actions were sustained by the arbitrator, in the face of a claim by the union that the company had never in its entire history disciplined an employee for poor work-performance. The arbitrator held that the company had not given up its disciplinary rights by mere inaction, and that its procedure of progressively stiffer penalties was more than enough to erase the effects, if any, of its past tolerance in this area.

Moreover, the company was positive in its actions in this situation. It knew what is wanted to do and issued positive instructions; it documented its case by having its foreman record each instance at the time, such that overwhelming and virtually irrefutable proof was brought to bear in support of its actions.

The Power of Positive Action

Clarity and positiveness are virtues that are as vital to the success of disciplinary measures as consistency.

Take the case of two employees who punched out early, knowing that the time clock was recording erroneously. The errant time clock was actually not the one they were supposed to use; it belonged to another section. Unfortunately for them, their own foreman came by just as they were leaving and said: "Hey, aren't you guys going to continue work?" Without replying, the pair disappeared into the locker room and on out of the plant as planned. A five-day suspension was given them for insubordination.

The arbitrator, in a rather literal ruling, reversed the company on the ground that the foreman had given the men no order but merely asked a question. Since he had not insisted on an answer, given no orders, and had not been positive in his supervisory role, there was nothing for the employees to be insubordinate to, and hence they had not disobeyed orders as alleged.

Another arbitrator upset a company's denial of holiday pay to an individual who left the plant two hours early on the day before a holiday. The contract required eight hours of work on the last scheduled day before the holiday before an employee would be given holiday pay, unless excused by the company. The testimony showed that the foreman knew the man had gone (because the latter had told him what he was going to do) but took no steps to prevent his leaving. Later, however, he called the man at home and told him he would get no holiday pay.

The arbitrator ruled that the foreman had effectively condoned or excused the absence by inaction;

that the belated phone call was no help; and that the man was, by the foreman's actions, contractually entitled to his holiday pay. The arbitrator stated that he would have upheld not only deprival of holiday pay but a suspension if the foreman had actually required the man to stay but had been refused.

Investigation and Screening

It was pointed out earlier that discipline is in the nature of a last resort—supervisors should not regard it as a panacea for all their personnel problems. When discipline does have to be dealt out, clarity, positiveness, consistency, objectivity—and, above all, thorough investigation to uncover and evaluate all the facts—are cornerstones of a fair and effective policy.

There is nothing whatever wrong with the practice followed in many plants of suspending an employee for an infraction that the supervisor discovers, pending complete investigation. There is also nothing wrong with the foreman collecting the full results of his investigations and checking his proposed action through with higher levels of management, and especially with industrial relations personnel.

Avoidance of hasty, impulsive decisions is highly desirable in all labor matters, but especially so in discipline cases, where emotions tend to run high. A cooling-off period will frequently generate more objectivity than many foremen can muster on the spot; it will also serve to get general agreement among management people as to final action, with lessened likelihood of later reversal of the foreman by his superiors.

Because personalities are obviously involved, grievances involving discipline are much more tautly contested than general contract-interpretation cases. Moreover, they are commonly much more difficult to handle by either side in an informal fashion. Evidence, especially the direct oral testimony of those involved, is usually demanded by both sides—even by parties who conduct other types of arbitration hearings through informal arguments. Cross-examinations can get heated and personal.

Hence it is extremely important that each case be investigated and screened before arbitration—not at the hearing table. There is little point in thinking that an experienced arbitrator can be talked into sustaining as unprovable case, or that it is necessarily better for him to reverse an ill-conceived company decision than for the company to reverse itself voluntarily.

And the same applies to the unions, who, perhaps because of the basically political nature of their organization, appear more susceptible to pressures in this field. Too often, neither side realizes that it is better in the long run to do one's own corrective job than to drag in an outside party to do it for you. Sole reliance on this outside type of crutch is inevitably debilitating; the type of precedent thus set up is all too likely to become an inflexible pattern—one in which the first- and second-step grievance meetings become mere formalities.

Of course, the foregoing is not meant to disparage honest, strongly-held differences of opinion. It is, however, intended to emphasize the necessity of full and objective investigation and screening of all cases, with the basic reasons for having a disciplinary policy fully in mind at all times.

Unvarnished Facts About Paint Dispute

Let's look at an example of what happens when these precepts are ignored; it involves a case of insubordination that began when a foreman ordered two employees to paint the floor. The employees protested; this work was not in their classification, and they couldn't be required to do it. The foreman insisted; other operations were done and the men were idle at the moment; the company was paying for their time, and was entitled to use it productively. Moreover, the painting was unskilled, undemanding work that anyone could do by merely slopping the paint on with a brush.

The dispute grew heated. The foreman said, "Do the job or you're fired!" He went to his office to write out the slips. The union committeeman was called in; the discharge slips were handed out to him; nothing could now reverse the action despite some desperate table-pounding by the steward. The ensuing grievance meetings took place in short order; they were rather brief, emotional and unproductive, with neither witnesses nor the two reluctant painters being called in.

But at the arbitration hearing, someone remarked that it was strange that the employees hadn't been in the foreman's office when they were being discharged. Belatedly, the emotions were put aside and the facts ascertained. It then became quite apparent that while the foreman and steward were discussing the firings at white-hot pitch, the employees had simply and quietly gone out and done the painting job they had been told to do. ∎

Meet the Author

John W. Whittlesey is labor relations attorney for Union Carbide Corp., in New York City. His previous two articles for *Chemical Engineering* were "The Role of Foremen in Labor Grievances and Arbitration" (July 22, 1963), and "The Foreman vs. the Union Organizer" (Jan. 6, 1964).

His 11 years at Union Carbide have involved him in all legal aspects of corporate labor relations, including arbitration, and were preceded by ten years as labor attorney for the U.S. Chamber of Commerce.

Automation–how to cope with its impact on men and equipment

by Dr. John Drake, Drake-Beam & Associates, Inc., Psychological Consultants, New York

Before management can prepare employees for the advent of automation, they must know what to expect themselves. As I see it, machine innovations are not expected, in the near future, to be very dramatic. In craft-type industries, as opposed to process industries, the revolution is in the development of controls that can steer, direct, and adjust the work of conventional machines, as opposed to the development of new machines.

Automation *might* mean fewer workers, but I doubt it. Many companies have found that automation often leads to a higher level of employment. For one thing, with automation, you often have more efficiency, hence, lower prices and a greater volume of work. So, even though people may "lose" their present jobs, they may be needed for other assignments; they won't be out of work.

For another, you may be able to turn out more work than you did before. If you are able to print more sheets of paper, you will probably need more sheets of paper, you will probably need more people in packaging, shipping, billing, maintenance, and warehousing.

Workforce composition

Once automation is introduced, the profile of your workforce will change. Here's what you might expect:

- **Fewer middle managers.** Production planners, inventory controllers, schedulers, etc., will be eliminated. Their knowledge, skill and decision-making can be programmed into the computer.
- **More professionals.** You will require more engineers—particularly those that are concerned with maintenance and design of auxiliary equipment.
- **Additional maintenance workers.** The cost of a breakdown of high speed automated equipment is too expensive for it to be down any period of time.
- **Fewer line workmen.** Individuals engaged in handling, packaging, and similar routine tasks that lend themselves readily to automation will be replaced first.
- **More professional managers.** As industry becomes more technologically sophisticated, competition will be keener and better management will be required. Professional managers will be characterized by advance degrees in business administration and will bring skill in motivation, planning, and organizing.

Nature of work to change

Significant changes in the nature of the work can be expected once automation is introduced. For instance, production workers will have different attitudes towards their work. In the automated situation, they will often feel more responsible for their work assignment.

Instead of concentrating on small, relatively meaningless tasks, the production worker will now control a vast array of machinery. Their role will change from providing the skills to accepting the responsibility.

Supervisors also gain in power with automation. Automated equipment is expensive, hence, for profitable operation it must be kept running. This leads to more direct consultation between foremen and management.

However, with this increase in power, the supervisor also works under increased pressure. Supervisors will be more concerned about the people-aspects of their jobs.

The emphasis will not be so much on expediting and getting the work out, because the machines will take care of this. But rather, on building team cohesiveness. It will be very important for the supervisor to keep the team together. Absenteeism and tardiness are prime problems in an automated situation.

Positive program to implement automation

The most common reactions to automation is that of

Engineering impact. *Maintenance personnel might be "switched off" when they see that new piece of automated equipment come in the door.*

The maintenance crew might have some doubts that they can maintain, service, and troubleshoot the new equipment. These men need assurance that their jobs aren't at stake. You can supply this assurance by alerting the maintenance staff of the new equipment.

Also, by training them, you update their skills to cope with the new demands. Others, who cannot measure up will have to be relegated to lower skilled maintenance jobs, or transfer them to other tasks.

Facilities engineers also have a big stake in this. Since the new equipment requires an ideal atmosphere, it should not be located in areas where radiant heat, sunlight or poor air conditioning exits. The machine just won't be capable of maintaining proper accuracy.

fear. Fear of unemployment or fear that the robots and machines might eventually tell us what to do.

And from fear three different reactions will occur. In one, the employees will fight. They will increase their grievances, complaints. Or they will utilize passive resistance, slowdown, utter strike threats.

If they don't fight, they might run. The employees will up-and-quit before you're ready to have them leave. One problem here is that those who leave might not be the ones that you want to leave.

The third reaction is rigidity. Under these circumstances, the employee will display less willingness to discuss things and become unreasonable.

However, the negative impact of automation can be reduced through seven basic principles:

1. Change is more effective when it is understood. Provide a careful explanation for the reason for the change, the objectives of the change and the mechanics of the change. These are essential. Post all persons involved well in advance of the change. Show them pictures of the new equipment and if possible, display the equipment in the plant. Also tell them how many people are likely to be laid off in each department.

2. Change is more effective when those affected have helped create it. Wherever feasible, automation changes should be discussed with the union, shop stewards and others concerned. Make the men part of the change.

3. Change is more effective when it is implemented after another change has been assimilated. Do not introduce automation right after completion of another major revision. Allow an adjustment period so that everyone has an opportunity to get things stabilized.

4. Change is more effective when it follows a series of successful changes. If you have made some changes recently and they have not worked out particularly well, avoid further change for awhile.

5. Change is more effective when those affected can see personal gain in the change. Management must see to it that they explain the benefits of the change to the employees. Often overlooked is the fact that there can be many positive attributes for the workers with the advent of automation. Some of these plus signs might be better distribution of the workload, interesting jobs, greater opportunity, more opportunity for training.

6. Change is more effective with new people. Therefore, make very careful provisions and reassure the old ones that privileges of seniority will be honored.

7. Change is more effective when there are appropriate degrees of trust and confidence between you and those affected by automation.

Case in point

A major U.S. company was mechanizing a packing line and the new equipment was going to replace 80% of the workforce—some 30 individuals (mostly women).

Eight months prior to installation, management and the union agent reviewed reasons for automating.

The reasons basically centered about the unit costs for packaging the item as opposed to their competitors' cost (who were using automation). The figures also cited probable business loss.

The company detailed a schedule which indicated how many people would be laid off at each point in time. Management included some proposals for seniority changes which would allow long-term personnel to be placed with other departments based upon seniority.

Three days later the business manager and the business agent met with all the shop stewards and reviewed the same information, then with the shift workers.

After eight months, the major automation was put on-stream with minimal conflict and disruption. ■

What's your management's rating? Try this mini-audit

A comprehensive audit program provides factual and realistic data about management strengths and weaknesses. The management audit, part of the overall appraisal or performance evaluation programs, actually consists of a series of sub-audits on specific areas (quality, manufacturing, engineering, etc.).

In essence, each sub-audit contains questions on the specific area. The questions are basic and necessary for successful operation. An important by-product, gathered by noting the weaknesses, is a list of goals to meet in the following operating period.

One advocate of audits, A. O. Smith Corp., Milwaukee, has a management audit program which rates 10 functional areas in each of 12 operating divisions.

Audit teams consist of two corporate staff men and one or two operating managers from the plants. Each team member, competent in his specialty, can make meaningful judgments and evaluations for the audit.

The team visits and tours each plant, interviews personnel, examines documents and records, and makes preliminary ratings. Team members discuss their findings with management before their final report.

Urban T. Kuechle, A. O. Smith president, believes audits provide a quicker evaluation. He notes that companies often ignore qualitative factors in favor of traditional numerical values (profit and loss, and budgetary controls) to evaluate management performance. By relying on numbers, it's usually two or three years too late when you realize you're in trouble.

Mini-audit gets you on your way

As an example, A. O. Smith and MODERN MANUFACTURING developed a manufacturing/engineering mini-audit. It acquaints you with the audit content and guides you in developing an audit program. The mini-audit also gives you a general readout of your operations. The questions are samples only. You should expand the list of questions, paying particular attention to meeting your own specific operating conditions.

Additionally, each question has a maximum point value assigned to it. While these values are examples, they can establish the importance of each question relative to the other questions.

Also, the point value is a means of ranking the operations. When the interviewer is auditing, he will determine the percentage of operation proficiency for each question and establish a percentage of point value to be scored. Hence, the rating score reflects the effectiveness in applying a certain procedure.

To see how your operations rate in general, why not run through the mini-audit? Then compare your rating against the ratings below to see where you stand.

(Turn page for audit questions)

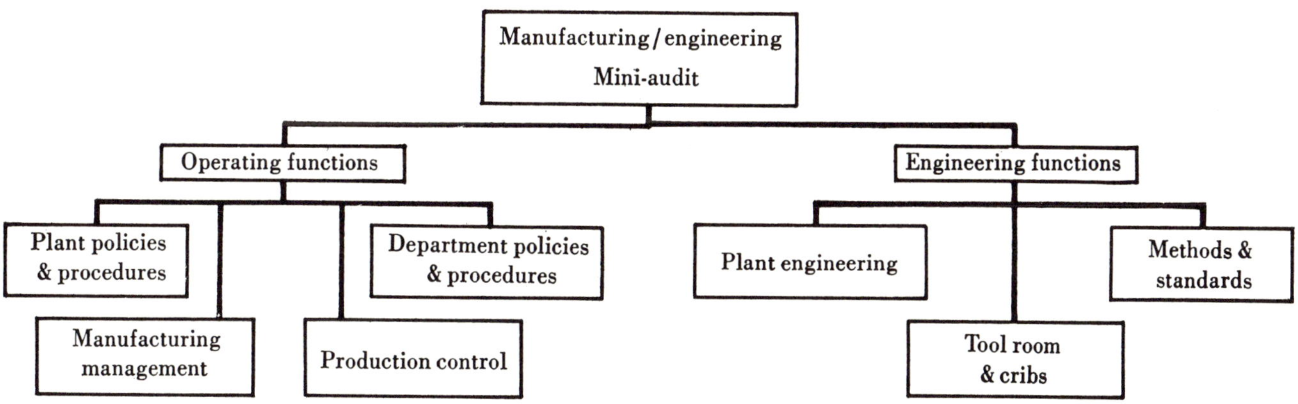

Operations rating review

Point total	Rating
78-82	Excellent (superior performance)
70-77	Good (adequate performance)
62-69	Fair (just getting by)
Below 62	Poor (not meeting minimums)

Engineering rating review

Point total	Rating
67-70	Excellent (superior performance)
59-66	Good (adequate performance)
52-58	Fair (just getting by)
Below 52	Poor (not meeting minimums)

Operations functions' *mini-audit covers four areas. Space is also provided for comments and point evaluation.*

Plant level policies and procedures	Max. rating	Your rating
1. Do you have an organization chart that fills plant requirements? Is it reviewed and updated?	4	
2. Do you have back-up men for key supervisors?	5	
3. What training programs do you have for hourly and exempt personnel?	4	
4. What procedure is used to evaluate and develop personnel? What is the frequency of this appraisal?	4	
5. Do you maintain a current age, condition, and machine utilization report?	4	
Departmental level policies and procedures		
1. Are there manning tables for various levels of production? How is manpower, workload established?	4	
2. What equipment records are kept (maintenance costs, utilization, preventive maintenance)?	4	
Manufacturing management (operations information)		
1. What plant and departmental budgets exist?	5	
2. What expense controls and reports are issued? Are they useful and relevant to controlling?	5	

Engineering functions' *mini-audit delves into three areas. Compare your score with the table on the opposite page.*

Plant engineering and maintenance	Max. rating	Your rating
1. How are maintenance costs controlled, reported?	4	
2. What is backlog of work in maintenance department? Is downtime reported for critical operations?	2	
3. What training programs exist for maintenance personnel?	4	
4. How is maintenance work planned and scheduled?	3	
5. What is the trend in maintenance cost as a function of direct to indirect labor hours, or cost per dollar of sales?	4	
6. Do you compare inside with outside costs for certain moving, building, or maintenance costs?	2	
Tool room and cribs		
1. What is tool withdrawal procedure?	3	
2. Do you have a record of standard tools for operator use, designers?	3	
3. What system is used for stocking, indexing tools?	3	
4. What training programs are in effect for crib attendants?	5	

Operations

	Max. rating	Your rating
3. What type system is used for data collection in labor reporting?	4	
4. What are output goals and standards for labor control?	4	
5. How do you determine indirect labor requirements at varying levels of production?	3	
6. Do you have a cost reduction program? How is the program administered?	4	
Production planning and control		
1. How are lead times and delivery determined? Are they realistic and published?	3	
2. What are capabilities of critical lines, machines?	5	
3. Is production control concerned with set up costs? What are costs as percent of direct costs?	4	
4. How are inventory levels determined? Are the inventory levels met?	4	
5. Is inventory balanced? What is ratio of active to inactive parts?	5	
6. Are storage areas well protected, orderly, clean? What is the percentage of inventory error?	4	
7. Does the shop produce only specified quantities and follow production control dates, schedules?	3	
Total manufacturing operations point score	82	

Engineering

	Max. rating	Your rating
5. What is tool inventory value and turnover rate?	3	
6. Are tooling components standardized? How is this program administered?	4	
Methods and standards		
1. What is workload and how is it scheduled? How are assignments made?	3	
2. What are the policies and procedures relating to the standards program (labor, tooling, methods)?	4	
3. What is the quality of routings and manufacturing instructions?	4	
4. What is the scope of the methods improvement program? How is it administered?	3	
5. How is the machine replacement program controlled and administered?	4	
6. What is the tool design and development program? How is the program administered and evaluated?	4	
7. Is there a plant layout? How was it developed?	3	
8. What training is provided for time study men and industrial engineers?	5	
Total manufacturing engineering point score	70	

Managing Engineering Projects

An experienced project manager explains how a new project is taken from the lab . . .

. . . and turned into a profitable operating plant

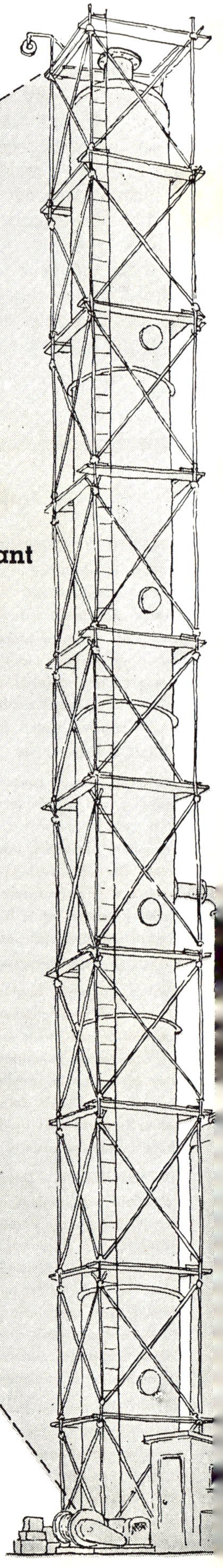

J. M. McLELLAN,
Procter & Gamble Co.

Construction projects are much like any growing business in the problems they encounter. But whereas most businesses grow and develop over a period of years, the life span of a \$500,000 to \$5,000,000 project is compressed into six months to two years.

During its brief life, one or more key people must make commitments, plan, supervise and finally be judged for success in achieving project goals. The problems, tools and techniques of project management are the subject of this article.

The project manager has a cradle-to-grave responsibility much like that of a task-force commander. He has an assigned objective, the nucleus of an organization, and usually a budget of dubious accuracy. It is his job to map out a plan of attack, secure the needed personnel and materiel, and see the job through to a successful conclusion. Once embarked, his control must be such that potential threats to the project are recognized sufficiently in advance to permit corrective action.

Defining the project manager's authority is difficult. Most often, he and his project engineers work through other departments of the company and through engineering and construction firms, so the organization chart may give little indication of the span of his authority. Actually, his authority is in proportion to project's importance to his company.

In a tight spot, he can borrow and use the authority of any one in the combined organizations to accomplish whatever is needed. If an individual's poor performance threatens the project, the manager need only point this out to the appropriate supervision to obtain improvement or a replacement. If delivery of key equipment is a major stumbling block, the remedy may lie in higher cost or perhaps in a personal friendship between someone in his organization and the vendor's president.

The project manager's authority in matters involving balance between scope, schedule and cost is more apparent. Anyone who has ever built a house is all too familiar with the insidious effects of a slight change here and a small addition there. Such unrecognized extras often eat up not only contingency allowances, but also furnishing and decorating allowances as well.

As with the task-force commander, credit for a successful undertaking is shared with all the participants. The onus of failure, however, is the project manager's alone, except in the case of faulty technical concept. Inevitably timing, control and balance are his prime concerns.

A project may involve widely varying technical specialties at the same plant and may take in a number of different plants. Examples for this article are drawn from the hypothetical Centerville expansion of the XYZ Chemical Co. This project includes a combination of actual situations, common to the industry, that help bring the project manager's role into focus.

The Centerville unit is to produce a product new to the XYZ Co.—one that is expected to have a revolutionary impact on its field. The product, synthesized from well-known chemicals, is aimed at the home-gardener market. Code-named "UL," it is to be in flake form and packed in bags.

Test-market results have been excellent, so the company is anxious to get into volume production. There is considerable anxiety over the possibility that established operators in the field might take the bloom off the market by rushing a hasty copy of UL into national sale.

The process, however, is far enough from the competitors' normal methods that they will probably need one to two years to catch up. Such a market lead would assure XYZ a rosy profit picture if the product lives up to predictions extrapolated from pilot operation. Truly profitable full-scale production, however, will require completion of R&D work on improved yield and byproduct disposition.

Studies of market locations, raw-material supplies, and labor costs have led to consideration of a new plant. However, the Centerville plant offers an attractive compromise between optimum location and the advantages of a going plant that is competently staffed and equipped with a full set of service facilities.

This Centerville expansion is to be built around the addition of a major synthesis unit, which includes a number of major liquids-processing systems, a control building, a tank farm, rail and truck unloading facilities, and a substation.

A separate unit will produce one of the ingredients, "ingredient B," in a separate area, using one of three available patented processes licensed to firms accustomed to engineering, building, and starting up the units on a turnkey basis. The increase in personnel leads to added offices, locker rooms, and eating facilities in the office area. A warehouse addition, although not originally included, proves justified and is added to the project scope during basic engineering.

Project Phases

FEASIBILITY DETERMINATION—This is the gestation period, in which the project grows from an idea (or a laboratory development) through practical considerations of market and profit potential, to a selection of optimum size and site. After a process has been selected, studies are made of various plant sizes and locations (taking into account labor, materiel, transportation and investment cost). When the optimum combination is clear, an investment go-ahead is given, based on a rough estimate.

BASIC ENGINEERING—This establishes the scope of the project in terms of capacity, detailed flow charts, plot plans, building and equipment layouts, estimates, and schedules. A nucleus of engineers, operating and research and development personnel, selects from among the basic alternatives. The engineering contractor is chosen and his key personnel are incorporated in the project team. Long-delivery equipment is bid and ordered, and decisions on contracting approach are made.

DESIGN AND PROCUREMENT—This comprises the conversion of basic engineering into completed drawings, models and specifications; procurement, expediting and inspection of equipment and major materials. The engineering complement is multiplied four or five times by the addition of designers, draftsmen, model makers, byers, etc. Equipment design is detailed in vendors' plants, and fabrication is performed by hundreds of people in dozens of shops across the country.

CONSTRUCTION—This includes letting contracts, construction, field procurement and field inspection. Construction begins long before engineering is complete, with early releases for site clearance, underground sewers and piping, foundations, etc. Field-force numbers reach a high plateau when the major effort in the fabricating shops comes to a close. Civil construction, equipment installation, piping, wiring and ductwork are completed in sequence. Inspection of equipment and construction begins well in advance of completion to minimize startup problems.

STARTUP—This involves training, trial operation, initial operation and trouble-shooting. Operator training begins during construction and is aimed at trial operation prior to startup. Engineering attention focuses on "getting the bugs out". After shakedown, a post-startup list is prepared to assure the orderly windup of all lose ends.

Project Start

The project manager is assigned early in the planning stage. His talent for organization may be applied to the coordination of the early studies and estimates, or he may merely work up the engineering and construction schedules consistent with each of the plans. His participation increases as the time of decision approaches until, when company management gives the project the green light, he has taken a leading part in defining the project commitments and is thoroughly familiar with its objectives.

The range of possible paces and the resulting range

Project phases and monthly manpower needs—Fig. 1

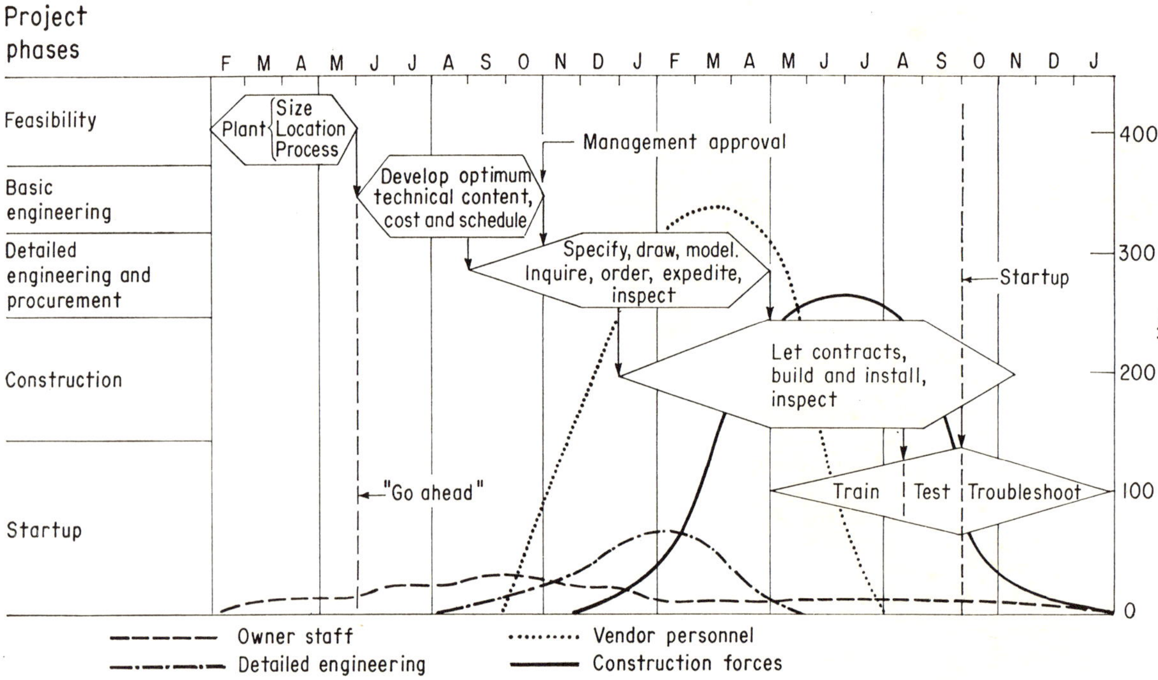

of costs have received varying recognition. In one company, the semantics of pace definition has been resolved by the establishment of "Normal, Rush and Blitz" as standard paces. (For definitions, see box.)

The Centerville basic schedule in Fig. 2 makes it clear that a "normal" pace is adequate for all but the synthesis unit. And even in the synthesis unit itself, if all goes well, most of the work can proceed at a normal pace. Only the synthesis-unit reaction and recovery systems stand out as possible trouble zones because of continuing R&D; so funds are included in the estimate to "rush" or even "blitz" these systems should the need arise.

Performance control and coordination begin as soon

Basic schedule for Centerville expansion—Fig. 2

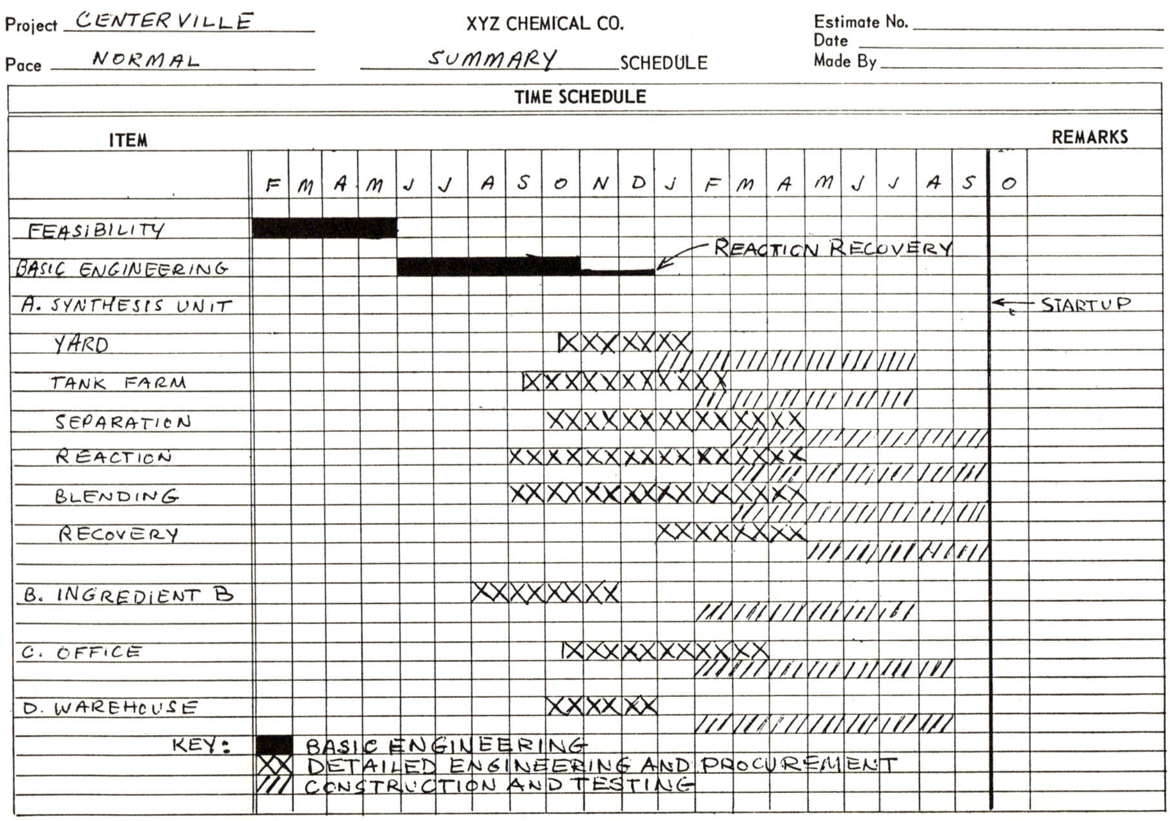

Project paces

NORMAL—Minimum capital costs consistent with minimum operating cost. Adequate time provided for optimum development, engineering, design, procurement and construction.

RUSH—Sacrifice of some capital cost (3% to 8%) to gain time. Compromises made principally in procurement and construction by spending limited extra amounts to speed items on critical path.

BLITZ—Sacrifices made wherever time can be gained. Development and engineering limited to the "sure quick way," and design studies for cost reduction dropped. Procurement from vendors who can deliver on a time rather than a cost basis. Construction department drops most competitive bidding and works overtime. Higher capital costs (10% to 20%) result, but project time may be cut drastically.

The range and acceptability of cost that are associated with the different paces varies as widely as do the engineering content and the objectives of the project themselves. A proposal to blitz a pilot project might be accepted even though it increased the costs 20%; the same proposal could be prohibitive for a plant aimed at the national market.

Project security techniques

Security techniques commonly applied are:

MISDIRECTION—Including code names for all revealing ingredients; purposeful misidentification of the process and physical characteristics with some other plausible product; wrong destinations for materials and equipment, which are then reshipped to project site.

AVOIDING THE "TRADE"—Dealing with design firms, fabricators and constructors not normally associated with the process or product, and who would be unlikely to have any contact with the competition. Privately designing any revealing equipment, for separate fabrication and assembly at owner's plant.

PHYSICAL CONTROLS—Burning wastebasket contents, using remote offices, sight barriers, badges, guards, etc.

as commitments have been made and accepted. Those interdepartmental commitments on which project commitments are based must be confirmed, defined and amplified.

Detailed schedules must be set up to control the early phases of the work. To this end, one or more project engineers—men who know the ropes and have the ability to get things done—are assigned to the project manager. Their basic responsibility is for detailed coordination of the kaleidoscopic activities of the project and for control of scope, schedule and cost. Later their efforts may be complemented by those of contractor project engineers who coordinate the work within their own firms and, after the project is further along, may even coordinate owner personnel.

The project plan adjusts scheduling and management to the types of contracts to be used later in the project. How the job is to be contracted governs how it will be engineered and the management controls to be used. Thus, for a lump-sum construction contract, engineering is brought to a sharp conclusion before the contract is let and, once the price has been established, cost control is limited to "extras."

With a cost-plus contract, engineering may greatly overlap construction, but extensive field planning and cost control may be needed.

Principal factors affecting the choice of a plan are cost, schedule and availability of personnel. Other considerations may include need for secrecy, the likelihood of continuing change due to overlapped R&D, and the state of the construction market. Often, the future state of the construction market is so obscure that flexibility to choose later among field contracts becomes a key feature.

A typical variety of project plans appear applicable in the four subprojects of the Centerville expansion.

The synthesis-unit project plan will dictate the startup date. The high-stakes objective justifies extensive overlap among R&D, engineering and construction. Personnel and know-how for much basic engineering are available in the XYZ Co. but outside help will be needed in some R&D areas and for detailed engineering. An engineering firm experienced in high-pressure-fluid chemical plant design is called for. Independent consultants and research institutes will also be phased in.

Local contractors are inexperienced in this type of construction, but a number of national concerns are well qualified. XYZ Co. has, on smaller jobs, coordinated a number of specialty contractors without using a general contractor. The size, complexity and pace of this job, however, point to a general contractor equipped with personnel and union contracts such that he can manage construction in whatever manner seems best at the time. Firm-price bidding of subcontracts may be attractive, so engineering is scheduled with this objective.

The ingredient-"B"-area project plan is aimed at turnkey contracting. Thorough definition of owner requirements prior to bidding and a thorough evaluation of bids are needed early in the project.

Thereafter, XYZ's contribution will be limited to inspection of engineering and construction, for conformity with specifications and for coordination with other areas of the plant.

The office-area project plan involves still other specialties and alternatives. Architects and engineering firms in the Centerville locale are anxious to participate in this phase of the job, and local construction contractors are accustomed to bidding competitively on their designs.

This portion of the project might be "spun off" for the plant itself to manage, using entirely local facilities. Alternatively, the synthesis-unit engineering firm could be given the design of the office area, and the construction could be managed by the synthesis-unit constructor. The latter arrangement is preferred by the engineering group because it eases the task of technical coordination and gives promise of economy both from a standpoint of fees and of combinations of utilities systems.

Company management also likes the second proposal. For an area that might be used for "empire building," it prefers to keep responsibility in the hands of the project manager. The work is given to the synthesis-unit engineering and construction contractors with instructions to invite lump-sum bidding by local contractors.

In the warehouse area, there are several engineering possibilities:

1. Continuing use of the same structural design.
2. Using one of a number of more modern designs.
3. Inviting contractors to bid on their own choice of structure on a turnkey design-and-construct basis.

A survey of current warehouse construction knocks out further consideration of 1, but points up the complexity of 2. Any of a variety of combinations of tilt-up, block or metal walls, with steel, timber or concrete structural systems seem to satisfy XYZ's needs.

The proponents of each combination claim cost and performance advantages over the others. XYZ is unwilling to invest the technical effort required for a thorough evaluation. It concludes that 3 will yield best results, with minimum effort. A Centerville engineering consultant is employed to propose a performance-type inquiry, for bidding on a design-and-construct basis.

Effect on Cost

Reams have been written and read by project managers about the effect on cost and schedule of a wide range of contracts in varying situations. Any number of compromise arrangements may be suited to individual situations. Discussion of all the possibilities is beyond the scope of this article. However, the basic characteristics of the extremes of the range, "lump sum" and "cost plus," are so different in their effects on a project that they deserve review.

Lump-sum contracts based on competitive bidding have a tremendous attraction. The cost is established early in the course of the work and the contractor shoulders all of the worries and risks. In the cost-plus contract, at the other extreme, the owner doesn't know what the cost is going to be until the job is done.

If there is inefficient management, labor slowdown, bad weather or bad luck, the owner pays. Often the best craftsmen are kept more or less permanently on the payrolls of the better local contractors to work on their lump-sum jobs so that the poorer producers gravitate to the cost-plus projects.

Under ideal conditions, with no changes, no collusion, plenty of time, and hungry bidders, the lump-sum contract would be the obvious choice.

As conditions move away from the ideal, however, the owner is more and more willing to accept the risk and the uncertainty in preference to paying the contractor. Neither form of contract offers an ironclad solution to the schedule problem. Even the "penalty clause" is little used because of the difficulty of proving responsibility for schedule failure. Generally, cost-plus gives the owner greater opportunity to try to control schedule than does lump-sum, but that is all.

In one case, a lump-sum electrical contractor on a process job was moving much too slowly. When checked he promised to soon bring a crew from another job in order to catch up. A month later—same pace, same promise. The contractor finally confessed his real problem was that his "good crew" was tied up on a power plant job that was behind schedule. He couldn't drop the power-plant job but neither could he afford to work a second-rate crew on this tightly estimated contract. The owner, faced with delayed startup, changed to a cost-plus contract with an oversized crew brought in "from the hall."

Cloak-and-Dagger Treatment

The need for secrecy can have a profound effect on the project plan, so security plans must be developed in the infancy of the project. The requirement may be as brief as the time needed to prepare an adequate publicity release or as long as the necessity to keep the entire project hidden from the competition.

The major sources of security leaks are from suppliers of ingredient materials, packaging materials and equipment, and from the casual conversation of employees. The various types of security measures are well known from their military applications. In private industry, however, there is less motivation of employees and suppliers toward support of security and penalties are practically impossible to apply.

The most common situation involves a reasonably high level of security in the early phases of the project after which the curtain is lifted with broad public release of project information. Following the public release, there may still be small areas subject to security, but the problems of communication with suppliers, governmental bodies, unions, contractors, the press, and the like are greatly eased.

Sources of Specialized Personnel

One source of personnel is the giant organization with the proved capacity to engineer, build and start up any type of plant anywhere. Fees for a group of specialists from such a company may run to 150% added to the actual salaries of the individuals. At the other end of the range is the manpower broker who furnishes lightly screened engineers and draftsmen to work under the owner's supervision for as little as 50% of salary.

The medium-sized owner, with a number of qualified people available for the project, may prefer to assemble a project team to his own taste from the great middle-ground of specialized personnel suppliers. He will draw from the groups below:

Basic Engineering and R&D—The "right" consultant may be found in one of the big engineering companies, in a research institute, in a university, or in private practice. Here, clear definition of the problem and of timing requirements are the most important factors from the standpoint of project management. If the consultant is a member of the firm that will do the detailed engineering, the transition into that phase is greatly eased.

Detailed Engineering—For process plants, complete engineering by one firm is most common. For service facilities (such as plant buildings), design may be by a group of firms including architectural, civil engineering and mechanical and electrical engineering firms. Process plants can be similarly split between engineering firms if arrangements are made to assure thorough technical coordination. Poor coordination, however, creates havoc.

Some engineering organizations lean heavily on equipment vendors for detailed engineering while others design in extreme detail. The effect on enginering cost can be profound, as can the effect on procurement and construction cost.

While performance specifications are usual for standard equipment such as pumps and compressors, they can also be used for piping systems, hangers, tanks, vacuum systems, conveying systems, and novel machinery. The problems and advantages of vendor design of special systems and equipment are discussed later.

Construction contractors will furnish all or any part of the field supervision required for the project. Consultants specializing in construction management also provide owner representation on a fee basis. Fees in either case, when converted to percent of salary, are comparable to fees for engineering personnel.

The choice between the one extreme of employing a full general-contractor organization and the other of rounding out an owner team with, perhaps, a rented accountant, purchasing agent and field engineer, is dictated by the available resources of the owner.

The Project Team

Putting together the project team is one of the trickiest steps in its management. It is the project manager's job to select the right organizations and individuals and weld them into an effective unit. Overemphasis on low overhead and elimination of fees can lead to cost and schedule disaster. One of the main reasons for use of established engineering and construction organizations is that it is their business to put effective teams into operation on short notice.

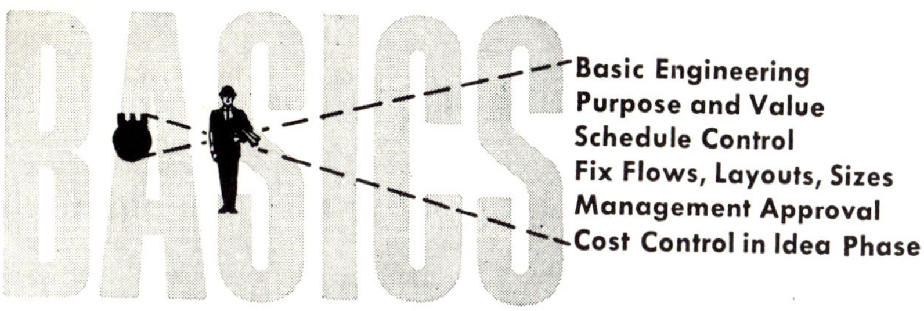

This can be the most interesting and profitable phase of the project. In it, brainstorming and important engineering developments can produce big breakthroughs, the major management decisions are made, and plans are laid for subsequent phases.

A novel or technically complex project may contain great opportunities for improvements over the initial concept if the time and talent are made available for their development. Often the greatest gains come from the byproducts of initial investigations. A study aimed at a more productive process may yield capital cost savings, or one aimed at loss control may reduce maintenance problems.

Regardless of development potential, thorough coordination of basic engineering pays for itself. The payoff comes from the savings produced in the detailed engineering and construction phases. Basic engineering allowed to drag behind schedule always provokes the familiar cry from engineering and construction contractors: "Your changes are running up my costs." The resulting confusion and cost may be the culmination of a well-thought-out calculated-risk plan and may be well justified. More commonly, they are simply the result of poor control.

The value of a good basic engineering phase defies factual analysis because the yardsticks change from project to project. A look at what bad basic engineering can do to a project is revealing.

Extensive and late changes can raise engineering costs 10 to 20%, which (if engineering is 10 to 15% of project cost) raises totals 1.0 to 3.0%. Engineering always tries to absorb such changes, but if it can't, "extras" in procurement and construction are inevitable.

These extras can amount to 5% of job cost, of which 2 to 3% is waste. Competent project engineering in this period, on the other hand, costs only 0.2 to 0.4%. This is a low price for safeguarding against a possible extra cost of 1.0 to 6.0%, and controlling schedules besides!

Schedule Control

Schedule control during basic engineering calls for intuition, teamwork, and a willingness to work down a number of different avenues simultaneously. Interlocking developments on the Centerville project illustrate the timing problem.

The initial plan calls for a new steam boiler to meet the heat demand of the synthesis unit. In addition, a new well may be required for make-up water. Since some spare capacity is available in the existing boilers, a smaller local heat source such as a flash boiler or a Dowtherm unit may suffice. A bigger investment in heat-exchange equipment can reduce demand, too. The right combination may eliminate the need for the well.

Part way through basic engineering, the reaction system is changed from batch to continuous to gain the better yields proved out in pilot-plant operation. Heat demands level out and the whole heat-balance structure changes.

Meanwhile the power-supply picture is being developed for a range of possible requirements.

How to control this scheduler's nightmare? At the start, development lists are made up, listing subjects, potential benefits and time required. The potential profitability of time spent in basic engineering is measured against the project objectives and a cut off date for developments is established. Priorities and manpower are assigned. Schedule commitments are made. The work begins. As progress is made, the project engineer interrelates developments, reevaluates profitability, forces arbitrary decisions in the less-profitable studies and, finally, combining force with diplomacy, brings the studies to a conclusion in time to make a unified presentation of the project to management.

Bar diagrams with plenty of room for notes and revisions are generally used for schedule control in this phase. Critical-path schedules may also be effective if there aren't too many changes. The important thing in this period, when new ideas are continually being generated, examined, and expanded or dropped, is to have meetings of the minds on timing commitments. The timing commitments are confirmed with schedules and lists. Personal coordination is essential, however, if the schedules are to be kept alive and useful.

Fix Flows, Layouts, Services, Sizes

The biggest bone of contention in basic engineering is the question of when it is complete. R&D sees no reason why, with the process "practically" settled, detailed engineering shouldn't start at once. Process engineers and specialists want time for equipment sizing and selection but see no reason why detailed engineering shouldn't get started on the few settled areas while the remainder of the project is whipped into shape. Supervision responsible for engineering cost and schedule says: "Why start nibbling at non-critical-path details when the bulk of the job is so far from ready?"

Controlled overlap between R&D, and basic and detailed engineering is part of the answer, to be sure. If costs are to be kept down, however, there must be a meeting of the minds on the level of detail required before detailed engineering can "take off."

The detail to be available at the end of Centerville basic engineering is spelled out as follows:

Flow charts show all equipment, controls, process pipelines and ducts, with flow rates and materials of construction. Steam, air and condensate return lines are omitted, but vital service flows such as cooling water are shown.

Equipment arrangement is shown to scale, in rough models or layouts accurate to less than a foot. Underground services, as well as main pipe and conduit banks and bridges, are shown, as are access platforms and stairs.

Service facilities such as control houses, offices, locker rooms, shops and boiler houses are laid out similarly to scale.

Basic engineering, no matter how well modeled or drawn, is useless as a foundation for detailed engineering unless it is firm. Real firmness is achieved by getting everyone involved, with a thorough understanding of the what, where, how and why of the whole as well of the component parts. Layout development is led by the principal engineers and coordinated by the project engineer, but active participation by the various specialists and plant personnel is also essential. As the preliminary models reach completion, everyone is invited to put in his two cents, from R&D to maintenance and constructors.

Management Approval

The final approval is secured in a full-dress review. The entire project group is assembled in one room, often for the first time, for presentation to company management. For the Centerville project, the walls are hung with flow charts, architectural renditions, building plans, and aerial photographs. Styrofoam models show the entire property in small scale and the synthesis unit in larger scale. Cost comparisons and outline schedules are on poster boards.

Formality of presentation should not be a stumbling block. Free-hand flow charts and sketches, chalk on blackboards and casual speech are generally the rule. Pictures and models can be worth thousands of words without being masterpieces.

Management knows that if changes are to be made this is the last chance to make them. From here on, the spending rate will accelerate as engineering and procurement spread out. Accordingly the review gets down to essentials. The project may be sent back to the drafting board for any number of reasons. Costs may need paring. Adaptability for a future formula change may be lacking. Ideas may be generated in the over-all review that are just too good to pass up. If, on the other hand, the project survives intact it will probably live out its normal life.

After management acceptance, the material from the management review is put into a "scope summation" booklet, which will be the project bible for the duration. The scope summation condenses the technical content of the project in written, flow-chart, layout, and model-photo form and includes scope-summation estimate and summary schedules. The booklet is the prime educational tool for the ever-widening circle of participants in the ensuing stages. More importantly, perhaps, it is always available as a definitive statement of the scope of the project.

Cost Control in the Idea Phase

Cost control in the idea phase of the project must be both flexible and imaginative. Control in the accounting sense is of little importance since relatively little money is being spent. The decisions and plans being made, however, will trigger big expenditures.

Control in this stage calls for recognition of what proposals will do to total cost. A proposal to change one area may interact surprisingly with other areas. If a schedule is altered, the whole contracting plan may be affected. Worse, with engineers and managers the optimists they are, cost increases may be ignored in the expectation of compensating decreases.

Two tools help to keep costs in focus:

Scope-Change Authorization—This is a very brief document, describes the change in project scope and, in the baldest terms, its effect on cost and schedule. Its apparent simplicity belies its value. It demands sharpness of focus despite the fast pace of the project. This focus eliminates misunderstandings and sometimes eliminates the changes themselves. Approval authorizes a change in the basic commitments, which are commitments that it is project management's duty to protect.

Capital-Cost-Estimate Reconciliation — This is a "whodunit" document that gets the facts out into the open when costs have gone awry. If they are unpleasant, it is all the more important to establish their causes. Spotting the causes of cost trouble gives management the opportunity to manage. Increases may be entirely acceptable when the values gained are understood. On the other hand, misunderstanding may be bared that has led to the inclusion of unneeded, costly items. In any case, with the facts in the open, management has the basis for decisions.

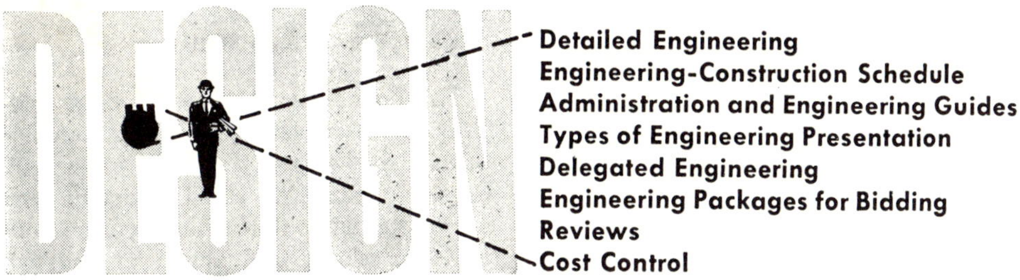

The detailed engineering and procurement phase is comparable to the logistics stage of a military task-force operation. In this phase, scores to hundreds of people select, design, buy and fabricate the thousands of items to be used in the construction phase. Detailed planning, good communications and sensitive progress reporting are musts if the project is to stay on course.

Up to this point, this article has been concerned with the relatively broad aspects of project management. From here on the discussion, like the project manager, concentrates on details. The reader, however, like the neophyte in a project, may wonder at the seeming disparity of the subject matter. Actually each subject must be settled in its proper time for any project if the work is to flow smoothly. Then, with timing and procedures firm, management need be concerned only with the crises inevitable to fast-paced one-of-a-kind operations.

Engineering-Construction Schedule

The engineering-construction schedule, prepared early in the detailed engineering phase, integrates all of the subsequent activities of the project. Until basic engineering is complete, lack of fixed scope makes detailed scheduling impractical. Thereafter the schedule determines the timing of all activities.

Several types of scheduling are in common use—bar charts, critical-path schedules, or combinations of the two. Bar-chart scheduling is a familiar technique. Critical pathing on the other hand has both advantages and limitations when applied in this field. Some of its characteristic differences from bar charting are worthy of comment:

The apparent complexity of the system may be its biggest stumbling block. However it need be no more complex than a bar chart at the same stage of project development. Furthermore its chief advantage lies in its clear statement of interdependencies as opposed to the hazier implications of the bar system.

For the big job, a computer and specially trained personnel may be justified. For the small to medium project, however, a compromise technique may be used

Activities in detailed engineering and purchasing

In moving from the general to the specific, the following activities must be developed more or less simultaneously:

DETAILED ENGINEERING

Scheduling	Bidding packages
Guides for engineering and administration	Periodic reviews
Design-presentation methods	Cost control
Delegated engineering	

PURCHASING

Bidding, discounts, mass buying	Expediting
Qualifying of fabricators	Detailed follow-up
Flexible price purchasing	

showing activities on a time scale without losing the dependency feature. Such a system is outlined in the CHEMICAL ENGINEERING article of Dec. 10, 1962, on "Critical Path Scheduling Simplified." Similar adaptations have been used successfully by other companies on entire projects.

Revision is more the rule than the exception in project scheduling, so flexibility is an important characteristic. The bar chart, because it leaves so much unsaid, is easier to revise. This ease, however, can be a trap for the unwary scheduler who may overlook an important dependency. The critical-path schedule, on the other hand, does not have to remain a thing of beauty to be useful. It can stand a great deal of freehand revision before redrawing becomes necessary.

The writer's point of view is that though the critical path approach is more demanding it is well worth the effort. The usual comment from the project engineer using the system for the first time is, "It certainly makes you learn the job." Once scheduled, the project engineer's grip on the job and timing is unshakable.

Reappraisal and compromise are inevitable in making up the engineering schedule for a tightly overlapped project. Early underground construction to beat bad weather may necessitate issuance of civil drawings uncomfortably early. Late R&D work on a byproduct-recovery system may dictate blitz tactics in a limited area. Early ordering of equipment may be needed, not so much for delivery but to get firm dimensions for civil and piping layouts. Every problem must be resolved before the engineering-construction schedule can be called firm.

When complete, the engineering-construction schedule is fanned out into subschedules, often in list rather than chart form. Typical subschedules, which may assign responsibility as well as timing, include:

Subject	Content
Detailed engineering...	All drawings, models, and their specifications.
Procurement..........	All equipment specifications, inquiries, orders, vendor prints, deliveries.
Construction.........	By areas, interrelating crafts, deliveries and contracts. (Usually detailed in latter stages of engineering.)
Inspection............	Pre-startup inspection, systems testing, practice runs.
Estimates............	By types: for field-cost control, bid check, final funding.
Cost reports..........	Engineering, purchases, construction.

Types of Design Presentation

The choice of design presentation (specification, models, drawings) can have far-reaching effects on the success of the project. The choice is made with the following objectives: engineering control, easy owner visualization and clear bidding details for builder comprehension.

Selection of the presentation method for most parts of the project is fairly routine. The choice between models and orthographic drawings, however, is one in which habit, prejudice, and unfamiliarity, play their parts. Lack of competent model builders can be a deciding factor. Nevertheless good modeling means fewer mistakes—fewer by management, plant, engineers and builders, because they can visualize better,

Administration and engineering guides

Getting acquainted is the first task of the new team of engineering firm and owner. The process can be speeded by an outline of the administrative and technical practices to be followed. This outline, in memorandum or booklet form, covers:

PROJECT ORGANIZATION —Organization charts and addresses.

ENGINEERING ADMINISTRATION—Who originates, approves, issues specifications, drawings, models, etc., in each specialized engineering field. (Note that the time taken for these steps may significantly affect project schedule.)

PURCHASE AND CONSTRUCTION AUTHORIZATION— Who originates, approves, and authorizes bid comparisons, vendor and contractor selection. Who authorizes field expenditures.

ENGINEERING STANDARDS—Owner, government, industry, and technical institute standards to be used in the project.

ENGINEERING GUIDES—Rules, examples, or payout formulas to illuminate the hazy areas in which the owner's taste of investment policy is the controlling factor. Gray zones such as the following can substantially affect owner satisfaction as well as cost:

Maintenance ease	Comfort and safety
Cleanout ease	Heat loss
Appearance	Downtime elimination
Noise, dust, fume control	Level of automation

Design presentation methods

The common methods of design presentation for process plants are (roughly in order of engineering cost):

DESIGN PRESENTATION	APPLICATION
Written specifications.	Catalog equipment. Novel equipment fully engineered by vendor. Complete manufacturing units fully engineered by vendor.
Rough drawings or sketches with specifications.	Equipment tailored to the project with detailed design by vendor.
Schematic drawings with specifications.	Flow charts, wiring diagrams. Usually used as a step toward presentation to scale.
Photodrawings or marked photos with specifications.	In cluttered existing areas. For equipment location, pipe and conduit routing.
Scale models with specifications.	In new areas where complexity makes two-dimensional presentation and visualization difficult.
Orthographic drawings.	For same uses as above if modellers not available. For single-plane layout. For equipment too complex for sketching or to be fabricated by shops without engineering departments. For civil engineering. For fabricating details of piping, ducts, chutes, etc., taken in field from models, photodrawings and installed equipment.
Isometric drawings with specifications.	Replaces any of the above in special cases.

and fewer due to dimensional conflicts because the model provides an automatic check against interferences. With costs of the two methods near a standoff, the popularity of modeling has grown simply because it produces better all-around projects.

Delegated Engineering

Engineering is delegated to vendors and constructors to tap added sources of specialized talent. The practice often, but not always, saves time and money. A tank builder, for instance, can detail his tanks for fabrication cheaper and better than the engineering firm. On the other hand, if the design of an extensive fume-removal system is delegated to the supplier, the over-all cost will probably be higher despite lowered engineering costs—because bidding will be limited to the high-overhead operators that are equipped with engineering departments.

Delegated engineering is inevitably accompanied by added problems in dimensional coordination. The engineering of the fume system, for example, will require careful checking against interference with piping, access ways and conduit banks that are being designed simultaneously by the engineering firm. If routing of conduit, on the other hand, is delegated to the contractor, there may be no opportunity to check details before construction is far along.

Actually, delegated engineering is normal practice in many specialties. Its possible drawbacks merit special attention, however, if the owner or the engineering firm is venturing into new fields or dealing with unfamiliar suppliers.

Engineering Packages for Bidding

The groupings of engineering best suited to mass purchasing and contract bidding vary from project to project. Often an early review reveals the need for changes from normal engineering sequences in order to secure effective bidding free from subsequent extras. Sometimes the early review points up the impracticality of lump-sum bidding, so planning is based on cost-plus contracting.

In the Centerville project, group equipment buying is obvious for packages involving tanks, vessels, pumps, exchangers, controls, electrical equipment and structural steel. The field engineers should be given the opportunity to secure bids on civil work, underground piping and plumbing, insulation, wiring and sheet metal. Piping is eliminated from the lump-sum list because of the probability of changes.

New interdependencies are imposed on the engineering schedule by the need to complete all engineering of a given type at the same time. Fast work in the synthesis area may be required, for example, to issue plans for bids for underground piping and sewers as soon as is necessary to meet the office-area schedule. It may be possible, however, to make a preliminary issue for bidding, which is correct within a few feet, with final dimensions following in ample time for economical construction.

Early review of bidding packages makes it possible to weigh the disadvantages in engineering against the potential advantages in the field. Lack of such review produces the familiar comment, "If we could have built these two together, we could really have saved a pile of money!"

Reviews by Plant, Constructors, Management

Theoretically the engineering firm should be able to detail the project on its own, once it has received the scope summation, the design instructions and the engineering guides. Actually, continuing consultation is essential if the work is to go smoothly.

As detailed engineering spreads out, the number of decisions multiplies. The designers will inevitably make some selections that will clash with the taste or the pocketbook of the owner or with the facilities of the constructor. To keep the key parties in harmony, periodic reviews are set up for plant personnel, constructors and management, as subjects of interest to each come into focus. Differences in viewpoint are apparent at such reviews:

The engineers' prime concern is that the units work well. They also want the plant-operating people to like their handiwork.

The plant personnel assume that the unit will work well. They are more concerned with less tangible subjects such as access, process flexibility, ease of maintenance, appearance, and convenience.

The constructor looks for the details of design and specification that will make trouble in the field. Less tangible, but equally important, is the improvement in field planning that results from the constructor's early understanding of the job.

Management's interest in this phase may focus on any subject, from the effect of appearance on community relations to the adaptability of automatic controls to an upstream formula change.

Project management concentrates upon staying on course and on schedule. The ideas enthusiastically generated in these reviews must be fitted into the budget and the schedule to be acceptable.

A final review, similar to the scope-summation review, may also be held. This gives all concerned an opportunity to speak before final release to the field.

Cost Control Geared to Activity

Cut-and-dried systems of cost control seem to be applied successfully in the construction of large dams, petrochemical plants and power plants. The possibilities of duplication are both greater and more frequent in such construction than in medium-size projects in the chemical industry. These demand more flexible cost-control systems to accommodate the wide variations encountered. Systems adequate for plant-managed small projects, on the other hand, bog down when multiplied ten- or a hundred-fold.

The project plan must provide a cost-control program sensitive enough to detect trouble as soon as it starts. The basic control is the scope-summation estimate. This estimate, though far from refined, is the first to reflect the detailed scope to be used for engineering. The pattern thereafter is one of refining each portion of the estimate as details become available, and of reporting and extrapolating from actual costs as they develop.

The three phases in which major spending occurs are: detailed engineering, procurement, and construction. Different cost-control systems are used for each phase.

The detailed engineering phase, which is least subject to accurate estimating and control, fortunately accounts for only 10 to 15% of the total cost. If engineering is done by the owner, normal practices will prevail. If an outsidê engineering firm is on the job, its normal cost-control system can in most instances be adapted to the project. Here the project manager will choose the line of least resistance as the line of least confusion.

Equipment purchases, which total 25 to 40% of the job cost, are readily correlated against the estimate. Although the number of items is large, the accounting is simple and the reporting sensitive.

Since field costs amount to 50 to 60% of the job cost, the system for their control requires early planning and continuing attention. Refined estimate packages are planned early in detailed engineering to provide yardsticks that will accompany bidding packages for lump-sum contracts and to aid in managing cost-plus contracts. The use of these yardsticks is discussed further under "Construction."

Monthly reports of engineering, purchasing and field costs are begun early for the obvious purpose of applying the yardsticks. A less appreciated reason for the early start is the invariable need to smoke out bugs, mistakes and misunderstandings. By the time these have been eliminated, the margin for error has disappeared and accurate reporting is essential.

Effective buying, although hard to measure, probably saves 5 to 10% of material and equipment cost (for the average project). The project manager, knowing that planned buying costs little more than unorganized buying, sets up the project accordingly. He can count on the buyer to take the lead in developing good bidder lists, checking financial stability, and handling freight allowance, credits for early payment, invoice checking, progress payments and similar routines. It is up to him, however, to see that the project derives maximum possible benefit from the competitive forces at work in the construction-materials market.

Qualifying Vendors

The engineers must verify the technical qualifications of equipment suppliers just as the buyer must check their financial stability. The easy way is to include only nationally ranked firms on the bidder list. The hard but often profitable way is to include the small and medium-sized firms that can demonstrate the necessary ability. While short on overhead, these firms are often long on specialized talent and low in cost.

Placing a critical-path order with such a vendor ranks as a calculated risk calling for the judgment of the project manager.

Sometimes, firm-pricing bidding isn't feasible because of blitz scheduling, the certainty of changes that would make bids meaningless, or a design too complex for tight bidding. In such cases, flexible pricing is needed. Hourly rates for shop equipment time, for differing labor classifications, or for shop time as a whole may be used.

None of these systems provide really good cost control since there is no yardstick for judging proposals or performance. Usually the shops under consideration have machinery that is different in capacity and accuracy, different accounting practices and different rate structures. Costs often exceed the estimate—however, no better approach is available.

Expediting

A common view is that once an order has been placed it is the vendor's problem to make good his delivery promise. But if delivery is at all critical it is the purchaser who must worry about all the contingencies not covered by the vendor's promise. Slow approval of vendor drawings, changes by the customer, late deliveries of parts, gas pockets in castings—all are good excuses for late delivery from the vendor's point of view. It is the expediter who guards against such contingencies.

Detailed Follow-Up

Detailed follow-up is an essential, though routine, part of good procurement practice. The hundreds of inquiries, bids, orders, vendor drawings, approval letters, expediting reports, and inspection reports must move smoothly if confusion in the engineering depart-

Buying market categorized

The buying market for chemical engineering projects falls into three rather loose categories, each subject to differing buying techniques:

MASS-PRODUCED MATERIALS AND EQUIPMENT—This category includes standard cataloged and price-listed items such as piping, electrical and building materials; motors, common instruments and controls; and other standardized equipment. Pricing of such items can be negotiated, based on approximate quantities, without detailed take-offs. Discounting practices vary widely. But jobbers, wholesalers and contractors generally get, and can pass on to the customer, discounts related to their volume of business.

EQUIPMENT SUBSTANTIALLY ENGINEERED BY VENDOR—Prices of equipment bought on performance specifications are determined primarily by technical content and secondarily by manufacturing facilities and pricing policies. Heat exchangers, vacuum and refrigeration systems, cooling towers, centrifuges, filters and the less-standardized applications of pumps, fans, agitators and controls fall into this category. Inquiries for such equipment should be grouped to invite volume pricing, but the costs of individual units should also be spelled out so that the best fit of quotation to specification can be recognized.

EQUIPMENT LARGELY ENGINEERED BY THE PURCHASER—This category includes tanks, vessels, structural steel, and special machinery in which the purchaser does the engineering. Here, prices are dependent on shop facilities and the vagaries of feast-or-famine pricing policy, and vendor estimating. In this field, the specialized low-overhead little fellows as well as the big operators should be given a chance, by inviting bidding on packages within their respective capacities. Thus, for the Centerville vessels, the big operators might bid on the entire lot while smaller operators bid on only shop-fabricated atmospheric tanks or pressure vessels.

ment and in the field are to be avoided. Absence of the inevitable lost or strayed papers must be detected before serious harm is done.

The project clerk, reporting to the project engineer, uses many-columned lists to keep track of every procurement step for each item to be purchased. He compares progress with target dates from the engineering-construction schedule and warns the project engineer about items heading for trouble in time for him to take corrective measures.

This section is not concerned with the right way to build a plant, but rather with what the project manager does to assure that the plant is built right. His part in the construction phase is divided into two distinct steps: first, setting up effective field management; second, making sure that it works.

If the first step is successful, the second becomes largely a matter of application by the project engineer and resident engineer who handle week-to-week developments.

There is no all-purpose management system suited to the conditions of every project. In the Centerville project, for example, different management techniques were required for the general contractor cost-plus synthesis unit, the turnkey lump-sum ingredient B and warehouse areas, and the office area ideally suited to lump-sum subcontracting.

For any project, there are certain points that must be pinned down to answer the question, "How will the job be managed, and who will do it?" The project manager, having established the competence of the general or specialty contractors under consideration, will go into the following points before awarding the principal contracts, and again in the formative stages of the field organization:

- Contract Management.
- Field-work planning and cost control.
- Schedule control.
- Project cost control and reporting.

Contract Management

The biggest decision is whether to have a general contractor at all. The fee paid to the contractor is not the major consideration. Rather the question is which company or group of companies can get the job done satisfactorily on time at the lowest cost. If the job can be effectively broken down into lump-sum specialty contracts, if the competitive bidding situation is good, and if the owner is staffed to coordinate the specialty contractors competently, then the general contractor is not needed.

The project manager, of course, likes the competitive aspect and the certainty of the fixed-price contract. Since lump-sum negotiating is a lot of work for the buyer, as well as for the bidder, he wants it used only where the scope is firm and bidding looks profitable. A contracting plan is set up early, spelling out the types of contracts to be used in each specialty, and the timing of negotiations.

Lump-sum contracts begun in an atmosphere of brotherly love can deteriorate into pitched battles with profits, reputations, individual's jobs, and even bankruptcy at stake. Regardless of the fiscal conclusions, such fiascos can only harm a project. Real or claimed misunderstandings can be minimized using the following steps:

Technical content of the contract should be defined with care, giving particular attention to fringe areas.

Bidder lists, while broad enough to discourage collusion, should exclude contractors of doubtful qualifications.

Bid-check estimates should be prepared from the bid drawings so that if bids are all out of line they can all be rejected.

Pre-contract conferences should be held with the low bidder or bidders to assure a real meeting of the minds. Detail of scope, site conditions, and schedule should be reviewed. Procedures for authorizing and pricing extras should be settled.

The contract should be let with each party retaining a set of documents and drawings signed by both parties.

As the work progresses, field supervision should be organized to stay on top of:

- Coordination with other contractors.
- Quality inspection.
- Authorization of extras.

Cost-plus contracts, in which the purchaser carries all the risk, call for different handling. The following considerations apply in the selection of both the prime contractor and his subcontractors.

The individuals to be provided, and their qualifications, are a key point in the negotiations. The good field organization is built on good "followings." The good job superintendent can attract a following of good craft superintendents and field staff. The good professional foreman can bring with him a following of productive craftsmen. These people make up effective construction teams. A pickup team, while it may be effective, has the potential drawbacks of any such team, compounded by union and contractual conflicts of interest.

Overhead and fee charges are not as simple as they might appear. Definition is required as to whether items such as small tools, rental equipment, accountants, receiving clerks, office facilities and offsite personnel are in overhead or in base cost.

Buying connections, discounts and warehousing facilities may differ between contractors.

Planning, scheduling and cost-control systems for the subcontractor are fitted into the over-all job pattern.

Guaranteed maximum and unit price contracts are variations that must be handled much like lump-sum contracts if their competitive value is to be retained.

Field-Work Planning and Cost Control

Money that is saved through efficient work on cost-plus contracts reduces owner costs. On lump-sum contracts, on the other hand, it is the contractor who profits from field savings. The owner's interest in field efficiency consequently focuses on the cost-plus contractors.

Systems of work planning and cost control are as varied as the contractors and the work they do. Some owners have developed systems applicable to their own construction, but most look to the contractors for this essential ingredient of good management.

If his own company has a fully developed system, the project manager needs only assure himself that the contractors can and will use it effectively. If he is not so fortunate, he is on his own in judging between claims of those whose supervisors carry their controls in their heads and those who go to the other extreme.

An example of what can happen with poor field controls occurred on a piping job on the top floor of a four-story building. Activity checks indicated that everyone was busy all the time but nevertheless costs were skyrocketing. During a project engineer's visit, he and the resident engineer decided to spot-check a busy pipefitter.

The fitter picked up a length of pipe at the east end of the fourth floor, walked to the east stairwell, down three flights to the ground floor, out the door, around the roadway four hundred feet to the other end of the building, up the west stairwell to the fourth floor, and over to a pile of pipes.

Here, an easy hundred-foot walk from where he had started, he put down the pipe. There were some changes made, of course, but the job was too near completion and too far out of control for a full recovery. It ended up badly over its budget, and late in starting up.

Schedule Control

If field management is to control the job, it must develop its own schedule, taking into account the realities of weather, design, deliveries, contracting plans, labor quality and availability, and so on. The engineering-construction schedule seldom reflects these realities, and never is in the detail needed for real field control. Good scheduling, like good planning, is necessary for the elimination of crises and for the low costs that go with balanced, well-directed crews and coordinated use of construction equipment.

If detailed planning is used, critical-path scheduling with computer support is ideal. If less-sophisticated controls are applied, simplified critical-path or bar-chart scheduling is adequate. Regardless of technique, detailed scheduling and progress reporting are essential.

An important dividend of good field-schedule control from a project management standpoint is that schedule jeopardy can be cured relatively cheaply if recognized early. Crews can be expanded or even worked overtime for brief periods without rousing thirst for overtime in the entire workforce. Off-site fabrication may also help.

Late in the job, an attempt to save a schedule leads to exorbitant overtime costs and may even defeat its own purpose. The crews, simultaneously sensing panic and the end of the job, may slow down enough to nullify the added hours worked.

For cost-plus contracts, the schedule can be manipulated with relative ease. No contractual conflicts are involved if the owner makes minor changes in job conditions. For lump-sum contracts, however, the time to effect schedule control is before the contract is let.

If the lump-sum bidder is uncertain of his ability to meet schedule requirements, he should not be awarded the contract. Once awarded, field management must continue to check his progress. If he fails to reach major check points on time, it may be necessary, as in the case of the wiring contractor mentioned earlier, to face the facts with him early enough to save the project.

Another case of a lump-sum contractor behind schedule developed when his piping superintendent ran afoul of the local union powers. A slowdown ensued but the superintendent kept the fact from the home office. Finally, job management, noting failure to meet schedule check points, called in the owner of the piping firm. He promptly replaced his superintendent, made peace with the union, and brought the job in on time (apparently profitably).

Project-Cost Control and Reporting

Project-cost control extends back to the home office, whereas field-cost control is for the use of field management in controlling field performance.

Project-cost control is concerned with subjects that can be expected to significantly affect the success of the project in meeting its commitments.

Periodic reporting (usually monthly) of field-controlled costs is basic to project-cost control. A simple project, such as Centerville's ingredient B unit, requires only simple routines to compare actual costs to the estimate and to extrapolate to a predicted cost.

In a complex and changing project such as the synthesis unit, cost reporting can grow from an exasperation to an accounting nightmare. If the job runs into cost troubles, a poorly conceived cost system can become so time-consuming that field management becomes office-bound just when it is needed outside.

A few rules are basic to good project cost reporting:

Accounting should conform to logical blocks of work consistent with field-management plans.

Estimates made before field plans had been jelled should be regrouped to match field coding.

Changes in job content should be accompanied by estimate changes to keep the estimate applicable to the work.

Reporting should be started early so that by the time the job is in full swing the system will be in good working order.

"Spot" cost reporting also is a valuable tool. The project manager often requests spot reporting to judge performance in areas such as pipe-shop-fabrication, steel erection, or formwork. Contract bids compared to the estimate also offer valuable information. Such spot reporting is usually informal, current, sharply focused and more sensitive than routine monthly reporting.

Construction Labor Relations

Labor relations are a vital consideration in letting field contracts. Jurisdictional squabbles, slowdowns and strikes can hamstring an otherwise healthy project. The contractor's relations with local and national unions can be more important than any management techniques. In some areas, notably in the southeastern states, open-shop contractors can operate effectively on multimillion-dollar projects. In some other regions, however, craftsmen needed for chemical plant projects are so organized that contractors must have union agreements to man jobs adequately.

Union agreements can be local or International (meaning national). Most national general contractors have International agreements with the civil trades, many with the piping union, and some with the electrical union. These International agreements have two substantial advantages. First: work may continue through a strike on the basis that any gains from the strike settlement will be made up retroactively. Second: the contractor may employ labor directly without employing and paying the fee of a specialty subcontractor.

Nevertheless the International contract is no guarantee of labor harmony. The national contractor moving into a new city or area may hit it off well with the local unions. He may, on the other hand, see his job used as a battlefield by unions who pay little attention to their national headquarters.

In letting prime contracts for a multi-area project like the Centerville expansion, the project manager keeps construction labor relations in mind. One of his criteria in selecting prime contractors for the ingredient B and the warehouse turnkey jobs, as well as the synthesis and office work, is their ability to maintain labor harmony with the other contractors on the job.

Once the job is under way, the owner's representatives steer clear of contractor labor problems. Tests of strength leading to firings or walkouts may occur at any stage. The project manager, knowing the rough and ready nature of construction labor relations accepts the judgment of his contractors up to a point. If, however, labor relations and productivity go steadily downhill he may require changes in personnel, or even of contractors, to get the job back on the track.

The Centerville project was the scene of a common form of jurisdictional conflict. The ironworkers, whose local was in the city fifty miles east, were observed to be setting the handrail for the synthesis-unit stairs and platforms. The pipefitters, who came from a local in another city seventy-five miles west, knew that setting handrails made of pipe was their work. The fitter steward ordered his men off the job.

The contractor superintendent brought the stewards of the two crafts together to arrange a settlement. He found them both firmly entrenched behind contracts which because union practices differed in the two cities, were absolutely in conflict. With piping on the critical path, the superintendent took the only practical course. He suspended all work on handrails, pending a settlement, thus opening the door for the fitters to return to work, which they did reluctantly a day latter.

With the crisis averted, the superintendent called his home office, which in turn called the International headquarters of both unions. The International officers called their respective locals with instructions, but the locals refused to make peace. Again the work suffered, as the men took more and more time to discuss the merits of the case.

The home office persuaded officers from the Internationals to visit Centerville to settle the issue. At Centerville they visited the job, reviewed the drawings, and then repaired to a hotel room to debate the issue. The next day, the argument was over. The superintendent was informed that the fitters would install the handrails.

The resident engineer never did learn just what combination of logic and power determined the outcome. He did know, however, that the job was lucky to have had only a one-day walkout of a principal craft union rather than during the whole controversy.

Community Relations

New construction means welcome new jobs for the community but it may also mean problems calling for careful handling by the owner. Accepting the welcome is easy. Handling the problems, however, may demand extraordinary planning and foresight.

The project manager's part depends on the situation and its timing. Top management usually deals with the city fathers and sets the ground rules. The project manager sees that engineering and construction conform. Conformity with codes and regulations is a normal part of the engineering for the project, requiring early investigation and review with the pertinent authorities.

The project manager is particularly concerned with the time required for legal proceedings and permit approvals. These, though routine and uncontested, can still be so slow as to wreck a schedule. If relations are cordial, however, engineering and construction may be allowed to move ahead on the basis of conditional releases, with the owner accepting the risk of later disapproval.

Plant Relations

"You can't live with 'em and you can't live without 'em" also describes a common feeling of the resident engineer regarding plant personnel in the latter stages of in-plant construction. Plant cooperation is essential to economic construction but construction interference is a headache to operating personnel. Further, as the new installation takes shape, the operating people frequently see new ways to improve the design.

The project manager, to ease the pressures on a resident engineer caught between a cooperative but demanding plant organization on the one hand and a tight schedule and budget on the other, meets with plant and field management early in the construction phase.

Channels are set up to route improvement ideas through the project engineer for technical and fiscal control. Agreement is reached on locations of field

Performance criteria

In judging capabilities and performance in this field, the project manager can apply the following criteria:

Lead-time between work planning and work performance should be sufficient to encourage orderly crew-balancing and procurement of materials.

Method and sequence should be developed in enough detail that supervision need not improvise the day the work starts and can be assured that equipment will be available.

Performance yardsticks and a performance reporting system should provide reactions promptly, so that bad performance can be corrected early in the game.

Cost of planning, measuring and reporting should be commensurate with benefits gained.

Controlling costs

The project manager and the project engineer participate in:

SCHEDULE VS. COST DECISIONS

Early or late threats to on-time completion may seem to call for overtime, or overmanned crews. If the price tag is too high, postponement of startup of a system, or of the job as a whole may be the right solution.

Premium expenditures to minimize interference with going production may appear justified at the plant level. If, however, sales requirements have slackened or if another plant can pick up the load, the extra cost may be avoided.

OUT-OF-LIMITS-COSTS, ACTUAL OR POTENTIAL

If total cost commitments are threatened, drastic action may be needed. A change in field supervision or even in contractors may have to be made if labor costs are out of control. Shortage of funds might lead to postponement of an area like the Centerville recovery system, which though valuable was not essential to the operation.

A segment of the project may be less profitable than the rest. If the cost of the Centerville warehouse, for instance, were to increase too much, rental warehousing would be more profitable. If so, the warehouse would be dropped regardless of whether the project as a whole was financially solvent.

Creeping inflation can be prevented by field recognition even though it has escaped the home office. Careful correlation of engineering and the estimate, before the work is done, can show up items for which no money has been provided.

FIELD-ORIGINATED CONTRACTS

Contracts on which field management takes bids may rank as major segments of cost. Project management normally takes part in all such major contracts.

JUDGMENT OF FIELD-MANAGEMENT EFFECTIVENESS

Everyone from the gate guard to the company president passes judgment on construction-management effectiveness. The direct responsibility, however, falls on the project manager and his project and resident engineers. This calls for continuing attention to the investment vs. value received from the combination of field overhead staff, field shops, construction equipment and field labor.

offices and shops, security and safety procedures, use of plant service facilities, sources of construction utilities, and liaison to minimize interference. Thereafter, operating and construction supervisors work together to spot and resolve conflicts before they become serious.

STARTUP

Successful startups are no more the product of chance than any other phase of the project. As in other phases, successful startups require early planning, with particular emphasis on personnel assignment and training. The techniques of planning, scheduling and controlling this phase of the job are as important as any already discussed. These have, however, been developed in such excellent detail in four articles by J. E. Troyan in CHEMICAL ENGINEERING 1960 and 1961 that any elaboration here would be redundant.

It is the project manager's function to assure that the time, money and people are available to carry out the programs outlined in Troyan's articles.

Post-Startup Management

After startup, the loose ends must be neatly tied down before the project can be closed. During the critical phase of initial startup, the men on the site have concentrated all their energies on getting the plant into production.

Anything not immediately essential to operation has been put aside for later consideration and action.

Consequently, not until about a month after startup is attention turned to the details of adjustment and polish necessary to the steady high efficiency production that is the aim of the project team.

Training of operating personnel is carried out by operating supervision, concurrent with construction. Plans are laid early to provide opportunities for familiarization with equipment and to permit practice operation without disrupting the project.

At this point, the construction force has shrunk to a few men in each craft doing finish-up work. The technical staff is getting ready to disperse, and the operating forces are in full control. Now, before the team breaks up, the tailing-out phase is organized.

Post-startup lists are made of all the ideas or problems that have come to light during initial operation and that, though not critical, should be taken care of before the job can really be called satisfactory and complete.

Many of these are easily corrected once recognized. Others require home-office engineering. Some suggested changes for improved appearance or operating convenience may cost more than they are worth.

All are listed and priced. The questionable items are settled by the project manager or by company management. The construction force finishes its few remaining items and breaks camp. Responsibility for items on the rest of the list is transferred to the plant and to the regular engineering force, and funds are set aside out of the project budget to cover them.

Now in the calm after the battle of startup, the success of the entire operation can be judged. If startup was on schedule, if costs were within the estimate, and if efficiencies are rising fast, the calculated risks, the pressures and the painstaking attention to detail have paid off. ■

Meet the Author

J. M. McLELLAN *has taken a leading part in developing project-management techniques for Procter & Gamble. Having an M.E. degree from Cornell, his background includes experience in plant operations and maintenance, personnel relations, and a teaching stint at Cornell.*

Since World War II, he has been project engineer or manager for construction totaling more than $100 million.

He is a licensed professional engineer in Ohio, and has an M.E. degree from Cornell. His hobby is sailing—he is Vice Commodore of his sailing club, and an active racer.

Managing an Engineering Project in Europe

As U.S. companies become increasingly involved in overseas plants, participation of American engineers in design and construction becomes greater. Here are some timely observations on the differences and similarities between U.S. and foreign procedures.

RICHARD L. LEWIS, Process Plants Corp.

The project manager building his first plant or facility in a foreign country is faced with a multitude of new and sometimes unexpected situations. These invariably add to the usual problems in engineering and economic decisions, communicating effectively, meeting schedules, and staying within a budget.

How well and rapidly the manager, staff and all others involved in the project become aware of and familiar with local customs, rules, laws and ways of doing business can have a major impact on the success of the completed project in terms of time, money, operability and profitability. This is particularly true where neither the individuals nor the company have had previous experience in the country or area.

Problems With Language

How often are the problems we encounter not only in business and engineering but in everyday life only problems of communication? The ambiguous specification or contract, the conference from which participants depart with diametrically opposite understanding of conclusions reached, the misunderstood phone call—all these are far more likely to occur when there are language differences and barriers. Thus, it is even more important than ever to write and speak clearly, confirm all important verbal agreements and instructions promptly in writing, and summarize and recapitulate all decisions before ending a meeting.

The use of English as the "official" language for the project presents no insurmountable problems. In fact, English is becoming the language of international commerce and engineering. It would be safe to say that most American companies engaged in engineering and construction in Europe use English for specifications, drawings, purchase orders, contracts and correspondence. The same may be true, though to a lesser extent, for European engineering contractors particularly if they are engaged in worldwide ventures. Also, any major chemical or industrial project in a European country will have international aspects, and English often becomes the common language.

It is sometimes embarrassing to the American to realize the extent to which written English is understood, and perhaps to a lesser degree spoken, by professional, technical and business people. Generally the ability to read and understand a foreign language surpasses the ability to write and converse in it. For example, an inspector may fully understand specifications and drawings in English but will not have sufficient command of the language to be able to write a report in it. Similarly, construction people have little difficulty following drawings and speci-

Overseas: A Different World

The observations in this report are based on the author's own experiences in Europe and the U.S. as the owner's Project Manager. Every project is a unique undertaking, and no two projects are ever alike. Nevertheless, an attempt is made to highlight those aspects of a European project and its management that would be different from what could be expected in the U.S. Where specific experiences are cited, these may not always be typical, but conversely, many general observations can be used as a guide on any project outside the U.S.

fications but may have difficulty in communicating in English. A certain amount of translation is necessary and interpreters will occasionally be needed.

The problem of communicating in spoken English as opposed to written English is a very real one and must be recognized. The most important thing to remember and practice is to always speak slowly and distinctly. Until you know the ability of an individual to follow you, never assume that what you have said has been completely understood.

It is also important to review the wording of specifications and contracts for clarity and accuracy, and to ensure that there is no ambiguity. Corrections if tactfully made and explained will generally be willingly accepted.

It would be helpful if the project manager were to try to learn the foreign language. However, there may be more than one foreign language involved, and even if one takes an intensive language course before starting on the project, fluency may not come without a fair amount of practice.

Almost from the project conception and concurrently with negotiations for raw-materials supply and site selection, the project manager must concern

Checklist of basic information—Table I

(Additional to usual utility and site data)

Government regulations

What are the legal requirements covering the following items? How are they administered, and by what government agency?

- ☐ Permission to construct a plant.
- ☐ Import permits or restrictions.
- ☐ Currency restrictions.
- ☐ Permission to operate a plant.
- ☐ Design and testing of boilers and storage tanks.
- ☐ Design and operation of hoists and mechanical equipment.
- ☐ Design and testing of electrical equipment and distribution systems.
- ☐ Design and testing of structures.
- ☐ Safety requirements and fire protection.
- ☐ Water and air pollution, sewage and waste disposal.
- ☐ Liability for defective design or construction.
- ☐ Zoning and local building laws, and jurisdiction of local government.
- ☐ Working conditions and social facilities.
- ☐ Insurance requirements, liability and accident.

Taxes and duty

- ☐ Sales taxes; invoice tax, and business taxes.
- ☐ Import taxes and duty tariffs.
- ☐ Duty exoneration, if any.
- ☐ Importation procedures, where and how cleared and documentation required.
- ☐ How are applicable duty tariffs determined and can the determination be appealed.
- ☐ Restrictions on temporary importation (construction equipment).

Labor

- ☐ Laws relating to employment of nationals.
- ☐ Laws relating to employment of foreigners.
- ☐ Working hours, overtime, legal holidays, vacations and benefits.

Local and national code standards and practices

(Other than regulatory)

- ☐ Structural design.
- ☐ Pressure-vessel design.
- ☐ Electrical equipment and distribution systems (particularly common voltage levels).
- ☐ Standards, if any, for pipe, valves and fittings. Commonly available to what standards?
- ☐ Commonly used structural and architectural materials.
- ☐ Underground piping, both utility and sewer.
- ☐ Reinforced concrete design and placement.
- ☐ Paving and road-surfacing materials and practices.

Procurement and commerce

- ☐ Commercial practices such as payment terms, supplier financing, escalation and penalties.
- ☐ Price stability.
- ☐ Capabilities of local fabricators.
- ☐ Local supply houses and stock carried.
- ☐ What are the representations or service facilities and parts supply on behalf of major equipment manufacturers (compressors, pumps, instruments)?

Transportation

- ☐ Access to site by road, rail and water. Limitations on load or size.
- ☐ Port facilities, rail and road transportation from potential foreign-supply sources.
- ☐ Transportation costs and port charges.
- ☐ Contacts with reputable forwarding agencies.

Construction

- ☐ Labor availability both skilled and unskilled.
- ☐ Local construction contractors, their experience and capability.
- ☐ Availability of construction equipment and rental rates.
- ☐ Temporary hiring, manpower agencies, cost and availability of locally hired construction staff.
- ☐ Customary insurance carried by contractors.

General

- ☐ Contacts with local industries for general information on their experiences.
- ☐ Availability of soils, concrete, quality-control, and other testing facilities.

himself with a thorough investigation and accumulation of information on local government, labor, codes, safety laws, taxes, social laws, building practices, material availability, and import and monetary restrictions (if any) to the extent that all of these may affect the design, construction, cost and operation of the plant. This is in addition to the normal site and utility data that must be developed. There are specific areas that are potential pitfalls if they are not recognized early enough. The check list in Table I covers many of the key questions.

In seeking this information, there are innumerable sources to which the newcomer can turn. This deserves a word of caution because frequently information from two different sources will be contradictory and perhaps, even with the best of intentions, erroneous. The project manager must always be prepared to use his best judgment in such instances.

A company setting up a business venture in a new country will undoubtedly retain the services of a law firm. Frequently, this can be an excellent source of information and guidance in matters relating to governmental regulations and procedures, legal and business practices and tax laws. A law firm can also provide introductions to officials in governmental departments having jurisdiction over such matters as building, safety and operating permits, foreign exchange and import restrictions. It can also provide the background and guidance in the important area of politics and protocol that one must observe particularly when dealing with officials of local government who may have a degree of autonomy from a higher level of government.

Needless to say, legal advice should always be sought on all contractual matters, and the project manager should take full advantage of the availability of such advice at all times, bearing in mind that this advice will often lean in a conservative direction.

A local firm of accountants if retained by your company can be a good source of information on taxes and general business practices. In tax matters, the advice of accountants will generally be in the direction of what is accepted practice. This must often be weighed against the legal advice on tax law that may tend in the direction of interpreting the letter of the law. Often, there can be a big difference.

Accountants can also provide background information on firms being considered for important orders or contracts. Such information is far more difficult to obtain in Europe than in the U.S. The corporate structures and interests behind many European companies can frequently be extremely complex, leading in many unexpected directions.

Established local plants, either foreign or nationally owned, particularly if they are in the petrochemical field, are another good source of information. Generally speaking, such companies will be cooperative and will quite freely give you the benefit of their experience providing that it does not involve questions of a proprietary nature.

Contacts with engineering contractors (dealt with in more detail later) can be helpful. Local contractors, fabricators and utility companies will welcome early contacts and are another source of information.

Government agencies, including those of the U.S. Government, can provide a wealth of information. In some countries, particularly those that are encouraging foreign investment, there are agencies specifically set up to assist foreign companies. Such agencies may even intercede on behalf of the foreign investor with other Government departments in such matters as approvals, licenses and permits, and can frequently cut much of the red tape. This is particularly true in Belgium where the agency, "Direction des Investissements Etrangers," is a department of the Ministry of Economic Affairs.

The assistance, information and good will of local Government officials—especially those with whom you will quite likely have many dealings—will prove valuable. It is well to acquire a little knowledge of local politics, the organization, interrelation and jurisdiction of local and higher governmental departments and officials with whom you may have to deal. A good starting point at the local level should perhaps be the mayor of the town or municipality in which the plant will be located. Then, through the mayor, contact could be made with the town engineer, and officials involved with health, safety and labor. While

Change in Belgian tax law—Table II

As of Jan. 1, 1970, Belgium will adopt a tax on value added (T.V.A.) system in place of the present transmission tax. At the time of this writing, only the principles of the new law have been established by the government. Specific details and procedures have yet to be formulated.

Eventually the new tax law may have an important effect on the cost of plant and equipment, because it now seems likely that T.V.A. paid on capital expenditures will be credited against T.V.A. on product sales.

In any event, there will be a four-year transition period during which a progressively greater portion of the tax will be credited. This period will present many difficulties and pose some interesting problems relating to capital expenditures and plant construction. Ultimately the adoption of T.V.A. will greatly simplify and eliminate the tax problems described elsewhere in this report.

The status of T.V.A. and the normal T.V.A. tax rate in various European countries is:

	Effective Date	Normal Rate
France	1940's	20%
Denmark	July 3, 1967	12½%
Germany	Jan. 1, 1968	11%
The Netherlands	Jan. 1, 1969	12%
Belgium	Jan. 1, 1970	20%
Luxemburg	Jan. 1, 1970	?
Italy	1972?	10%
Norway	?	?

Information requirements and procedures provide the basis for successful project management

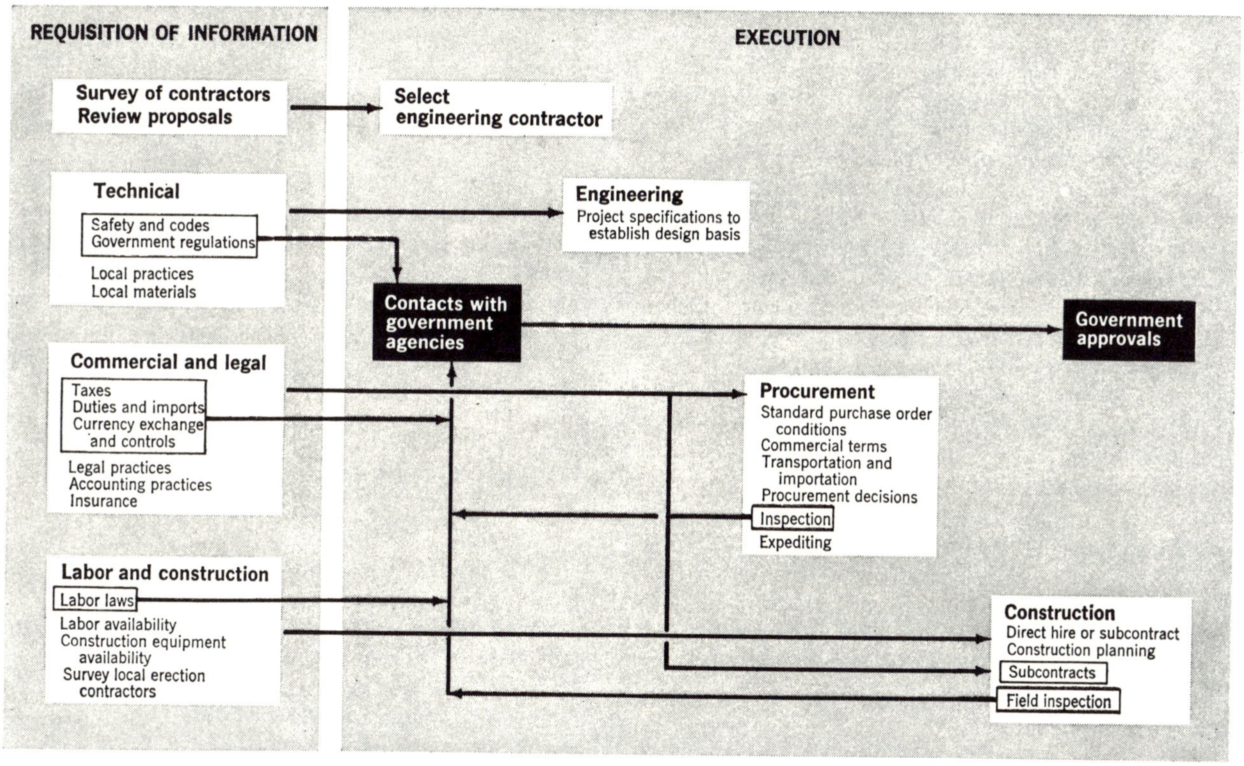

some higher government official whom you may already have contacted may have ultimate jurisdiction in approving your plants, a good relationship with local officials can be invaluable. A word of caution—the observance of accepted formalities may be different from what we are accustomed to, and recognition of hierarchies and personal sensitivities should not be overlooked.

Taxes

Unlike the U.S., most if not all European countries have some form of direct business or excise tax. For example, in Italy it is a gross business tax (approximately 6%) imposed on the seller and included in the seller's price. In Belgium, it is an invoice or transmission tax, (currently 7%) added separately to each invoice, and paid ultimately by the buyer in the form of tax stamps affixed to the invoice.

Another form of direct tax, used in France and one that may eventually be adopted throughout the Common Market, is the so-called tax on value added (T.V.A.). Rather than taxing the entire value of a transaction, the T.V.A. tax is applied only on the value added between any two successive transactions (e.g., difference between cost of raw material and finished product). It tends to be self-regulating and eliminates many of the complexities and inconsistencies of other forms of taxation, particularly the transmission tax. See Table II for the recently announced change in the Belgian law, and typical T.V.A. taxes in Europe.

On a typical project in Europe, direct taxes and duties are a significant item in the total cost of a project, ranging anywhere from 10 to 20%. Familiarity with the tax laws is essential to avoid paying double and even triple taxes on certain transactions. A brief description of the Belgian tax laws will illustrate this.

All commercial transactions for goods (either domestic or imported) and services performed inside the country are subject to tax. In the case of services, it is a 7% invoice tax. In the case of materials, it is either a 7% transmission tax paid on each transfer, or in some special instances a one-time tax called a "forfeitary tax," between 12 and 20%, paid at the first transfer when the material is sold by the producer.

The intent of the law is that goods will be subject to the forfeitary tax, or in the normal course of distribution from producer to ultimate owner, at least two successive transmission taxes. It does not take much imagination to conjure up a situation where this tax could be paid many times over. This is easily illustrated by considering a lump-sum turnkey contract as it might be conducted in the U.S. The owner would pay a 7% tax on the lump-sum price paid to the prime contractor, and this lump-sum price would include a 7% transmission tax on all invoices paid by the contractor for material and labor. In addition, any subcontractor who purchased material would pay a 7% tax on such purchases. This already totals three tax levies and does not include similar ones paid in transfer of material from producer to distributor.

Every transaction involving an invoice and payment is taxable and such tax will be collected no matter how many previous transmission taxes have been paid. An exception in the law that permits the elimination of a transmission tax concerns contractors or a general contractor and subcontractor working together on a specific building contract. They may legally form a "Temporary Association" for the execution of such a contract, and invoicing between members of the association for this work is not taxed. A single transmission tax is applied to the total cost of the work. Such an arrangement is practical for a commercial building or similar project with only two or three contractors but has only limited value for a major petrochemical project.

Reduction of Multiple Taxation

There are many steps that can be taken to reduce multiple taxation, which if carefully executed are clearly within the law. However, there are many gray areas that ultimately will be subject to the interpretation of the tax authorities.

The 7% tax on the total project cost can be eliminated if the owner contracts directly with suppliers and subcontractors rather than through the engineering contractor. This is relatively simple. It requires that all purchase orders and contracts must be in the name of the owner and signed by the owner's legally authorized representative. Furthermore, all invoices must be addressed to the owner and be paid by him. The responsibilities and function of the engineering contractor are, of course, not affected. This saving cannot be made on a lump-sum contract.

The second area in which double taxation can be avoided involves orders or contracts covering material and erection labor. The simplest illustration of this would be a cooling tower purchased on an erected basis. The cooling tower manufacturer would be requested to quote a total erected cost, with the cost of the lumber shown separately. Two orders would then be written, one to the tower manufacturer for the complete tower less lumber, and the second from the owner to lumber supplier, essentially written by the tower manufacturer but on the owner's purchase order form and over the owner's signature. The owner will eventually, on the tower manufacturer's authorization and with his own verification of material received, pay the lumber supplier directly.

Similar situations where this type of savings can be effected would be erection contracts involving the supply of material; orders for alloy vessels in which the material may be a substantial part of the cost; pipe fabrication and structural steel fabrication (provided the labor is performed in Belgium).

Considerable care must be taken in the terms and working of such split orders. First, to ensure that the owner is adequately protected and the supplier or contractor's responsibility is not diminished, and second, to ensure that the contract or orders will stand the scrutiny of the tax authorities. This last factor plus the administration of additional orders will determine the extent to which such split orders will be written.

It is also interesting to note that services performed outside Belgium are not subject to the transmission tax. This applies to all engineering performed outside Belgium, and it is possible for any such order that is not a standard item (a control panel or an extensive conveying system, for example) to write a separate order for engineering services. In this instance, there is an even more substantial saving to be made in duties, because such services would not be subject to duty, as will be explained later.

Import Duties and Taxes

As of July 1968, customs duties on trade between members of the Common Market were eliminated. However, it is not generally appreciated that there are still border taxes and formalities.

Customs duties vary from zero (for imports from within the Common Market) to as high as 20-25% depending on the nature of the item and the source. There are generally no restrictions on imports into Common Market countries. Duties are levied on the basis of landed cost, which includes packing, insurance and freight to point of importation.

In addition, all imports regardless of source are subject to the transmission tax of 7% (in Belgium) plus a border tax variously called an internal tariff, supplementary tax or import tax, which will vary from 6% to 11%. These two taxes are applied on the landed cost plus customs duties.

The import tax can be varied at any time by a simple government decree. This tariff generally but not necessarily corresponds to the export bonus paid to the supplier by the government of the exporting country. The bonus in turn is supposed to be an exoneration of excise taxes normally levied in the exporting country as a device to encourage exports. The buyer will never be directly aware of this bonus, because it is paid to the supplier, although it may be allowed for in the supplier's price if he must be competitive.

The schedule of duties by commodity classification is extensive. It is often possible to classify an item under more than one category, particularly when it is a component of a piece of equipment or a system. A common example would be a motor or an auxiliary pump for a compressor. If it is imported as an integral part of a compressor, it may be subject to a lower duty than as a separate item. Normally, the duty classification is determined by the customs officer at the time of importation, and his decision is usually not subject to subsequent review.

In order to obtain the most favorable rate, it is

necessary to employ a competent customs broker and to attempt to ship components either separately or together (whichever may result in the most favorable classification) with invoices appropriately written with an eye to the description that will fit the most desirable classification.

In Belgium and probably elsewhere, there is an alternative to depending on the interpretation of the customs inspector. This involves the application for a special import status that permits classification and payment of duties on a provisional basis at the time of importation, plus a 10% guarantee. Upon completion of the project, the customs officials will make a complete onsite review of the equipment and make appropriate rate rulings. The problem with this alternative is that it ties up cash in the form of the 10% guarantee, and the final determination may not be made for a year or longer after plant completion. This leaves the matter in suspense, and there is always the risk that the final determination may be less favorable than the provisional one.

Border formalities have a nuisance value, the likes of which are never experienced on a domestic U.S. project. First, documentation that includes the original invoice and bill of lading must be presented before a shipment can be cleared through customs. There seems to be no certain way of ensuring that these documents arrive in the right hands before the shipment. Lost documents can sabotage a schedule more effectively and in a more frustrating way than any strike or failure on the part of a supplier. With air-shipments (particularly if urgent), it is essential that documents travel with the shipment. In Europe import licenses and consular invoices are generally not required, but where these are necessary they can contribute significantly to delays and longer schedules.

A further hurdle imposed by border formalities involves temporary importation, as in the case of a component that will be incorporated into a piece of equipment and then reexported, or the exportation of an item previously imported, as in the case of a defective component returned to a manufacturer in another country for repair. The formalities and red tape involved can result in unbelievable delay. If the situation warrants, it is well worthwhile having someone hand-carry small items to be returned for repair by plane, train or car. There are also frequent problems with replacement of defective parts on imported equipment unless one is willing to pay the duty a second time.

Safety Regulations

The extent and nature of regulatory control of industrial and civil design and construction differs vastly from country to country, and even within a country. At one extreme is a minimum of regulation, either by design or by default, and a reliance on voluntary industry standards with a legal liability, both civil and criminal, for defective design and construction. At the other extreme is very close and detailed control of all aspects of design and construction.

In some instances, the regulations have their origins in what is, or perhaps what was at one time, the predominant industry in the country. But in most instances growth and diversification of industry since World War II have outdated many of the regulations, although they are changing rapidly.

Mining and steelmaking were until recent times the predominant industries of Belgium. Consequently, Belgium has very stringent regulations concerning hoisting equipment of every description including cranes. Regulations covering pressure vessels apply only to vessels containing steam. Recently, provisional regulations were introduced to cover LPG storage vessels, but there are no specific requirements of any description for pressure vessels containing hydrocarbons regardless of size or pressure.

On the other hand, Holland has a government agency called "Stoomwezen" that has control over all safety aspects of plant design and construction. This extends not only to pressure vessels but to piping systems, relief systems and structural design. For example, the agency requires review and approval from a safety aspect of all flowsheets. They have even been known to send an inspector (at the owner's expense) to the U.S. to witness the testing of critical rupture disks and large pressure vessels even though the latter were designed and tested in accordance with the ASME code.

Switzerland also has very stringent regulations for industrial plants covering esthetic as well as safety aspects. A common sight in certain areas of Switzerland is the wooden posts that must be erected, and left standing for up to six months, to outline the extremities of any proposed structure. This provides an opportunity for neighboring property owners to take exception to the esthetics of the proposed structure. Atmospheric and water-pollution standards are extremely exacting, but this is becoming as common throughout Europe as in the U.S.

Regardless of the extent of regulation, it can be safely assumed that in comparison with a U.S. project, the whole area of government controls and formalities will assume considerably more significance in terms of time and effort. Further, a failure to fully investigate and comply with such regulations can be disastrous.

Belgian Practice

A brief description of the regulations and their administration in Belgium will serve to illustrate how a typical chemical plant project would be affected.

In Belgium, laws affecting safety and standards in industrial installations are under the jurisdiction of the Ministry of Labor and Economic Affairs and are contained in what is officially known as "The General Regulations for Protection of Labor." However, with exception of the actual stamping of boilers or steam vessels, the administration of these laws and regulations is delegated to approved private inspection agencies.

In practice, the services of one of these agencies is engaged on a time and expense basis. During engineering, this agency will review the overall design and particularly the electrical design to advise of any noncompliance with regulations. For those pressure vessels that do fall under the regulations, they review and accumulate documentation on all design calculations, relief sizing, etc.

They, or affiliated agencies in other European countries, will conduct any legally required shop inspections and collect material certifications and a substantial number of material and weld-test coupons.

Finally, when construction is nearing completion, they will conduct field inspections and tests on the basis of which they will certify that the plant conforms to the regulations. These field tests and inspections include extensive testing of the electrical system (insulation, grounding, relaying), and load tests on all hoisting equipment.

At this time, steam boilers and steam pressure vessels are inspected, hydrotested and stamped by a government inspector after review of all documentation collected and verified by the inspection agency.

In general, the authorized agency, as an intermediary between the owner and various levels of government bureaucracy, is an advantage particularly because the rules are quite broad and open to interpretation. Because the governmental agency has an ultimate legal responsibility, it will refer doubtful points to the officials for clarification, or ultimately to a body consisting of government engineers and representatives of the inspection agencies that advise on new or revised regulations.

Belgium has very detailed and specific regulations covering the distribution and use of electric power. However, they are in some aspects so restrictive that their application to a petrochemical plant can present real problems. For instance, 3,000-v. motors and switchgear are classified as high voltage equipment and must be located in separate buildings with restricted access (originally having mine winding-gear in mind). This presents problems, for example, in the conventional compressor house. Fortunately, the officials administering the regulations are flexible in their interpretations and are willing to accept compliance in the form of a sign at all entrances to the compressor house designating it as an "electrical equipment room with access restricted to authorized persons."

There is a formal procedure for requesting approval for deviations of the regulation. If the application is properly documented and the request based on accepted and good practice, a variance will be granted. The process, however, can be slow—often too slow to permit holding up important engineering decisions. In such a case, a calculated risk may have to be taken by proceeding on the assumption that approval will be granted. In those many areas that the regulations are silent (pressure vessels containing hydrocarbons, and structures of all types), the laws, which provide criminal liability for negligence in design or construction and civil liability for defective work for a period of 10 years, are a relatively good assurance that sound practices will be followed.

Finally, there are two permits that must be obtained directly from the responsible government agencies. One is a permit to construct the plant, and the second is a permit to operate the plant. The first requires a presentation of the plant layout showing all buildings and equipment, listing rated horsepower, sizes and pressure rating of steam vessels, together with a written description of the plant and process. The operating permit is granted mainly on the basis of the acceptance reports of the authorized agency.

Codes and Standards

Other than mandatory regulation, it is necessary to determine what optional codes and practices will be followed to ensure that the plant will meet the owner's standards. In making these decisions, local practices should be followed wherever possible. However, when there is any question as to the adequacy of local codes, it is best to start with the standard specification one would use in the U.S. and modify it where and when necessary. This can be onerous and sometimes time consuming, but is the safest way to avoid surprises when the the plant begins to take physical shape.

For pressure vessels, the following source is an excellent guide, although a little outdated, to the various codes, their legal standing and their administration: ASME paper 61-SA-40, "A Comparison of United States, European and British Commonwealth Codes for the Construction of Welded Boilers and Pressure Vessels" by J. F. Lancaster.

The decision as to which code to use will be determined by local legal requirements, and by the owner's need to protect and assure himself of an acceptable quality standard. In some instances, local codes may satisfy both requirements while in others a combination of local and ASME codes may be followed.

The ASME code is frequently used in Europe, wholly or in part. Many fabricators are familiar with its requirements. Although the code may be followed in design and fabrication, code stamping cannot be obtained, and compliance must be checked by owner's or contractor's inspection.

The most widely used and the most comprehensive European code is in the German A. D. Merkblatter. This code and most European codes are less conservative than ASME in allowable stresses. Most are based on 2/3 of yield stress. Further, most European codes permit progressively higher stress at temperatures below 650 F. However, where failure by creep becomes a consideration, the European codes are more conservative than ASME in limiting allowable stress. The new Div. 2 of ASME Code Section VIII probably comes quite close to many European codes as regards vessel-wall thickness.

The German DIN standards for materials are extremely comprehensive, and it is generally possible to find German equivalents for ASME materials. In general, European codes are not as detailed as ASME in fabrication requirements. Weld-quality control methods vary considerably. Where circumstances permit, a workable compromise on pressure-vessel codes is the use of materials and allowable stresses based on the German codes but with fabrication and design detail on ASME. Welder qualification in accordance with ASME presents no particular problem.

Piping, Valves and Fittings

German DIN piping standards are probably the most comprehensive and widely used in Europe. Pipe and forged fittings to USAS dimensions and ASTM material specifications are readily obtainable (although not necessarily from stock) and widely used particularly by American companies and contractors. Forged and cast-steel valves to USAS standards are also obtainable. In fact, a number of U.S. manufacturers have licenses or manufacturing facilities in Europe. However, for special valves such as lip-seal bonnets, high-temperature or high-pressure (2,500 psi. rating), there may be no acceptable equivalent (to USAS standards) available in Europe. Purchase of these may be necessary in the U.S. ASME standards for piping fabrication and welding as well as for welder qualifications can readily be met.

Cast-iron valves with USAS flanges and face-to-face dimensions, particularly in the large sizes, are not readily available even on special order from European sources. These can be obtained, at a price, in the U.K. Furthermore, there appears to be very little standardization among European manufacturers, or even within countries, on cast-iron valves. This presents a real problem when a valve must be replaced. If a compromise must be made, it is best to retain the USAS flange dimensions and accept non-standard face-to-face dimensions.

Malleable-iron fittings are not available from European sources. If imported from the U.S., they may be more expensive than European forged-steel fittings.

Metric dimension pipe and tubing in many sizes closely approximate English sizes, i.e., 25 mm. and 1 in. In small tubing, particularly instrument tubing, and steel tubing using compression fittings, it is probably better to use metric sizes, e.g., 6 mm. (1/4 in.), 9 mm. (3/8 in.). European manufactured equipment will use metric tubing. Although the metric and inch-size tubing are very nearly equivalent, the difference is sufficient that the fittings are not interchangeable.

Electrical Codes

For electrical codes, DIN standards for electrical equipment are comprehensive and widely used. DIN equivalents are available for most NEMA enclosures. Some manufacturers will also offer motors to NEMA enclosure standards. Generally, application of typical U.S. specifications based on NEMA standards and API practices for hazardous areas presents few problems.

Unfortunately, there is no standardization of motor-frame sizes as in the U.S. Therefore, it is wise to specify a single manufacturer for all but very small or special motors, or better still, to buy all motors directly from a single vendor. The difficulties of coordinating engineering details are outweighed by eliminating a very real maintenance problem and by a saving in cost.

Conduit and conduit fittings can present problems. Although Crouse-Hinds and similar fittings as well as conduit are available in inch sizes, there are unfortunately a number of thread standards—U.S., metric and UNI threads. The latter is a unified metric thread system that probably will become the standard for all metric threads in Europe but is not yet widely used.

For electrical-wiring standards, it is well to base specifications on the National Electric code modified where necessary. Local wiring practices and codes, both mandatory and otherwise, vary widely. Direct-burial wire-armoured cable is common for underground service. Underground conduit is rarely used except as a convenience for pulling underground cable after construction is "out-of-the-ground." In this case, regulations may require the use of a direct-burial cable for higher voltages (e.g., 3 kv.), even though it is in conduit.

It is also wise to try to standardize electrical devices such as relays, pressure switches, and fuses. Many such devices of American type are available in Europe, and there are a number of quality European manufacturers. However, a simple item such as a fuse of a type common, for example, in Italy may be unobtainable in Belgium. The same is unfortunately true of simple things like plugs and receptacles. An attempt at standardization on components commonly available in the country in which the plant is being constructed will to some extent, but probably not entirely, eliminate these small problems that can become especially troublesome at startup time.

Structural and Architectural

Structural codes and practices, both steel and concrete, present probably the biggest difference from the U.S., and there are even big differences between European countries.

Most countries have well-defined codes. In Belgium, they are called "Norms." Frequently these are not mandatory but almost invariably structural designs have to be certified by an engineer or architect licensed in the country in which the structure is to be erected. Furthermore, there is (in Belgium and elsewhere) a criminal liability on the part of the engineer and/or builder if any structure should collapse due to negligence in design or construction. While the Norms are not mandatory in Belgium, they would represent an acceptable minimum standard in a court of law. For these reasons, structural design is often done in the country in which the plant is being built even when the engineering is done elsewhere.

A comparison of ACI standards for reinforced concrete design and placement should be made with local codes. To the extent that local codes are less restrictive than ACI, or are silent, the ACI standard should be used. For example, Belgian Norms permit the use of plain bars for reinforcement.

Concrete work is a frequent cause of complaint from U.S. engineers in Europe. In part, this is due to poor placement practices and a lack of awareness of and attention to detail. Insufficient vibration and improper slump are typical problems. Even the need for casting important foundations in a single pour is overlooked. The common practice of finishing floors and paving with a cement topping is unsatisfactory in most industrial plants, and the techniques of finishing a monolithic slab floor are not well known.

There are also significant differences in the design and fabrication of structural steel. The choice of standard shapes is quite limited as compared to the U.S. In channels and I-beams there is only one standard weight. This makes it impossible, for example, to use a U.S. standard stairway detail employing 10-in. lightweight channel for the stringers. Instead, a 150-mm. (6 in.) channel would have to be used. It is heavier than the 10-in. lightweight U.S. channel and does not fully cover the width of a normal tread. This is not a major problem but in many instances it results in less-economical choice of section.

Another problem is that standards for structural connections comparable to AISC do not exist in Italy or Belgium. They are frequently designed individually. A more important problem relates to the widespread use of field-welded structures rather than bolted. This occurs because of a tendency to avoid detail design and the fabrication accuracy required for structures in which field connections are bolted, plus unawareness of the economics of bolting. To appreciate this, one has only to see two cranes tied up for an hour or two holding a beam in position while the structure is being aligned, and then the connections welded. Welding presents a problem if galvanized steel structures are contemplated. Much of the advantage of galvanizing is lost if the connections are welded.

Standards and practices as well as mandatory local codes for plumbing, domestic water and sewers vary to such an extent that any generalization is impossible except to say that any attempt to adhere to U.S. practices will be largely futile. Sewer pipes, for example, whether cast iron or concrete, are generally made to the manufacturer's own standards. Copper pipe is rarely used for plumbing, and lead pipe is still common.

The same difficulties will also be encountered in architectural details. Doors and windows, for example, are generally not standard. Even a particular manufacturer will have few standard items.

U.S. manufacturers of metal siding are well represented in Europe. With the possible exception of this material, it is best to follow local practices and use local materials. For example, on a small building in Belgium, brick walls may be less expensive than metal siding. Concrete block, when it is available, may also be more expensive than brick and often the dimensional quality is not good enough to permit its use without plastering.

Particular attention should be paid to lighting, heating and ventilation in critical working areas, because European standards are not as strict as U.S. ones.

An area of architectural design that should receive special attention is that relating to fire prevention. Many widely used materials, particularly insulating and roofing materials, have a very poor fire rating, and would not be considered acceptable by U.S. standards. The rule of outward opening doors is almost universally violated and fire escapes are frequently overlooked. In some parts of Europe, there are rigid fire-safety standards, but in many there are none whatsoever. Insurance companies have been very reluctant of late to insure chemical plants, particularly in Europe. It is wise to work closely with the insurers during the design and construction phases to meet their requirements and improve the insurability of the installation.

Engineering Practices And Methods

American contractors in Europe generally follow the same engineering and drafting practices as the parent organization. They would also use the same project-management systems and computer programs for piping take-offs, cost control, exchanger ratings and the like. Practices among European firms will vary widely. Some firms have to a lesser or greater extent, through working for or with American companies, adopted our practices and even in some instances gone one better.

Generally, there is a tendency to over-elaboration in drafting detail, and conversely a lack of detail in essentials such as piping isometrics. Specifications tend to be too loose (partly a language problem) and too much faith is placed in the manufacturer or supplier. In this regard, there is sometimes a reluctance on the part of suppliers to provide the type of detailed information about their products that is asked of and provided by a U.S. supplier. This is changing quite rapidly. Most vendors now provide reliable certified drawings although this was not the case 10 or 15 years ago. European suppliers are no better and no worse than the average U.S. supplier in providing essential after-order information like spare-parts lists and operating manuals.

Beware of one pitfall with vendor drawings. European machine drawings use the first-angle projection whereas the U.S. standard is the third-angle projection. Confusion between the two can result in embarrassing mistakes such as incorrect directions of rotation.

For vessels and exchangers, it is quite common for the engineering contractor to provide the vendor with a complete vessel design, including what would in the U.S. be considered shop details. The reason for this is that many qualified fabricators do not have the capability to do detail design. More important, engineering can proceed without any holdup for procurement and certified prints from the vendor. For the same reason an engineering contractor will sometimes rate the exchangers.

Detail piping-fabrication drawings will generally be prepared by the engineering contractor, a practice sometimes but by no means always followed in the U.S.

Both structural steel and concrete drawings prepared by the contractor will be considerably more detailed than is the U.S. practice. This is because of lack of widely followed, uniform standards for steel-fabrication details and reinforcing-bar detail in concrete design. Concrete drawings will also frequently show a bill of materials for reinforcing bars, concrete quantities, and area of formwork. This is necessary if concrete work is subcontracted on the basis of unit prices.

Symbols on electrical drawings, particularly schematics, invariably follow one or another of the European standards. This is not a problem if a complete key is provided on each drawing or at least on a separate standard symbol drawing. Details of electrical wiring and schematics is as good as, if not better than, in the the U.S.

Instrument-engineering details as well as symbols will, particularly in the case of an American engineering company, follow U.S. practices. This, perhaps more so than any other discipline related to the chemical or petrochemical industry, is American in concept, design and overwhelmingly so in availability of equipment.

The question that may well be asked is: How do European companies handle their engineering since there are relatively few major engineering contractors of purely European origin? The answer probably is that engineering of chemical and industrial plants is an "in-house" function of either large chemical companies or the large machinery and equipment manufacturers. In fact, many of the European engineering firms have evolved from such beginnings and still retain the corporate tie to the parent organization.

Selecting an Engineering Contractor

Many of the major U.S. engineering companies have competent, experienced organizations in Europe, and there are several European engineering companies that are capable of handling multimillion-dollar projects. There is a decided advantage in keeping the engineering and procurement functions in the same geographical location, and both preferably close to the market in which most of the purchasing will be done. Hence, there should be little reason to consider engineering a European project in the U.S. Communications by telephone, telex and even between computers across the Atlantic are now quite good but unfortunately mail (as in the U.S.) is quite erratic and dependence on written communication between the U.S. and Europe will put a definite schedule disadvantage on the U.S.-engineered project. Even doing the so-called basic engineering in the U.S., and detailed engineering in Europe can only lengthen the overall schedule and add to the difficulty of effective project management.

Generally speaking, engineering costs in Europe, particularly in Italy and the U.K., are lower than in the U.S. At present, comparable average hourly costs including overhead and burden would be in the range of \$10 to \$14 in the U.S. and \$6 to \$7 in Italy. Part of this saving would be offset by a lower efficiency—European engineering productivity is considered to be about 90-95% that of U.S.

In selecting an engineering contractor in Europe, aside from the obvious cost considerations and whether you are intending a cost-plus or lump-sum project, there are a few areas to which particular attention should be paid.

First, joint ventures that seem to proliferate in Europe should be avoided. This warning applies not only to engineering but also to fabrication and construction. There may be instances where a joint venture will offer an advantage or where they may be politically expedient, but they should always be looked at critically. At best, they pose problems in communication and definition of responsibility; at worst they do nothing but add another "cut" to the overhead and profit. Early in any contact with people or firms offering goods or services, ascertain whether they are principals or agents.

Second, talk to recent clients of the contractors you are considering. Use the information you obtain in reviewing their organization and capabilities but with discretion and due allowance for possible personality conflicts that can give a distorted picture.

Give considerable weight to a company's experience in the area or country that you are building in. Knowledge of local conditions and requirements from past experience can be a decided advantage.

Look very critically at the size of the organization in relation to the size of your project. There is practically no pool of experienced designers and draftsmen in Europe such as exists in the U.S. In fact many companies are obliged to run training programs to build their staff. There are also few experienced design and drafting job-shops such as exist in the U.S., to whom peak-load work can be "farmed out." Consequently, you should have some idea of the total engineering hours your project will require, and make an estimate of the peak manpower requirements, particularly in designing and drafting. If this represents a substantial portion of available manpower, you may run the risk of delays in engineering even if the contractor is at the time looking for work.

The offices of American engineering companies in Europe tend progressively to reduce American staff even at the top levels in favor of local nationals. Many of these people are competent and well qualified but it is important to obtain a roster of proposed project staff with a resume of their experience and to arrange for face-to-face meetings. Look for fluency in English and their knowledge and experience in American engineering and project-management practices. Follow up on references from previous clients and eventually obtain a commitment from the contractor on specific key people. Assure yourself that qualified alternates would be available in event of illness or resignation of a key man.

Investigate in some detail the systems, procedures and general organization of each company. Tour the offices, look at work on the drawing boards, look at schedules, reports and planning of current or recent projects. Look at data-processing facilities and the manner in which they are used. Specifically try to determine how the contractor proposes to organize and run your project. A more thorough review may be advisable in the case of a Europen engineering company.

Pay particular attention to the procurement organization, especially inspection and expediting. Find out how many inspectors and expeditors are on the payroll, where they are located and their experience. Look at inspection and expediting reports.

Finally, review the construction organization and examine the qualifications of available people. While there are a growing number of European engineers experienced and competent in Americans methods of engineering management, there appears to be a shortage of good construction managers. Many of the top field men are still Americans. Bear in mind that an American construction manager will be facing an entirely new situation in his first European project. The best way to make an evaluation of the type outlined above (assuming that the number of potential contractors has been narrowed down to a few likely candidates) is to prepare a questionnaire or request for proposals that outline the nature, scope and size of the project, and asks for specific information in the areas you wish to evaluate. The replies themselves will give a good idea of the competence of the organization, and after a detailed review can form the basis for a one- or two-day visit to the offices of each company to meet the people and see the organization in operation.

Purchase-order commercial terms—Table III

Payment.
- Currency.
- Form of payment (letter of credit or cash).
- Terms (payment schedule).
- Where to be made.

Escalation.
Delivery penalty and/or bonus.
Cancellation.
Export bonus (in favor of buyer or seller?).
Point of final acceptance and inspection (should always be jobsite).
Court of jurisdiction (preferably country of importation).
Delivery quotation—generally implies agreement on responsibility for:
- Export fees, taxes and documentation.
- Transport costs and arrangements.
- Insurance.
- Importation costs, duties and taxes.

Procurement and Subcontracting

International commercial transactions invariably involve a number of details that are absent in a purely domestic transaction in the U.S. Any purchase or contractual commitment must carefully spell out payment terms, currency, escalation, delivery point, penalties or bonus, if any, for ontime delivery, and court of jurisdiction in case of dispute. See checklist for purchase orders in Table III.

In Europe, payment terms invariably involve a down payment with an order, progress payments and a guarantee holdback. An American company selling to a European company, even if an affiliate of a U.S. firm, will require an irrevocable letter of credit with an order, or some other form of guarantee. Funds tied up in a letter of credit earn no interest.

It is essential to specify the currency in which payment will be made. Europe, in contrast to other foreign areas, has relatively few restrictions on transfer of funds, but any transfer of funds involves bank commissions and the normal spread between buying and selling rates. These can add up to significant amounts. Therefore, consideration should always be given to paying in the currency of available funds. Aside from potential devaluations, there is also a normal fluctuation in exchange rates that adds some element of risk whenever currency exchange is involved.

Escalation clauses are the rule rather than the exception in all proposals. If an escalation clause is included in the final contract or order, it must of course specify the index to be used. Where different indexes are used for material and labor, it must specify the percentage of labor and material in the total order and in some cases the percentage of engineering that may or may not be subject to escalation. Further, the effective period of escalation must be defined. This could be up to the time of shipment (an unfavorable basis from the

Simplified definitions of quotation terms—Table IV

Term	Responsibility for— Export Taxes, Fees and Documents†	Import Taxes and Duties	Transport Cost and Arrangements†	Clean Bill of Lading	Insurance Cost and Provision†
"Ex Factory"	Buyer	Buyer	Buyer	N.A.*	Buyer
"F.O.B." (Named carrier, named inland point of departure), or (Named inland carrier at named point of exportation)	Buyer	Buyer	See notes	Seller	Buyer
"F.O.B. Vessel" (Named port of shipment)	Buyer	Buyer	See notes	Seller	Buyer
"F.O.B." (Inland point in country of importation)	Seller	Seller	See notes	N.A.*	Seller
"F.A.S. Vessel" (Named port of shipment)	Buyer	Buyer	See notes	Seller (Dock receipt)	Buyer
"C. & F." (Named point of destination)	Seller	Buyer	See notes	Seller	Buyer
"C.I.F." (Named point of destination)	Seller	Buyer	See notes	Seller	Seller

* N.A. means not applicable.

† General Notes

Seller bears costs and risks up to time at which buyer is obliged to take delivery. Seller renders assistance at buyer's request and expense in obtaining export and import documents.

Buyer must take delivery as soon as goods are at his disposal per delivery terms.

Provision for and payment of transportation to point of delivery is by seller. From point of delivery, it is arranged by buyer, except for: "FRT. PREPAID TO..." and "FRT. ALLOWED TO..." Seller's responsibility for risks ends at loading point upon receipt of bill of lading.

For comprehensive definition see: "Revised American Foreign Trade Definitions 1941," National Foreign Trade Council Inc., New York.

buyer's point of view) or on a progressive basis tied to actual progress. In construction contracts, it would be figured monthly for the work done in that month. Escalation clauses tend to complicate cost control, and can add significantly to the accounting work load. Often, it is advantageous to negotiate either a limit on escalation, or better still, to negotiate a flat increment to the contract price in lieu of escalation. The gamble that is taken can often be offset by the fact that fixing the escalation is an incentive for the supplier to deliver as early as possible.

Delivery point along with responsibility for freight, taxes, duties and insurance must be clearly specified. The "Revised American Foreign Trade Definitions" issued by National Foreign Trade Council, Inc., New York, provides a good but not absolute guide to the generally accepted meaning and implication of the terms F.O.B., F.A.S., C. & F., C.I.F., and the many variations of each. For a simplified list of terms, see Table IV that shows buyer and seller responsibilities.

It is quite common in Europe to include a delivery penalty clause, and more often, both bonus and penalty in orders and contracts. Penalties are sometimes limited by law to a maximum of around 5-10% and are usually applied at ½-1% per week, up to the maximum. The penalty clause generally excludes force majeur (strikes are frequently considered force majeur) and must be specific as to its application when partial deliveries are involved. In practice, recovery of penalties is often difficult and frequently not worth the time and effort that will have to be expended. Unless the penalty can be very simply applied and where delivery is critical, it is best to use the penalty as a negotiating point and trade it off for a price cut before placing the order, or against the inevitable disputed extras at the time of final settlement.

One aspect of European commerce that will be somewhat disturbing to an American is the latitude for negotiation in competitive bidding. This goes beyond the usual adjustments to the initial bid resulting from review and reconciliation of details and variations in each offering. In effect, bids are sometimes submitted with the intent that prices are negotiable. The extent to which this practice is followed varies from country to country, from industry to industry, and with the level of economic activity.

Monopoly practices and cartel agreements are another aspect of commerce unfamiliar in the U.S. For example, the steel cartel restricts the marketing of mill products so that an Italian steel mill will not sell mill products such as pipe or shapes for delivery in Belgium. They will sell to a fabricator or processor whose ultimate product will be shipped to Belgium, although the fabrication may be as simple as coating and wrapping pipe or threading conduit. If the price advantage is significant, and particularly in the case of large tonnage quantities, ways can often be found to get around cartel agreements.

Paralleling the tendency to negotiate initial prices is the negotiation of the final settlement. Invariably the claims will exceed the expected recovery. It requires firmness on the part of the purchaser and in no way does it reflect on the honesty or integrity of the seller. It is a way of doing business.

Suppliers in Europe do not carry anywhere near the stock of common items as is the case in the U.S. This is especially true of pipe and fittings. Therefore, it becomes essential that bulk purchase orders be placed early and in sufficient quantity. Small quantities, if available from stock, will command an appreciably higher price than larger quantities fabricated against an order. To further complicate the problem, restocking of surplus by a supplier is very rare and if accepted at all, the charge will be in excess of 50% of the purchase price. If import duties are involved they would generally have to be written off. Much of the surplus material has often to be sold for scrap prices.

An accurate bid comparison can be quite difficult to make when comparing prices from two or more countries. Currency exchange, duties, taxes and freight introduce an element of uncertainty. Since duties and taxes can be several times more than the usual difference between competitive bids, it is important to determine the correct customs classification, the applicable duty and taxes. Where currencies are relatively stable, it is best to establish a standard rate of exchange for comparison purposes during the life of the project.

Packing costs are often quoted separately, sometimes as a percentage (anywhere from ½-2½%) of the purchase price. It is important to agree on the price and type of packing. This is even more important with ocean shipments. For critical items that could be damaged by saltwater corrosion or in handling, it is well worthwhile writing a packing specification. MIL standard DSAM-4145-2 December, 1963, can be useful as a guide.

A common practice in Europe is pricing of commodity items by delivered weight. In many instances, this is of relatively little importance except that delivery to the field is checked by weighing rather than by count or by length. For heavy shipments, this may be impossible, and the seller's weights have to be accepted. However, for some items, particularly pipe, this can be to the buyer's disadvantage. USAS standards permit a 12½% tolerance on nominal wall thickness. With pricing on delivered weight, there is no incentive for the pipe manufacturer to replace worn dies. Consequently, the wall thickness tends to be on the heavy side, resulting in a higher price per unit length. Further, in the case of pipe, requirements are measured by length, not weight, and the order must specify the length required, not the weight. Suppliers normally quote prices per kilo. During negotiation, it is possible to insist that this price be converted to a unit price per meter based on nominal weight, but frequently the supplier will want to increase his price by 2-3%.

When considering procurement of machinery and equipment that will require parts and service, give consideration to proximity of a qualified service organization and the availability of spare parts on short notice. This problem can be much more serious than in the U.S.

Again, a word of caution on joint ventures and third parties. With manufacturers, determine in advance who his major subsuppliers will be. When in doubt, investigate. It is important with vessel fabricators and the like to obtain from the contractor a detailed survey of the vendor's facilities. If one is not available, a survey should be made. Check references with other customers. For contracts for erection or labor include a clause prohibiting subcontracting without owner's approval.

Schedules and Scheduling

The schedule for a project in Europe will probably be from 20-30% longer than a comparable U.S. project. It could be even longer than this. Prevailing economic conditions will have a big influence on the overall schedule, but among the reasons for anticipating a longer schedule are red tape and governmental formalities, somewhat longer deliveries of heavy machinery and equipment (although this is improving), a comparative lack of stock of "shelf" items, and inefficient construction practices. But more than anything else, the relation of time and money is still not as well appreciated in Europe as in the U.S.

The inertia in following up or taking positive action with regard to the dictates of a schedule is probably the biggest single battle that must be waged on a project in Europe. The reasons for this relative lack of aggressiveness are probably founded in social systems and customs. Class distinctions and deference to rank still govern attitudes and behavior in Europe. It is almost unthinkable, for example, for a schedule man who may not have a degree to press a professional man such as an engineer or a person of higher rank. This is particularly true with the outside expediter. His contacts with suppliers will often but not always be on the basis of rank. He is often not considered a "representative of the purchaser" but a "representative of the purchaser having a certain rank." Consequently, his contacts are often confined to the lower echelons of the supplier's staff. When the expediter chooses to be aggressive, he may not be taken seriously by the supplier's management; this is not always true but it is definitely an underlying factor in most relationships. Thus, high-level contacts are often necessary to obtain action both within an organization and between organizations.

Scheduling systems are generally variations of the conventional bar chart. American-affiliated engineering companies generally follow the same systems as the parent company. A number of companies claim to use critical path methods, but in practice the use of such methods is in its infancy and is frequently used only on customer insistence. Its value in such cases is questionable.

The project manager must expect to insist on, and take an active part in, regular and frequent schedule reviews—more so than on a U.S. project. He must be prepared to continually push and goad for action.

One factor that must not be overlooked in a European schedule is holidays and vacations. Depending on the country, there may be more than twice as many annual public holidays as in the U.S., and the actual days vary widely from country to country. National vacation periods, generally two to three weeks long, are also the custom in Europe. These vary from country to country, stretch from the latter part of June through mid-August, and during this entire period a slowdown in all activity must be expected. There is also another unavoidable slowdown over Christmas and New Year often extending through the first week of January, during which time business virtually comes to a standstill.

Cost Control

Multiple currencies and variable exchange rates can present a problem in cost control and reporting. For consistency throughout the life of the project, it is advisable to report all costs in a single currency. To avoid additional complications, it is best to use the currency in which the bulk of the project funds will be available. On the basis of exchange rates prevailing at the start of the project, establish "standard" exchange rates for conversion of all commitments and expenditures to the "official" reporting currency.

Invoices when received and approved for payment should be reported as expenditures converted at the standard conversion rate. When the invoices are paid and a statement is received from the bank showing the actual conversion rate and payment plus bank commission, accumulate the differences between actual total payment and the reported expenditure and show this difference as an adjustment to the total expenditures. In this manner the cost-control report will reflect underruns or overruns that can be accounted for in terms of changes, engineering, or procurement factors and still provide a record of actual cash expenditures. Exchange rates will normally fluctuate within the official price range. Even within this range, fluctuations can become significant if large sums of money are involved. Therefore, it is wise even while using the "standard" exchange rate for bid comparisons, to follow the trends in exchange rates and take them into account when deciding any close bids that are quoted in different currencies. If during the course of a project, it appears that actual rates have changed appreciably from the "standard," it may be necessary to estimate the effect of this change on commitments and show it on the cost report as an adjustment to the total commitments. An actual devaluation of one or more currencies during the course of the project can be handled in one of two ways, depending on how far the project has progressed and the magnitude of the effect of the devaluation. If the project is still in the active engineering and procurement stage, it may be best to change the "standard" rate of exchange and correct the commitments and cost forecast for all affected items. If most of the procurement commitments have been made, it is best to retain the original "standard" exchange rate for individual cost items and show an adjustment to total commitments and expenditures.

Aside from these problems, cost-control procedures will be much the same as on a domestic project. American or American-oriented contractors follow accepted cost-control procedures, some computerized.

Inspection and Expediting

If schedules and delivery promises are to be maintained, active expediting is even more important than in the U.S. To be effective, it must extend also to subsuppliers.

Generally, expediting reporting is good in that considerable detail will be recorded. Frequently lacking in the report is the assessment based on fact or judgment of progress relative to the schedule target. This can often be ascribed to lack of experience—expediting in the American sense is relatively new in Europe.

Experienced European contractors tend to put much less faith in the supplier in the matter of quality than is the U.S. practice. This is in part based on experience, plus the fact that exported or imported items are less conveniently returned if defective. As a result, a far wider range of items will be inspected, and in many cases the contractor will insist that no shipments be made without a shipping release.

Inspection is extended to pipe, fittings, fabricated items of all types, relief valves, pumps, electric motors (frequently tested before shipment) and switchgear, in addition to the usual items like pressure vessels. To cover this wide range of items requires a large inspection force and one that must have some specialists in fields other than in vessels and similar welded fabrication.

Whether this wider and more costly inspection can be justified is debatable. Even though it is not uncommon to find pipe fittings, cable and even motors legitimately rejected by this type of inspection, many defects will and do pass unnoticed until after receipt in the field. There is a definite law of diminishing returns. Costs on a wide inspection basis can become high. It is advisable at an early date to clearly define inspection requirements as to what items are to be inspected and how.

There are a number of inspection agencies in Europe that can be used to supplement the contractor's inspection staff, but it is essential to be selective in the individual who will be carrying out the work. It is advisable to have these individuals report directly to the contractor's inspection department rather than through their own organization. Outside agencies are frequently used for field testing and inspection in the areas of weld, X-ray, concrete and earthwork.

Transportation

The services of a competent forwarding and customs clearing agent are essential because unlike a purely domestic project, shipments will have to pass through one or more borders, and there is little or no uniformity in rates or in regulations governing the transport industry. Ultimate responsibility for collecting the proper documentation and having it available for prompt clearance of shipments through customs will rest with the clearing agent. The forwarding agent may also be the clearing agent, but frequently they are different concerns who nevertheless must work closely together.

The primary function of a forwarding agent is to arrange or contract for transportation on behalf of the owner at the most favorable rates and timing. Depending on individual circumstances, this can be extended to a responsibility for co-ordinating all shipments, checking on shipping dates directly with suppliers, and arranging transportation accordingly. There is no set basis on which such services are provided except for ocean freight where the agent receives a commission from the steamship company, and this should be his total compensation.

Overland shipments can be handled on the basis of the forwarder quoting firm freight rates for the complete project, or he can act as the owner's agent contracting for transportation in the name of the owner and receiving a fee ostensibly only from the owner. It is the obligation of the forwarder to obtain the best possible rates for ocean and/or overland shipments. These rates can vary widely, particularly between conference and nonconference shipping lines, and on shipments between European ports.

Even when the forwarding agent is charged with responsibility for arranging and co-ordinating all shipments, constant followup is required by the engineering-contractor's traffic department. Rates should also be spot-checked.

Rail shipment in Europe is slow, unreliable and limited as to loads and clearance. Consequently, most overland shipments are by truck. In Northern Europe, there are extensive inland waterways permitting heavy or large items to be shipped by barge. From the Mediterranean area to Northern Europe, ocean freight is generally slower and more expensive than truck shipments over the Alps. Surprisingly such shipments of direct, full truckloads are quite fast, and are only occasionally delayed by bad weather. Wherever possible through-trucking should be used. Some trucking concerns are licensed to truck through countries without border formalities. Such trucking is called "T. I. R." Unloading and reloading for customs clearance in the country of destination should be avoided. If necessary, arrange for clearance at the jobsite.

Insurance

The project manager will be concerned with two aspects of insurance, one involving liability and accident insurance during construction, and the other the insurability of the completed plant.

Most European contractors do not carry the high limits of third-party liability coverage that is customary in the U.S. As a result of statutory liability, many do carry coverage for defective work. If possible, the owner should obtain third-party liability coverage that includes all contractors as insureds. This may result in duplication of coverage, in some instances, but it can avoid later difficulties and confusion. Motor-vehicle liability insurance coverage in Europe is generally at quite low limits.

Insurance for the completed plant will undoubtedly be placed in the American or U.K. insurance market. The high loss-experience in the chemical industry in the last few years has made insurance companies reluctant to underwrite the risks. Unfortunately, this is especially true of Euroepan installations where the recent loss-experience has been particularly poor. Therefore, it is wise to have, if possible, a firm commitment on insurance at an early stage of the project.

Construction

For many reasons, a direct-hire construction job in Europe is the exception rather than the rule. Firstly, there is in most instances no pool of skilled construction labor as in the U.S. Skilled craftsmen and particularly the good ones tend to be retained in permanent employment. Secondly, the labor laws make temporary hiring difficult and costly.

On the other hand, there is little restriction on the mobility of labor across borders within and often even from outside the Common Market. There is a general shortage of skilled and unskilled labor in Northern Europe. A sizable portion of the labor force, both skilled and unskilled, in Germany, Belgium, France, and to a lesser extent Switzerland, are from Spain, Italy and North Africa. Although there are required formalities such as working papers and affidavits from employers, these (except perhaps in Switzerland) are not so much restrictive as of nuisance value. It is common for erection contractors and their crews to work relatively freely in different countries within the Common Market.

A further big difference between the European and U.S. construction job is labor unions. They do exist and are quite strong in Europe but they operate more as political organizations with government, employers and the unions as three participants in negotiations. The results of these negotiations are embodied in labor laws and regulations rather than as contracts between union and employers. Consequently there are no closed shops and no jurisdictional disputes. However, very restrictive labor laws offset these advantages.

Without being too specific because laws do vary from country to country, the following illustrate some of the laws of Belgium and their effect on a construction job.

There are very stringent restrictions on overtime work of any description. Specific authorization for overtime must be obtained on a day-by-day basis from the Dept. of Labor. Such permission is usually only granted if the employer can prove extenuating circumstances. Such overtime is worked at premium rates. In addition, the worker must in the following weeks be given equivalent time off on normal pay.

As indicated earlier, there are up to twice as many national holidays as in the U.S. Work during the normal vacation period generally comes to a halt. In any event, the worker must be given his full vacation—generally three weeks per year.

Termination of a worker requires a minimum of one month's notice, a minimum of two weeks' severance pay, plus time off on full pay to seek other employment. Disciplinary dismissal is very difficult to enforce, and is subject to review by the labor department.

Transportation must be provided daily to and from the jobsite.

Payroll benefits add between 50 and 60% to direct hourly wages, and the resulting skilled labor cost is on the order of \$3-\$3.50 per hr. On a subcontract basis including subcontractor's overhead and profit, it is closer to \$4.50-\$5 per hr.

Invariably almost all construction work is subcontracted either on a lump-sum or unit-price basis with the engineering contractor providing overall construction management. His supervisory field staff will consist of permanent employees in key-specialist positions with the remainder locally hired, preferably through a manpower agency of which there are an increasing number. This method of hiring local staff, although more expensive, avoids all of the administrative and legal details. There are also a number of agencies, particularly in the U.K., that specialize in construction staff and supervisory personnel.

Because of the added accounting burden on a subcontract job (particularly if unit priced), the engineering contractor's field staff will be larger than on a comparable U.S. project. Also, the subcontractors will have their own field staffs. Hence, the overall ratio of supervision and staff to productive workers will be much higher on a European job than on a comparable U.S. project.

ALUMINUM storage silos are hoisted into place atop forest of reinforced-concrete supports.

There are many large well-established civil contractors in business in most countries. Unfortunately even the largest of these are forced to subcontract plumbing, plastering, heating and similar building work to the small local contractor. This area of the project frequently presents problems out of all proportion to its importance.

At least two very well-qualified and competent instrument-installation contractors operate internationally in Europe, and generally it is not difficult to find qualified electrical contractors.

Piping fabrication, piping, and mechanical erection are a problem on a sizeable project. Local contractors will generally have neither sufficient equipment nor manpower to handle a large job. In areas where there has been a heavy concentration of petrochemical construction, local concerns have expanded into the field. In the Antwerp area, some of the shipyards have moved into both the construction and contract-maintenance fields, but the fluctuation in workload frequently puts a strain on available manpower, supervision, and equipment. Often as a result, this part of a construction project is handled by joint ventures between several contractors, or by a contractor with a number of subcontracts. Neither of these arrangements is ideal.

An exception to this general situation is in Italy where a number of companies have specialized in international mechanical and piping erection work on a big scale. They are well equipped and organized with relatively large pools of skilled craftsmen, and they are capable of moving men, equipment and facilities to almost any jobsite. Their workmanship in welding and pipefitting is good.

There are very few specialist pipe-fabrication shops

in Europe. Quite frequently all piping is field fabricated. This is really not a disadvantage because it is possible to obtain quality work, and with good scheduling it makes for efficient utilization of manpower.

Except for isolated instances, a reasonably tight schedule will preclude lump-sum erection contracts. They will be either entirely unit price or a combination of lump-sum and unit price. This adds quite appreciably to the field staff, in that field-staff specialists will be involved in work measurement as well as direction and control of the work program.

Units of work measurement for unit-price work are quite different from the U.S. Pipe fabrication and erection are on a weight basis—generally calculated—and include all valves and fittings. It is an entirely empirical method of pricing based on experience factors. Unit prices are generally broken down into a number of categories—by size of piping, whether yard piping or in a unit; by material of construction, either alloy or carbon steel; and by elevation above grade. Labor efficiency, particularly in pipe fabrication and erection, is also measured by weight output per manhour.

Changes to completed work are usually charged on time and material basis or on a schedule of charges for unit operation, such as cuts and welds by pipe size, schedule and alloy. This is the unit measurement generally used in the U.S. for pipe fabrication. Some attempts have been made by American contractors to introduce this system but with limited success. As always, there are unforeseen situations where these experience factors do not fit the situation and result in gross inequities. In such a case, negotiation must be resorted to.

Mechanical erection including machinery and steel structures is also priced on a weight basis, but it is important to spell out precisely what operations are included and which are extra. Alignment and assembly is best done on a time and material basis.

Civil work is generally contracted on a unit-price basis when lump sum is impractical. There is a reluctance to hire men and equipment by the hour because this would relieve the contractor of all responsibility for performance. In concrete and structural work, the system of unit prices, yards of concrete, square feet of form work, kilos of reinforcing steel, is complicated but workable. However, earthwork based on yardage can be an endless point of disagreement even when considerable detail is spelled out as to how measurements are to be made. There will have to be upwards of 50 or 100 different unit prices depending on the precise nature of the operation—for example, excavation and backfill both by hand and machine with variations depending on depth, existence of groundwater, distance the material must be moved, etc.

Unit-price basis for electrical and instrument work tends to be somewhat simpler and more readily rationalized by circuits, instrument loops, etc.

Construction methods, in general, and civil work, in particular, are much less efficient than U.S. experience, due in part to less mechanization and (by U.S. standards) many outmoded practices. Lower labor rates (although after benefits, not much lower) probably more than offset the inefficiencies and additional supervision. The resulting total cost is perhaps 5% less than the same job in the U.S.

There is a general tendency to rely on vendor specialists to assemble, align and check out almost all mechanical equipment. In some instances this is good, but if carried out to extremes it can add appreciably and unnecessarily to the cost of the job. There is no good reason, other than a reluctance to assume responsibility, why much common equipment like small pumps cannot be installed, aligned, and checked by the contractor's field engineers.

The success of a European construction project will depend on the ability of the construction management to co-ordinate the activities of the various contractors, to keep intercontractor disputes to a minimum, and to pressure contractors into meeting commitments as to manpower and supervision. Quality control is also more difficult on subcontracted work than on direct hire. Generally, it is necessary to employ outside testing agencies—especially for civil work. A frequent complaint and perhaps a universal one is the difficulty of getting unfinished items completed at the tail-end of the job. The only solution to this, and not often successful, is a substantial hold-back with the contractual right to have the work completed by others at the contractor's expense. Unfortunately no third party is interested in the leftover headaches.

Jobsite housekeeping is another problem that can only be dealt with on the basis that the owner has the right, after due warning, to have site cleanup during the course of the job and after completion done by outsiders at the subcontractor's expense.

Good administration and sound systems are needed to control and authorize extra work and/or extra cost items.

Thorough testing and inspection must be maintained throughout the work, and particularly at the end prior to acceptance. All contractors should be contractually required to retain sufficient supervision and manpower to rectify all defects uncovered during the startup period. This applies particularly to electrical, piping and instrument work. ■

Meet the Author

Richard L. Lewis is Director of Engineering with Process Plants Corp., 112-20 14 Ave., College Point, N. Y. 11356. Until recently, he was project manager for over 12 years for National Distillers and Chemical Corp. He came to the U.S. in 1953 from South Africa and has managed many projects in the U.S. for several of the leading engineering contractors. He became involved with foreign projects in 1965. He holds a degree in mechanical engineering from Witwatersrand University.

Author Index

Subject Index